Biomathematics

Volume 18

Simon A. Levin Thomas G. Hallam
Louis J. Gross (Eds.)

Applied
Mathematical Ecology

With 114 Figures

Springer-Verlag
Berlin Heidelberg New York
London Paris Tokyo

Simon A. Levin
Section of Ecology and Systematics
and Environmental Research Center
Cornell University
Ithaca, NY 14853-0239 USA

Thomas G. Hallam
Louis J. Gross
Department of Mathematics
and Graduate Program in Ecology
University of Tennessee
Knoxville, TN 37996-1300 USA

Mathematics Subject Classification (1980):
92-01, 92-02, 92A15, 92A17

ISBN-13: 978-3-642-64789-5   e-ISBN-13: 978-3-642-61317-3
DOI: 10.1007/978-3-642-61317-3

Library of Congress Cataloging-in-Publication Data.
Applied mathematical ecology / Simon A. Levin, Thomas G. Hallam, Louis J. Gross (eds.)
p. cm. – (Biomathematics; v. 18)
Proceedings of the Second Autumn Course on Mathematical Ecology,
held Nov. – Dec. 1986 at the International Centre for Theoretical Physics in Trieste, Italy.
Includes bibliographies and indexes.
ISBN 0-387-19465-7 (alk. paper)
1. Ecology – Mathematical Models – Congresses. I. Levin, Simon A. II. Hallam, T.G.
(Thomas G.) III. Gross, Louis J. IV. Autumn Course on Mathematical Ecology (2nd:
International Centre for Theoretical Physics: 1986) V. Series.
QH541.15.M3A66 1989 574.5'072'4–dc19 88-37326 CIP

© Springer-Verlag Berlin Heidelberg 1989
Softcover reprint of the hardcover 1st edition 1989

Typesetting: Thomson Press (India) Ltd., New Delhi
2141/3140-543210   Printed on acid-free paper

Dedication

This work is dedicated to the families of all the faculty and participants of the Second Autumn Course on Mathematical Ecology who tolerated extended absences, sacrificed companionship, or utilized finances so that we could gather in Trieste. The editors would particularly like to dedicate these efforts to their wives and children. Special recognition—to Marilyn Kallet and Heather Gross—to Rachel, Guy, and Bethany Hallam—to Carole, Jay, and Rachel Levin—for their love and understanding.

Acknowledgments

The Second Autumn Course on Mathematical Ecology was produced through the efforts of many. We appreciate the most gracious support of Professor Abdus Salam, Director of the International Centre for Theoretical Physics (ICTP) and Professor Luciano Bertocchi, Associate Director of ICTP, who provided financial, logistical and moral support for the Course. Professor Giovanni Vidossich, who suggested that we organize the First Autumn Course on Mathematical Ecology in 1982, again supported and coordinated the current efforts with his usual care and expertise. Gabriella DeMeo served as the Course Secretary and ably assisted with the organization and the many problems thrust in her direction by the directors, lecturers, and participants. The attitude and help of the ICTP staff was conducive to a most pleasant learning experience and working environment.

The energetic group of scientists who came to Trieste eager to learn and the scientists who served as faculty found time to get acquainted, to discuss scientific problems of mutual interest, and to appreciate the academic needs and situations of their colleagues. We are grateful to all for their interest in the Course and their success in meeting its objectives.

The efforts of the speakers and participants are greatly appreciated by the directors and by all who were able to survive the rigorous schedule. Many scientists provided thoughtful reviews of the manuscripts in this book. The drafts also were influenced by comments from the faculty, participants, and research scholars who were present in Trieste.

To everyone involved with the Course, we acknowledge your contributions with pleasure. The Course was a delightful experience for each of us and we thank you for your efforts in making it so enjoyable.

T.G. Hallam   *L.J. Gross*   *S.A. Levin*

Preface

The Second Autumn Course on Mathematical Ecology was held at the International Centre for Theoretical Physics in Trieste, Italy in November and December of 1986. During the four year period that had elapsed since the First Autumn Course on Mathematical Ecology, sufficient progress had been made in applied mathematical ecology to merit tilting the balance maintained between theoretical aspects and applications in the 1982 Course toward applications. The course format, while similar to that of the first Autumn Course on Mathematical Ecology, consequently focused upon applications of mathematical ecology. Current areas of application are almost as diverse as the spectrum covered by ecology. The topics of this book reflect this diversity and were chosen because of perceived interest and utility to developing countries.

Topical lectures began with foundational material mostly derived from Mathematical Ecology: An Introduction (a compilation of the lectures of the 1982 course published by Springer-Verlag in this series, Volume 17) and, when possible, progressed to the frontiers of research. In addition to the course lectures, workshops were arranged for small groups to supplement and enhance the learning experience. Other perspectives were provided through presentations by course participants and speakers at the associated Research Conference. Many of the research papers are in a companion volume, Mathematical Ecology: Proceedings Trieste 1986, published by World Scientific Press in 1988.

This book is structured primarily by application area. Part II provides an introduction to mathematical and statistical applications in resource management. Biological concepts are interwoven with economic constraints to attack problems of biological resource exploitation, conservation of our natural resources and agricultural ecology. Part III consists of articles on the fundamental aspects of epidemiology and case studies of the diseases rubella, influenza and AIDS. Part IV addresses some problems of ecotoxicology by modelling the fate and effects of chemicals in aquatic systems. Part V is directed to several topics in demography, population biology and plant ecology, with emphasis on structured population models.

The themes of the book — resource management, epidemiology, ecotoxicology, demography, and population ecology — are indicative of some of the Course developments. There were other areas of application presented in the course but not included here primarily because of other obligations of the speakers. Professor Peter Hammerstein addressed problems of behavioral ecology from an evolutionary game theoretic approach. Professor Mark Harwell directed a workshop on ecological consequences of nuclear war and gave several lectures on ecosystem processes. Professor J. Alberto Leon talked on aspects of population genetics. Professor G.P. Patil discussed topics of diversity and statistical ecology. Professor Luigi Ricciardi introduced stochastic population models and the associated analysis. We are pleased to acknowledge, with thanks, their efforts during the course.

Knoxville, Tennessee, and   *T.G. Hallam, L.J. Gross and*
Ithaca, New York   *S.A. Levin*

Table of Contents

List of Authors

Roy M. Anderson, Department of Pure and Applied Biology, Imperial College, London University, London SW7 2BB, England

Joan L. Aron, Department of Population Dynamics, Johns Hopkins School of Hygiene and Public Health, Baltimore, MD 21205, USA

Carlos Castillo-Chavez, Center for Applied Mathematics, Section of Ecology and Systematics, and Biometrics Unit, Warren Hall, Cornell University, Ithaca, NY 14853, USA

Colin W. Clark, Institute of Applied Mathematics, University of British Columbia, Vancouver B.C., Canada

Andrew P. Dobson, Department of Biology, University of Rochester, Rochester, NY 14627, USA

Louis J. Gross, Department of Mathematics and Graduate Programs in Ecology, and Plant Physiology and Genetics, the University of Tennessee, Knoxville, TN 37996-1300, USA

William S.C. Gurney, Department of Physics and Applied Physics, University of Strathclyde, Glascow G4 ONG, Scotland, UK

Thomas G. Hallam, Department of Mathematics and Graduate Program in Ecology, University of Tennessee, Knoxville, TN 37996-1300, USA

Herbert W. Hethcote, Department of Mathematics, University of Iowa, Iowa City, IA 52242, USA

John Impagliazzo, Department of Computer Science, Hofstra University, Hempstead, NY 11550, USA

S.A.L.M. Kooijman, Biologisch Laboratorium, Vrije Universiteit, Amsterdam, The Netherlands

Ray R. Lassiter, Environmental Research Laboratory, US Environmental Protection Agency, College Station Road, Athens, GA 30613, USA

Simon A. Levin, Ecology and Systematics, E 347 Corson Hall, Cornell University, Ithaca, NY 14853-0239, USA

Wei-min Liu, Ecology and Systematics, Corson Hall, and Center for Applied Mathematics, Cornell University, Ithaca NY 14853, USA (current address: Department of Mathematical Sciences, Indiana University Purdue, University at Indianapolis, Indianapolis, IN 46205, USA)

Marc Mangel, Department of Zoology, University of California, Davis, CA 95616, USA

Robert M. May, Department of Biology, Princeton University, Princeton, NJ 08544, USA (current addresses: Zoology Department, University of Oxford, OX13 PS, England and Department of Pure and Applied Biology, Imperial College, London SW1 2BB, England)

Robert McKelvey, Department of Mathematical Sciences, University of Montana, Missoula, MT 59812, USA

J.A.J. Metz, Institute for Theoretical Biology, University of Leiden, Groenhovenstraat 5, 2311 BT Leiden, The Netherlands

Roger M. Nisbet, Department of Physics and Applied Physics, University of Strathclyde, Glascow G4 ONG, Scotland, UK

Robert V. Thomann, Environmental Engineering and Science, Manhattan College, Riverdale, NY 10471, USA

Part I. Introduction

Ecology in Theory and Application

Simon A. Levin

Fraser Darling (1967) wrote that "Ecology.... was a bigger idea than the initiators grasped." Darling and others, writing in the 1960's, addressed the new demands placed upon ecologists as public awareness of an environmental crisis grew. The science of ecology had, until that time, been developed to satisfy purer, more abstract, objectives—a search for understanding and explanation. It was ill prepared to do more than provide anecdotes in support of the need to guard nature's treasures and the "balance of nature." Only in the last two decades have serious efforts been directed to developing the theoretical basis we need to manage natural systems from a sound ecological basis.

The notion of the balance of nature was evident even in the writings of St. Thomas Aquinas. Such ideas pervade the nonmathematical literature; and thus, it is not surprising that they were also represented in early theoretical work. John Graunt, Thomas Malthus, and Charles Darwin all recognized that population growth must carry with it the roots of its own diminishment, and this led naturally (but not inescapably) to the notion of equilibrium. Sadler, Quetelet, and others, mesmerized by parallels from physics, introduced models of population regulation built upon vague forces underlying the system's "resistance to growth" (Hutchinson, 1978). Verhulst introduced the logistic, the simplest and most commonly used nonlinear growth model; and other such efforts were little more than variations on the logistic theme. Characteristic of all of this early work were the concepts of balance and equilibrium, of homogeneity, and of determinism. The predominant emphasis was on individuals and their adaptations, and on the populations they comprised.

The early 20th century saw a growth of theoretical ideas, with increased attention to dynamics and change. The interrelatedness of species became a central subject of investigation in plant community theory, where Gleason's individualistic and stochastic concepts lost out to Clements' notion of the plant formation as a holistic superorganism that developed over time to a climax state determined by the local climate (Whittaker, 1975; McIntosh, 1985). Again, homogeneity and determinism prevailed.

Meanwhile, Grinnell, Elton, and other animal ecologists were developing and refining the intertwined notions of niche and community, elucidating the role of the species within its community. Mathematical models assumed center stage; and the investigations of Volterra, Lotka, and Kostitzin became recognized for their power in aiding understanding, in explanation, and in developing and exploring hypotheses. Volterra, motivated by his son-in-law d'Ancona's discussion of the fluctuations of the Adriatic fisheries, demonstrated that it was at least possible that such fluctuations were simply the consequence of the interactions between predators and their prey. His related work in competition theory led to fundamental insights concerning the coexistence of species, and inspired the young Russian ecologist Gause to conduct his classic experiments on competition between species of microorganisms. It was the "Golden Age of Theoretical Ecology" (Scudo and Ziegler, 1978). Yet homogeneity and an only slightly modified emphasis on equilibrium, determinism, and asymptotic behavior remained powerful aspects of the theory.

In the middle of the 20th century, the concepts of spatial and temporal variability as essential properties of a healthy ecosystem were placed on firmer theoretical footing by community ecologists such as A.S. Watt, R.H. Whittaker, G.E. Hutchinson, and Robert MacArthur. Simultaneously, the concept of the *ecosystem*, as introduced by Tansley (1935), received attention as the context for studying nutrient and energy flows and biogeochemical cycling.

Following Watt's prescient presidential address to the British Ecological Society (Watt, 1947), appreciation grew for the importance of variability in space and time as a factor structuring communities, and as a key to coexistence and coevolution. As Watt's work and much that followed showed, natural biotic and abiotic disturbance recycles limiting resources, developing mosaics of successional change that allow species to subdivide resources temporally. Woody Allen has said, "Time is nature's way of making sure everything doesn't happen at once." The explicit incorporation of disturbance, variability, and stochasticity as part of the description of the normative community is thus an imperative. As Robert Paine and I have argued, local unpredictability is for many species globally the most predictable aspect of systems.

Theoretical and experimental investigations into the importance of gaps and mosaic phenomena have demonstrated the inseparability of the concepts of equilibrium and scale. As one moves to finer and finer scales of observation, systems become more and more variable over time and space, and the degree of variability changes as a function of the spatial and temporal scales of observation. Such a realization has long been part of the thinking of oceanographers, who observe patchiness and variability on virtually every scale of investigation. A major conclusion is that there is no single correct scale of observation: the insights one achieves from any investigation are contingent on the choice of scales. Pattern is neither a property of the system alone nor of the observer, but of an interaction between them.

The importance of scales also becomes apparent from an examination of population models, both in terms of their general dynamic properties and in terms of their applicability to real populations. Much recent mathematical work has demonstrated that even the simplest models of population dynamics can exhibit oscillatory and even chaotic behavior; as a consequence, it is impossible to predict accurately the precise dynamics of populations governed by such equations.

To some extent, such observations render moot the classical debate over whether populations are controlled by density-dependent or density-independent factors. Close to the theoretical equilibrium, the dynamics of populations may be indistinguishable from those of appropriately chosen stochastic density-independent models; near equilibrium, density dependence is very weak, and will be obscured by any overriding density-independent variation. On the other hand, far from equilibrium, density-dependent factors assume more importance because the nonlinearities are stronger. Thus, density dependence is the primary mechanism constraining major excursions in population density and keeping populations within bounds; but within those bounds, density independent phenomena predominate. Concepts of stability that rely on asymptotic return to an equilibrium state are seen to be irrelevant on many scales of interest, and more general concepts such as boundedness and resiliency replace them.

The major conclusions of such studies are that there are inherent limits to predictability, that predictability depends on the scale of investigation, and that there is no single correct scale of inquiry. This recognition is of central importance in the current revitalization of mathematical theory, inspired by applied imperatives arising from renewable and nonrenewable resource management, epidemiology, and environmental protection. Through these needs and studies came a clarification of what ecological theory could and could not do. Francois Jacob (1976, 1977), in his insightful discussions of evolutionary theory, made clear the distinctions between explanation and prediction, and the degree to which prediction of evolutionary events is hampered by the importance of stochasticity, historical contingency, and constraints. Such ideas apply equally to prediction of ecological responses to stress.

In the late 1960's and early 1970's, as a result of environmental activism and concern over the effects of chemicals in the environment, a suite of environmental laws were passed that, in one form or another, called for the protection of ecosystems. In general, those laws tended to be vague when it came to establishing criteria for measuring ecological effects, since they were designed to address broad objectives without being hampered in applicability by specific references to particular systems. Much legislation, and certainly the conceptual and mathematical models that they spawned, still emphasized ideas such as equilibrium, constancy, homogeneity, stability, and predictability. This in turn led to the development of simple-minded and misguided predictive approaches, built upon large-scale and highly detailed models. Finally, as the need for tools and answers grew, this necessity led to the flagrant abuse of mathematical theory. Ideas developed for explanation and pedagogical purposes were recruited for prediction, without the warnings that belonged on the label. The message of Francois Jacob, that explanation and prediction were quite different animals, was ignored; and the distinctions between what was basic and what was applied became blurred.

One of the conclusions from the early frustrating experiences in predicting the responses of ecological systems to stress was the need to develop a deeper theory of how ecosystems operate. The description of the behaviour of any system can be carried out blindly, or it can be based on some degree of understanding of the basic mechanisms underlying system dynamics. The search to understand any complex system is a search for pattern; that is, for the reduction of complexity to a few simple rules, principles that allow abstraction of the essence from the noise. This is also the key to management: pattern exists at all levels and at all scales, and recognition of this multiplicity of scales is fundamental to describing and understanding ecosystems. It is essential to strip away what is irrelevant detail, and to determine those processes and components that are central to the integrity of the system: the keystone species of Robert T. Paine (Paine, 1966), or the factors controlling recruitment and larval survival. The emphasis on equilibrium and homogeneity obscures the search for pattern, and represents the baggage of historical tradition.

Moreover, overly detailed and reductionistic models of populations and systems obscure any pattern by introducing irrelevant detail, often on the specious premise that somehow more detail and more reduction assures greater truth. This point of view is predicated in part on the fallacious notion that there is some exact system description possible, and that any model is an approximation to that exact description. In reality, there can be no such ideal, because there can be no "correct" level of aggregation. The taxonomic species, for example, is an imperfect tool of classification, and ignores the differences among the individuals of the species with regard to demographic and phenotypic properties; thus it is just one possible grouping within a particular nested hierarchy. More importantly, the particular hierarchical decomposition that arises from a taxonomic classification system bears much less relevance for some ecological descriptions than would one that was functionally based.

What society values, and what environmental laws are designed to protect, are not the intricate details of systems, but rather the essential features that make those systems recognizable to biologists and that are somewhat predictable over time. We should not expect to make systems more predictable in our models than natural variability dictates; in fact, variability, such as that associated with seasonality, gap phase dynamics, or fire, is one of the most predictable features of many systems, and is essential to the maintenance of resiliency and to the persistence of most species. Moreover, it is well recognized that the climatic and atmospheric systems driving biotic systems are very limited in predictability. In 1985, Neil Frank, Director of the National Hurricane Center, said "All our weather planes and radars create a false impression that we can do a great job forecasting. That's a myth. What we're doing is a great job observing."

Furthermore, predictability is inextricably intertwined with variability, and with the temporal and spatial scales of interest. Thus, a central challenge in ecological theory must be an elaboration of the understanding of how scales relate, how systems behave on multiple scales, and how the measurement and dynamics of particular phenomena vary across scales. The inherent limits to predictability on long time scales emphasize the importance of monitoring, and of coupling any management action with some mechanism for modification based on analysis of the data obtained from monitoring. Such adaptive management recognizes explicitly the limits to predictability, and places emphasis on short-term prediction in which nonlinear phenomena have diminished importance.

In this regard, environmental management presents a number of problems that are generic in the consideration of the responses of systems to quite distinct stresses. The prediction of the dynamics of natural populations has never been resolved adequately, even for renewable resources such as fisheries, which have been the objects of scientific scrutiny for decades. The difficulties relate to our inability to identify and predict changes in the factors controlling dynamics, to the spatial and temporal variability of parameters and even mechanisms of control, and to the inherent propensity of nonlinear dynamical models to exhibit turbulent dynamics that make parameter estimation a daunting challenge. When communities and ecosystems are considered, with the consequent multiplication of pathways of interaction, the problems are similarly multiplied. Some relief can be achieved by judicious simplification that properly recognizes the shift in detail appropriate to the shifts in levels of organization, but a core of irremovable uncertainty will always remain. Scientific advisers must make clear to managers the levels of uncertainty, and must not give in to the temptation to seem to present the certainty that their clients seek. Model outputs must be presented not just in terms of averages, but with associated variances in relation to stochastic effects and uncertainties in parameter estimation. Finally, as Crawford Holling has emphasized (Holling, 1986), an inescapable conclusion from the existence of such uncertainty is that there will be surprises associated with virtually any management action, and that any management strategy must have some potential for (adaptive) modification when experience and monitoring so dictate.

As we enter the last decade of the 20th century, we face environmental challenges greater than ever before as the scale shifts from local problems to global and regional ones. We are confronted with changes in the distributions and exchanges of elements on broad scales, with the alarming loss of biotic and habitat diversity, with the consequences of species invasions, with toxification and contamination of our aquifers and other systems, with the need for alternatives for waste disposal, with the collapse of resource systems, and even with the risk of global thermonuclear war. As never before, these challenges mandate a better integration across disciplines, especially across the physical and biological sciences, and an integration set in a holistic perspective encompassing ecosystems, regions, landscapes, and the biosphere. We must recognize explicitly the multiplicity of scales within ecosystems, and develop a perspective that looks across scales and that builds upon a multiplicity of models rather than seeking the single "correct" one. We need to couple system-level testing that allows identification of emergent phenomena with mechanistic studies designed to provide understanding and the basis for extrapolation.

In dealing with the effects of chemicals in the environment and their local and global consequences, the control of agricultural pests and human disease, the management of scarce resources and other problems facing society, we must develop more sophisticated quantitative tools for prediction, for aiding understanding, and as guides to management. This volume, built upon the foundations laid in Hallam and Levin (1986), is dedicated to those objectives. The chapters explore current approaches to problems in demography and epidemiology, resource management, and ecotoxicology, and apply classical and modern mathematical developments in ecology and epidemiology to a variety of case studies.

Finally, we must acknowledge the limits of our ability to predict—the ecologist's uncertainty principle—and be prepared to manage in the face of that uncertainty. That will require the development of more sophisticated and flexible approaches to risk assessment and management. It also will require an increased public awareness that there are limits to predictability, and that the fuzzy boundary between science and policy justifies, even necessitates, public involvement in the decision-making process. There are few scientific absolutes in environmental decision-making; rather, environmental management must be an expression of the values and needs of society, as manifest in the statutes the people's representatives enact and in societal participation in public discourse on environmental issues.

Acknowledgments. The author's research is sponsored by the U.S. Environmental Protection Agency and by the National Science Foundation. The views expressed herein are those of the author, and do not necessarily represent those of the sponsors or of Cornell University. The first part of this article is the text of a lecture originally developed as the 1985 Charles A. Alexander lecture at Cornell University. Parts of this text have appeared in various other publications.

References

Darling, F.F. (1967) A wider environment of ecology and conservation. Daedalus *96*, 1003–1019.
Hallam, T.G., Levin, S.A. (eds.) (1986) Mathematical Ecology—An Introduction. Biomathematics, vol. 17. Springer Berlin Heidelberg New York London Paris Tokyo, 457 pp.
Holling, C.S. (1986) The resilience of terrestrial ecosystems; local surprise and global change. In: W.C. Clark, R.E. Munn (eds.) Sustainable Development of the Biosphere. Cambridge University Press, Cambridge, pp. 292–317
Hutchinson, G.W. (1978) An Introduction to Population Ecology. Yale University Press, New Haven, Conn., 260 pp.
Jacob, F. (1976) The Possible and the Actual. Allen & Unwin Ltd., London, 72 + viii pp.
Jacob, F. (1977) Evolution and tinkering. Science *196*, 1161–1166
McIntosh, J.P. (1985) The Background of Ecology: Concept and Theory. Cambridge Studies in Ecology. Cambridge University Press, Cambridge, 383 pp.
Paine, R.T. (1966) Food web complexity and species diversity. Am. Nat. *100*, 65–75
Scudo, F.M., Ziegler, J.R. (1978) The Golden Age of Theoretical Ecology: 1923–1940. Lecture Notes in Biomathematics, vol. 22. Springer, Berlin Heidelberg New York, 490 pp.
Tansley, A.G. (1935) Use and abuse of vegetation concepts and terms. Ecology *16*, 284–307
Watt, A.S. (1947) Pattern and process in the plant community. J. Ecol. *35*, 1–22
Whittaker, R.H. (1975) Communities and Ecosystems, 2nd edn. Macmillan, New York, 162 pp.

Part II. Resource Management

Bioeconomic Modeling and Resource Management

Colin W. Clark

These notes developed from a series of lectures given originally at the University of Bremen in June 1986. They provide an introduction to the modeling of biological resource exploitation. Some of the material is covered in considerably greater detail in my books (Clark 1976a, 1985), but other topics represent more recent research. A numerically oriented problem book (Conrad and Clark, 1987) is also available.

The research opportunities in resource modeling are great, particularly in such topics as: multispecies systems, age-structured populations, genetics of exploited populations, stochastics and uncertainty, and regulatory systems. Publication of good research is virtually guaranteed by the existence of several quality journals, including Journal of Mathematical Biology, Marine Resource Economics, Journal of Environmental Economics and Management, and Natural Resource Modeling.

Contents

Lecture 1

Introduction to Bioeconomic Modeling

1.1 A General Production Model

We consider a renewable resource stock, whose size at time t is denoted by $x = x(t)$. The example we have in mind is a certain population of fish, with $x(t)$ representing the total biomass. The population has a natural growth rate $F(x)$:

$$\frac{dx}{dt} = F(x) \tag{1.1}$$

Here $F(x)$ is assumed to be a concave function, with

$$F(0) = 0 \quad \text{and} \quad F(K) = 0 \quad \text{for some } K > 0$$

(see Fig. 1.1). The most commonly used example is the *logistic* growth model, in which

$$F(X) = rx(1 - x/K) \tag{1.2}$$

Here $r > 0$ is called the *intrinsic growth rate* and $K > 0$ is the *carrying capacity*. Alternative functional forms for $F(x)$ have been described by Pella and Tomlinson (1969), Shepherd (1982), and others.

From Eq. (1.1) it is clear that population biomass $x(t)$ will increase whenever $x < K$ (since $dx/dt > 0$ for $x < K$), and will decrease if $x > K$. Hence $x = K$ is a *stable equilibrium* population:

$$x(t) \to K \quad \text{as} \quad t \to \infty$$

For the case of the logistic model (1.2), Eq. (1.1) can easily be solved in closed form:

$$x(t) = \frac{K}{1 + ce^{-rt}}, \quad c = \frac{K - x_0}{x_0} \tag{1.3}$$

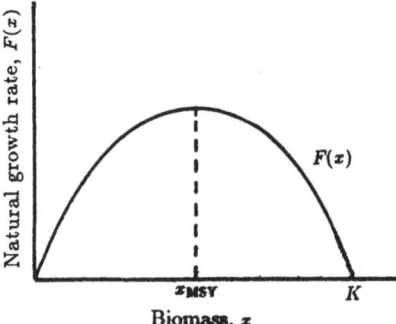

Fig. 1.1. The natural growth rate $F(x)$ of a biological resource stock x; x_{MSY} denotes the biomass corresponding to maximum sustainable yield.

Harvesting

We next modify the biological model (1.1) to allow for continuous harvesting at a (variable) rate $h = h(t)$:

$$\frac{dx}{dt} = F(x) - h(t) \tag{1.4}$$

This equation immediately raises the question: what is the "optimal" rate of harvest? As we shall see, this question is more subtle than it may first appear. Let us begin with a naive, but popular approach.

Namely, assume that $h = $ constant, and ask the question, what is the *maximum sustainable harvest*? The dynamics of Eq. (1.4) for h constant are depicted in Fig. 1.2. For $h < \max F(x)$, the equation has two equilibria, x_1 (unstable) $< x_2$ (stable). On the other hand, for $h > \max F(x)$ there are no equilibria, and $dx/dt < 0$ for all x. The special case

$$h = \max F(x) = F(x_{\mathrm{MSY}}) \tag{1.5}$$

gives rise to a single semistable equilibrium at $x = x_{\mathrm{MSY}}$. This case, which is of fundamental importance in resource management, is referred to as *Maximum Sustainable Yield* (MSY). MSY is the management objective of many resource management agencies. Taken literally, MSY implies both the perpetual preservation of the resource stock, and the maximization of the total long-term yield. These are both obviously desirable objectives.

There are, however, a number of problems with the MSY approach. In particular:

1. MSY is unstable: an arbitrarily small error may lead to overexploitation, and ultimately in principle to extinction of the resource.
2. MSY ignores all economic aspects of resource exploitation.
3. MSY cannot be applied directly to the exploitation of ecologically inter-dependent species.

We will turn therefore to the economics of resource exploitation. Our analysis

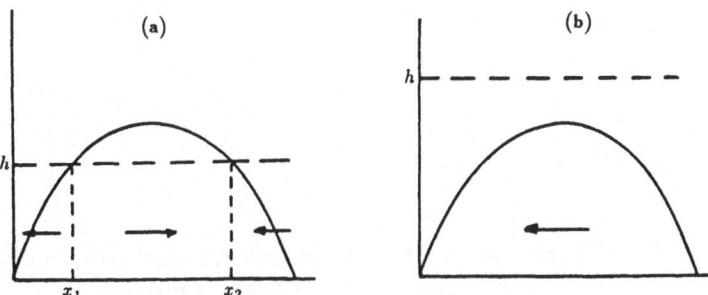

Fig. 1.2. Population dynamics with constant-rate harvesting: (a)$h < MSY$; x_2 is a stable equilibrium, x_1 unstable; (b)$h > MSY$; no stable equilibrium exists.

will also be of use in overcoming objection 1 (and is also a necessary component for dealing with objection 3).

Effort

The concept of harvesting *effort* arose originally in the case of fisheries. Fishing effort is a measure of the intensity of fishing. Typically, fishing effort is represented by the total number of vessel days spent fishing per year. On a continuous-time basis, effort E can then be thought of as the number of (standardized) vessels actively fishing at a given time.

The relationship between catch (rate) and effort is often modeled by the catch equation (Schaefer, 1954):

$$h = qEx \tag{1.6}$$

where q is a constant, called *catchability*. The genesis of Eq. (1.6) is fairly evident: for a given fish stock x, catch rate is proportional to effort. Likewise, for fixed effort E, catch rate is proportional to stock abundance. Equation (1.6) can be rewritten as

$$\frac{h}{E} = qx$$

which says that catch per unit effort (CPUE) is proportional to stock abundance. In situations where both catch and effort data are collected, the CPUE statistic often provides a good preliminary indication of the state of the fish stock.

The combined model

$$\left.\begin{array}{l} \dfrac{dx}{dt} = F(x) - h \\[2mm] h = qEx \end{array}\right\} \tag{1.7}$$

is often referred to as a *general-production* model. It was first developed for use in the tuna fisheries of the Eastern Tropical Pacific Ocean (Schaefer, 1954).

For the logistic growth model (1.2), the equilibrium relationship between effort E and *sustained* catch h is easily found by setting $dx/dt = 0$ in Eq. (1.7) and eliminating x:

$$h = \bar{h}(E) = qKE(1 - qE/r) \tag{1.8}$$

The values of x, E, and h at MSY are

$$x_{MSY} = K/2$$
$$E_{MSY} = r/2q$$
$$h_{MSY} = rK/4$$

Observe that the above model employs three parameters: r, K, and q. If, however, it is possible to estimate the two parameters a, b of a linear regression

$$\frac{h}{E} = a - bE$$

then h_{MSY} is given simply by

$$h_{MSY} = a^2/4b$$

Unfortunately, this approach seldom works very well in practice for statistical and other reasons (Ludwig and Hilborn, 1982).

1.2 Economic Models

Schaefer's general production model was extended to a bioeconomic model by Gordon (1954). Let p denote the price of the resource (\$/tonne) and let c denote the cost of effort (\$/SDF, where SDF = standardized day's fishing). Then net revenue flow is given by

$$R = ph - cE = (pqx - c)E \qquad (1.9)$$

For the case of sustainable yield, this becomes by (1.8)

$$\bar{R} = p\bar{h}(E) - cE \qquad (1.10)$$

(see Fig. 1.3). According to Gordon, the *optimal* level of effort is $E = E_0$, at which sustainable net revenue $\bar{R}$ is maximized. This is usually referred to as Maximum Economic Yield (MEY). Gordon predicted that a sole owner of the fishery would manage effort so as to achieve MEY; this prediction was criticized by Scott (1955).

A second prediction of Gordon's model is that, in an unregulated "common-property" fishery, effort will reach an equilibrium level E_∞ at which $\bar{R} = 0$. The simple argument behind this prediction is this: for $E > E_\infty$ fishing costs exceed fishing revenues (Fig. 1.3). Hence effort will decrease due to the departure of money-losing vessels. Likewise if $E < E_\infty$ then positive profits can be made and E will therefore tend to increase.

The situation predicted by Gordon's model—overfishing combined with low incomes to fishermen—is an all-but universal phenomenon in fisheries. The interesting question is, if by exerting less effort E_0 fishermen can make more money, why don't they learn this and benefit thereby? The answer was given

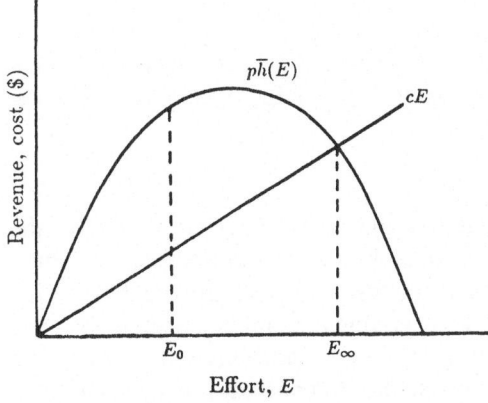

Fig. 1.3. Gordon's (1954) fishery model: E_∞ is the bionomic effort equilibrium of the common property fishery; E_0 is the level of effort that maximizes sustainable economic rent, referred to as "maximum economic yield" (MEY).

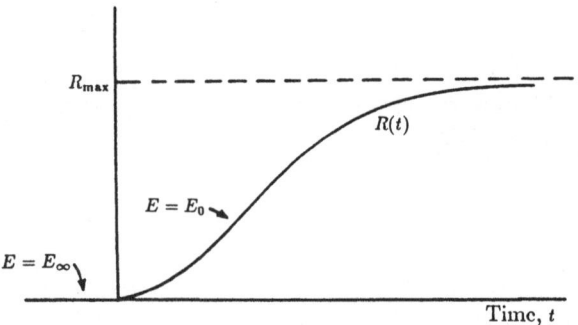

Fig. 1.4. Net economic revenue $R(t)$ as a function of time t, following a sudden decrease in effort from E_∞ to E_0 (see Fig. 1.3).

colorfully (if not quite accurately) by Hardin (1968), who used the phrase "The Tragedy of the Commons." Acting in his own best interest, the individual fisherman is forced to exert an excess of effort in order to maximize his share of the profits (or "economic rents," to use the economists' phrase). In conjunction with everyone else, this action leads inevitably to excessive effort and ultimately zero (or near-zero) net revenues. Some form of cooperation, or as Hardin says, mutual coercion, is needed in order to overcome the Tragedy of the Commons.

But there is more to the story than that! Consider a fishery in equilibrium at E_∞ (called *bionomic equilibrium* by Gordon). Agreement is reached to reduce E to E_0 somehow. What happens?

In fact, profits do *not* go up immediately to $R_{max} = \bar{R}(\bar{E}_0)$, because it necessarily takes time for the new equilibrium to become established (Fig. 1.4). Mathematically, R only approaches R_{max} asymptotically as $t \to \infty$. The rate of convergence depends primarily on the intrinsic growth rate r of the resource. In the case of slowly growing resources (whales, some fish species, forests), recovery of the resource may take years or decades.

So now we are led to yet another problem—the "cost of time." This brings in some interesting mathematics.

1.3 A Dynamic Model

The standard economic device for reflecting the value of time is called *discounting*. Namely, the discounted *present value* of a payment $V(t)$ due t years from today is defined to be

$$PV = e^{-\delta t} \cdot V(t) \tag{1.11}$$

where δ is the (instantaneous) annual discount rate. Clearly δ is also the (instantaneous, or continuous) rate of compound interest: a current deposit equal to PV will, by accumulation of compound interest at rate δ, grow to have value $V(t)$ in t years' time. (For simplicity we here assume zero inflation; otherwise one must distinguish between real and inflationary interest and discount rates).

Returning to our resource model, suppose that the harvest rate $h(t)$ is applied from time t to $t + dt$. This will produce a net payment of

$$V(t)dt = R\,dt$$

where by (1.9) and (1.7)

$$R = R(x, h) = \pi(x)h \tag{1.12}$$

where

$$\pi(x) = p - \frac{c}{qx}$$

The total present value of this harvest "policy" is given by

$$PV = \int_0^\infty e^{-\delta t}\pi(x)h\,dt \tag{1.13}$$

which is subject to

$$\frac{dx}{dt} = F(x) - h, \quad x(0) = x_0 \tag{1.14}$$

$$x(t) \geqq 0 \tag{1.15}$$

According to standard economic theory, the private owner of a resource stock would be motivated to select a production strategy $h(t)$ which maximizes his present value PV.

Let us also assume that the harvest rate is constrained:

$$0 \leqq h(t) \leqq h_{max} \tag{1.16}$$

[One can reformulate the above model to allow effort $E(t)$ as the "control variable," replacing (1.16) by

$$0 \leqq E(t) \leqq E_{max} \tag{1.17}$$

The results are analogous to those described in the following.]

The problem of determining the function $h(t)$, $0 \leqq t < \infty$, that maximizes (1.13) subject to the conditions (1.14)–(1.16) is a problem in "optimal control," or classically speaking, in the calculus of variations. The problem can be solved by means of the Pontrjagin maximum principle (see Sec. 2.1 below), but because of its special form, it can in fact be solved by completely elementary methods.

Namely, define

$$Z(x) = \int_{x_\infty}^x \pi(y)\,dy \tag{1.18}$$

where

$$\pi(x_\infty) = 0$$

Then

$$\frac{dZ}{dt} = \pi(x(t))\frac{dx}{dt}$$

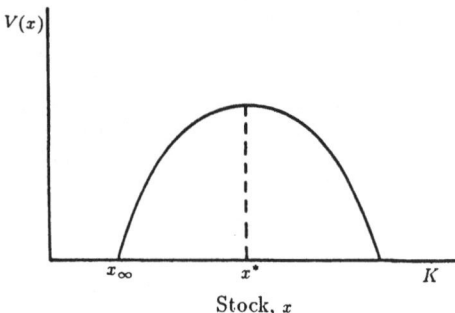

$V(x)$

x_∞ x^* K

Stock, x

Fig. 1.5. The discounted value function $V(X)$; optimal harvesting consists of the most rapid approach to equilibrium at $x = x^*$.

and an integration by parts yields

$$PV = \int_0^\infty e^{-\delta t} \pi(x)[F(x) - dx/dt]\, dt$$

$$= \int_0^\infty e^{-\delta t} [\pi(x)F(x) - \delta Z(x)]\, dt + Z(x_0) \tag{1.19}$$

Define

$$V(x) = \pi(x)F(x) - \delta Z(x) = \left(p - \frac{c}{qx} \right) F(x) - \delta Z(x) \tag{1.20}$$

When $F(x)$ is logistic, $V(x)$ is a concave function (Fig. 1.5). In order to maximize (1.19) one should clearly try to maximize $V(x)$. Because of the discount factor $e^{-\delta t}$, this maximization should be achieved as rapidly as possible. This implies that the optimal harvest policy is given by

$$h(t) = \begin{cases} h_{\max} & \text{if } x(t) > x^* \\ F(x^*) & \text{if } x(t) = x^* \\ 0 & \text{if } x(t) < x^* \end{cases} \tag{1.21}$$

where x^* maximizes $V(x)$.

The technical phrase for (1.21) is that $h(t)$ is a "bang–bang, singular" control policy. The equilibrium solution $x \equiv x^*$, $h \equiv h^* = F(x^*)$ is the singular solution (of control theory); when $x(t) \neq x^*$ one of the bang–bang controls $h_{\max}$ or 0 must be utilized. This policy is also sometimes called the most rapid approach policy (MRAP). In the Operations Research literature the phrase *myopic* control problem is used, indicating that one only needs to know the current stock level x, and current costs and prices p, c, δ, in order to deduce the optimal current control. [If p, c, δ are time dependent it turns out that one also needs to know dp/dt and dc/dt; see Lecture 2.]

By differentiation of (1.20) we obtain the following necessary condition for the optimal equilibrium biomass level x^*:

$$F'(x^*) + \frac{\pi'(x^*)F(x^*)}{\pi(x^*)} = \delta \tag{1.22}$$

I call this formula the Fundamental Theorem of Renewable Resource Economics. It shows up, in various guises, in many different problems of optimal resource management.

Some special cases are of interest. First suppose $c = 0$ and $\delta = 0$. Then (1.22) becomes simply

$$F'(x^*) = 0$$

which is the MSY solution. If both exploitation costs and the "value of time" are zero, then MSY is optimal.

Next suppose $c = 0$ but $\delta > 0$. Then (1.22) becomes

$$F'(x^*) = \delta \tag{1.23}$$

This formula, which says that marginal productivity equals the discount rate, has a distinguished economic history associated with such luminaries as Ramsay, Keynes, Samuelson, and Koopmans. It is often referred to as the (modified) Golden Rule of Capital Accumulation.

Equation (1.23) suggests that *discounting is an anticonservationist factor*, and this is true also for Eq. (1.22). Namely, the larger the discount rate, the smaller the optimal biomass x^*. In fact, in the case of (1.23) we see that

$$\delta \geq r \text{ implies that } x^* = 0 \quad (\text{if } c = 0)$$

Under certain circumstances, *extinction* of the resource population can be the profit-maximizing policy. This somewhat unnerving result is easily understood: whenever the intrinsic growth rate of a resource stock is less than the discount (i.e. interest) rate, then profits are maximized by liquidating the resource asset— provided that it is profitable to do so in the short run. While this may seem to be an unlikely special case, I am prepared to argue that time discounting is one of the main factors underlying the widespread destruction of forests, soils, and wildlife populations, and is also eminently displayed in society's current attitude towards acid rain, nuclear and toxic wastes, oceanic pollution, ozone destruction, and many other side effects of "progress."

Finally, suppose that $\delta = 0$ but $c > 0$. Then (1.22) can be written as

$$\frac{d}{dx}(\pi(x)F(x)) = 0 \tag{1.24}$$

Note that $\pi(x)F(x) = R(x) = $ sustained economic rent at stock level x. Thus (1.22) is equivalent to MEY, in the case that $\delta = 0$.

It is historically interesting to note that some economists have criticized the biologists' notion of MSY for its failure to incorporate costs of resource exploitation. In its place, MEY was put forth as a new desideratum (with the additional advantage that MEY was even more conservationist than MSY). It was not at first realized that MEY overlooks the value of time. Nowadays the acronym OEY (optimal economic yield) is sometimes used to describe the equilibrium in the general case $c > 0$, $\delta > 0$; whether OEY is more or less conservationist than MSY depends on parameter values. Whenever p/c and δ are both large, OEY is in fact less conservationist than MSY.

Antarctic Baleen Whales

A notorious example of overexploitation is provided by the Antarctic whaling industry in this century (Clark and Lamberson, 1982). Parameters for this fishery (lumping blue, fin, and sei whales together into "blue whale units" BWUs) are:

$r = .05$ per annum

$K = 400,000$ BWU

$q = 1.3 \times 10^{-5}$ per catcher day

$p = \$7000$ per BWU

$c = \$5000$ per catcher day

(costs and prices refer to ca 1965). This yields

$x_\infty = 55000$ BWU

which is about 13% of the pre-exploited biomass.

Straightforward algebra provides the explicit value for x^* for the logistic model:

$$x^* = \tfrac{1}{4}\{x_\infty + K(1 - \delta/r) + [(x_\infty + K(1 - \delta/r))^2 + 8Kx_\infty\delta/r]^{1/2}\} \qquad (1.25)$$

and Table 1 shows x^* as a function of the discount rate δ.

What are the conclusions to be drawn from this analysis? First, the optimal equilibrium stock level (for profit maximization) depends sharply on the discount rate. As noted earlier, discount rates are very important features in resource conservation. What is the "optimal" discount rate, then? This topic has been argued at length by economists, with no clear agreement. What discount rates do firms or consumers actually use? Some published reports have estimated private discount rates up to 40% per annum! "Real" interest rates in the economy, as represented by bank rate of interest minus rate of inflation, are more stable, having perhaps ranged from + 5% to − 5% per annum recently. (Negative real rates are fortunately rare; they occur during uncontrolled inflation.)

Table 1. Influence of the rate of discount on profit maximizing exploitation of Antarctic baleen whale stocks

Discount rate (per annum)	Optimal stock level (BWU)	Optimal sustained yield (BWU per annum)	Initial profit (10^9)
0%	227500	4905	0.99
3%	151200	4700	1.37
5%	119500	4190	1.50
10%	85300	3355	1.61
15%	73400	3015	1.63
20%	68600	2845	1.64
25%	65600	2745	1.65
∞	55000	2370	1.65

Source: United Nations Food and Agriculture Organization, Mammals in the Sea, FAO, Rome, 1978.

In resource industries, discount rates probably also reflect "risk discounting"—see Lecture 2. It seems likely that resource firms expect returns above 10% on their investments. Table 1 suggests that conservation motives could be minimal in such cases. In the case of whaling (a *very* lucrative business in the early postwar years), even the advent of the International Whaling Commission failed to prevent depletion of the whale resource. Current estimates of Antarctic whale stocks are something like 10,000 blue whales, 70,000 fin whales, and 200,000 sei whales, a total of 78,000 BWU. Conservationist pressure has forced the IWC to adopt a moratorium on these species, but active whaling states (Japan, USSR) still threaten to leave IWC and resume whaling.

High interest rate (and low capital availability) are, unfortunately, characteristic of developing countries. If these rates are employed in resource decisions, the outcome may be a temporary boom in certain resource industries, followed by a bust as resource stocks are depleted. The problem will be exacerbated if, as often seems to happen, the boom years lead to overoptimism and a failure to manage overexpansion and overexploitation in resource industries.

Biological resources are economically important. Mismanagement is inexcusable. Good management requires understanding of the entire resource system. Mathematical models can greatly aid in improving our understanding—provided no particular model is taken too seriously.

Lecture 2

Extensions of the Basic Model

In the first lecture we showed that the optimal harvest policy was a bang–bang singular policy, or MRAP. Such control policies are easy to compute and easy to implement. It is therefore worthwhile to ask under what conditions optimal policies are of this nature.

A more general formulation of our optimal harvest problem is as follows:

$$\text{maximize} \int_0^\infty e^{-\delta t} \pi(x, h, t)\, dt \tag{2.1}$$

$$\text{subject to } \frac{dx}{dt} = G(x, h), \quad x(0) = x_0 \tag{2.2}$$

$$0 \le h \le h_{\max} \tag{2.3}$$

In performing the simplification in Eq. (1.19), it was important to have dx/dt appearing linearly in the integrand, so that an integration by parts could be performed. For this to occur, it must be the case that both $\pi(x, h)$ and $G(x, h)$ contain the control variable h *linearly*. In such cases (2.1)–(2.3) is said to be a *singular* control problem. The optimal control for a singular control problem is always a combination of singular and bang–bang controls. We next explain how this works out.

2.1 Myopic Control Problems and Blocked Intervals

We now drop the assumption that the parameters p, c, and δ are constants, and allow any or all of them to be functions of time t (but *not* of h). First it is easy to see that the discount factor corresponding to a time-varying instantaneous discount rate $\delta(t)$ is

$$\alpha(t) = \exp\left(- \int_0^t \delta(s)\,ds \right) \tag{2.4}$$

Our optimal harvest problem now becomes

$$\text{maximize} \int_0^\infty \alpha(t)\pi(x,t)h\,dt \tag{2.5}$$

$$\text{subject to } \frac{dx}{dt} = F(x) - h \tag{2.6}$$

$$0 \le h \le h_{\max} \tag{2.7}$$

(In fact F can also be allowed to depend on t.)

One can again use integration by parts, but it is quicker and cleaner to invoke the Pontrjagin maximum principle (PMP) (see e.g., Clark, 1976a, Ch. 4). Introduce the Hamiltonian expression

$$\begin{aligned}\mathcal{H} &= \alpha(t)\pi(x,t)h + \lambda[F(x) - h] \\ &= (\alpha\pi - \lambda)h + \lambda F(x)\end{aligned} \tag{2.8}$$

where $\lambda = \lambda(t)$ is the so-called adjoint variable. According to PMP, the optimal control h must at all times maximize the Hamiltonian $\mathcal{H}$. This immediately implies that

$$h(t) = \begin{cases} h_{\max} & \text{if } \lambda(t) < \alpha(t)\pi(x(t),t) \\ 0 & \text{if } \lambda(t) > \alpha(t)\pi(x(t),t) \end{cases} \tag{2.9}$$

and these are the bang–bang controls.

Singular control arises when

$$\lambda(t) \equiv \alpha(t)\pi(x,t) \tag{2.10}$$

over an open time interval. The adjoint equation of PMP is

$$\frac{d\lambda}{dt} = -\frac{\partial \mathcal{H}}{\partial x} \tag{2.11}$$

A necessary condition for singular control is obtained by equating $d\lambda/dt$ in Eqs. (2.10) and (2.11):

$$\frac{d\lambda}{dt} = \dot{\alpha}\pi + \alpha\pi_x[F(x) - h] + \alpha\pi_t$$

$$-\frac{\partial \mathcal{H}}{\partial x} = -\alpha\pi_x h - \lambda F'(x)$$

$$= -\alpha\pi_x h - \alpha\pi F'(x)$$

Hence we obtain (since $\dot{\alpha} = -\delta\alpha$)

$$F'(x) + \frac{\pi_x F(x)}{\pi} = \delta(t) - \frac{\pi_t}{\pi} \tag{2.12}$$

This is an implicit equation for x, which determines the *singular* solution $x = x^*(t)$.

Note that for the case of constant parameters, (2.12) reduces to our previous result (1.22) for the optimal equilibrium biomass x^*.

In the general case, the optimal control must be obtained by combining singular and bang–bang controls. An example involving slowly varying parameters is illustrated in Fig. 2.1. Here an initial bang–bang adjustment reduces x from x_0 to the singular path $x^*(t)$. Subsequently singular control $h^*(t)$ is used, where

$$h^*(t) = F(x^*(t)) - \frac{dx^*}{dt} \tag{2.13}$$

The singular solution is sometimes called a *myopic* solution, reflecting the fact that $x^*(t)$ can be calculated from purely "myopic" data, namely the current values of economic and biological parameters, and their current rates of change.

Blocked Intervals

It may happen however that, because of rapid parameter shifts, the singular path $x^*(t)$ changes so rapidly that the singular control (2.13) violates the control constraints $0 \le h \le h_{max}$. In such a situation the singular (myopic) solution is not feasible, and a (temporary) bang–bang solution has to be used. The time interval during which bang–bang control becomes optimal is called a "blocked interval."

As a simple example, suppose all parameters are constant, except for δ, which jumps from δ_1 to δ_2 at a predictable future time $t = T$. Then there are two (constant) singular biomass levels x_1^*, x_2^*, but the transition from x_1^* to x_2^* cannot be made instantaneously. The optimal solution appears as in Fig. 2.2: a blocked interval, during which $h = h_{max}$, surrounds the switch point $t = T$. Here we assumed $\delta_1 < \delta_2$ so that $x_1^* > x_2^*$.

The intuition behind this example is as follows. We know that x^* represents

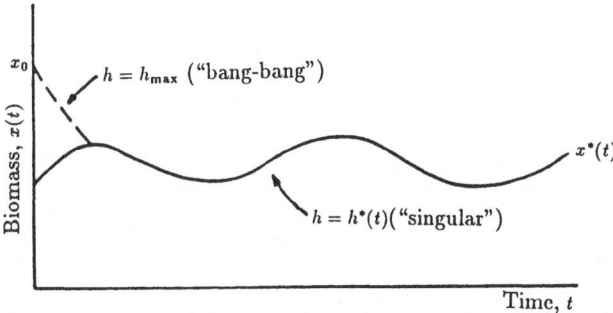

Fig. 2.1. A singular biomass trajectory $x^*(t)$ with "bang-bang" approach.

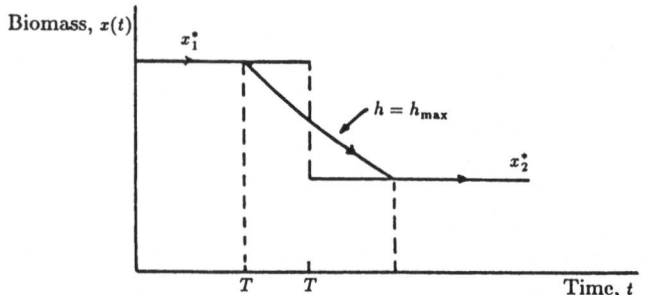

Fig. 2.2. A blocked interval surrounding the time T at which the discount rate switches from δ_1 to a larger value δ_2.

an "optimal level of conservation" of the resource stock x. Also x^* is a decreasing function of the discount rate δ. But if we know that δ will be higher in the future, then today's conservation incentive may also be reduced. Therefore we start harvesting at h_{max} prior to the time T. (As an exercise the reader may wish to describe the optimal policy for the case of a sudden price change, say $p_1 > p_2$. The solution is quite different from that of Fig. 2.2.)

2.2 Discrete-Time Models

In resource modeling it is extremely useful to be able to choose either a continuous-time or a discrete-time model, according to need. In fact, one often wishes to employ both continuous and discrete-time components in the same model (see Lecture 4).

In this section we therefore describe a discrete-time version of the renewable resource model of Lecture 1. To maintain the reader's interest we will in fact consider a stochastic model. It will turn out that the optimal harvest policy is again "myopic."

We shall consider a harvested population with nonoverlapping generations; the five species of Pacific salmon are a well known example. The system dynamics are illustrated in Fig. 2.3. Adult fish, called *recruits*, enter the fishery, where a

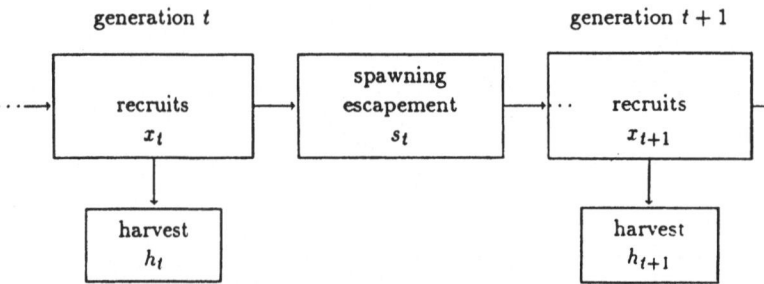

Fig. 2.3. A stochastic spawner-recruit model.

harvest is taken. The remaining recruits escape to spawn, giving rise to the subsequent generation (possibly after a delay of several years to reach maturity).

The equations of population dynamics are simply

$$x_{t+1} = F(s_t, w_t) \tag{2.14}$$

$$s_t = x_t - h_t \tag{2.15}$$

where w_t is a random "noise" term. Thus recruitment is determined by previous escapement, but with stochastic environmentally-induced variation. Let $\pi(x_t, h_t)$ denote net revenue derived from the tth harvest. Our problem is to determine the harvest policy which maximizes the expected present value of future harvests:

$$\text{maximize } E\left\{ \sum_{t=0}^{T} \rho^t \pi(x_t, h_t) \right\} \tag{2.16}$$

subject to (2.14) and (2.15) and

$$0 \leq h_t \leq x_t \tag{2.17}$$

Here $\rho \in (0, 1)$ is the discount factor.

Because of the stochastic element, however, it is necessary to specify our problem in greater detail. We define the *value function*

$$J_T(x) = \max E\left\{ \sum_{t=0}^{T} \rho^t \pi(x_t, h_t) \mid x_0 = x \right\} \tag{2.18}$$

The optimal harvest policy can then be computed by an iterative procedure known as *stochastic dynamic programming*.

For $T = 0$ we have

$$J_0(x) = \max_{0 \leq h \leq x} \pi(x, h) \tag{2.19}$$

and this is an elementary optimization problem, with solution $h = h^*(x)$, say.

Next, for $T = 1$ suppose that some harvest $h \leq x$ is taken in the first period ($t = 0$). This yields a return $\pi(x, h)$, and leads to a second period recruitment equal to $x_1 = F(x - h, w)$ with probability $\phi(w)$, say. An optimal (one-period) harvest is then taken in the second and last period. The total discounted expected return is therefore

$$\pi(x, h) + \rho E_w\{J_0(F(x - h, w))\}$$

An optimal first-period harvest must maximize this expression. In other words,

$$J_1(x) = \max_{0 \leq h \leq x} [\pi(x, h) + \rho E_w\{J_0(F(x - h, w))\}] \tag{2.20}$$

Since we already know how to compute $J_0(\cdot)$, we can therefore also compute $J_1(x)$ and $h_1^*(x)$.

Proceeding inductively we obtain

$$J_{k+1}(x) = \max_{0 \leq h \leq x} [\pi(x, h) + \rho E_w\{J_k(F(x - h, w))\}] \tag{2.21}$$

Eq. (2.21) is the dynamic programming equation (DPE), or Bellman equation, for our model. It specifies an iterative algorithm for computing the optimal T-horizon harvest policy $h_T^*(x)$ for any T, as a function of the current stock level x. Such a rule is called a *feedback* control policy.

In general the DPE must be solved numerically. By making additional simplifying assumptions, however, we can obtain analytical results. We now consider the specific problem (Reed, 1979):

$$J_T(x) = \max E\left\{\sum_{t=0}^{T} \rho^t \pi(x_t, h_t) \mid x_0 = x\right\} \tag{2.22}$$

subject to $x_{t+1} = Z_t F(x_t - h_t)$ (2.23)

$$0 \leqq h_t \leqq x_t \tag{2.24}$$

where π has the form

$$\pi(x, h) = \int_{x-h}^{x} (p - c(y)) dy \tag{2.25}$$

and $\{Z_t\}$ is a sequence of i.i.d. random variables with mean 1. Note that this form (2.25) for the net return from one period's exploitation would follow from the following submodel for harvesting within a single season:

$$\frac{dy}{dt} = -qEy, \quad y(0) = x, \quad y(1) = x - h, \quad c(y) = \frac{c}{qy}$$

The assumed form (2.25) for net revenue π is sufficient to imply the optimal harvesting is bang–bang-singular. Let $Q(x)$ be an indefinite integral of $p - c(x)$, so that

$$\pi(x, h) = Q(x) - Q(x - h) \tag{2.26}$$

For $T = 0$ we have

$$\begin{aligned} J_0(x) &= \max_{0 \leqq h \leqq x} \pi(x, h) \\ &= \max_{0 \leqq h \leqq x} [Q(x) - Q(x - h)] \\ &= Q(x) - Q(x_\infty) \quad \text{if } x \geqq x_\infty \end{aligned}$$

where $p - c(x_\infty) = 0$ (and $J_0(x) = 0$ if $x < x_\infty$). The optimal policy in the terminal period is obviously to harvest all of the resource that is profitable.

For $T = 1$ we now obtain

$$\begin{aligned} J_1(x) &= \max_{0 \leqq h \leqq x} [Q(x) - Q(x - h) + \rho E_w\{J_0(wF(x - h))\}] \\ &= \max_{0 \leqq h \leqq x} [Q(x) - Q(s) + \rho E_w\{Q(wF(s)) - Q(x_\infty)\}] \\ &= \max_{0 \leqq h \leqq x} [V(s)] + Q(x) - \rho Q(x_\infty) \end{aligned}$$

where

$$V(s) = \rho E_w\{Q(wF(s))\} - Q(s) \tag{2.27}$$

Assume (as in Lecture 1) that $V(s)$ is concave and has a unique maximum at $s = s^*$. Then (2.27) implies that the optimal escapement is s^*, except when $x < s^*$, in which case $s = x$ is optimal. The optimal harvest policy for $T = 1$ is therefore given by

$$h = \begin{cases} x - s^* & \text{if } x \geq s^* \\ 0 & \text{if } x < s^* \end{cases} \tag{2.28}$$

and we have

$$J_1(x) = \begin{cases} Q(x) + V(s^*) - \rho Q(x_\infty) & \text{if } x \geq s^* \\ Q(x) + V(x) - \rho Q(x_\infty) & \text{if } x < s^* \end{cases} \tag{2.29}$$

Let us adopt the additional assumption that s^* is *self-sustaining*, in the sense that

$$\Pr(zF(s^*) \geq s^*) = 1 \tag{2.30}$$

The case $x < s^*$ can then be eliminated. Since $J_1(x) = Q(x) + \text{const}$ it is easy to see that

$$\begin{aligned} J_2(x) &= \max_{0 \leq h \leq x} [V(s)] + Q(x) + \rho V(s^*) - \rho^2 Q(x_\infty) \\ &= (1 + \rho)V(s^*) + Q(x) - \rho^2 Q(x_\infty) \end{aligned}$$

and the optimal harvest policy is again given by (2.28). By induction we conclude that (2.28) is optimal for all $T \geq 1$, and hence also in the limit as $t \to \infty$.

The policy (2.28) is called a "fixed-target escapement policy," since each harvest returns the escapement stock level s_k to the target s^*. This policy is completely analogous to the bang–bang singular (MRAP) policy derived in Lecture 1. The only difference is that yield now fluctuates from year to year, in response to fluctuations in annual recruitment. [If in fact s^* is not self-sustaining, then there will be years in which $x < s^*$ and no harvest is taken. Actually in this case $s^* = s_T^*$ depends on the time horizon T, but s_T^* converges to a stationary equilibrium s^* as $T \to \infty$; see Reed 1979.]

By differentiation of (2.27) we see that the necessary condition for the optimal escapement s can be written as

$$F'(s) \cdot \frac{p - E_w\{wc(wF(s))\}}{p - c(s)} = \frac{1}{\rho} \tag{2.31}$$

This is the stochastic, discrete-time analog of Eq. (1.22).

2.3 Nonlinearities

We have remarked that the validity of the simple MRAP harvest rule depends upon linearity of the harvesting model with respect to harvest rate h. In particular, this requires net revenue flow $\pi(x, h)$ to be a linear function of h. In the original continuous-time model we had

$$\left. \begin{aligned} \pi &= ph - cE \\ h &= qxE \end{aligned} \right\} \tag{2.32}$$

i.e. renenue is a linear function of harvest and cost is a linear function of effort, at least for $0 \leq h \leq h_{max}$.

Are these linearity assumptions reasonable? This depends largely on the relative *scale* of the resource industry being modeled. In the case of a small, local fishery, the rate of fish harvest is likely to have little effect on the market price of fish. But for larger fisheries, annual catch rates can have serious effects on price levels. An individual firm exploiting such a resource might well be motivated to limit catches simply to maintain a satisfactory price level.

Let $p(h)$ denote the inverse demand function; $p(h)$ is the price of fish when the harvest rate is h. We assume $p'(h) \leq 0$. The resource owner (if one exists) then wishes to

$$\text{maximize} \int_0^\infty e^{-\delta t} [p(h)h - c(x)h] \, dt \qquad (2.33)$$

$$\text{given } \frac{dx}{dt} = F(x) - h, \quad x(0) = x_0 \qquad (2.34)$$

To keep things simple, let us first ignore costs and consider the following problem:

$$\text{maximize} \int_0^\infty e^{-\delta t} U(h) \, dt \qquad (2.35)$$

$$\text{subject to } \frac{dx}{dt} = F(x) - h, \quad x(0) = x_0 \qquad (2.36)$$

where $U(h) = hp(h)$. We apply PMP to this problem, with Hamiltonian

$$\mathcal{H} = e^{-\delta t} U(h) + \lambda [F(x) - h]$$

Since h must maximize $\mathcal{H}$ we have

$$\lambda = -e^{-\delta t} U'(h)$$

$$\frac{d\lambda}{dt} = e^{-\delta t} \left[\delta U'(h) - U''(h) \frac{dh}{dt} \right]$$

By the adjoint equation

$$\frac{d\lambda}{dt} = -\frac{\partial \mathcal{H}}{\partial x} = -\lambda F'(x) = -e^{-\delta t} U'(h) F'(x)$$

These equations imply

$$\frac{dh}{dt} = [\delta - F'(x)] \frac{U'(h)}{U''(h)} \qquad (2.37)$$

and we still have the state equation

$$\frac{dx}{dt} = F(x) - h \qquad (2.38)$$

The phase-plane topology for Eqs. (2.37) and (2.38) is shown in Fig. 2.4. The optimal trajectory (x, h) must be one of the solution trajectories of this system.

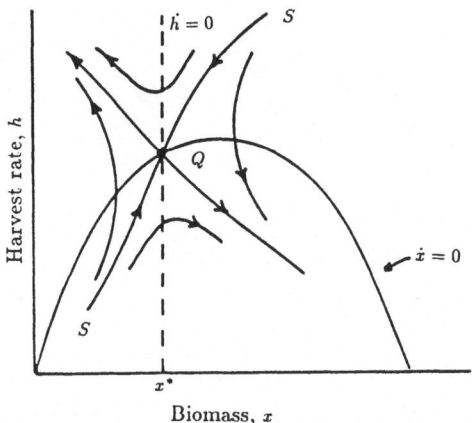

Fig. 2.4. Phase-plane diagram for optimal harvesting in a nonlinear control model. With an infinite time horizon the optimal harvest $h(x)$ is determined by the convergent separatrices S.

With an infinite time horizon, clearly only the convergent separatrix trajectories, labeled SQS in the figure, qualify. These curves completely specify the optimal harvest policy $h = h(x)$ in feedback form. The optimal policy converges towards an equilibrium biomass x^* and sustained yield h^* given by

$$\left. \begin{array}{r} F'(x^*) = \delta \\ h^* = F(x^*) \end{array} \right\} \tag{2.39}$$

Recall that this is the same formula obtained earlier, for the case of zero costs—see Eq. (1.23).

Although our nonlinear problem has an optimal equilibrium solution, the dynamic adjustment policy is no longer a bang–bang policy. Instead, it is an asymptotic approach policy. (In growth economics, this solution has been called an "asymptotic turnpike.")

When costs are included, the objective integral can be expressed as

$$\int_0^\infty e^{-\delta t} \pi(x, h) \, dt$$

The resulting equilibrium (x^*, h^*) is now determined by the equations

$$\left. \begin{array}{r} F'(x) + \dfrac{\partial \pi / \partial x}{\partial \pi / \partial h} = \delta \\ h = F(x) \end{array} \right\} \tag{2.40}$$

This includes both (2.39) and (1.22) as special cases.

Stochastic Model

In the stochastic case, nonlinearity may arise for a different reason, namely *risk aversion*. Thus one can formulate the discrete-time problem with risk aversion as:

$$\text{maximize } E\left\{ \sum_{t=0}^{T} \rho^t U(\pi(x_t, h_t)) \right\} \tag{2.41}$$

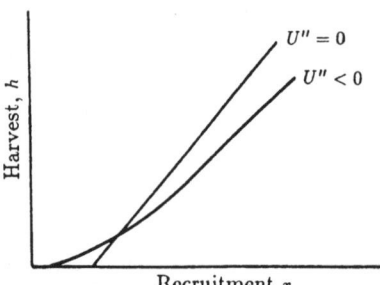

Fig. 2.5. Optimal feedback harvest $h(x)$ for linear utility $(U'' = 0)$, and for decreasing marginal utility $(U'' < 0)$.

subject to $x_{t+1} = F(s_t, w_t)$ (2.42)

where U is a concave (nonlinear) utility function. In this situation, the constant target escapement policy (2.29) is replaced by a curvilinear feedback harvest policy (Fig. 2.5), which must be determined by numerical methods.

2.4 Forestry

The classical economic model of forest exploitation concerns the optimal "rotation" period: at what age should a forest be harvested and replanted? The solution goes back to a German forester, M. Faustmann, in 1849.

Let $V(t)$ denote the net value of timber in a certain forest stand, as a function of age t. Let c denote the cost of logging and replanting. If the forest is logged at times $t_1 < t_2 < \cdots$ then the total present value of net revenues is given by

$$PV = \sum_{k=1}^{\infty} e^{-\delta t_k}[V(\Delta t_k) - c]$$ (2.43)

where $\Delta t_k = t_k - t_{k-1}$ $(t_0 = 0)$. We wish to determine the sequence $\{t_k\}$ that maximizes (2.43).

Because we assume that parameters do not change over time, it is apparent that in fact the optimal sequence is determined by a single optimal rotation period T, so that $t_k = kT$. Hence

$$PV = \sum_{k=1}^{\infty} e^{-k\delta t}[V(T) - c] = \frac{V(T) - c}{e^{\delta T} - 1}$$ (2.44)

The optimal rotation period T is obtained by differentiation:

$$\frac{V'(T)}{V(T) - c} = \frac{\delta}{1 - e^{-\delta T}}$$ (2.45)

This is the Faustmann formula. Notice that as $\delta \to 0$ the right side of (2.45) approaches $1/T$, and (2.45) then becomes the necessary condition for maximizing average net economic yield per unit time.

For typical temperate forests, it turns out that the optimal rotation period T, and the average long-term timber yield, $V(T)/T$, depend strongly on the discount rate δ. Clark (1976a, p. 261) quotes the following example for fir forests in British Columbia:

Annual discount rate	Optimal rotation age	Average annual yield
0	100 yr	$9.10/acre
0.03	70	7.10
0.05	63	4.20
0.10	49	2.80

The appropriate discount rate for use in resource management has long been a matter of controversy. (Faustmann's results apparently led to great arguments regarding German forest practice, even reaching the popular press of the day; Scott, 1972). At present, most government forest services seem to employ zero discounting with an objective of maximizing sustained yields. This could be highly suboptimal, as we will now see.

The Risk of Forest Fires

Now suppose that there is a constant risk λ per unit time that the forest will be destroyed by fire (see Reed, 1987). Thus forest fires are a Poisson process:

$$\Pr(\text{fire in } t, t+dt) = \lambda\, dt \tag{2.46}$$

Let $\tau_1, \tau_2, \ldots$ denote the times between successive destructions of the forest, whether due to fire or logging; we assume that the τ_i are i.i.d. random variables. As before, there exists an optimal age T for logging the forest. We have, for the random variable τ

$$F(t) = \Pr(\tau \le t) = \begin{cases} 1 - e^{-\lambda t} & \text{for } t < T \\ 1 & \text{for } t \ge T \end{cases} \tag{2.47}$$

The net economic return at each forest destruction is

$$Y = \begin{cases} -c_2 & \text{if } \tau < T \\ V(T) - c_1 & \text{if } \tau \ge T \end{cases} \tag{2.48}$$

where c_1, c_2 are the costs of logging and replanting, or of clearning and replanting a burned forest.

The expected present value is given by

$$PV = E\left\{ \sum_{n=1}^{\infty} e^{-\delta(\tau_1 + \cdots + \tau_n)} Y_n \right\} \tag{2.49}$$

Using the independence of the τ_i, we have

$$PV = \sum_1^\infty E\{e^{-\delta(\tau_1 + \cdots + \tau_n)}\} E\{e^{-\delta\tau_n} Y_n\}$$

$$= E\{e^{-\delta\tau} Y\} \sum_1^\infty E\{e^{-\delta\tau}\}^{n-1}$$

$$= \frac{E\{e^{-\delta\tau} Y\}}{1 - E\{e^{-\delta\tau}\}}$$

Now using (2.47) it is easy to show that

$$E\{e^{-\delta\tau}\} = \frac{\lambda + \delta e^{-(\lambda+\delta)T}}{\lambda + \delta}$$

$$E\{e^{-\delta\tau} Y\} = (V(T) - c_1)e^{-(\lambda+\delta)T} - \frac{\lambda c_2}{\lambda + \delta}(1 - e^{-(\lambda+\delta)T})$$

Hence we obtain

$$PV = \frac{(\lambda + \delta)(V(T) - c_1)e^{-(\lambda+\delta)T}}{\delta(1 - e^{-(\lambda+\delta)T})} - \frac{\lambda c_2}{\delta}$$

By differentiation it follows that the optimal T satisfies

$$\frac{V'(T)}{V(T) - c} = \frac{\delta + \lambda}{1 - e^{-(\delta+\lambda)T}} \tag{2.50}$$

In other words, the risk of fire is simply *added on* to the discount rate, a procedure which is often referred to as *risk discounting*.

It seems that this result is not yet well known among professional foresters. Even if the monetary discount rate δ is set equal to zero, it is still necessary to include risk discounting. Failure to do so can lead to a mis-specification of management policy, and to significant reductions in forest productivity (Reed and Errico, 1985).

Lecture 3

Investment Strategy

In the bioeconomic model of Lecture 1, net economic return was represented as

$$\pi = ph - cE$$

(Some generalizations were discussed in Lecture 2.) The costs of harvesting were assumed directly proportional to harvesting effort E. Thus the model only considers *variable* costs.

In actual resource industries fixed costs may be as important as variable costs. For example, fishing vessels costing in excess of $10 million per vessel are not exceptional in oceanic fisheries, e.g., for tuna or cod. Processing plants constitute another major fixed investment, as do sawmills and logging roads in the forest industry.

A simplistic analysis would treat fixed costs as once-and-for-all investments made at $t = 0$. In real life, however, fixed costs may be only relatively fixed. Capital equipment depreciates through wear and obsolescence. An additional complication arises in resource economics, since the resource base may itself vary over time. Indeed, probably one of the justifications underlying MSY as a management objective is stability, so that capital investments can be efficiently utilized.

In this lecture we describe a renewable resource model in which both the rate of exploitation and the rate of investment in equipment used for exploitation are treated as control variables (Clark, Clarke, and Munro, 1979). We continue to use the simple fishery model of Lecture 1:

$$\frac{dx}{dt} = F(x) - qEx \tag{3.1}$$

$$0 \leq E(t) \leq E_{max} \tag{3.2}$$

As noted previously, effort E could be quantified in terms of the number of fishing vessels. Thus E_{max} equals the size of the fishing fleet, which we will treat as "capital stock" K:

$$E_{max} = K \tag{3.3}$$

The dynamics of K are:

$$\frac{dK}{dt} = I - \gamma K \tag{3.4}$$

where I = rate of investment, and γ = rate of depreciation. Finally, the present value of net economic revenues is given by

$$PV = \int_0^\infty e^{-\delta t}(pqxE - c_1 E - c_2 I)\,dt \tag{3.5}$$

where c_1 denotes the (*variable*) cost of effort, and c_2 is the cost of investment in fishing vessels.

The mathematical problem, then, is to determine the optimal effort and investment policies, maximizing (3.5) subject to (3.1)–(3.4), with $x(0)$, $K(0)$ given. This problem is much more difficult than the simple problem of Lecture 1, primarily because there are now two "state variables," x and K. The complete solution will be described here, together with the underlying intuition. Mathematical details of the proof can be found in the cited reference.

Case 1. Reversible Investment
Suppose first that investment I is unconstrained:

$$-\infty \leq I \leq +\infty \tag{3.6}$$

(The case $I = +\infty$ corresponds to a "pulse" investment, i.e. an instantaneous jump in the level of capital K.) This means that fishing vessels can be purchased or sold in any amount, at any time, and at the same price c_2 for both purchase and sale.

By integration by parts,

$$\int_0^\infty e^{-\delta t} I\, dt = \int_0^\infty e^{-\delta t}\left(\frac{dK}{dt} + \gamma K\right)dt$$

$$= (\delta + \gamma)\int_0^\infty e^{-\delta t} K\, dt - K_0$$

Hence the objective (3.5) becomes

$$PV = \int_0^\infty e^{-\delta t}(pqxE - c_1 E - c_2(\delta + \gamma)K)dt + c_2 K_0$$

Furthermore, if K can be instantaneously adjusted, it is obvious that one will never retain unused fishing capacity, i.e. $E(t) \equiv K(t)$. Hence we have

$$PV = \int_0^\infty e^{-\delta t}(pqx - (c_1 + c_2(\delta + \gamma)))E\, dt$$

$$= \int_0^\infty e^{-\delta t}(pqx - c_{\text{total}})E\, dt \qquad (3.7)$$

with $c_{\text{total}} = c_1 + c_2(\delta + \gamma)$. This has a simple interpretation: since capital K is completely reversible, all costs of fishing can be treated as variable costs. Total cost c_{total} consists of operating cost c_1 plus interest and depreciation on capital, $\delta c_2 + \gamma c_2$.

Now (3.7) is identical with (1.13), the objective PV for our original model.[*] Hence the optimal policy has an equilibrium (singular) solution x_{total}^* given by Eq. (1.22), with $\pi(x) = p - c_{\text{total}}/qx$. The optimal controls are bang–bang: at $t = 0$ a maximum fleet of vessels $K_{\max}$ should be purchased, with $E = K_{\max}$ reducing x_0 to x_{total}^*, and then all except $E_{\text{total}}^* = F(x_{\text{total}}^*)/qx_{\text{total}}^*$ vessels should be sold off. Clearly the assumption of completely reversible investment is a bit unrealistic. However, this case helps to understand the more interesting cases of irreversible investment.

Case 2. Irreversible Investment

Making the opposite assumption to case 1, we now suppose that disinvestment is completely impossible:

$$0 \leq I \leq +\infty \qquad (3.8)$$

(A less severe assumption might be that disinvestment is possible, but at a price smaller than c_2; this case is fully analyzed in the reference.)

We will use PMP to obtain the singular solutions to our problem. Write

$$E(t) = \phi(t)K(t)$$

---

[*] Actually we should also include the constraint $0 \leq E \leq E_{\max} = K_{\max}$.

so that (3.2) now becomes

$$0 \le \phi(t) \le 1 \tag{3.9}$$

We treat ϕ and I as the control variables. The Hamiltonian is then

$$\mathcal{H} = e^{-\delta t}[(pqx - c_1)\phi K - c_2 I] + \lambda_1[F(x) - qx\phi K] + \lambda_2[I - \gamma K]$$
$$= \phi K[e^{-\delta t}(pqx - c_1) - \lambda_1 qx] + I[\lambda_2 - e^{-\delta t}c_2] + \lambda_1 F(x) - \lambda_2 \gamma K \tag{3.10}$$

Since $\mathcal{H}$ is linear in both control variables, various possibilities of singular control arise.

(a) I Singular

If the coefficient of I in Eq. (3.10) vanishes, we have

$$\lambda_2 = e^{-\delta t}c_2 \tag{3.11}$$

$$\frac{d\lambda_2}{dt} = -\delta e^{-\delta t}c_2$$

By the adjoint equation,

$$\frac{d\lambda_2}{dt} = -\frac{\partial \mathcal{H}}{\partial K} = -\phi[e^{-\delta t}(pqx - c_1) - \lambda_1 qx] + \lambda_2 \gamma \tag{3.12}$$

There are three possibilities for the control ϕ: $\phi = 0$, $\phi = 1$, or ϕ singular. If ϕ is singular, then

$$e^{-\delta t}(pqx_1 - c) - \lambda_1 qx = 0$$

so that

$$\frac{d\lambda_2}{dt} = \lambda_2 \gamma = \gamma e^{-\delta t}c_2$$

But this contradicts Eq. (3.11). Hence both I and ϕ cannot be simultaneously singular (it cannot be optimal to be investing but only utilizing part of the fleet simultaneously).

Similarly, $\phi = 0$ leads to a contradiction.

For $\phi = 1$, Eqs. (3.11) and (3.12) imply that

$$\lambda_1 = e^{-\delta t}(p - c_{total}/qx), \quad c_{total} = c_1 + c_2(\delta + \gamma) \tag{3.13}$$

Now use the adjoint equation $d\lambda_1/dt = -\partial \mathcal{H}/\partial x$ to conclude (as in Sec. 2.1) that

$$F'(x) - \frac{c'_{total}(x)F(x)}{p - c_{total}(x)} = \delta$$

where $c_{total}(x) = p - c_{total}/qx$. Thus $x = x^*_{total}$ in this case. We will see later that x^*_{total} is the unique long-term optimal equilibrium solution to our model of irreversible investment. The optimal controls for this equilibrium solution are $\phi = 1$, and $I = \gamma K^*_{total}$ where $K^*_{total} = f(x^*_{total})/qx^*_{total}$.

(b) $I = 0$

This is the only remaining possible case (except for $I = +\infty$, which can only occur instantaneously). If ϕ is singular, then by (3.10) we have

$$\lambda_1 = e^{-\delta t}(p - c_1/qx)$$

The same calculation using the adjoint equation, now yields

$$F'(x) - \frac{c_1'(x)F(x)}{p - c_1(x)} = \delta$$

where $c_1(x) = c_1/qx$. Thus $x = x_1^*$, the optimal sustained biomass corresponding to variable costs c_1 only. Since $c_{total} > c_1$ we have

$$x_1^* < x_{total}^* \tag{3.14}$$

From the above calculations we see that there are *two* constant singular solutions $x_1^* < x_{total}^*$. It only remains to show how bang–bang controls for ϕ and I are used to coverage to one of these singular solutions. The result, which is a bit complicated to describe (and *very* complicated to prove!) will be described with reference to Fig. 3.1.

Observe first that x_1^* and x_{total}^* have the following interpretations: x_1^* is the optimal biomass for the case of zero capital costs. In an unnecessarily large fleet K has been purchased (e.g. by mistake), then capital costs would be irrelevant to present management decisions, and x_1^* would be the optimal resource biomass. On the other hand, whenever further investment is under consideration, its cost must be considered, and the optimal equilibrium biomass in then x_{total}^*, which is determined by *total* costs, $c_1 + (\delta + \gamma)c_2$. Investment will never be optimal when $x < x_{total}^*$.

Figure 3.1 is a feedback control diagram, specifying the optimal controls ϕ and I as functions of the current state variables x and K. There are three control regions R_1, R_2, and R_3, separated by two switching curves S_1, S_2. Region R_1 corresponds to an underexploited resource, $x > x_{total}^*$, with undercapitalization.

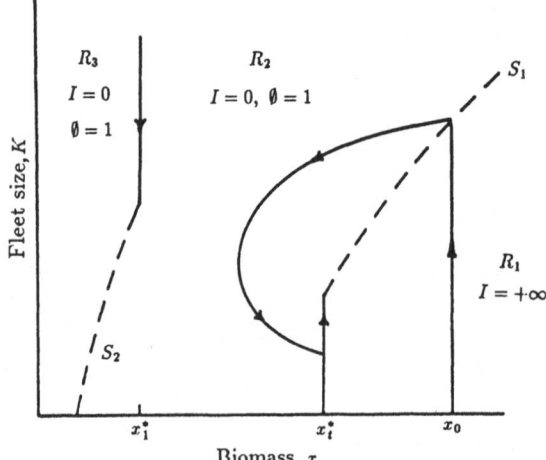

Fig. 3.1. Feedback control diagram for optimal investment I and harvesting effort E. An optimal trajectory, starting at (x_0, K_0) with $x_0 > x_{total}^*$ and $K_0 = 0$ is also shown.

An infinite investment rate is used to increase K to its optimum value as given by the switching curve S_1. Note that the larger the initial biomass x_0 the greater the level of investment.

Harvesting the resource at full capacity now begins, and the trajectory (x, K) lies in region R_2 $[\phi = 1, I = 0]$. Since $I = 0$, fleet size K decreases exponentially due to depreciation ("wear and tear," sinking of vessels,...). The resource stock x is at first rapidly reduced, but eventually the depreciating fleet is no longer capable of making such large catches, and the resource eventually begins to increase again. Ultimately x reaches x^*_{total}, at which time an additional pulse investment becomes optimal, raising K to $K^*_{total} = F(x^*_{total})/qx^*_{total}$. Here finally a steady state is reached, with $\phi = 1$ and with singular investment $I = \gamma K^*_{total}$.

To recapitulate, the profit-maximizing strategy for renewable resource exploitation with irreversible investment consists of three phases:

1. initial "overinvestment," in order to generate large initial revenues from the unexploited resource,
2. resource exploitation using existing (depreciating) capacity, and
3. an ultimate sustained yield phase, with continuous investment to make up for depreciation.

While this result may seem unduly complex, it is worth remarking that "boom-and-bust" cycles in resource development are actually extremely common. Some authors have described such cycles as highly irrational; it can be shown that an unregulated common-property resource industry will experience a more drastic cycle, with larger initial investment and a greater degree of resource stock reduction (and finally a lower equilibrium biomass) than is optimal (see below).

The above model has been fitted to data from the Antarctic whaling industry by Clark and Lamberson (1982). With an unexploited initial biomass of 400,000 BWU, the optimal policy initially employs 13 factory fleets ($20 million per fleet), which reduce whale stocks to 80,000 BWU in the first 15 years of whaling. Equilibrium is reached after an additional 16 years of whaling, with $x^*_{total} = 11500$ BWU and $K^*_{total} = 1$ factory fleet.

In actual fact, following World War II, some 26 new whale factory fleets were constructed, whereas the optimum would have been 9 (whale stocks in 1946 were only about 250,000 BWU). Whale stocks were reduced to around 75,000 BWU before whaling was finally stopped in 1979. (Blue and fin whales were protected prior to 1979.)

We have not yet discussed the control region R_3 in Fig. 3.1. This is a "moratorium" region, with both $\phi = 0$ and $I = 0$. A profit maximizer would never deliberately drive the resource stock into this region, but unregulated exploitation might well do so. The boundary between regions R_2 and R_3 is formed partly by the switching curve S_2, where the fishing effort control ϕ switches from 0 to 1, and partly by a segment of the vertical line $x = x^*_1$, where the singular control $\phi(t) = F(x^*_1)/qK(t)x^*_1$ is used. This latter situation corresponds to optimal sustained yield without investment costs.

The two switching curves S_1 and S_2 are determined as follows. Let $S(x, K)$ denote the present value of future revenues, starting with $x(0) = x$, $K(0) = K$, and using the controls $E(t) \equiv K(t)$, $I \equiv 0$ until $x(t)$ returns to x^*_{total} from below (see the trajectory in Fig. 3.1), thereafter jumping from K to K^*_{total} and remaining at the long-run equilibrium $(x^*_{total}, K^*_{total})$. The curve S_1 is then determined by the equation

$$\frac{\partial S(x, K)}{\partial K} = c_2 \tag{3.15}$$

This equation has the interpretation that the marginal return to capital should equal the cost of capital; this is the standard optimal investment rule.

Similarly, the curve S_2 is determined by the equation

$$\frac{\partial S(x, K)}{\partial x} = p - \frac{c}{qx} \tag{3.16}$$

which can be considered as an optimal rule for "investment in fish," $p - c/qx$ being the opportunity cost of a unit of fish. Equations (3.15) and (3.16) can be used as the basis of a numerical algorithm for computing the switching curves. [There is a slight additional complication in cases where the (x, K) trajectory crosses both switching curves, but this can be handled in a fairly obvious way—see the reference.]

The Unregulated Case

The investment cycle in an unregulated, common property fishery has been studied by McKelvey (1985). The theoretical predictions are qualitatively similar to the profit-maximizing case discussed above, with the difference that x^*_1 is replaced by $x_{1\infty} = c_1/pq$ and x^*_{total} by $x_{total\,\infty} = c_{total}/pq$. These are the two bionomic equilibria for variable and total effort cost, respectively.

The prediction of two different bionomic equilibria disagrees, of course, with the original Gordon model, which did not attempt to distinguish fixed and variable costs. What the new model predicts is an original phase of expansion of fishing capacity, resulting in overcapitalization. This phase is followed by an intermediate phase with vessels neither leaving nor entering the fishery, except possibly through bankruptcy. If regulation of the fishery were to be initiated at this time, temporary overfishing (while vessels depreciate) might well be optimal. Thus a complete moratorium on fishing whenever $x < x^*_{total}$ (as suggested by the simple variable-cost model) is not generally the optimal recovery policy.

Stochastic Aspects

A model of optimal investment for a stochastically fluctuating resource stock has been developed by Charles (1983). In this model, investment (including replacement of depreciated capital) is optimal when resource abundance is higher than usual, and is not optimal when resource stocks are low. Similar results would presumably prevail under price fluctuations.

One aspect of resource investment which is often important is the fact that the *size* of the resource stock may be highly uncertain before exploitation commences. If so, one cannot determine the optimal initial level of capacity (K_0)

by the methods described above. One way to approach the problem is to make an initial estimate of the resource stock (and its productive potential), and then to proceed as before, on the assumption that the estimate is correct. The consequences of an error in the estimate are asymmetric: an overestimate of resource potential can lead to wasteful overexpansion, and also to overexploitation of the resource. An underestimate will result in underexpansion, but this can be corrected later, much more easily than overexpansion. The asymmetry of this is due to the irreversibility of investment.

Problems of decisions under uncertainty are studied in the literature on Decision Theory, and Operations Research. Most of the theory, however, applies only to static problems, and is not much use in resource situations. A dynamic Bayesian decision model of optimal initial fishery investment with uncertain stock size is described by Clark et al. (1985).

Lecture 4

Theory of Resource Regulation

The principal reason for developing models of resource economics is to obtain policy implications for resource management. In the first three lectures, however, we only addressed policy matters indirectly, comparing the undesirable results of non-regulation with the supposedly optimal results of complete centralized control. The latter situation really only pertains to an individual owner of the resource stock—an unlikely situation for many renewable resources. For practical purposes it therefore becomes important to extend these models to cases involving the regulation of multiple resource users.

The ideal solution to the common-property problem is the establishment of inviolable individual resource rights. In socialist countries this ideal is achieved by fiat: the state owns all resources and presumably manages them for maximum social benefit. Whether economic rent is still dissipated by the massive bureaucracy is not within the scope of these lectures.

In capitalist countries, private ownership of resources is normal in cases where rights are easily defined and secured—for example, in agriculture. It should also be the case that externalities are minor, and easily controlled. Thus forests may remain in the public domain because of significant externalities related to water flow, recreational use, etc.

In the case of fishery resources, private ownership is unusual, except for sedentary species, such as shellfish, which are readily protected from poachers. In some countries, such as Japan, coastal fishing communities may possess recognized exclusive rights to fish close to their shores. However, most marine fisheries are still exploited by competing fishermen and fishing companies. Under these circumstances, the dual problems of overfishing and overexpansion of fishing capacity remain paramount.

In order to be explicit, this chapter will describe models of regulation of commercial fisheries; further details appear in Clark (1980).

4.1 A Behavioral Submodel

In modeling fishery regulation it is first necessary to develop a submodel of fishermen's behavior. By one "fisherman" we will mean an individual vessel owner/captain, who makes continual decisions on how intensively to fish. For simplicity we ignore many other types of decisions, such as search strategy (Mangel and Clark 1983), investment strategy (Clark 1980, McKelvey 1985), discard strategy (Clark 1982), and so on.

We will suppose that the fisherman can continuously control the rate of effort, $E(t)$, exerted by his vessel. This might be achieved by fishing more hours per day, or more days per month, or by hiring more helpers, cruising at greater speed, etc. Catch rate is still given by

$$h = qE(t)x \tag{4.1}$$

but we now assume a typical cost-of-effort curve (Figs. 4.1, 4.2), with marginal cost at first decreasing, then increasing.

The individual fisherman takes price as fixed, so that his net revenue flow is given by

$$\pi = ph - c(E) = pqxE - c(E) \tag{4.2}$$

In the absence of any regulation, we assume that the fisherman simply maximizes his net revenue flow; hence

$$\begin{cases} c'(E) = pqx & \text{if } pqx > r \\ \quad E = 0 & \text{if } pqx < r \end{cases} \tag{4.3}$$

where r denotes minimum marginal cost ($=$ minimum average cost). In economics, the segment of the marginal cost curve to the right of E_r is called the *supply curve* for (individual) effort.

Different types of fishery regulation will have different effects on Eq. (4.3), as will be explained in the sequel.

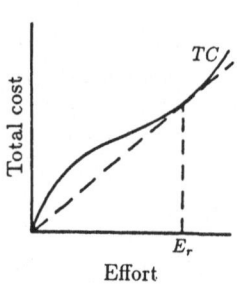

Fig. 4.1. Total cost of effort per vessel.

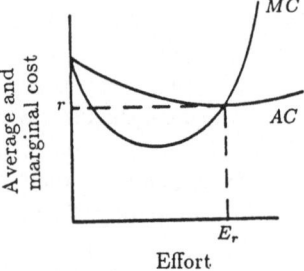

Fig. 4.2. Average and marginal effort cost per vessel. The marginal cost curve MC lying to the right of E_r is the vessel's effort supply curve.

4.2 The Unregulated Common-Property Fishery

Note that the value of E given by Eq. (4.3) is an increasing function of x (fishermen work harder when fish are more aboundant). Write this function as

$$E = E(x)$$

Then individual net revenue flow is

$$\pi = \pi(x) = pqxE(x) - c(E(x)) > 0 \text{ provided } x > x_\infty = \frac{r}{pq} \tag{4.4}$$

Suppose now that N identical vessels enter the fishery. Using our basic model of population dynamics (Lecture 1), we then have

$$\frac{dx}{dt} = F(x) - NqxE(x) \tag{4.5}$$

Bionomic equilibrium is reached when $\pi = 0$ and $dx/dt = 0$, i.e. when $x = x_\infty$ and $N = N_\infty$, where

$$N_\infty = \frac{F(x_\infty)}{qx_\infty E(x_\infty)} \tag{4.6}$$

where

$$\pi(x_\infty) = 0 \quad \text{and} \quad E(x_\infty) = r \tag{4.7}$$

(A more detailed analysis would also include fixed costs, see Clark (1980) and McKelvey (1985). Also see Clark (1980) for the case of nonidentical vessels.)

4.3 Optimization

Suppose next, as an extreme example, that a management authority can control both the number of vessels, $N = N(t)$, and the effort $E(t)$ of each vessel. The authority wishes to

$$\underset{N,E}{\text{maximize}} \int_0^\infty e^{-\delta t} N[pqxE - c(E)]\, dt \tag{4.8}$$

$$\text{subject to } \frac{dx}{dt} = F(x) - qxNE, \quad x_0 \text{ given} \tag{4.9}$$

The Hamiltonian for this problem is

$$\mathcal{H} = e^{-\delta t}\{N[pqxE - c(E)] + \mu[F(x) - qxNE]\} \tag{4.10}$$

where $\mu = e^{-\delta t}\lambda$, which is called the "current" adjoint variable.

This model is linear in N (including fixed costs would improve this aspect of the model), so there is a singular solution for N, which occurs when

$$c(E) = (p - \mu)qxE \tag{4.11}$$

We must also have $\partial \mathcal{H}/\partial E = 0$, which implies that

$$c'(E) = (p - \mu)qx \qquad (4.12)$$

These two equations imply that marginal cost $=$ average cost, i.e.

$$E = E_r \qquad (4.13)$$

It is interesting to observe that the optimal level of effort per vessel, E_r is the same as the level of effort used in the unregulated fishery. Because the adjoint variable μ turns out to be positive, however (it represents the "shadow price" of the resource stock x, which is positive unless the resource is economically valueless), the facts that

$$c'(E_r) = (p - \mu)qx \quad \text{for optimum}$$
$$c'(E_r) = pqx \qquad \quad \text{for nonregulated case}$$

show that (for singular N) the optimal biomass x is always larger than the bionomic equilibrium.

The optimal equilibrium values of x and N can be obtained from the maximum principle, which readily yields the necessary condition (with $d\mu/dt = 0$)

$$F'(x) + \frac{rF(x)}{qx^2(p - r/qx)} = \delta \qquad (4.14)$$

$$F(x) = qxNE_r \qquad (4.15)$$

Letting x^*, N^*, E^* denote optimal equilibrium values, and x_∞, N_∞, E_∞ denote bionomic (unregulated) equilibrium, we therefore have shown that

$$E^* = E_\infty = E_r$$
$$x^* > x_\infty \qquad (4.16)$$
$$N^* < N_\infty$$

Also, of course, $\pi^* > \pi_\infty = 0$. These results occur with the original Gordon (1954) model, which predicted overexploitation of unregulated common-property resources.

4.4 Regulation

Taxes

Is there some way to achieve the optimal fishing policy, other than by completely centralized control? One suggestion immediately follows from Eq. (4.11), namely a *tax* τ, equal to the adjoint variable μ, levied on all landed fish. For in this case the net revenue to the individual fisherman is simply reduced by the tax:

$$\pi_\tau = (\rho - \tau)qxE - c(E) \qquad (4.17)$$

If the fisherman chooses E so as to maximize π_τ, then

$$c'(E) = (p - \tau)qx \qquad (4.18)$$

which is the same as (4.12) if $\tau = \mu$. Since the other model equations are unchanged, the taxed fishery has the same equilibrium solution as the optimal fishery. (Since we have not included fixed costs, it is not so clear how the taxed fishery's dynamics will compare with the optimum; see McKelvey, 1985, for a detailed analysis of this question.)

It is a standard result of welfare economics that externalities can be corrected by means of taxes. Unfortunately, however, taxes are unpopular with resource industries. While one might argue for the fairness of taxes (sometimes called royalties, from the days when resources belonged to the king) on the exploitation of publicly owned resources, it is still worthwhile to ask whether other approaches to management of common-property resources might also work.

Licenses

Recall from Eq. (4.16) that the equilibrium levels of effort are the same for the optimal and the unregulated cases, but that the unregulated fishery has an excess of vessels: $N_\infty > N^*$. This might suggest that one could achieve the optimal result by simply restricting vessel numbers to N^*, by a system of issuing fishing vessel licenses. Let us show that in fact this method does not work.

First we recall the equilibrium equations in the unregulated case:

$$c'(E) = pqx$$
$$\pi = pqxE - c(E) = 0$$
$$F(x) = qxNE$$

These equations determine the bionomic equilibrium x_∞, N_∞, and E_∞.

Now suppose $N = N_0$ is fixed by licensing, with $N_0 < N_\infty$. Then only two equations hold:

$$c'(E) = pqx$$
$$F(x) = qxN_0E$$

These equations determine x_0 and E_0; we have $\pi_0 = pqx_0E_0 - c(E_0) > 0$: each licensed vessels earns a positive net income.

Recall, however, that for the optimal solution we have Eq. (4.11):

$$c'(E) = (p - \mu)qx$$

where $\mu > 0$. Consequently the licensing method cannot achieve optimality—the licensed vessels exert an excess of effort.

In the past decade, limited vessel licensing has been introduced in many commercial fisheries. What has invariably happened is that licensed vessel owners have then spent large sums of money to increase the fishing power of their vessels, for example by upgrading engines, adding sophisticated electronics, building in freezing plants, and so on. In other words, the license holder is motivated to increase his level of effort, as predicted by our simple model. (A more detailed model allowing for individual investment decisions does indeed predict increased investment.) The ultimate total fishing power may be as great as, or greater than, the level that would be reached without licenses.

Why is it that fisherman behave so "perversely"? After all if all licensed fishermen simply utilized the optimal effort level E^*, then all would profit handsomely. But then any *given* individual fisherman can further increase his profits by increasing E—i.e. we still have "The Tragedy of the Commons" as described by Hardin (1968).

Total Catch Quotas

Another popular type of regulation in fisheries has been the Total Allowable Catch (TAC) system. In this approach, an annual quota for a given species is determined on the basis of biological studies of productivity and current stock levels. The management authority keeps track of the total catch as the fishing season progresses, and closes the fishery when the target quota has been taken.

From the economic point of view, this is probably the most disastrous regulation system, even if combined with vessel licensing. Each individual vessel must compete viciously for a share of the catch before the fishery is closed. Fish quality becomes of secondary importance. Safety precautions are ignored. Markets and processing facilities are saturated during a brief annual season of feverish fishing activity.

Of course total catches must be controlled, but this is a necessary condition, not a sufficient condition for rational management.

Allocated Catch Quotas

The situation changes completely if we assume that the annual catch quota is *allocated* to individual vessels. Let us assume that the ith fisherman has a quota Q_i, which allows him to fish at the rate Q_i:

$$h_i = qE_i x \leq Q_i \tag{4.19}$$

If a given fisherman does not choose to use his entire quota, we assume that he can put it up for sale to other fishermen. Conversely, he can attempt to purchase additional quota units from other fishermen. Let m denote the unit price at which quotas are traded. The *total* quota is fixed:

$$\sum_{i=1}^{N} Q_i = Q \tag{4.20}$$

Net revenue is given by

$$\pi_i = ph_i - c(E_i)$$
$$= ph_i - c\left(\frac{h_i}{qx}\right) \tag{4.21}$$

Each fisherman attempts to maximize π_i:

$$\underset{0 \leq h_i \leq Q_i}{\text{maximize } \pi_i(h_i)} \tag{4.22}$$

Now suppose the maximum is at $h_i = Q_i$. Then the fisherman will wish to purchase additional quota units if and only if

$$\frac{\partial \pi_i}{\partial h_i}(Q_i) > m$$

This means that the equation

$$\frac{\partial \pi_i}{\partial h_i} = m \tag{4.23}$$

determines the ith fisherman's *demand function* $D_i(m)$ for quota units. The total demand is

$$D(m) = \sum_{i=1}^{N} D_i(m) \tag{4.24}$$

and the market-clearing equation (supply equals demand) is then simply

$$D(m) = Q \tag{4.25}$$

Carrying our the differentiation in (4.23) and assuming market clearing (so that $h_i = Q_i = D_i(m)$), we obtain

$$c_i'(E_i) = (p - m)qx \tag{4.26}$$

Note that this equation has the same form as (4.11):

$$c'(E) = (p - \mu)qx \quad (\mu = \text{shadow price})$$

and also (4.18):

$$c'(E) = (p - \tau)qx \quad (\tau = \text{tax on catch})$$

We conclude that allocated, transferable quotas are mathematically equivalent, in terms of control of fishing effort, to taxes on catch. Both are capable of achieving optimal exploitation...in theory. I say "in theory" because quite a few assumptions underlie this result.

From the point of view of fishermen's incomes, taxes and allocated quotas obviously have quite different implications. In an optimizing tax system, all the economic rent from the fishery accrues to the government, while in a quota system it all accrues to the fishermen. But of course it is possible to combine taxes and quotas: suppose the tax is τ_1, and fishermen have allocated quotas. Redoing the above calculations results now in the condition

$$c_i'(E_i) = (p - \tau_1 - m_1)qx \tag{4.27}$$

where $\tau_1 + m_1 = m$. In this case the government receives τ_1 instead of τ, and the fisherman receives $\tau - \tau_1$.

Many governments are currently considering the introduction of allocated quota systems (Iceland has already done so). It should be pointed out that there are quite a few significant complications, including:

—monitoring and enforcement of quotas
—adjustment of quotas to stock fluctuations
—inducement to discard the less valuable sizes or species of fish.

The economic benefits of an allocated catch system can be large, however. My prediction is that countries that introduce successful quota (or quota plus tax) systems will derive substantial economic benefits from their fishery resources, whereas those that fail to do so will continue to find their fisheries a drain on the economy.

Lecture 5

Age-Structured Models

The forestry models discussed in Lecture 2 involved age structures, but in a trivial way (all trees in a forest were assumed to have the same age). In this lecture we describe two types of age-structured models used in fisheries. The second of these, the famous dynamic-pool model of Beverton and Holt, leads to unsolved optimization problems.

5.1 A Delayed Recruitment Model

The spawner-recruit model discussed in Lecture 2

$$x_{t+1} = F(s_t) \tag{5.1}$$

$$s_t = x_t - h_t \tag{5.2}$$

assumes either nonoverlapping generations, or else that fish reach sexual maturity after one period. This is realistic for a few species of fish (and for most species of insects), but not for many others.

A more general model, which allows for a lengthy period to reach sexual maturity, is the following:

$$x_{t+1} = \sigma s_t + F(s_{t-n}) \tag{5.3}$$

where x_t, s_t again denote recruitment and escapement of adults in year t, and where $\sigma \in (0, 1)$ is the annual survival rate, and $n \geq 0$ the period of sexual maturity. Only adults are assumed to be harvested; recruitment to the adult population x_{t+1} is determined by escapement (s_{t-n}) which occurred $n + 1$ years previously. This model has been employed in the study of marine mammal populations. The local stability properties are discussed by Clark (1976b).

What effect does the introduction of a delay term have on harvest policy? Consider the following deterministic optimization problem:

$$\text{maximize} \sum_{t=0}^{\infty} \rho^t \pi(x_t, h_t) \tag{5.4}$$

$$0 \leq h_t \leq x_t \tag{5.5}$$

$$\text{subject to } x_{t+1} = \sigma s_t + F(s_{t-n}) \tag{5.6}$$

$$s_t = x_t - h_t \tag{5.7}$$

where, as in Lecture 2

$$\pi(x,h) = \int_{x-h}^{x} (p - c(y)) \, dy \tag{5.8}$$

Proceeding formally, introduced the Lagrangian

$$\mathcal{L} = \sum_{t=0}^{\infty} \{\rho^t \pi(x_t, h_t) - \lambda_t [x_{t+1} - \sigma(x_t - h_t) - F(x_{t-n} - h_{t-n})]\} \tag{5.9}$$

We have the necessary conditions

$$\frac{\partial \mathcal{L}}{\partial x_t} = 0 \quad (t \geq 1) \qquad \text{and} \qquad \frac{\partial \mathcal{L}}{\partial h_t} = 0 \quad (t \geq 0) \tag{5.10}$$

Let us attempt only to find an equilibrium solution with escapement s:

$$\begin{aligned} x_t &= x = \sigma s + F(s) \\ h_t &= h = x - s \end{aligned} \tag{5.11}$$

Then a simple calculation yields the necessary condition

$$[\sigma + \rho^n F'(s)] \frac{p - c(F(s))}{p - c(s)} = \frac{1}{\rho} \tag{5.12}$$

for the optimal escapement s. This equation is similar to Eq. (2.31), and in fact the two equations become identical if the two models are specialized so as to agree, i.e. $\sigma = 0$, $n = 0$ in (5.12), and $w = 1$ in (2.31).

How much difference does the delay model make? Clark (1976b) calculated optimal harvest policies for Antarctic whale data, using both the delay model, and the simple continuous-time model of Lecture 1 (with intrinsic growth rate adjusted for delayed recruitment). The numerical results were the same to within about 2%. It is also shown by dynamic programming that the bang–bang approach policy is not optimal, but almost optimal, for the delay model. These results lend some credibility to the use of simplified models of population dynamics in resource modeling. However, if the age at which animals are harvested is important, then a more structured model is required.

5.2 The Dynamic Pool Model

The model which is most commonly used in fisheries management is the dynamic pool or cohort model, originally developed for North Sea fisheries by Beverton and Holt (1957). In this model, a fish population is depicted as consisting of a number of "cohorts," a cohort consisting of all fish of a given age. A cohort may be indexed by the year at which it originally becomes available or "recruits" to the fishery:

$N_k(t) =$ number of fish from year k recruitment
 alive at time $t \geq k$

Thus the (post-recruitment) age of fish in this k^{th} cohort is $t - k$. (Fish may not become recruits until their actual age is several years.)

Once recruited, fish are subject to both natural and fishing mortality, which are assumed to be additive:

$$\frac{dN_k}{dt} = -(M(t) + F_k(t))N_k \quad t \geq k \tag{5.13}$$

$$N_k(k) = R_k \tag{5.14}$$

or

$$N_k(t) = R_k \exp\left(-\int_k^t (M(u) + F_k(u))\,du\right) \tag{5.15}$$

If $w(a)$ denotes the average weight of fish at age a, then the expression

$$B_k(t) = N_k(t)w(t - k) \quad t \geq k \tag{5.16}$$

is the *cohort biomass*. The *unfished* cohort biomass is therefore (assuming $M =$ constant)

$$B_{k0}(t) = R_k e^{-M(t-k)} w(t - k) \tag{5.17}$$

A popular growth curve is the cubic exponential, or von Bertelanffy curve:

$$w(a) = w_\infty [1 - e^{-k(a-a_0)}]^3 \tag{5.18}$$

A typical unfished cohort biomass curve is then as illustrated in Fig. 5.1.

This curve suggests a management policy for maximizing yield: each cohort should be ultimately fished intensively near the time of maximum cohort biomass. If one assumes that

$$F_k(t) = q_k E(t)$$

and

$$0 \leq E(t) \leq E_{max}$$

then it is easy to formulate an optimal control model for maximizing total yield by weight, or total discounted net economic yield, from the given cohort. See Clark

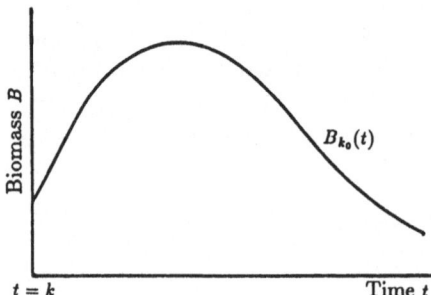

Fig. 5.1. Unfished cohort biomass $B_{k0}(T)$.

(1976a, Ch. 8) for details. Not surprisingly, the optimal policy can be quite sensitive to the rate of discount.

The major practical difficulties arise with this approach:

1. *Nonselectivity.* In most fisheries it is not possible (or certainly not economically feasible) to harvest each cohort separately. When nets are placed in the water, they capture all fish except those that are small enough to swim through the mesh. *Partial* selectivity may be obtainable by adjusting the mesh size of nets.
2. *Stock-Recruitment Relation.* The objective of maximizing the yield from each cohort ignores the relationship between adult fish stock and subsequent recruitment. When this effect is included, it is clear that a less intensive harvest policy becomes optimal.

It is clearly possible to construct an optimization model incorporating these two effects. Any such model, however, would suffer from several difficulties. First, computation of the optimal harvest policy would be extremely difficult, perhaps even impossible. (See Horwood and Whittle (1986) for a linearization approach to the solution of multicohort optimization models.) Second, the model would still be grossly unrealistic, because of the significant natural fluctuations exhibited by all fish populations. Third, some model components, especially the average stock-recruitment relation, are extremely poorly understood for most fish stocks.

Fishery managers are thus forced to make various compromises. First, it seems reasonable to try to separate the biological and economic objectives of management. The fishery biologist typically conceives of his mission as one of recommending annual catch quotas so as to achieve the maximum long-run yield (i.e. MSY). Very often this objective appears to be quite unsatisfactory to the fishing industry. Nevertheless, I argue that this is probably the best basic *biological* objective for fishery management. Economic objectives such as efficiency of production, quality control, optimal marketing, and so on, can be treated as being subject to the overriding biological objective. Most of the latter objectives would be encouraged by using allocated quota systems (possibly with royalties), as described in Lecture 4.

Let us conclude by considering how the cohort model is actually used in practice. Suppose that a constant fishing mortality F is applied to a given cohort, but only starting at age $a_c =$ age of first capture. We consider both F and a_c to be control variables.

Total yield from the cohort is then equal to

$$Y(F, a_c) = \int_{k+a_c}^{\infty} FN_k(t)w(t-k)\,dt$$

$$= FR_k e^{Fa_c} \int_{a_c}^{\infty} e^{-(M+F)u} w(u)\,du \tag{5.19}$$

One can then attempt to determine F and a_c so as to maximize $Y(F, a_c)$, treating R_k as an exogenous (i.e. not affected by F) variable. This is usually not satisfactory, since Y is only maximized as $F \to \infty$ and $a_c =$ age of maximum cohort biomass (Fig. 5.2). We can't leave out economics, after all!

In practice, a_c may not be subject to much fine control. If we fix a_c, then

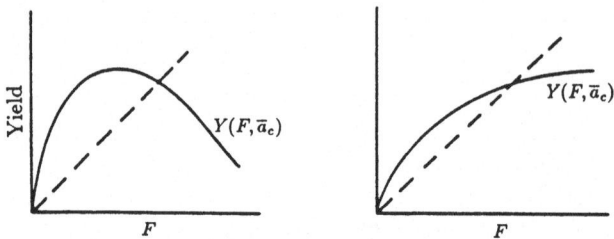

Fig. 5.2. Sustainable yield curves for values of $\bar{a}_c$ less than and greater than the age of maximum biomass.

$Y = Y(F, \bar{a}_c)$ will have one of the two shapes of Fig. 5.2. The first case occurs if $\bar{a}_c <$ age of maximum biomass, and vice versa. Now one is tempted to make Fig. 5.2 into an economic graph, by multiplying F by price p, and putting in a cost line cF (Gulland, 1968). Rather than using actual price-cost parameters, Gulland suggested a rule of thumb whereby F^* is determined by the condition that $Y'(F^*) = 0.1 \times Y'(0)$. This is now called the "$F_{0.1}$ principle," and is widely used in management. Although the calculation of $F_{0.1}$ ignores any stock-recruitment relation, the result is sufficiently conservative that most biologists seem to believe that in most cases it is unlikely to lead to undue depletion of fish stocks.

Pulse Fishing

In the 1970's, the distant-water fishing fleets of certain countries used a method that has been called pulse fishing. The fleets would enter a certain fishing area, and then fish extremely intensively, virtually removing all fish in the area. The fleet would then move to some other area, and eventually would cycle back to repeat the process.

Intuitively, one would probably think that pulse fishing was highly suboptimal. However, under certain conditions it can be shown by simple algebra (Clark, 1976a, p. 293) that pulse fishing can produce a higher average yield than any sustained-yield policy. The main condition required is that fishing is nonselective of cohorts.

It can also be shown that pulse, or periodic fishing, may be economically optimal because of economies of scale in effort costs (Clark, 1976a, p. 167). For example, consider the basic model of Lecture 1, but with nonlinear, nonconvex cost function $c(E)$ (Fig. 5.3):

$$\text{maximize} \int_0^\infty e^{-\delta t}[pqxE - c(E)]\, dt \tag{5.20}$$

$$\text{subject to } \frac{dx}{dt} = F(x) - qxE \tag{5.21}$$

Suppose that when one calculates the optimal equilibrium solution x^*, $E^* = F(x^*)/qx^*$ by the usual PMP technique, one discovers that $E^* < E_r$ (Fig. 5.3). Then it is easy to see that in fact this solution cannot be optimal. For one could

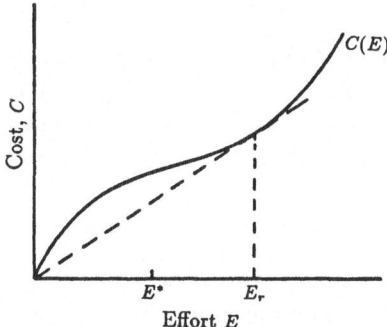

Fig. 5.3. Nonconvex total effort cost $C(E)$. The optimal sustained effort is a "chattering" control with average E^*.

reduce costs by using some combination of $E = 0$ and $E = E_r$. If effort is switched rapidly from 0 to E_r and back, the same biological equilibrium x^* can be arbitrarily closely approximated. (This is called "chattering" control.) Thus some kind of pulse fishing is economically cheaper than sustained-yield.

(The reason that PMP does not give the correct solution in this case, is that the validity of PMP depends upon certain implicit convexity conditions, which are violated in this example.)

Lecture 6

Uncertainty and Disequilibrium

Many of the models discussed in the previous lectures gave rise to optimal equilibrium solutions. Even in the case of a fluctuating resource stock, the simple model of Lecture 2 produced a constant-escapement policy, which stabilizes escapement (but certainly does not stabilize annual yield).

As soon as one begins to inject greater detail and realism into bioeconomic models, both the possibility and the desirability of equilibrium solutions become questionable. In the present lecture I will first discuss some aspects of uncertainty in resource modeling, and will then discuss other ways in which non-equilibrium solutions may arise. (See Mangel, 1985, for a more complete discussion of uncertainty in resource systems.)

6.1 Dynamic Decision Theory

Resource stocks are subject to many stochastic effects. The growth, mortality, and reproductive rates of fish, insect, and animal populations vary in a random manner. Fires, disease, and insect attacks may harm or destroy forests, and these phenomena are usually unpredictable.

Random changes in resource stocks have two very important bioeconomic consequences. First they lead to risk discounting, and secondly they result in both model and parameter uncertainty.

Risk Discounting

We have already seen risk discounting in its purest form in Section 2.4, where it was shown that the effect of including forest fire risk λ is the same as simply adding λ to δ, the annual discount rate. This is a straightforward result, which should be completely noncontroversial: failure to include this adjustment to the discount rate may lead to a significant loss of forest productivity. The same model can be used to assess the benefits of fire control (Reed, 1987).

Risk discounting can take on more perverse forms, however. In the case of a fish population, for example, it may be fully apparent to the fishing industry that fish are available, and can immediately be caught and marketed profitably. Future prospects are often more uncertain. Consequently the industry tends to discount the future, and attempts to harvest the resource intensively in the present. In meetings with management authorities, it is almost always the case that the industry wants higher catches than the managers wish to permit. In this setting, the fact that models always involve uncertainty gives the industry the upper hand in arguments with resource biologists.

Decision Theory

In principle, the methods of decision theory could be used to compute optimal management strategies in the presence of uncertainty. Unfortunately, dynamic decision theory (which is obviously required for resource decisions) is mathematically difficult. Numerical computation of optimal policies rapidly runs up against the "curse of dimensionality."

Figure 6.1 shows schematically the decision problem for an uncertain, controlled system (Clark, 1985). When applied to a fishery, the state X_k would refer to recruitment in year k, the probability distributions π_k would involve current estimates of parameters and of stock abundance, the controls U_k would include annual harvest h_k and also stock assessment activities. Updating of probability distributions is carried out using the Bayes formula.

In principle, the optimal controls for a system of this kind can be found by dynamic programming. In practice, the capacity of a large computer can be rapidly swamped by storage and CPU demands for all but the smallest examples. Here is a relatively simple example, which appears to become quite difficult if some of the restrictive assumptions are relaxed.

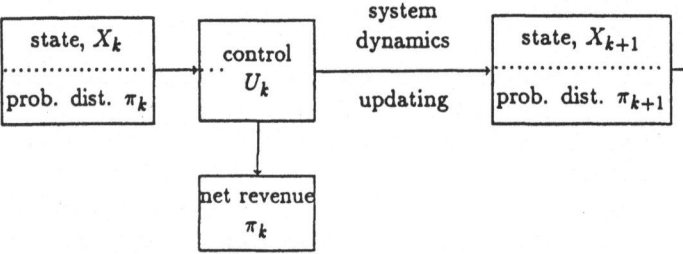

Fig. 6.1. Stochastic diagram of an uncertain, stochastic control system.

Uncertain Stock Size

We recall the stock-recruitment model described in Lecture 2:

$$x_{t+1} = Z_t F(x_t - h_t) \tag{6.1}$$

$$0 \leq x_t \leq h_t \tag{6.2}$$

$$\text{maximize } E\left\{ \sum_{t=0}^{T} \rho^t \pi(x_t, h_t) \mid x_0 = x \right\} \tag{6.3}$$

In this formulation it is tacitly assumed that the current size of the resource stock x_t is known exactly before the harvest control h_t is chosen. In the case of a fish population this may be an unrealistic assumption: the management agency may have to specify an annual quota on the basis of fragmentary information regarding current stock abundance. If the stock estimate is highly overoptimistic, the specified quota might even exceed the total stock! (There is some anecdotal evidence that this occurred in the Gulf of St. Lawrence herring fishery in 1983.)

Suppose, for example, that the recruitment level x_t cannot be monitored at all prior to the setting of a quota Q_t (see Clark and Kirkwood, 1986). Assume that the specified quota is actually caught, unless it exceeds x_t:

$$h_t = \langle Q_t, x_t \rangle = \min(Q_t, x_t) \tag{6.4}$$

Equation (6.1) describes the population dynamics.

We will assume that the final spawning escapement $s_t = x_t - h_t$ is known exactly; this assumption is roughly valid for anadromous species (e.g. salmon) which spawn in freshwater lakes and streams. If s_t has density $f(\cdot)$, then $x_{t+1} = Z_t F(s_t)$ has distribution density

$$\pi(x) = \frac{1}{g_t} f\left(\frac{x}{g_t}\right), \quad \text{where } g_t = F(s_t) \tag{6.5}$$

As in Sec. 2.2, let $J_T(s)$ denote the value function:

$$J_T(s) = \max E\left\{ \sum_{t=1}^{T} \rho^{t-1} h_t \mid s_0 = s \right\} \tag{6.6}$$

(here we assume zero costs for simplicity). The maximum is taken with respect to the quotas $Q_0, \ldots, Q_T \geq 0$. Then

$$J_1(s) = \max_{Q_1 \geq 0} E\{\langle zF(s), Q_1 \rangle\}$$
$$= E\{zF(s)\} = F(s) \tag{6.7}$$

and $Q_1^* = +\infty$ (no finite quota is needed when there is no future). The dynamic programming equation is

$$J_{t+1}(s) = \max_{Q \geq 0} E_\pi\{\langle x_1, Q \rangle + \rho J_t(F(x_1 - \langle x_1, Q \rangle))\} \tag{6.8}$$

where π is the density (6.5), with $g_t = F(s)$. Eq. (6.8) follows from the usual argument: suppose initial escapement s is known, and a quota Q is chosen. Then

$$x_1 = Z_1 F(s)$$

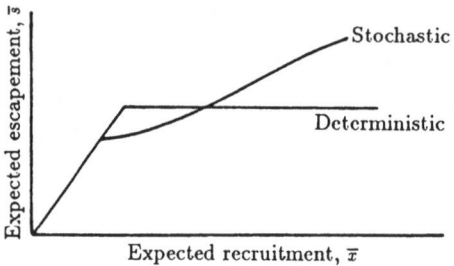

Fig. 6.2. Optimal expected escapement for a model with uncertain stock size.

with probability density π, and

$$h_1 = \langle x_1, Q \rangle$$

so that $s_1 = x_1 - \langle x_1, Q \rangle$. Expected total PV is therefore

$$E_\pi\{h_1 + \rho J_t(s_1)\}$$

and maximization over $Q \geq 0$ yields Eq. (6.8).

In Sec. 2.2 we showed, under the hypothesis that recruitment x_t is always known before the harvest h_t is specified, that the optimal harvest (= quota) policy is a constant-escapement policy. This is no longer true for the present model, and in fact the optimal *expected* escapement is not constant (Fig. 6.2). There is simply no analog of equilibrium sustained yield in this model. Recruitment, harvest, and escapement all fluctuate randomly. The possibility of a "disaster," with $h = x$, always exists. The implication is, of course, that harvesting a resource without estimating its abundance is a dangerous procedure. The above model can be used as a framework for estimating the economic value of stock estimates: see Clark and Kirkwood (1986).

Parameter Estimation

Another dubious assumption underlying our models is that the model parameters are accurately known. This is seldom the case. The estimation of stock-recruitment relationships for fish populations, for example, is widely described as the most important unsolved problem in fisheries management.

Consider first the case in which pre-harvest recruitment x_t cannot be assessed, but post-harvest escapement s_t can be accurately determined. Since catches $h_t = x_t - s_t$ are known, the fishery data produces an accurate time series of escapement and subsequent recruitment, with presumably a wide dispersion (Fig. 6.3a). This data may lead to reasonably good estimates of the stock-recruitment parameters, and also of the density f for Z_t.

But now consider the opposite case, where recruitment can be observed prior to harvest. According to our stochastic model of Sec. 2.2, the optimal harvest policy uses a fixed target escapement s^* (Fig. 6.3b). Under such a management policy there will be limited dispersion in the escapement data, and hence it may be difficult to obtain good estimates of the stock-recruitment function. The

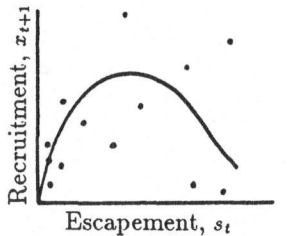
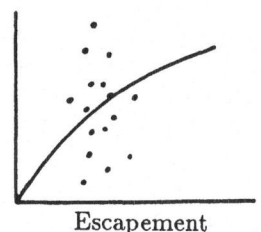

Fig. 6.3. Stock-recruitment curves generated from highly variable data (a) highly contrasted data; (b) low contrast.

managers may never know whether higher escapements would result in higher yields.

The problem of optimal harvesting under parameter uncertainty has been studied by Ludwig and Walters (1982). Whenever the existing parameter estimates are poor, the optimal harvest strategy involves a period of "probing" by means of deliberate variation in escapement levels. This again contradicts the received wisdom of sustainable equilibrium policies.

There are other ways in which uncertainty should mitigate against sustained yield policies, particularly MSY, but theoretical studies are still minimal. For example, herring and other pelagic schooling species are notoriously subject to population crashes, but it is not always obvious to what extent fishing is responsible. It would seem that a rational management policy would simply leave part of the stock completely unfished in case the remaining population unexpectedly crashes.

6.2 Disequilibrium Management Policies

Western technology seems addicted to equilibrium solutions. Occasionally our illusions are shattered by reality—by Chernobyls, by "100-year" floods that occur three years in a row, by the collapse of OPEC, by the sudden disappearance of the anchoveta. Equilibrium-dominated management is unable to cope with these surprises.

There follows a list of some of the ways in which equilibrium thinking in resource management may be inappropriate. Many of these have already been discussed in these notes, but in most cases no adequate theory exists.

1. *Boom-and-bust cycles.* These are very common in resource industries. They may in fact sometimes be "optimal" from the profit viewpoint. But they can also be socially disrupting.
2. *Natural resource fluctuations.* Fish stocks, particularly, undergo large-scale natural changes in abundance. While the implications for management are now fairly well recognized, there is still the danger that planners will assume that constant yields can be achieved.
3. *Monoculture and pest outbreaks.* The obvious examples here are in agriculture, but there are many parallel examples in resource management: suppression of

forest fires leads to occasional gigantic fires; spraying for spruce bundworm leads to an ever-increasing need for more spraying, as the whole ecosystem destabilizes (Holling 1978); aquaculture spreads fish diseases; apparent yield-maximizing exploitation results in loss of genetic diversity, which may have long-term depressive effects on yields.

4. *Global synergism of local bioeconomic phenomena.* The human species now pervades almost every habitable region on earth. Exploitation of renewable and exhaustible resources is rapidly expanding into the last wildernesses. On a global scale, ecosystems are being permanently altered by large-scale industrial developments and movements of populations. These developments, mostly the result of individual decisions, have potentially serious global implications, including climatic destabilization and the widespread destruction of genetic material. Concerns have been expressed regarding the sustainability of current and projected levels of economic development (Clark and Munn, 1986). The bioeconomic principles discussed here apply also to global resource issues, but the implications have not yet been investigated in any detail.

References

Beverton, R.J.H., Holt, S.J. (1957) On the dynamics of Exploited Fish Populations. Ministry of Agriculture, Fisheries and Food (London), Fish. Invest. Ser. 2(19)

Charles, A. (1983) Optimal fisheries investment under uncertainty. Can. J. Fish. Aquat. Sci. *40*, 2080–2091

Clark, C.W. (1976a) Mathematical Bioeconomics: the Optimal Management of Renewable Resources. Wiley Interscience, New York

Clark, C.W. (1976b) A delayed-recruitment model of population dynamics, with an application to baleen whale populations. J. Math. Biology *31*, 381–391

Clark, C.W. (1980) Restricted access to common-property fishery resources: a game-theoretic analysis. In: P.T. Liu (ed.) Dynamic Optimization and Mathematical Economics. Plenum, New York, pp. 117–132

Clark, C.W. (1982) Concentration profiles and the production and management of marine fisheries. In: W. Eichhorn (ed.) Economic Theory of Natural Resources. Physica-Verlag, Würzburg Wien, pp. 97–112

Clark, C.W. (1985) Bioeconomic Modelling and Fisheries Management. Wiley-Interscience, New York

Clark, C.W., Kirkwood, G.P. (1986) On uncertain renewable resource stocks: optimal harvest policies and the value of stock surveys. J. Envir. Econ. Manag. *13*, 235–244

Clark, C.W., Lamberson, R.H. (1982) An economic history and analysis of pelagic whaling. Marine Policy *6*, 103–120

Clark, C.W., Clarke, F.H., Munro, G.R. (1979) The optimal exploitation of renewable resource stocks: problems of irreversible investment. Econometrica *47*, 25–49

Clark, C.W., Charles, A., Beddington, J.R., Mangel, M. (1985) Optimal capacity decisions in a developing fishery. Marine Res. Econ. *1*, 25–54

Clark, W.C., Munn, R.E. (1986) Sustainable Development of the Biosphere. Intern. Inst. Appl. Systems Anal., Laxenburg, Austria

Conrad, J.M., Clark, C.W. (1987) Notes and Problems in Resource Economics. Cambridge Univ. Press, New York

Gordon, H.S. (1954) The economic theory of a common property resource: the fishery. J. Polit. Economy *62*, 124–142

Gulland, J.A. (1968) The concept of marginal yield from exploited fish stocks. J. du Cons. Intern. Expl. Mer *32*, 256–261

Hardin, G. (1968) The tragedy of the commons. Science *162*, 1243–1247

Holling, C.S. (ed.) (1978) Adaptive Environmental Assessment and Management. Wiley, New York

Horwood, J.W., Whittle, P. (1986) The optimal harvest from a multicohort stock. IMA J. Math. Appl. in Medicine and Biology *3*, 129–142

Ludwig, D.A., Hilborn, R. (1982) Management of possibly overexploited stocks on the basis of catch and effort data. Univ. of Brit. Col., Vancouver, Inst. Appl. Math. Stat. Tech. Rep. No. 82–3

Ludwig, D.A., Walters, C.J. (1982) Optimal harvesting with imprecise parameter estimates. Ecol. Modelling *14*, 273–292

Mangel, M. (1985) Decision and Control in Uncertain Resource Systems. Academic Press, New York

Mangel, M., Clark, C.W. (1983) Uncertainty, search, and information in fisheries. J. du Cons. Inter. Explor. Mer *41*, 93–103

McKelvey, R.W. (1985) Decentralized regulation of a common property renewable resource industry with irreversible investment. J. Envir. Econ. Manag. *12*, 287–307

Pella, J.J., Tomlinson, P.K. (1969) A generalized stock production model. Bull. Inter-Amer. Trop. Tuna Comm. *13*, 421–496

Reed, W.J. (1979) Optimal escapement levels in stochastic and deterministic harvesting models. J. Environ. Econ. Manag. *6*, 350–363

Reed, W.J. (1987) Protecting a forest against fire: optimal protection patterns and harvest policies. Nat. Res. Modeling *2*, 23–54

Reed, W.J., Errico, D. (1985) Assessing the long-run yield of a forest stand subject to the risk of fire. Can. J. For. Res. *15*, 680–687

Schaefer, M.B. (1954) Some aspects of the dynamics of populations important to the management of commercial marine fisheries. Bull. Inter-Amer. Trop. Tuna Comm. *1*, 25–56

Scott, A.D. (1955) The fishery: the objectives of sole ownership. J. Polit. Econ. *63*, 116–124

Scott, A.D. (1972) The Economics of Conservation 2nd edn. McLelland-Stewart, Toronto

Shepherd, J.G. (1982) A family of general production curves for exploited populations. Math. Biosci. *59*, 79–93

Common Property and the Conservation of Natural Resources

Robert McKelvey

Contents

1. Introduction

Biological resource modeling is a blend of population biology and resource economics. The choice of the word "resource" suggests that the emphasis is on use (presumably conservative and efficient use), rather than on permanent preservation.

Biological resources are, by their very nature, renewable. They also can be exhausted. To understand the economic forces that drive the clearcutting of a virgin forest, one may well compare the mining-out of a gold mine or an oil well. On the other hand, certain natural resources that are usually thought of as non-renewable (e.g. depleted soil or groundwater), over a sufficiently long time period may be restorable. Minerals are being concentrated at the mid-oceanic rifts. Thus, both conceptually and methodologically, it is best to examine biological resources in the broader context of natural resource systems.

Our focus here is on questions of public policy. Resource conservation raises profound public policy issues. It concerns intergenerational equity: the balance between consuming in the present and maintaining options for the future. Inherently, natural resource models explore dynamic tradeoffs.

Common property issues abound in natural resource policy analysis, partly because ownership often is held in the public domain, and partly because of the very nature of the resources. The ability of a body of water to absorb and neutralize pollutants is utilized by all in common. It is nearly impossible to devise an ownership scheme applicable to the tuna, which migrate annually across thousands of miles of ocean. No one can segregate the air that he breathes.

In Western economic thought, the common property character of natural resource utilization is most often regarded as a market defect, to be corrected by direct regulation or by indirect incentive schemes. This analysis has grown out of observation of the industrialized world, and its assumptions reflect the values and culture of that world.

Most natural resource modeling has emphasized the Western viewpoint, and these lecture notes certainly are no exception. But it should be remembered that in stable pre-industrial societies, rational common property resource utilization often has been achieved, through traditional institutions which balance rights and obligations.

In the following pages I will show how analytical models can be used to explore some important public policy issues in the management of our natural resource heritage.

2. Optimal Depletion of An Exhaustible Resource

2.1 A Dynamic Model

What determines the rate at which a natural resource is utilized? Consider the "mining" of a lode of an exhaustible natural resource: e.g. a gold mine, an oil well, or a stand of virgin timber. Let $Q(t)$ denote the rate of extraction (harvesting). The residual resource stock $x(t)$ will be depleted according to

$$dx/dt = -Q(t) \leq 0 \quad \text{for } 0 \leq t \leq T \leq \infty, \tag{2.1}$$

with the extraction terminating at time T. (In the case of the forest, this equation ignores regrowth). To assure that $x(t) \geq 0$ always, the extraction profile is restricted so that

$$\int_0^T Q(t)\, dt \leq x(0) = R. \tag{2.2}$$

Let $U(Q)$ be the utility of extracting at the rate $Q(t)$. In a first case utility is to be interpreted straightforwardly as market value:

$$U_M(Q) = Q \cdot P(Q). \tag{2.3}$$

Here the market price P depends on the rate at which the resource is being supplied (and consumed). In a second case utility is interpreted as value to consumers:

$$U_C(Q) = \int_0^Q P(q)\,dq. \tag{2.4}$$

This takes into account the consumers' differential willingness to pay. Both of these monetary utilities are subsumed in the general analysis. As is customary, we assume that price $P(Q)$ is convex and monotone decreasing from

$$P(0) = P_0 \leq \infty$$

(P_0 is the "choke price") to 0 as $Q \to \infty$.

Intertemporal utility comparisons require assigning a numerical value to a stream of utility flow over time, i.e. to the function $Q(\cdot)$. For monetary utility, neoclassical economics generally takes this value to be

$$\int_{t_0}^{\infty} \exp[-\delta(t-t_0)] \cdot U[Q(t)]\,dt,$$

the discounted current-value (at time t_0) of the flow of future returns.

This measure is highly controversial, certainly among biologists and environmentalists. It is easiest to accept when thought of as the monetary incentive driving a market economy or motivating a social planner. We shall accept this interpretation without debate, and go on. As an index of economic efficiency, it seems adequate for comparing alternative market structures, and in particular in assessing common property resource exploitation.

Monetary utility of extraction must be balanced against costs. We shall assume a unit extraction cost $c(x)$ which is independent of the rate of extraction Q, but rises as the residual stock $x(t)$ is depleted. Then, given that the residual stock at time t_0 is x_0, and that a particular extractive profile $Q(\cdot)$ is chosen, the discounted *net* present value achieved will be

$$V[x_0, t_0; Q(\cdot)] = \int_{t_0}^{T} \exp[-\delta(t-t_0)] \cdot [U(Q) - C(x)]\,dt. \tag{2.5}$$

2.2 The Maximum Principle

The problem of optimization is to determine the particular extractive profile that will maximize V, consistent with the differential equation (2.1) and constraint (2.2). The solution is standard in optimal control theory, a direct application of the "maximum principle". (Pontryagin et al., 1962).

Suppose that there exists an optimal extraction profile $Q^*(\cdot)$, and resulting depletion schedule $x^*(\cdot)$. Substituting these in the integral (2.5) will yield the optimal return

$$W(x_0, t_0) = V(x_0, t_0, Q^*).$$

To apply the maximum principle, we must introduce the dual dynamic variable

$$\Gamma(t) = D_x W(x, t). \tag{2.6}$$

(Here, D_x denotes partial derivative with respect to x).

$\Gamma(t)$ is the marginal value (or shadow price) of the current resource stock. With it, the *net* instantaneous rate of return to harvesting at rate $Q(t)$, taking account of depletion costs as well as extraction costs, is given by the Hamiltonian

$$H(t) = U(Q) - C(x)Q - \Gamma \cdot Q. \tag{2.7}$$

To optimize the extraction rate, $Q^*(t)$ must be chosen to maximize H at each point in time.

When $Q^*(t) > 0$, this requires that $D_Q H = 0$, i.e.

$$U'(Q) = C(x) + \Gamma. \tag{2.8}$$

(Here, $U' = dU/dt$). Equation (2.8) is the marginal rule for harvesting: marginal utility must balance total unit costs.

If the extraction process is completed in finite time, then termination occurs when instantaneous net return $H(t)$ has dropped to zero. Comparing (2.7), this means that

$$U'(Q_T) = U(Q_T)/Q_T.$$

Hence, by our assumptions on U, $Q_T = 0$ and $U'(Q_T) = P_0$. (This result would have to be modified if we were to allow for non-linear extraction costs).

The shadow price $\Gamma(t)$ can be thought of as the current market price for a resource *asset*. It evolves according to the differential equation

$$d\Gamma/dt - \delta\Gamma = -D_x H = C'(x)Q, \tag{2.9}$$

sometimes called the *arbitrage* equation.

When harvest costs are constant, so that $C'(x) = 0$, this equation says that $\Gamma(t) = \Gamma_0 \cdot e^{\delta t}$, i.e. shadow price rises at the rate of interest. Its present value discounted back to time $t = 0$, $\mu(t) = e^{-\delta t} \cdot \Gamma(t)$, remains constant. (This result of Hotelling (1931) remains valid when extraction cost is a non-linear function of Q alone). With depletion, (since $C'(x) < 0$), $\mu(t)$ drops over time: the discounted asset value of the marginal stock falls as the cost of its extraction rises.

One of two things happens, depending on the behavior of $C(x)$.

1. If $C(0) < P_0$, so that unit harvest costs always lie below the choke price, then the entire lode will be extracted in finite or infinite time. At terminal time,

   $$x_T = 0, \quad \text{and} \quad \Gamma_T = P_0 - C(0).$$

   This is the Malthusian case of physical exhaustion of the resource.
2. Otherwise extraction terminates when the lode is depleted to $x_T > 0$ for which

   $$C(x_T) = P_0. \quad \text{Then} \quad \Gamma_T = 0,$$

   i.e. the residual lode is of no economic value. This is the Ricardian case, of economic exhaustion of the resource.

2.3 Resource Scarcity

The above simple analysis allows us insight into the concept of resource scarcity, an issue which preoccupied the early conservation movement and motivated the limits-to-growth debate of the early 1970's. The view that scarcity is a purely physical phenomenon, i.e. represents the depletion of a fixed finite supply of material, is challenged by the economic perspective. That asserts that a physical material becomes a resource only when it acquires a value that exceeds the cost of its extraction. If its utility is low or the costs high, so that it is not economically recoverable, it is not a resource. In this sense, the resource base is not fixed, but may expand when technology lowers recovery costs or new demand raises prices, thereby making feasible the use of lower grade or geographically more remote stocks.

These considerations have bearing on the related question of quantifying resource scarcity. While current market price (reflecting demand) is one measure of scarcity, and current recovery costs is another, on the other hand, it is the shadow price (the asset price) that incorporates the stream of future costs and utility.

Of course, in reality, neither future costs nor future utility can be known with certainty at the present time. Empirical values of Γ can reflect only peoples' expectations of what the future will bring. Market measures of scarcity thus are only perceptions of scarcity. However, it is not clear that physical measures possess a more objective reality! (For further discussion, see Fisher 1981 and Smith 1979).

2.4 Social Manager vs. Monopolist

The optimization analysis enables us to begin examining the ways in which industry structure influences the pattern of resource utilization. Optimization is consistent with a resource industry that is dominated by a single central actor, maximizing net present value. He may be either a monopolist (with utility function U_M) or a social manager (whose utility function is U_S, and who will be maximizing the sum of "producer and consumer surpluses").

To compare, it is convenient to eliminate $\Gamma(t)$ in favor of $Q(t)$ and to plot phase plane trajectories in (x, Q)-space. Introducing $\Pi(Q) = U'(Q)$ and differentiating the marginal rule (2.8) totally with respect to time yields

$$\Pi'(Q)\cdot\dot{Q} = C'(x)\cdot\dot{x} + \dot{\Gamma} = - C'(x)Q + [\delta\Gamma + C'(x)Q] = \delta\Gamma = \delta[\Pi(Q) - C(x)].$$

Hence $\dot{x} = - Q$; $\dot{Q} = \delta[\Pi(Q) - C(x)]/\Pi'(Q)$. (Here $\dot{x} = dx/dt$).

Dividing the second differential equation by the first, the phase plane trajectory has the slope

$$dQ/dx = \delta[\Pi(Q) - C(x)]/Q\cdot|\Pi'(Q)| > 0,$$

since $\Pi'(Q) < 0$.

Now, by equations (2.3) and (2.4),

$$\Pi_M(Q) = d/dQ\cdot[QP(Q)] = P(Q) + QP'(Q) < P(Q) = \Pi_S(Q),$$

while similarly

$$|\Pi'_S(Q)| < |\Pi'_M(Q)|.$$

It follows that, should the trajectories of the social manager and the monopolist cross at a point (x, Q) of phase space, then

$$0 < (dQ/dx)_M < (dQ/dx)_S$$

there. Hence the trajectories cross at no more than one point.

But these trajectories have the same terminal point in phase space, namely $(x_T, Q_T = 0)$. Hence the monopolist's trajectory must lie entirely below the social manager's trajectory:

$$Q_M(x) < Q_S(x) \quad \text{for } x > x_T.$$

This means that the monopolist extracts the resource more slowly at each level of depletion than does the social manager, and so reaches the terminal depletion level at a later time.

Historically, conflicts have arisen between conservationists and the advocates of large scale public sector projects designed to provide resources cheaply, such as high dams on western rivers. Perhaps ironically, the monopolist is more of a conservationist than is the social manager. (Hotelling, 1931).

3. Competitive Exploitation of a Resource

The notion of central management by a present-value optimizer represents one extreme idealization of a natural resource industry. The opposite idealization is that of pure competition. In competitive conditions, as we shall see, exploitation patterns and the profitability of the industry can be very different from those of "socially optimal" management.

We shall for now maintain our focus on a non-renewable resource, being depleted according to equations (1) and (2). The total extraction rate $Q(t)$ is

$$Q(t) = \sum_{n=1}^{N} q_n(t),$$

the sum of the individual extraction rates of the N competitors.

3.1 Privately Controlled Resources

Our first case examines a resource which can be decomposed homogeneously to any degree. Thus, an individual competitor controls a portion $r = R/N$ of the resource, which is a microcosm of the resource body as a whole. Let $x_n(t)$ represent the individual's residual stock at time t. We are assuming that each producer's costs rise, as he depletes his share, in the same proportion as would occur for the resource body as a whole. Thus the individual's unit extraction cost is

$$C_n(x_n) = C(Nx_n),$$

consistent with the industry-wide cost function $C(x)$ of the preceding model.

Let us further assume that the number of competitors is sufficiently large that none expects to alter the behavior of the others significantly, and so each merely reacts to the environment in which he finds himself. Hence each individual n takes as given

$$Q_n(t) = \sum_{m \neq n} q_m(t),$$

the cumulative extraction rate of the others. Within this environment he is a profit-maximizer: he chooses $q_n(\cdot)$ to maximize

$$\int_0^T e^{n-\delta t}[P(Q) - C_n(x_n)] \cdot q_n dt,$$

where of course $Q = Q_n + q_n$ and $Q_n(\cdot)$ is given.

Proceeding as before,

$$H_n = q_n \cdot [P(Q) - C(Nx_n) - \Gamma_n] - \Gamma_n Q_n.$$

Hence, the marginal rule is that

$$q_n \cdot P'(Q) + P(Q) = C(Nx_n) + \Gamma_n,$$

and the dual dynamic equation is

$$d\Gamma_n/dt = \delta \Gamma_n + Nq_c C'(Nx_n).$$

Here Γ_n is the shadow price of the individual's own private reserve of the resource.

At the Nash equilibrium (G. Owen, 1968), all individuals are reacting identically so that $Q = Nq_n$, $x = Nx_n$, and $\Gamma_n = \Gamma$, for all n. Recalling the notation of Sect. 2, that

$$\Pi_M = d/dQ[QP(Q)], \quad \Pi_S = P(Q),$$

respectively "marginal revenue" and price, the individuals' marginal rules can be aggregated into an industry-wide marginal rule

$$[(N-1)/N] \cdot \Pi_S(Q) + [1/N] \cdot \Pi_M(Q) = C(x) + \Gamma, \tag{3.1}$$

determining Q. Here the dual dynamic equation is

$$d\Gamma/dt - \delta\Gamma = Q \cdot C'(x). \tag{3.2}$$

Comparing these with Eq. (2.8) and (2.9), we see that the dynamics of the N-competitor industry is intermediate between that of a monopolist and that of a social manager. The extreme case $N = 1$ reduces exactly to that of a monopolist, and the limit as $N \to \infty$ is identical with that of the social manager. Notice also that for this model, setting $N = \infty$ has the same effect as treating the firms as "price takers": that is, each firm ignores his own impact on price.

The limiting case of pure competition among price-taking firms is an instance of Adam Smith's "hidden hand": pure competition yields the socially optimal result. As we shall see shortly, this result breaks down when the resource is common property.

Also, in our simplified model, the resource stock is infinitely divisible into substocks, which then can be individually managed. Unit extraction costs are independent of extraction rate; that is, there are no economies of scale. Hence there is no optimal size of firm and so no tendency for the number of firms to be limited.

3.2 Common Property

The basic assumption underlying the proceeding analysis is the complete physical separation of the substocks being separately managed. Depletion of one substock has no effect on the cost of extraction of another. But there are many natural reserves, both renewable and non-renewable, for which this assumption is not appropriate. Examples are a commonly-exploited oil reservoir, an irrigation aquifer, open rangeland, a herd of big game animals, or a stock of marine mammals.

Consider first a non-renewable resource, obeying

$$dx/dt = - \Sigma q_n.$$

Drawing on the common property resource pool, each individual competitor will choose his extraction profile $q_n(t)$ to maximize

$$\int_0^\infty e^{-\delta t}[P(Q) - C(x)]\cdot q_n dt,$$

given the extraction profile Q_n of the others. Here $\dot{x}(t)$ is the residual stock in the common resource pool; extraction costs depend on this aggregate amount.

The Hamiltonian is

$$H_n = q_n\cdot[P(Q) - C(x) - \Gamma_n] - \Gamma_n Q_n,$$

where Γ_n is the shadow value to the nth operator of a unit of the resource stock. For each individual operator, q_n is determined by the marginal rule

$$q_n\cdot P'(Q) + P(Q) = C(x) + \Gamma_n,$$

and Γ_n satisfies the dual dynamic equation

$$d\Gamma_n/dt - \delta\Gamma_n = C'(x)\cdot q_n.$$

For identically reacting individuals these equations aggregate to

$$[(N-1)/N]\cdot \Pi_S(Q) + [1/N]\cdot \Pi_M(Q) = C(x) + \Gamma \tag{3.3}$$

and

$$d\Gamma/dt - \delta\Gamma = C'(x)/N \tag{3.4}$$

Comparing (3.3–3.4) with (3.1–3.2), the only difference one sees is the factor $1/N$ on the right hand side of the dual dynamic equation. It reflects that, in common property use, the asset value to the individual of the marginal unit of resource is depressed, since he must share it with $N-1$ others.

In the purely competitive limit, as $N\to\infty$, we have

$$P(Q) = C(x) + \Gamma, \quad d\Gamma/dt = \delta\Gamma.$$

In the Ricardian case (of economic exhaustion), extraction ceases when the stock is depleted to x_T at which $C(x_T) = P_0$. Since $\Gamma_T = 0$ otherwise $\Gamma(t) = 0$ identically; the individual operator places no value at all on the conservation of the resource asset.

In the Malthusian case (of physical exhaustion), $x_T = 0$ and $\Gamma_T = P_0 - C(0)$. Thus $\Gamma(t) = \Gamma_T \cdot e^{-\delta(T-t)}$: the asset retains a non-zero value to each individual.

In either case, the diminished level of the shadow price, as compared to optimal social management, means that the resource will be extracted more rapidly, and so will be brought to physical or economic exhaustion earlier than would be socially optimal. Further, the present value return will be lowered. This illustrates a general principle, that competitive exploitation of a common property resource tends to be economically inefficient and anti-conservationist.

3.3 Renewable Resources

The anti-conservationist tilt imparted by common property resource utilization shows additional facets when the resource is renewable, as for example a groundwater pool that is being recharged from surface runoff, or a self-renewing biological population. For these an appropriate stock-dynamic equation might be

$$dx/dt = F(x) - Q(t), \tag{3.5}$$

where the $F(x)$ represents the natural renewal.

Applying control theory methods as before, one arrives at identically the same marginal rule (3.1) for determining harvest. (Now $N = 1$ corresponds to a monopolist and $N = \infty$ to either the social manager *or* open access competition.) However the dual dynamic equation, which for either monopolist or social manager is

$$d\Gamma/dt = \Gamma \cdot [\delta - F(x)] + C'(x)Q, \tag{3.6}$$

becomes for common property exploitation

$$d\Gamma/dt = \Gamma \cdot [\delta - F'(x)] + C'(x)Q/N, \tag{3.7}$$

and in the open access limit

$$d\Gamma/dt = \Gamma \cdot [\delta - F(x)]. \tag{3.8}$$

The most prominent changes induced in the dynamics result from the renewal term $F(x)$ in (3.5): with it, the resource stock level can approach a steady state in which harvest Q exactly balances renewal $F(x)$. At this steady state, harvest rate, stock level, and shadow price all are constant.

Solving the equilibrium equations, $dx/dt = d\Gamma/dt = 0$, yields the equilibrium values $x^*, Q^* = F(x)$, and Γ^* for optimal management, and $x_N, Q_N = F(x_N)$, and Γ_N for common property exploitation. These are related by, respectively,

$$\Gamma^* = P(Q^*) - C(x^*) = -C'(x^*)Q^*/[\delta - F'(x^*)]; \quad Q^* = F(x^*),$$

and

$$\Gamma_N = P(Q_N) - C(x_N) = C'(x_N)Q_N/[\delta - F'(x_N)]N; \quad Q_N = F(x_N).$$

In the open access limit $N \to \infty$,

$$\Gamma_\infty = 0, \quad C(x_\infty) = P(Q_\infty), \quad Q_\infty = F(x_\infty).$$

A qualitative description of the trajectories is this: initially the stock is driven down, in a "mining" phase, but ultimately exploitation is moderated, to approach a steady state in which harvesting just balances renewal. In common property exploitation, the asset value is less than for social management. Consequently the early harvest is more rapid, and the ultimate steady state occurs at a lowered stock level ($x_N < x^*$). (For a more complete analysis, see Eswaran and Lewis 1984).

Thus, in the common property regime, the higher early consumption is paid for by greater depletion, and hence in the long run, higher steady state harvest costs and lower steady state harvest rates. In the open access limit $N = \infty$, the competitors put no value whatever on the asset (i.e. $\Gamma^\infty = 0$); hence their exploitation rate is limited only by current harvests. In a sense they are totally myopic, discounting the future completely (i.e. they might as well set $\delta = \infty$).

3.4 Incompletely Isolated Resources

It has been argued that the common property effect sometimes has been exaggerated, because, for example, local populations of an exploited biological resource often are more or less isolated, with little movement among them. Our model can be modified simply in order to explore this issue.

Suppose that there exist N self-renewing resource pools, with limited migration among them. Let us assume in fact that the migration rate between any pair of pools is proportional to their differential stock density. Thus

$$dx_n/dt = F(x_n) - q_n + \sum_{m \neq n} \sigma_{nm}(x_m - x_n) \quad \text{for } m, n = 1, 2, \ldots, N, \quad \text{with } \sigma_{nm} = \sigma_{mn}.$$

The present value return to harvest in the nth pool is

$$V_n = \int_0^\infty e^{-\delta t} \cdot [U(q_n) - C(x_n)] \cdot q_n dt.$$

For an industry-wide social manager, the task is to choose all harvest flows $\{q_n\}$, from $n = 1, \ldots, N$, so as to maximize $\sum V_n$.

The Hamiltonian for this problem is

$$\mathbf{H} = \sum_m \left\{ [U(q_m) - C(x_m)] \cdot q_m + \Gamma_m \left[F(x_m) - q_m + \sum_{k \neq m} \sigma_{km}(x_k - x_m) \right] \right\}.$$

Hence, the marginal rule for each q_m is

$$\Pi(q_m) = C(x_m) + \Gamma_m, \quad m = 1, 2, \ldots, N,$$

and the dual dynamic equations are

$$d\Gamma_m/dt = [\delta - F'(x_m)] \cdot \Gamma_m + C'(x_m) \cdot q_m - \sum_{k \neq m} \sigma_{km}(\Gamma_k - \Gamma_m).$$

Presuming symmetry among the pools (including initial conditions), all stock levels, flows, and shadow values will be the same:

$$x_n(t) = x(t), \quad q_n(t) = q(t), \quad \Gamma_n(t) = \Gamma(t).$$

Hence

$$dx/dt = F(x) - q, \quad d\Gamma/dt = \Gamma \cdot [\delta - F'(x)], \quad \text{and} \quad \Pi(q) = C(x) + \Gamma.$$

There will be no net flow between pools. At steady state,

$$q^* = F(x), \quad \Gamma^* = C'(x^*) \cdot q^* / [\delta - F'(x^*)] = \Pi(q^*) - C(x^*).$$

Now suppose that the pools are harvested separately by independent operators. Each will wish to choose his $q_n(t)$ to maximize V_n, given the harvest rates $\{q_m\}_{m \neq n}$ of the others. His Hamiltonian is

$$\mathbf{H}_n = [U(q_n) - C(x_n)] \cdot q_n + \sum_{m=1}^{N} \Gamma_m \left[F(x_m) - q_m + \sum_{k \neq m} \sigma_{km}(x_k - x_m) \right].$$

(Note that the individual operator must attach a shadow value to *all* stocks, not just the stock in the pool that he exploits directly). His marginal rule for q_n is

$$\Pi(q_n) = C(x_n) + \Gamma_n,$$

and his dual dynamic equations are

$$d\Gamma_n/dt = \Gamma_n[\delta - F'(x_n)] + C'(x_n) - \sum_{k \neq n} \sigma_{kn}(\Gamma_k - \Gamma_m),$$

and

$$d\Gamma_m/dt = \Gamma_m[\delta - F'(x_m)] - \sum_{k \neq m} \sigma_{km}(\Gamma_k - \Gamma_m) \quad \text{for } m \neq n.$$

Presuming symmetry, the Nash equilibrium will balance all pools at the same stock and flow levels: $x_m = x$ and $q_m = q$. Furthermore each harvester will assign shadow value Γ to his own pool and a common value μ to all other pools. The dynamics are

$$dx/dt = F(x) - q, \quad \Pi(q) = C(x) + \Gamma,$$
$$d\Gamma/dt = \Gamma[\delta - F'(x)] + C'(x) \cdot q + (N - 1)\sigma(\Gamma - \mu),$$
$$d\mu/dt = \mu[\delta - F'(x)] + \sigma(\Gamma - \mu).$$

In the limiting case as $N \to \infty$, assuming that the overall migration constant $N \cdot \sigma(N)$ tends to a finite limit Ω, the dual equations become

$$d\Gamma/dt = \Gamma[\delta + \Omega - F'(x)] + C'(x) \cdot q, \quad \mu \equiv 0;$$

and in the steady state

$$q^0 = F(x^0), \quad \Gamma^0 = C'(x^0) \cdot q^0 / [\delta + \Omega + F'(x^0)] = \Pi(q^0) - C(x^0).$$

Thus, at open access, the effect of allowing migration between pools is to change the effective discount rate from δ to $\delta^{\#} = \delta + \Omega$. If the migration constant Ω is zero, so that the pools are sealed off from one another, then $\delta^{\#} = \delta$ and the individual harvesters' operations coincide with the social manager's: i.e. there is no commons.

At the opposite extreme, where $\Omega \to \infty$ so that the pools are effectively merged, the harvesters put a zero value $\Gamma = 0$ on the resource asset. Then $\delta^{\#} = \infty$, and the operators' behavior is completely myopic.

Between these extremes, as the migration rate-constant Ω varies between 0 and ∞, there are a continuum of intermediate cases, exhibiting a degree of myopia which is reflected in the effective discount rate $\delta^{\#}$.

4. Capital Effects

4.1 Modeling Investment

To a population biologist, the renewal term $F(x)$ in our stock dynamics equation has a straightforward interpretation, as density dependent growth in biomass $x(t)$. Typically, it might be the logistic growth term in the "surplus production" harvest model. To the economist, $F(x)$ suggests self-renewal of a persistent capital stock.

A natural resource stock can be looked upon as just one of the input factors in producing the harvest output. In a dynamic model, inputs are of two kinds. Some ("labor" or "harvest effort") are thought of as providing their services instantaneously. Others ("capital stocks") persist over time, and produce a flow of services. Plainly the natural resource stock is of this second character.

From the mathematical perspective, capital stock variables must be modeled dynamically, say through differential equations, rather than instantaneously through algebraic relations. Their presence imparts an inertia to the natural resource system, since stocks change relatively slowly and their effects persist.

It follows, then, that to understand the dynamic performance of a natural resource system, one must take into account not only the natural resource stock itself, but also other capital inputs into harvest production.

As usual, we shall model the harvesting of a self-renewing natural resource stock $x(\cdot)$ by the differential equation

$$dx/dt = F(x) - Q. \tag{4.1}$$

But now we think of the output variable Q as being generated by the combined effects of a number of factor inputs. These include the current $x(t)$, but also other factors of production.

For simplicity, let us include in these a single capital stock $K(t)$, which depreciates proportionally but can be augmented by direct investment $I(t)$:

$$dK/dt = I(t) - \tau K. \tag{4.2}$$

Here, τ is the depreciation rate constant.

In the model, $I(\cdot)$ is treated as a control variable, to be determined by the harvesters. A second control variable is harvest effort $E(\cdot)$, which, along with current resource stock level, determines current harvest:

$$Q = \phi(x) \cdot E.$$

Furthermore, the effort that can be exerted is limited by the current capital stock level K:

$$0 \leq E \leq (\text{const.}) \cdot K.$$

(It is convenient to set the proportionality constant equal to one; however K is a *stock* while E is a *flow*.)

Combining these two conditions, we have that the stock levels $x(t)$ and $K(t)$ combine to limit current harvest to the range

$$0 \leq Q \leq \phi(x) \cdot K.$$

It will be convenient to suppress the explicit variable E in favor of Q itself as the second control variable. Thus Q is both input and output to production.

For a central social manager, the problem is to choose $Q(t)$ and $I(t)$ to maximize

$$\int_0^\infty e^{-\delta t} [U_S(Q) - C(x)Q - B(I)I] dt.$$

In the new term, $B(I)$ is the unit cost of investment at the level I, and is assumed to be a monotone increasing and convex function. (This optimization model first was considered by Clark, Clarke, and Munro (1979). For a streamlined analysis in the original formulation, see C. Clark's article in this volume.)

There is an intertemporal trade-off involved, in incurring current investment costs in order to permit increased harvest effort in the future. This parallels the trade-off between present harvest and future growth in the resource stock.

The Hamiltonian for this problem is

$$\mathbf{H} = U_S(Q) - C(x)Q - B(I)I + \Gamma \cdot [F(x) - Q] + \Theta \cdot [I - \tau K]$$

and the Lagrangian, which incorporates the inequality constraint, is

$$\mathbf{L} = \mathbf{H} + \sigma^+ \cdot [\phi(x) \cdot K - Q].$$

Here the multiplier σ^+ is zero unless the constraint binds.

From Mangasarian's generalized maximum principle, the marginal rule for harvest Q is

$$P(Q) = C(x) + \Gamma + \sigma^+;$$

that is, Q is set at the level at which price of the resource balances total costs of harvest. Thus

$$\sigma^+ = \max [0, P(Q) - C(x) - \Gamma],$$

the "positive part".

The marginal rule for investment is

$$\mu = B(I) + I \cdot B'(I);$$

i.e. investment is set to match its marginal cost to the shadow value of K. The dual dynamic equations for the shadow prices Γ and μ are

$$d\Gamma/dt = [\delta - F'(x)]\Gamma + C'(x)Q - \sigma^+ \cdot \phi'(x)K,$$

and

$$d\mu/dt = (\delta + \tau)\mu - \sigma\phi(x).$$

At equilibrium the capital capacity constraint does bind, so that

$$Q = F(x) = \phi(x)K; \quad \text{whence } K = F(x)/\phi(x), \quad I = \tau K.$$

From the equilibrium in μ, therefore

$$\sigma^+ = (\delta + \tau)\mu/\phi(x);$$

i.e. σ^+ equals the marginal cost of maintaining capital at the required level for equilibrium. From $d\Gamma/dt = 0$ together with the marginal rule for Q, we have

$$\Gamma = P(Q) - C(x) - \sigma^+ = [\sigma^+ \cdot \phi(x)K - C'(x)Q]/[\delta - F'(x)],$$

which determines the equilibrium value of x.

To examine the common property regime, one considers N individual harvesters, each investing in private capital

$$dK/dt = I_n - \tau K_n,$$

and having his harvest rate constrained by his current capital capacity

$$0 \le q_n \le \phi(x)K_n. \tag{4.3}$$

Each chooses harvest rate q_n and investment rate I_n to maximize his individual present-value return

$$\int_0^\infty e^{-\delta t}[P(Q) - C(x)q_n - B(I)I_n]dt, \tag{4.4}$$

subject to the harvest and investment profiles of the others.

The aggregate dynamics are governed by

$$P(Q) + P'(Q)Q/N = C(x) + \Gamma + \sigma^+,$$
$$\mu = B(I) + B'(I)I/N,$$
$$d\Gamma/dt = [\delta - F'(x)]\Gamma + C'(x)Q/N - \sigma^+ \cdot \phi'(x)K/N,$$
$$d\mu/dt = (\delta + \tau)\mu - \sigma^+ \cdot \phi(x).$$

In the open access limit,

$$P(Q) = C(x) + \sigma^+ = \max[0, P(Q) - C(x)], \tag{4.5}$$

$$\Gamma \equiv 0, \quad \mu = B(I), \tag{4.6}$$

$$d\mu/dt = (\delta + \tau)\mu - \sigma^+ \cdot \phi(x). \tag{4.7}$$

That is, the competitive open access industry will ignore the shadow value of the resource stock, and will set investment so as to match the shadow price of capital to the market price of investment, rather than to its marginal cost. This means that the competitive industry will invest more heavily and harvest more heavily than would a socially managed industry at the same state (x, K).

4.2 The Problem of Entry

All of our analysis has assumed a fixed number of harvesters who enter at initial time and remain thereafter. A truly dynamic analysis ought to take into account the entry and exit of harvesters through the course of time. In general this is difficult: it amounts to a differential game with a non-constant set of players.

However there is one case in which the analysis proves tractable. This is the case of *irreversible* entry: players enter over the course of time, but none may exit.

This interpretation may be placed on our open access investment model (Eq. 4.1–4.4) by ruling out disinvestment, i.e. by requiring that $I(t) \geq 0$. The industry is thought of as made up of a continuum of harvesters, and new investment at any instant of time represents the entry then of a new set of "atomic" firms.

Integration of the dual dynamic equation (4.7) yields

$$\mu(t_0) = \int_0^\infty \exp\left[-(\delta - \tau)(t - t_0)\right] \cdot [P(Q) - C(x)]^+ \cdot \phi(x) dt$$

It represents the accumulated flow of returns to a unit of stock acquired at time t_0, discounting both for time preference and for depreciation.

The aggregate entry at t_0 is determined (in 4.5) by equating $\mu(t_0)$ to the average cost of a unit of investment, namely $IB(I)/I$. In effect, atomic investors enter in aggregate to the level at which their eventual returns will just meet their initial investment costs: the result is that all profits will be dissipated. This is the expected aggregate behavior of the common, as described originally by Scott Gordon in his celebrated 1954 article. Here it appears in a fully dynamic setting. (For a more complete analysis, see McKelvey, 1985.)

5. Regulating the Common

One benefit of applying simple economic models to natural resource conservation issues is the heightened awareness of the way in which government intervention, through regulation or taxation, can impact decisively the pattern of natural resource development and conservation in a market economy. Even taxes imposed purely as revenue measures can have profound unintended effects.

On the other hand, an important feature of present-day environmental and natural resource policy has been the deliberate use of taxation and other financial incentives as a means of indirect and decentralized guidance of resource development. Related to this is the large body of economic theory that has grown up, concerning the positive use of economic incentives to correct market defects. Of these, our particular focus here is on the negative common property externalities that seem to be ubiquitous in natural resource utilization.

The particular common property externality that we have been examining throughout these notes is the so-called "stock externality": the circumstance that an industry's aggregate harvesting depresses the stock, and thereby brings about poorer future harvests for each individual operator. That is, common property harvesting exerts a persistent negative external effect upon the individual's "production function".

There are certainly other forms of common property externality. One traditionally examined is the crowding or congestion externality. In fisheries, for example, this could entail interference with a vessel's maneuvering in confined waters, the tangling of nets, and the like. As with the stock externality, the effect of crowding generally is negative.

Other externalities may be positive. In exploring for oil or in searching for a school of fish, a cooperative industry-wide information network might enhance everyone's success.

A notorious difficulty, in decentralized regulation of a common property resource industry, has been the design of physical controls and economic incentives that achieve "economic rationality" (i.e. efficiency) and at the same time are simple, inexpensive to institute and to monitor, and non-intrusive.

Fishery management provides many examples of how hard it can be to meet these conflicting goals.

For example, when the number of vessels in a fishery is limited by licensing, the fishermen may respond by building larger vessels. When vessel size is limited, they may compensate by installing more elaborate harvesting and processing gear.

If the season length is limited, fishermen will respond by installing more powerful engines in order to move more quickly to the grounds, or more sophisticated electronic search equipment. They even may hire spotter aircraft.

It would seem that regulating the common is rather like trying to squeeze a balloon!

It will be convenience to couch our discussion of economic incentives mainly in terms of taxes and subsidies (= negative taxes). This may seem inappropriately narrow. Indeed, a variety of alternative regulatory mechanisms have been advocated, either direct quantity controls, or the creation of markets for otherwise unpriced commodities or factors of production.

However, in the context of a deterministic environment and complete information, such mechanisms generally can be shown to be entirely equivalent in principle to the levying of certain taxes. (For a discussion of the case of the fishery, see Clark, 1988). Hence our analysis really applies quite broadly.

The issue to be explored here is of a different character. We will be concerned not so much with the particular forms of regulatory mechanism as with the question of where within the resource system they ought to be applied. We will be seeking out the pressure points within the system at which one can achieve regulatory control.

5.1 Controlling Multiple Inputs

It generally is regarded as impractical to regulate simultaneously all of the many input factors of harvest production. But what are the alternatives? More precisely, what is the minimum number of regulatory controls (either physical restrictions or economic incentives) that are necessary to achieve decentralized rationalization? Also, where in the system ought they to be applied?

A short answer is: one needs merely to control the independent externalities. I shall demonstrate this proposition, first for a simple static system, generalizing the classical paradigm of Scott Gordon (1954). In this way we shall gain a certain insight into the vastly more complicated dynamical systems.

Let us consider, then, a *single-product* renewable resource industry, in which each harvester's actions directly affect the production function of the others. (For joint production of several harvest outputs, see McKelvey, 1983).

We suppose that there are N harvesters, and that each achieves a harvest output y_n by applying a vector of inputs $\mathbf{x}_n$. He also contributes to one or more common property externality, his contribution $e_n^{(\alpha)}$ to the αth of these being determined by his primary input vector $\mathbf{x}_n$:

$$e_n^{(\alpha)} = E_n^{(\alpha)}(\mathbf{x}_n).$$

The individual contributions are assumed to aggregate additively to produce the industry-wide external effects:

$$e^{(\alpha)} = \sum_{n=1}^{N} e_n^{(\alpha)}.$$

(This simplifies things notationally, but is non-essential to the argument.)

For notational brevity, we shall combine the externalities into a vector

$$\mathbf{e}_n = (e_{n1}, \ldots, e_{n\alpha}, \ldots) = \mathbf{E}_n(\mathbf{x}_n)$$

and

$$\mathbf{E} = \sum_{n=1}^{N} \mathbf{e}_n.$$

Furthermore, relative to any particular harvester n, the external vector may be split into his own contribution plus that of the others:

$$\mathbf{E} = \mathbf{e}_n + \mathbf{F}_n, \quad \text{where } \mathbf{F}_n = \sum_{m \neq n} \mathbf{e}_m.$$

The individual operator's harvest output is a function of his primary inputs and the industry-wide externality vector:

$$y_n = G_n(\mathbf{x}_n, \mathbf{E}).$$

The payoff to the individual harvester is

$$V_n(\mathbf{x}_n, \mathbf{F}_n) = p \cdot G_n(\mathbf{x}_n, \mathbf{e}_n + \mathbf{F}_n) - \mathbf{c} \cdot \mathbf{x}_n,$$

where p is the price of the harvested resource and $\mathbf{c}$ is the vector of unit costs of the input factors.

In socially optimal management, the inputs $\{\mathbf{x}_n^*\}$, $n = 1, \ldots, N$, are chosen to maximize

$$V = \sum_{n=1}^{N} V_n;$$

the maximum value will be denoted V^*.

In a common property regime, by contrast, the inputs are assumed to constitute a Nash equilibrium $\{x_n^0\}$; $n = 1,\ldots, N$. That is, for every n and feasible input x_n',

$$V_n(x_n', F_n^0) \leq V_n(x_n^0, F_n^0), \quad n = 1,\ldots, N;$$

the equilibrium payoff configuration will be denoted $\{V_n^0\}$. Hence, each operator's payoff would suffer should be depart unilaterally from the equilibrium.

We now undertake to analyze the two systems (social optimum and common property) in a parallel fashion.

First we will need some notation. We shall say that an individual externality vector e_n is *feasible* if it is generated by at least one admissible input vector x_n. A set $\{e_1,\ldots, e_n\}$ of feasible individual externality vectors constitutes a *feasible externality configuration*. $E = \sum e_n$ is then a *feasible aggregate externality*, and (e_n, F_n) is a *feasible externality pair*.

For (e_n, F_n) a feasible pair, we shall define the nth harvester's *internal optimization problem*: It is to choose x_n consistent with $E_n(x_n) = e_n$ which maximizes $V_n(x_n, F_n)$. The solution (assumed unique) will be denoted $X^\#(e_n, F_n)$ and the corresponding maximum value by $V_n^\#(e_n, F_n)$. Thus, briefly,

$$V_n^\#(e_n, F_n) = V_n[X^\#(e_n, F_n), F_n] = \max_{(x_n|e_n)} V(x_n, F_n).$$

In terms of the functions $V_n^\#(\cdot)$, the *external social optimization problem* is to choose a feasible externality configuration $\{e_n^{**}\}$ which maximizes

$$V = \sum V_n^\#(e_n, F_n).$$

Similarly, the *external Nash equilibrium* is a feasible configuration of externalities $\{e_n^\infty\}$ satisfying the stability condition, for $n = 1,\ldots, N$, that

$$V_n^\#(e_n', F_n^\infty) \leq V_n^\#(e_n^\infty, F_n^\infty),$$

for any feasible e_n'.

It is now easy to show that the original system's overall social optimum and Nash equilibrium both result by composing the internal with the external calculations.

That is, for the social optimum,

$$x_n^* = X_n^\#(e_n^{**}, F_n^{**}), \quad V^* = \sum V_n^\#(e_n^{**}, F_n^{**});$$

conversely, $e_n^{**} = e_n^* = E_n(x_n^*)$.

This follows directly from the string of equalities

$$V^* = V(\{x_n^*\}) = \max_{\{x_n\}} \sum V(x_n, F_n)$$

$$= \max_{\{e_n\}} \sum_n \max_{(x_n|e_n)} V(x_n, F_n)$$

$$= \max_{\{e_n\}} \sum_n V_n^\#(e_n, F_n)$$

$$= \sum_n V_n(e_n^{**}, F_n^{**}).$$

For the Nash equilibrium,

$$x_n^0 = X_n^\#(e_n^\infty, F_n^\infty), \quad V_n^0 = V_n^\#(e_n^\infty, F_n^\infty);$$

conversely, $e_n^\infty = e_n^0 = E_n(x_n^0)$.

To see this, let $\{x_n'\}$ be feasible and let $e_n' = E_n(x_n')$. By definition of the external equilibrium,

$$V_n^\#(e_n', F_n^\infty) \leq V_n^\#(e_n^\infty, F_n^\infty).$$

Let $x_n^\infty = X_n^\#(e_n^\infty, F_n^\infty)$, defined through internal optimization. Then

$$V_n(x_n', F_n^\infty) \leq V_n^\#(e_n', F_n^\infty)$$
$$\leq V_n^\#(e_n^\infty, F_n^\infty) \leq V_n(x_n^\infty, F_n^\infty).$$

But this means that $\{x_n^\infty\}$ is the Nash equilibrium for the global problem, i.e. $x_n^\infty = x_n^0$ and $e_n^\infty = e_n^0$. $\qquad\square$

In summary, the configuration of external variables is determined by external market structure to conform either to the social optimum or to the Nash equilibrium. In each case, the individual harvesters adjust their (internal) primary variables to achieve these external values in the most efficient way.

We are now in a position to examine regulatory control. Simply, the regulator must provide economic incentives for the harvesters to adjust their external variables to the external social optimum. It is sufficient for the regulator to cause each harvester to "feel" the shadow cost of the externalities he produces. This can be achieved, in one way, by levying taxes on the externality vector. It is necessary to assess no more separate taxes than there are functionally independent components of this vector.

With the tax, the individual operator at open access has the modified objective function $V_n^\#(e_n, F_n) - \tau_n \cdot e_n$. Assuming differentiability, and the functional independence of the externalities, the first order conditions for a Nash equilibrium become

$$\operatorname*{grad}_{(e_n)} V_n^\#(e_n, F_n) = \tau_n.$$

On the other hand, the first order conditions for the optimum, i.e. to maximize $\sum V_m^\#(e_m, F_m)$, are

$$\operatorname*{grad}_{(e_n)} V_n^\# + \sum_{m \neq n} \operatorname*{grad}_{(F_m)} V_m^\# = 0,$$

for every n. Hence the regulator need only set the tax equal to

$$\tau_n^{(\alpha)} = - \sum_{m \neq n} \operatorname*{grad}_{(F_m)} V_m(e_m^*, F_m^*),$$

with the right-hand-side evaluated at the optimum.

5.2 Regulating Dynamical Systems

Turning now to decentralized management in a dynamic context, we shall illustrate the general principles through an example, a variant of our basic model in these lecture notes.

For the individual harvester, we shall assume an instantaneous production function,

$$q_n = R(x, E, S) \cdot e_n,$$

which determines the rate of harvest at any instant of time.

Here, as usual, x denotes the resource stock level and e_n the individual's harvest effort (rate). Effort e_n is constrained by the individual harvester's current capital capacity K_n:

$$0 \le e_n \le K_n.$$

As usual, K_n is a stock variable, evolving according to

$$dK_n/dt = I_n - \tau K_n.$$

The industry *aggregate* effort, $E = \sum e_n$ also enters this production function, as a "crowding externality": it has an immediate, probably negative, impact on each harvester's current output.

A second externality in the production function is $S = \sum s_n$. Here s_n is the individual's *search* capacity. Unlike e_n, s_n is taken to be a capital stock variable, which builds up over time according to

$$ds_n/dt = J_n - \tau \cdot s_n.$$

Of course, J_n stands for the rate of investment in this kind of capital. The externality S generally will be positive, since cooperative search enhances the individual's harvest effectiveness.

The third externality is the resource stock level x itself. Like S, the stock externality results from the cumulative past actions of all of the harvesters. Aside from the more complicated production function, the resource stock dynamic equation is unchanged:

$$dx/dt = F(x) - Q, \quad Q = \sum_{n=1}^{N} q_n.$$

The discounted present-value net return to a harvester is

$$V_n = \int_0^\infty e^{-\delta t} \cdot \{[P - C(x)]q_n - B(I)I_n - A(J)J_n\} dt,$$

taking into account investment costs that are assumed to rise with the industry-wide investment rates:

$$I = \sum_{n=1}^{N} I_n, \quad J = \sum_{n=1}^{N} J_n.$$

With central management (maximizing $\sum V_n$), our standard control theory procedures give the aggregate dynamics

$$dx/dt = F(x) - R(x, E, S)E, \quad 0 \le E \le K, \tag{5.1}$$

$$dK/dt = I - \tau K, \quad dS/dt = J - \tau S. \tag{5.2}$$

Dual dynamics for the shadow prices Γ, Θ, μ of x, S, K, respectively, are

$$d\Gamma/dt = \Gamma \cdot [\delta - F'(x)] + C'(x)E \cdot R - [P - \Gamma - C(x)]E \cdot D_x R \qquad (5.3)$$

$$d\Theta/dt = (\delta + \tau)\Theta - [P - \Gamma - C(x)]E \cdot D_S R, \qquad (5.4)$$

$$d\mu/dt = (\delta + \tau)\mu - \sigma^+, \qquad (5.5)$$

where

$$\sigma = [P - \Gamma - C(x)] \cdot [R + E \cdot D_E R], \qquad (5.6)$$

and $\sigma^+ = \max [0, \sigma]$ is the positive part.

Finally, investment is set by the marginal rules

$$B(I) + I \cdot B'(I) = \mu, \quad A(J) + J \cdot A'(J) = \Theta \qquad (5.7)$$

and harvest effort by the "bang–bang" rule

$$E = \begin{cases} 0 & \text{when} \quad \sigma > 0 \\ K & \text{when} \quad \sigma < 0. \end{cases} \qquad (5.8)$$

When $\sigma = 0$, effort may be determined by a marginal rule, reflecting the crowding externality:

$$R + E \cdot D_E R. \qquad (5.9)$$

By contrast, an open access common property regime $(N = \infty)$ is governed by (5.1–5.2), and

$$d\Gamma/dt = \Gamma \cdot [\delta - F'(x) + E \cdot D_x R], \qquad (5.3)'$$

$$d\Theta/dt = (\delta - \tau)\Theta + \Gamma \cdot E \cdot D_S R. \qquad (5.4)'$$

Equations (5.5) and (5.8) still hold for μ and E, except that now

$$\sigma = [P - \Gamma - C(x)] \cdot R - \Gamma \cdot E \cdot D_E R. \qquad (5.6)'$$

Finally, investment rules are modified to

$$B(I) = \mu, \quad A(J) = \Theta. \qquad (5.7)'$$

In comparing the two sets of dynamic equations, one can characterize the open access behavior as "defective" (relative to central management) in a number of ways:

1. Missing from the calculation of σ is the effect of the crowding externality

$$[P - C(x)] \cdot E \cdot D_E R.$$

Thereby, both current effort in (5.7)' and future effort capacity K in (5.5)' are distorted.

2. Investments I and J are set in (5.8) by equating asset values μ and Θ to investment costs. But in (5.8)' these are taken as average costs instead of as marginal costs in (5.8). These flaws relate to the open entry of capital, but disappear when unit investment costs are constant.

3. The shadow value Γ of the resource stock in (5.3)' is missing two terms which

relate to the stock externality. The first is

$$[P - C(x)] \cdot E \cdot D_x R,$$

which measures the thinning effect, i.e. the effect of stock depletion on harvest productivity. The second is

$$C'(x) \cdot E \cdot R,$$

which measures the effect of stock depletion on harvest costs.

4. The shadow value Θ of search capital in (5.4)' is missing the term

$$[P - C(x)] \cdot E \cdot D_S R,$$

which measures the external effect of industry-wide search capacity on harvest productivity.

The combined effects of these omissions is to alter substantially the system's state trajectories from those of central management. In particular, with an infinite time horizon, and a trajectory tending to a steady state, the open access system sets both $\Gamma(t) \equiv 0$ and $\Theta \equiv 0$. Thus at open access both resource asset value and search capital asset value are ignored. The first of these represents the "tragedy of the common"; the second, the "free rider" problem.

In this circumstance, the open access dynamics simplify even more:

$$dx/dt = F(x) - RE, \quad dK/dt = I - \tau K, \quad dS/dt = J - \tau S; \tag{5.10}$$

$$\Gamma \equiv \Theta \equiv 0, \quad d\mu/dt = (\delta + \tau)\mu - \sigma^+; \tag{5.11}$$

$$\sigma = [P - C(x)] \cdot R; \tag{5.12}$$

$$B(I) = \mu, \quad A(J) = \Theta; \tag{5.13}$$

$$E = \begin{cases} K & \text{if} \quad \sigma > 0 \\ 0 & \text{if} \quad \sigma < 0; \end{cases} \tag{5.14}$$

the case $\sigma \equiv 0$ on a time interval is inconsistent.

Economic regulation is designed to cause open access harvesters to "feel" the otherwise unrecognized costs and benefits of their actions. Since their primary inputs are their control functions $e_n(\cdot)$, $I_n(\cdot)$, and $J_n(\cdot)$, economic incentives must lead the harvesting firms to adjust these to the social optimum.

A straightforward approach is to levy taxes $\varepsilon(t)$ on effort e_n and $\beta(t)$ on investment I_n, and to pay a subsidy $\alpha(t)$ to encourage investment J_n. Thus, unit costs become

$$C(x) + \varepsilon(t), \quad B(I) + \beta(t), \quad A(J) + \alpha(t),$$

respectively. Notice that $\varepsilon(t)$ affects only equation (5.12) for σ, and that one could achieve the same effect there by taxing output harvest instead of effort: that is by decreasing net price to $P - \varepsilon(t)$.

Let $x^*(\cdot)$, $K^*(\cdot)$, etc. denote the components of the optimal trajectory. Consider the case where $\sigma^*(t) \neq 0$ along the trajectory, except perhaps for an instant in passing. Thus the bang–bang rule effectively determines $E^*(\cdot)$. We set ε, β, and α by

$$\sigma^*(t) = [P - C(x^*) - \varepsilon(t)] \cdot R[x^*(t), E^*(t), S^*(t)], \tag{5.15}$$

$$\mu^*(t) = B[I^*(t)] + \beta(t), \tag{5.16}$$

$$A[J^*(t)] = \alpha(t). \tag{5.17}$$

Assuming that the optimal system has been tracked accurately up to time t, then according to (5.12–5.13) these incentives will lead to a correct choice of $E(t)$, $I(t)$, and $J(t)$ at open access. Furthermore, by the dynamic equations (5.10–5.11), they will correctly project x, K, S, and μ into the future.

In fact, one can show that under these circumstances the tax β on I is redundant: by adjusting $\varepsilon(\cdot)$ appropriately, one may alter the evolution of $\mu(\cdot)$ so that

$$\mu(t) = B[I^*(t)],$$

and do this while retaining

$$\mathrm{sgn}\,[\sigma(t)] = \mathrm{sgn}\,[\sigma^*(t)],$$

hence while controlling $E(t)$ optimally.

On the other hand, when $\sigma^*(t) \equiv 0$ on an interval and $E^*(t)$ is determined by the marginal rule (5.9), then decentralized incentives seem to be ineffective.

References

Clark, C.W. (1988) Bioeconomic modeling and resource management. This volume, p. 11

Clark, C.W., Clark, F., Munro, G. (1979) The optimal exploitation of renewable resource stocks: problems of irreversible investment. Econometrica 47, 25–49

Eswaran, M., Lewis, T.R. (1984) Ultimate recovery of an exhaustible resource under different market structures. J. Environ. Econ. Manag. 11, 55–69

Fisher, A.C. (1981) Resource and Environmental Economics. Cambridge Univ. Press, Cambridge

Gordon, H.S. (1954) The economic theory of a common property resource: the fishery. J. Polit. Econ. 62, 124–142

Hotelling, H. (1931) The economics of exhaustible resources. J. Polit. Econ. 39, 137–175

McKelvey, R. (1983) The fishery in a fluctuating environment: coexistence of specialist and generalist fishing vessels in a multipurpose fleet. J. Environ. Econ. Manag. 10, 287–309

McKelvey, R. (1985) Decentralized regulation of a common property resource industry with irreversible investment. J. Environ. Econ. Manag. 12, 287–307

Owen, G. (1968) Game Theory. W.B. Saunders Co., Philadelphia

Pontryagin, L.S., et al. (1962) The Mathematical Theory of Optimal Processes. Wiley-Interscience, New York

Smith, K. (ed) (1979) Scarcity and Growth Reconsidered. Johns Hopkins Press, Baltimore

Information and Area-Wide Control in Agricultural Ecology

Marc Mangel

Contents

Introduction and Motivations

Agricultural ecology is a topic which could take the entire period of time available for the Course in Mathematical Ecology offered at the ICTP. For that reason, one must carefully select topics in the lectures. The topics chosen for these lectures are motivated by questions concerning agricultural productivity in developing countries. Productivity is often hampered by pest insects, which may cause enormous crop losses during outbreaks. There is considerable need for predicting where and when outbreaks are likely to be severe and to be able to implement management strategies that are effective but not excessively costly. These sentiments are echoed in the United Nations Africa Relief Program, as reported in the *New York Times* on 2 June 1986. The UN General Assembly adopted an agreement on African recovery that included the following points concerned with agricultural development:

> The immediate objective will be to cope with future emergencies and catastrophes through the following measures:
> —To create and sustain national emergency preparedness;

—To institute effective early warning systems;
—To establish flexible and efficient regional networks of crop protection
                                          (*New York Times*, 2 June 1986, page 4).

Developing nations do not have the luxury of widespread prophylactic use of insecticides because of the large financial expense (there are, of course, other costs as well), so that early warning systems and effective control are especially important.

This chapter is concerned with effective early warning systems for monitoring pest populations and regional networks of crop protection through control programs. The implementation of regional monitoring and control programs is rare in agriculture, but essential since insect populations do not recognize international political boundaries. A number of dramatic events in recent years have clearly demonstrated the need for a global view.

Early warning systems lead to the study of trapping for information in pest control (rather than trapping for control). Regional networks of crop protection lead to consideration of area-wide effects during control efforts, particularly application of pesticide and the use of the sterile insect methods.

The next section contains a discussion of the principles which will guide the analysis presented here. In particular, an "operational" approach is used to study the questions of interest. Since the underlying laws that govern the system of interest are unknown, we must attempt to learn these laws or at least construct some model of them that can be used for the purposes of interest. The third section contains a discussion of models for pest distributions and movement. The fourth section is concerned with trapping, particularly trapping for information (early warning). The fifth section contains a treatment of area-wide control of pests, particularly the management of pesticide resistance and the effective use of sterile insects in a management or control program.

Guiding Principles

When the phrase "operations research" (OR) is mentioned, most people think of a set of techniques such as linear and nonlinear programming, optimal control theory, game theory etc. This is not what the founders of OR in the United States, Philip Morse and George Kimball, intended at all. Their original objective is worth repeating here because it helps guide thinking about the role of analysis in operational problems. Morse was approached by the US Navy for aid in thwarting the operation of German submarines off the Eastern Coast of the US. He assembled a group of outstanding physical scientists and mathematicians into the Anti-Submarine Warfare Operations Research Group (see Morse, 1977, for his own description, or Mangel, 1982) for work on operational, rather than scientific, problems. The major difference between to two kinds of problems is that operational problems are typically so complex that we cannot deduce the dynamics of the system from first principles. Instead, we must develop a model of the system of interest. Once the model is developed, however, the approach of basic science is still used. Thus, we should think of operations research as the scientific method applied to operational problems.

As part of their work, Morse and his colleagues developed what they came to call "hemibel thinking". A hemibel is the logarithm of 3 and the general objective of hemibel thinking is to look for big improvements through the introduction of analysis into an operational problem. Morse and Kimball (1946, p. 38) describe hemibel thinking as follows:

It is well to emphasize that these constants which measure the operation are useful even though they are extremely approximate; it might almost be said that they are more valuable *because* they are very approximate. This is because the successful application of operations research usually results in improvements in factors of 3 or 10 or more. Many operations are ineffectively compared to their theoretical optimum because of a single faulty component... when the "bottleneck" has been discovered and removed, the improvements in effectiveness are measured in hundreds or even thousands of per cent. In our first study of any operation we are looking for these large factors of possible improvement. They can be discovered if the constants of the operation are given only to one significant figure, and any greater accuracy simply adds unessential detail... Having obtained the constants of the operation under study in units of hemibels (or to one significant figure), we take our next step by comparing these constants. We first compare the value of the constants obtained in actual operations with the optimum theoretical value, if this can be computed. If the actual value is within a hemibel (i.e., within a factor of 3) of the theoretical value, then it is extremely unlikely that any improvement in the details of the operation will result in significant improvement. In the usual case, however, there is a wide gap between the actual and theoretical results. In these cases a hint as to the possible means of improvement can usually be obtained by a crude sorting of the operational data... In many cases a theoretical study of the optimum values of the constants will indicate possibilities of improvement.

Rephrased for the agricultural pest control problem, this quotation takes the following form. We are not particularly interested in describing the crop-pest interaction in excruciating detail. Instead, the objective is to identify the key processes and major strategies that will quickly improve yield. Rather than fine-tuning models, we are looking for large differences between theory and data, using the analysis to show how to increase the chance of success in the battle against agricultural pests.

The principles of hemibel thinking are as valid today as they were 40 years ago. Scientists who are aiding decision makers involved in problems of agricultural ecology should keep the hemibel principle in mind when developing models. It is worth noting, too, that hemibel thinking is at odds with much of the current philosophy concerning Integrated Pest Management (IPM). Models in the current IPM approaches are typically highly detailed computer models, in which insight concerning interactions can be obtained only after intensive computational expense. It is always helpful to try a simple model before developing such a complex model. In addition, the cost of such detailed models may be prohibitive in developing nations.

In the majority of pest problems, we gather information not for its own sake, but to decide if an action such as some kind of pest control should be taken. Determining the threshold for this action, particularly when there is uncertainty associated with the decision, is not an easy job. Plant (1986) gives a good discussion of uncertainty and the economic threshold for action in pest management problems. In general, in these lectures it will be assumed that the threshold for action has already been determined. The importance of proper government policy as a means of avoiding disasters (see, e.g. the recent article on famine by Mellor and Gavian, 1987) can not be overstressed. Clark's lectures in this volume provide an example of how proper policy can be developed. Information transfer between the scientist and policy-maker is crucial, if scientific principles are going to be used in setting policy. It is essentially impossible to take too much care in insuring the effective transfer of scientific information.

Although it is very tempting to do so, one should not work on applied questions without a particular agricultural system and pest in mind. Most of the material presented in this chapter is concerned with methodology and the development of analytical tools, so that it will often appear that the problems are completely abstract ones. My own experience in problems of agricultural ecology comes from work on fruit flies of economic importance, such as the Mediterranean fruit fly or the apple maggot, and control of pests of cotton, particularly spider mites and lygus bug.

Modelling Pest Distributions

When considering a pest problem, the first thing that we need to know is the distribution and density of the damaging insect. To address these questions, consider a large region, of the order of perhaps hundreds of square kilometers, that is divided into cells and let A_i denote the area of the ith cell. The cells themselves might be of the order of square kilometers. For example, in California there are currently traps placed throughout the state for fruit flies of economic importance at a density of about one trap every 2.5 km^2. Let $N_i(t)$ denote the number of pests in cell i at the start of period t (if a discrete time formulation is used) or at time t (if a continuous time formulation is used). We are interested in the probability distribution of the vector $N(t) = \{N_i(t)\}$. This is defined by

$$p_i(n, t) = \text{Prob}\{N_i(t) = n\}. \tag{1}$$

Perhaps the simplest model is the Poisson distribution

$$p_i(n, t) = \exp(-\lambda(t)A_i)(\lambda(t)A_i)^n/n! \tag{2}$$

where $\lambda(t)$ is a parameter. The single parameter λ completely specifies the probability distribution, so that once it is known the entire distribution is known. The Poisson distribution has a very nice infinitesimal interpretation (which can also be used to derive Eq. (2)). Consider a small region $\Delta a \ll A$, where A denotes the area of a typical cell. Then Eq. (2) is equivalent to

$$\text{Prob}\{\text{no pest in } \Delta a\} = 1 - \lambda(t)\Delta a + o(\Delta a)$$
$$\text{Prob}\{\text{one pest in } \Delta a\} = \lambda(t)\Delta a + o(\Delta a)$$
$$\text{Prob}\{\text{more than one pest in } \Delta a\} = o(\Delta a) \tag{3}$$

where $o(z)$ represents terms such that $o(z)/z$ approaches 0 as z approaches 0. The parameter $\lambda(t)$ in the Poisson distribution can thus be interpreted as a proportionality constant relating the probability that a small region contains a pest and the size of the region. An explicit time dependence is included in this parameter, but for many of the applications discussed in this chapter one can view $\lambda(t)$ as a constant, as if a snapshot of the pest distribution at time t were taken.

The mean and variance of the Poisson distribution are easily shown to be

$$E\{N_i(t)\} = \lambda(t)A_i$$
$$\text{Var}\{N_i(t)\} = \lambda(t)A_i . \tag{4}$$

Here $E\{N_i(t)\}$ denotes the expectation or mean, recall that it is defined by $E\{N_i(t)\} = \sum n \Pr\{N_i(t) = n\}$, with the summation extending from 0 to ∞. The variance is $\text{Var}\{N_i(t)\} = \sum (n - E\{N_i(t)\})^2 \Pr\{N_i(t) = n\}$.

It is often quite useful to measure the variance in units of the mean and the way to do this is to use the coefficient of variation defined by

$$CV\{N_i(t)\} = \sqrt{\text{Var}\{N_i(t)\}}/E\{N_i(t)\}$$
$$= 1/\sqrt{\lambda(t)A_i} . \tag{5}$$

Eq. (5) has an important operational interpretation: As the region that is being sampled increases in area, the mean and the variance of the number of pests in the region will also increase (as in Eq. (4)) but the relative size of the variation (when the standard deviation is measured relative to the mean) decreases. Indeed, according to Eq. (5), as A_i increases, the coefficient of variation approaches 0.

The following computational algorithm is a useful way of computing terms in the Poisson distribution. First, set

$$p_i(0, t) = \exp(-\lambda(t)A_i)$$

and then note that

$$p_i(n + 1, t) = \exp(-\lambda(t)A_i)(\lambda(t)A_i)^{n+1}/(n + 1)!$$
$$= [\lambda(t)A_i/(n + 1)]p_i(n, t).$$

Thus, once $p_i(0, t)$ is computed, all of the other probabilities can be found iteratively.

Negative Binomial Distribution

It is unlikely, particularly for problems in which the area of interest is large, that the parameter $\lambda(t)$ in the Poisson distribution is constant over all of the cells. How, then, should we deal with the variation of the parameter? For simplicity in what follows, the time dependence of λ will be suppressed.

Imagine that the parameter of the Poisson distribution is itself random. The parameter will now be denoted in boldface by $\boldsymbol{\lambda}$, to indicate that it is also a random variable. One picture would be this: "Nature" picks the value of the parameter in a particular cell from some distribution and, once that value is picked the number of pests in the cell is given by the Poisson distribution. In other words, the Poisson distribution is now the conditional distribution of the number of pests in a cell, given the value of $\boldsymbol{\lambda}$.

In order to work with conditional probabilities, a digression to recall the definition of conditional probability and Bayes's formula is required. Suppose that $\mathscr{A}$ and $\mathscr{B}$ are two possible outcomes of a probabilistic process. The conditional probability of $\mathscr{A}$ given that $\mathscr{B}$ has occurred is then defined by

$$\text{Prob}\{\mathscr{A}|\mathscr{B}\} = \text{Prob}\{\mathscr{A},\mathscr{B}\}/\text{Prob}\{\mathscr{B}\} . \tag{6}$$

In this equation, $\text{Prob}\{\mathscr{A},\mathscr{B}\}$ is the probability that both events occur and it is implicitly assumed that the probability that the event $\mathscr{B}$ occurs is greater than 0. Rearranging Eq. (6) leads to

$$\text{Prob}\{\mathscr{A},\mathscr{B}\} = \text{Prob}\{\mathscr{A}|\mathscr{B}\}\,\text{Prob}\{\mathscr{B}\} \tag{7}$$

and then noting that the roles of $\mathscr{A}$ and $\mathscr{B}$ can be interchanged in these equations leads to Bayes's theorem

$$\text{Prob}\{\mathscr{B}|\mathscr{A}\} = \text{Prob}\{\mathscr{A}|\mathscr{B}\}\,\text{Prob}\{\mathscr{B}\}/\text{Prob}\{\mathscr{A}\} . \tag{8}$$

Eq. (8) will turn out to be extremely important when we consider the analysis of information in pest control problems.

Suppose now that the parameter λ has a density function $f(\lambda)$ so that $f(\lambda)d\lambda$ is the probability that $\lambda \leq \lambda \leq \lambda + d\lambda$. The probability that $N_i(t)$ takes a particular value n is then found combining the conditional Poisson distribution with the density of λ so that

$$p_i(n,t) = \int [\exp(-\lambda(t)A_i)(\lambda(t)A_i)^n/n!]f(\lambda)d\lambda . \tag{9}$$

In order to easily implement Eq. (9), we want to choose a density $f(\lambda)$ that will easily integrate against the Poisson distribution. An appropriate density will be described momentarily. Before doing that it is worthwhile to briefly consider a sampling problem, in order to show how the framework developed thus far can be employed to obtain useful information about pest populations.

Suppose that an insect is distributed according to the Poisson distribution, but that the value of the Poisson parameter λ is unknown. In order to learn about the value of the parameter, we assume that cells are sampled and that the insect counts obtained through sampling are used to make inferences about the Poisson parameter. In particular, assume that in the ith cell λ has density $f(\lambda)$ and that when this cell is sampled, the number of insects discovered equals to n. (Assume for the time being that this sample information is perfect; it will be seen that this assumption is not crucial to the following argument.) We wish to compute the *posterior* (i.e., after sampling) density for λ, given the data that n pests were discovered in the sample. Let $f(\lambda|n)$ denote this density. Use of Bayes's theorem shows that

$$\begin{aligned} f(\lambda|n)d\lambda &= \text{Pr}\{\lambda \leq \lambda \leq \lambda + d\lambda | N_i(t) = n\} \\ &= \text{Pr}\{\lambda \leq \lambda \leq \lambda + d\lambda, N_i(t) = n\}/\text{Pr}\{N_i(t) = n\} \\ &= f(\lambda)[\exp(-\lambda A)(\lambda A)^n/n!]/\int f(\lambda)[\exp(-\lambda A)(\lambda A)^n/n!]d\lambda . \end{aligned} \tag{10}$$

Although it somewhat abuses notation, it is very helpful to adopt the convention that $\lambda \approx \lambda$ for the more exact relationship $\lambda \leq \lambda \leq \lambda + d\lambda$. This will be done throughout the rest of the Chapter. From the viewpoint of implementation of these formulas, we should pick a density $f(\lambda)$ that will allow the integrals to be done easily. The density that is chosen is defined on $[0, \infty)$ for the range of values of the parameter. One choice for the density is the gamma density with parameters v and α. This density will be denoted by $f(\lambda; v, \alpha)$ and is

$$f(\lambda; v, \alpha) = \exp(-\alpha\lambda)\lambda^{v-1}\alpha^v/\Gamma(v) .\tag{11}$$

In this equation, $\Gamma(v)$ is the gamma function. For biologists who are not familiar with it, you can think of the gamma function in the same way that one thinks about the functions $\sin(x)$ or $\log(x)$. That is, the gamma function has one or more definitions, arises in particular situations, and has certain computational properties. The gamma function is defined by

$$\Gamma(v) = \int_0^\infty \exp(-t)t^{v-1}dt .\tag{12}$$

From the definition, we can show that the following recursion relationship holds (try it as an exercise!)

$$\Gamma(v+1) = v\Gamma(v)\tag{13}$$

so that for integer values of its argument, $\Gamma(v+1) = v!$. For values of the argument less than 1, Abramowitz and Stegun (1964) give an extremely accurate formula (page 256, paragraph 6.1.34). For values of the argument greater than 1, use Eq. (13) and the formula in Abramowitz and Stegun (which, by the way, is a book well worth owning!).

The information that $f(\lambda; v, \alpha)$ is a probability density provides an easy guide to doing what appear to be complicated integrals. That is, since a probability density must integrate to 1, one can integrate Eq. (11) over $[0, \infty)$ and multiply by the appropriate constants to show that

$$\int \exp(-\alpha\lambda)\lambda^{v-1}d\lambda = \Gamma(v)/\alpha^v .\tag{14}$$

The mean and variance of λ can be found by directly applying Eq. (14). Using this equation gives

$$E\{\lambda\} = v/\alpha$$
$$E\{\lambda^2\} = (v/\alpha)^2 + (v/\alpha^2)\tag{15}$$

from which it follows that the variance of λ is v/α^2 and the coefficient of variation is $CV\{\lambda\} = 1/\sqrt{v}$. Note that the coefficient of variation approaches 0 as v increases. Also, note that the mean of the distribution is constant when v/α is constant, so that one can hold the mean constant while varying the shape through changes in v. Finally, note that the density peaks at $\lambda^* = (v-1)/\alpha$, a value less than the mean. The reader is encouraged to verify all of these statements by sketching the shape of the gamma density for a few values of the parameters. If that's done, one can see the robustness of the distribution in terms of different shapes.

Using the gamma density in Bayes's formula leads to

$\Pr\{\lambda \approx \lambda | n \text{ pests present}\}$

$$= \frac{[\exp(-\alpha\lambda)\lambda^{v-1}\alpha^v/\Gamma(v)][\exp(-\lambda A)(\lambda A)^n/n!]}{\int[\exp(-\alpha\lambda)\lambda^{v-1}\alpha^v/\Gamma(v)][\exp(-\lambda A)(\lambda A)^n/n!]d\lambda}$$

$$= \exp(-(\alpha+A)\lambda)\lambda^{n+v-1}(\alpha+A)^{n+v}/\Gamma(n+v)$$

$$= f(\lambda|v+n, \alpha+A) . \tag{16}$$

This equation is an example of an *updating rule*: we begin with a prior idea about the distribution of the parameter, obtain information, and update the idea about the distribution of the parameter to obtain a posterior distribution. In this case, the prior idea is that λ follows a gamma distribution with parameters v and α, the information is that n pests were in area A, and the posterior distribution of the parameter is a gamma distribution with parameters $v+n$ and $\alpha+A$. Note that as long as some "positive" information is obtained, in the sense that $n > 0$, there is a reduction in uncertainty since the coefficient of variation of the posterior distribution $(1/\sqrt{(n+v)})$ is less than the coefficient of variation of the prior distribution $(1/\sqrt{v})$.

The main justification for choosing a gamma prior is that it works: we start with a gamma prior, use Poisson sampling, and end with another gamma distribution. Thus, we only need to keep track of the parameters of the distribution, rather than the distribution itself. Statisticians call the gamma a *conjugate prior* for the Poisson distribution (see, e.g. Berger, 1980 or DeGroot 1970 for a fuller discussion of Bayesian decision theory). There is a small "biological" story for the choice of a gamma prior: Suppose that the parameter of the Poisson distribution is proportional to the number $B(t)$ of pests and that the number satisfies a stochastic differential equation (SDE) of the form

$$dB = B[r(1 - B/K)dt + \sigma dW]$$

where r, K and σ are parameters and dW is the increment in Brownian motion (see the Chapter by Riccardi, in vol. 17 of this series). Then the equilibrium distribution, defined as the $\lim_{t \to \infty} B(t)$, is often a gamma distribution. This kind of model would suggest that a gamma prior for the Poisson parameter is a reasonable choice.

The denominator in Eq. (16) is the probability that n pests are present in the region with area A. Doing the integrals and a little bit of algebra leads to the result

$\Pr\{n \text{ pests present in region of area } A\} = p(n, A)$

$$(\Gamma(v+n)/n!\,\Gamma(v))(A/(\alpha+A))^n(\alpha/(\alpha+A))^v . \tag{17}$$

This is the negative binomial (NB) distribution. It is computed using an algorithm similar to the one given for the Poisson distribution. That is, first set

$$p(0, A) = (\alpha/(\alpha+A))^v$$

and then use the iteration formula obtained directly from the definition in Eq. (17)

$$p(n+1, A) = ((v+n)/(n+1))(A/(\alpha+A))p(n, A) . \tag{18}$$

If N has a NB distribution given by Eq. (17), then the mean and variance of N are

$$E\{N\} = (v/\alpha)A = m$$
$$\text{Var}\{N\} = (v/\alpha)A + v(A/\alpha)^2$$
$$= m + m^2/v . \tag{19}$$

The parameter v, which is often denoted by k in the ecological literature, can thus be interpreted as an "over-dispersion" parameter in the following sense: When v is large, the mean and variance of the NB distribution are approximately equal so that NB distribution is essentially a Poisson distribution. (The more mathematically inclined reader may wish to consider the behavior of the gamma density when $v \to \infty$ and v/α is held constant.) When v is small, on the other hand, the variance of the NB distribution can be much larger than the mean. Because ecological data often involve variances that far exceed means, the NB distribution has enjoyed immense popularity in ecology. It is not without its problems, however, and some of these are discussed below.

From Eq. (19), the coefficient of N is given by

$$\text{CV}\{N\} = ((1/m) + (1/v))^{0.5} . \tag{20}$$

This equation should be compared to Eq. (5). Since the mean of the NB distribution is proportional to the area A, as the sampled area increases, the mean m approaches ∞ and the coefficient of variation approaches a limit $v^{-0.5}$ which is non-zero and may be large when v is small. Thus, whereas for the Poisson distribution there was no limit on the relative accuracy of the sampling, for the NB distribution there is an inherent limit, determined by the over-dispersion parameter.

If Eq. (18) is rewritten in terms of m and k, a little bit of algebra shows that

$$p(n, A) = p(n; m, k)$$
$$= [\Gamma(k + n)/n! \, \Gamma(k)](m/(k + m))^n (k/(k + m))^k . \tag{21}$$

The $p(n; m, k)$ form will be used in the rest of this chapter. A remarkable feature of the NB distribution is the preponderance of zeroes when k is small. From Eq. (21),

$$p(0; m, k) = \text{Pr}\{N = 0 \text{ when } m \text{ and } k \text{ are the parameters of the distribution}\}$$
$$= (k/(k + m))^k \tag{22}$$

and this probability may be considerable, even if m is immense, if k is sufficiently small. For example, if $k = .09$ we obtain the following results:

$\log(m)$	$\text{Pr}\{N = 0\}$
0	0.799
1	0.734
2	0.672
3	0.614
4	0.562
5	0.513

The interpretation of these numbers is the following: If N has a NB distribution with parameters $m = e^5$ and $k = 0.09$ then the probability that a particular region may have 0 pests is slightly larger than $1/2$, even though the mean of the distribution is about 150! This kind of behavior is highly desirable when modelling pest insects, because pests are often totally absent from most regions (thus giving many zeroes) yet are abundant "on the average" because of a few highly intense, localized outbreaks.

Estimation of the parameters m and k is important in any applied problem. It is easy to show that if we have collected a data set and **m** and s^2 are the sample mean and variance, then **m** is the maximum likelihood estimate (MLE) for the mean of the NB distribution. Estimation of k is a little bit trickier. Kendall and Stuart (1979, page 78) show how MLE estimates for k can be determined. Two simple methods, which may be sufficient in many applied problems, are the following. We can simply use the moments of the empirical distribution and match them to the moments of the NB distribution. Thus, set $m = $ **m** and, in light of Eq. (19), set

$$k = m^2/(s^2 - m) \tag{23}$$

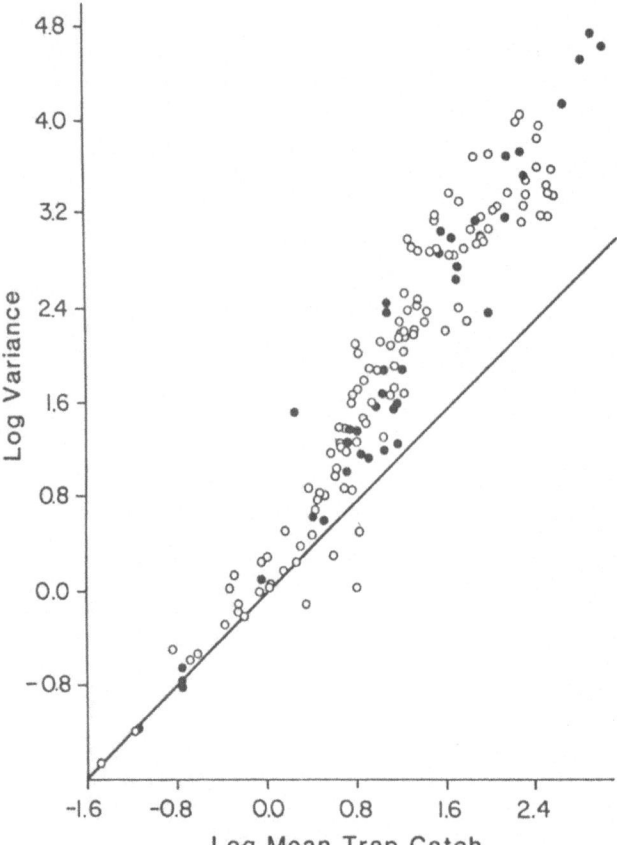

Fig. 1. P. Baker's data on medfly (*Ceratitis capitata* Wied) trapping. Open circles are Jackson traps, closed circles are delta traps

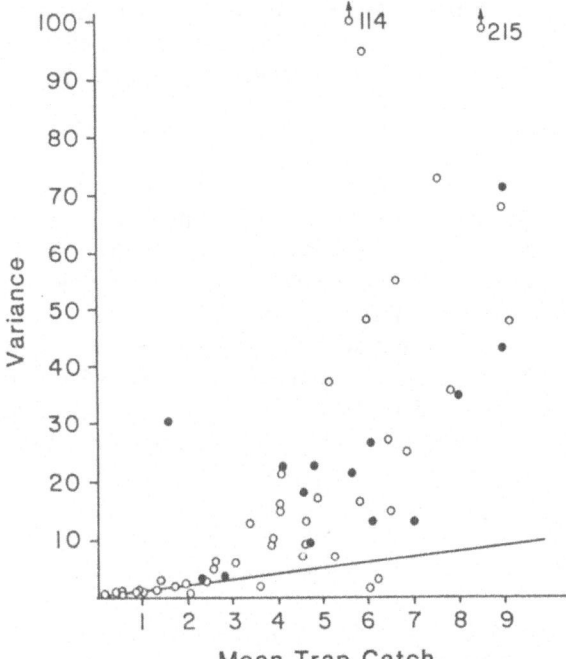

Fig. 2. P. Baker's data for small values of mean trap catch. Everything else is the same as in Fig. 1

as long as this equation makes sense (negative values of k, for example, are not allowed but may occur). Another approach is to rewrite the variance of N as

$$\log[\mathrm{Var}\{N\}] = \log[(m^2/k)(1 + (k/m))] \tag{24}$$

so that for values of $k/m \ll 1$ one obtains

$$\log[\mathrm{Var}\{N\}] = \log(m^2/k) + \log(1 + (k/m))$$
$$= 2\log(m) - \log(k) + o(k/m) \tag{25}$$

where $o(z)$ represents a term such that $o(z)/z$ goes to 0 as z goes to 0. A plot of the logarithm of the variance against the logarithm of the mean of the distribution will lead to an estimate of k from Eq. (25).

An example of how actual pest count data appear, and how the NB distribution might be used is shown in Figures 1 and 2, using data collected by Peter Baker (1985) who was trapping fruit flies in Mexico. These figures show the variance of the trap catch as a function of the mean. We clearly see an increase in the variance as the mean increases; use of a simple NB model gives a value of $k \approx 2$.

Other Models for Aggregated Distributions

The NB is only one of a family of distributions that can be used to model aggregated populations. It is sometimes misused, perhaps because simple over-enthusiasm and the ease with which it can be applied. A number of authors (e.g. Taylor et al. 1979 or Perry and Taylor 1986) have chastened individuals who use the NB distribution too glibly. In particular, the value of k often depends upon the mean. Taylor et al. (1979) and Perry and Taylor (1986) suggest that k is a function $k(m)$ of the mean, given by

$$k(m) = m/[am^{b-1} - 1]$$

where a and b are parameters. Perry (1981) discusses parameter estimation for this relationship (The motivation for this choice of functional form, and the interpretation of the parameters will be discussed below.) Another problem with the NB distribution is that the value of k may depend upon the sampling scale and structure of the population, so that blind use of the NB (or any other distribution!) could lead to silly results. For example, the reader may wish to consider a problem with about 50–100 cells in which $N_i = i^\beta$ where β is a parameter (e.g. try $\beta = 0.5, 1, 1.5$). Simply computing the mean and variance leads to a picture of considerable overdispersion, but it is not random overdispersion at all. Rather, there is a clear pattern to the data. Further discussion of this point is given by Debouzie and Thioulouse (1986). The essential point, however, is to use the NB distribution thoughtfully and to be aware that there are a number of other models that can be used with equal facility as the NB, especially with the accessibility and easy use of modern, desktop computers.

Some of these alternate models will now be discussed. They are all computationally more complicated than the NB distribution. None of them, however, is so complex that it can not be used with a small, desktop microcomputer. Rapid advances in computer technology are likely to make it even easier to use these distributions.

The Neyman Type-A distribution is obtained by compounding one Poisson distribution with another. If N follows a Neyman type A with parameters λ and θ then

$$\Pr\{N = n\} = \sum_{j=1}^{\infty} [\exp(-\lambda)\lambda^j/j!][\exp(-j\theta)(j\theta)^n/n!] . \tag{26}$$

The probability of a 0 in this model is

$$\Pr\{N = 0\} = \exp[-\lambda(1 - e^{-\theta})] \tag{27}$$

and the mean and variance of N are

$$E\{N\} = \lambda\theta$$
$$\mathrm{Var}\{N\} = \lambda\theta(1 + \theta) \tag{28}$$

so that it is clear that we can, by appropriate choice of the parameters, make the variance greatly exceed the mean.

A second class of contagious models are *urn models*. For pest control problems, they can be phrased as models involving an "occupancy approach". The typical urn problem would be phased as follows. We start with a mixture of "white" balls and "black" balls. Let $W(t)$ and $B(t)$ denote the number of white and black balls after t "drawings" or "samplings". The sampling rules are the following:

1. A single ball is removed. This is the sample.
2. One adds $\alpha + 1$ balls of the same color and β balls of the opposite color.

Although this is apparently a simple problem, its analysis is actually quite difficult. Feller (1968) provides a good introduction to urn problems. The paper by Bernard Friedman (Friedman 1948) is a gem of applied mathematics and still well worth reading. Two properties of urn models of special note are:

1. For the simple urn model just formulated, initial fluctuations drive the final outcome. That is, if one considers the proportion of white balls $p(t) = W(t)/(W(t) + B(t))$ as t increases, the distribution of $p(t)$ approaches a uniform distribution on $[0, 1]$. The behavior of a particular sample path is virtually completely determined by the first 10 or 20 samplings (subject to the values of α and β, of course). There is thus no "equilibrium" in the sense that all points in the (W, B) phase space are ultimately attracted to the same point.
2. A special case of the general urn problem is the Polya urn, in which $\beta = 0$. Under appropriate conditions, the Polya urn has a NB limiting distribution. This limit applies to the situation in which

$$W(0)/(W(0) + B(0)) \to 0$$
$$(\alpha + 1)/(W(0) + B(0)) \to 0$$

and the number of samples $T \to \infty$ in such a way that

$$TW(0)/(W(0) + B(0)) \to \theta \neq 0$$
$$T(\alpha + 1)/(W(0) + B(0)) \to \rho \neq 0 .$$

Under these conditions, the limiting distribution of the Polya urn is negative binomial (see Johnson and Kotz 1969 for further discussion).

The limiting distribution actually has a nice biological interpretation for the development of aggregation. As an example, consider how an aggregated distribution of insect eggs might arise. Suppose that as insects incounter possible habitats for their eggs they follow the rule of sampling a small volume of the habitat and, if they discover other eggs, adding some of their own eggs to those already present. (The basic idea here is that if other eggs are present, a previous female has decided that this is an acceptable habitat.) With the exception of having to add a description of the first insect that lays its eggs in the habitat, the description just given follows the spirit of a Polya urn model.

A third, and quite general model, for aggregation is known as *Taylor's power law*. It arises out of the statistical analysis of the relationship between the mean and variance of pest sampling data. If $\mathbf{m}$ and $\mathbf{s}^2$ denote the sample mean and variance, we fit a relationship of the form (see Taylor et al. 1979 for more details)

$$\mathbf{s}^2 = a\mathbf{m}^b \tag{29}$$

where a and b are parameters. These parameters are then used with a particular probability distribution. For example, if the pest distribution follows a NB model, then the mean and variance are related by the formula Variance $= m + (m^2/k)$. Using equation (29) gives a method for finding a functional form for the overdispersion parameter k as a function of the mean. This is the equation given previously.

Pest Growth and Movement

The picture established thus far is a large region, divided into cells with $N_i(t)$ denoting the number of pests in cell i at the start of period t. In the absence of any control actions, the values of $\{N_i(t+1)\}$ will be determined by growth and movement of the pest populations. These are both complicated processes and each has been modelled in several different ways (see, e.g. Hargrove 1981, Minogue and Fry 1983, Rogers 1979, Sawyer and Haynes 1985, 1986 or Taylor 1986). The purpose of this section is simply to state the assumptions that will be used in the next two or three sections on trapping and control.

In most applied problems, we want to be able to detect and control the pests at a relatively low population level, so that density dependent effects can be ignored. This leads to mathematically simple relationships (which are thus somewhat uninteresting to mathematicians!), simplifies much of the further analysis, and makes parameter estimation easier. Thus, in the absence of any movement, we might assume that

$$N_i(t+1) = r_i(t)N_i(t) \tag{30}$$

where $r_i(t)$ is the growth rate for the pest population in cell i during period t. We could easily incorporate more complicated growth models, with density dependence, if there is evidence for their need.

Movement is a little bit trickier to characterize. Small scale or local movements of the pest can often be effectively described by some kind of random walk or diffusion model (see, e.g. Broadbent and Kendall 1953, Gillis 1956 for "classic" work or Kareiva and Shigesada 1983, Root and Kareiva 1984 or Sawyer and Haynes 1985, 1986 for more recent work). Sometimes large scale motion can also be modelled using diffusion models. Although random walk models provide nice qualitative pictures for insect movements, there are even some difficulties with these models as descriptions of short scale movement. For example, Kareiva and Shigesada (1983) show that the actual movement of butterflies is wider ranging than the movement predicted by a random walk model. That is, the actual movement is more diffuse than the random walk.

In agricultural pest control problems, we are more likely to be interested in large scale movements of pests. Large scale movements are often driven by factors such as wind or the movements of animals and people that carry the pest (intentionally or not). In this case, wind patterns and roadway maps may provide the best information about the movement of pests. There is a paucity of data on the large scale movements of pests of interest in agricultural problems. Taylor (1986) provides one of the best descriptions of large scale movement. In general, however, we must tailor the movement model to the problem of interest.

Trapping and Information

In this section, a variety of trapping models are developed for the analysis of trapping for information, rather than control. (Also see Janesen and Metz, 1979;

Mangel, 1985, 1986; Mangel et al., 1984; McClendon et al., 1976; Metz et al., 1983, and Plant and Wilson, 1985).

As a motivation for the development of mathematical models, it is worthwhile to consider the experiments of Cunningham and Couey (1986) in which large numbers of marked Mediterranean fruit flies were released at differing distances from a single trap in an orchard. The proportion of flies released r feet from the trap and captured by the trap was measured. Cunningham and Couey found an almost perfect fit of their data to the formula

$$q(r, t) = A(t) \exp(-B(t)r) \tag{31}$$

where $q(r, t)$ is the proportion of flies released r feet from the trap and captured by time t, and $A(t)$ and $B(t)$ are parameters. Cunningham and Couey found that $A(24\,\text{hrs}) = 0.6527$, $B(24\,\text{hrs}) = 8.637 \times 10^{-3}$ and that $A(\infty) = 0.6970$, $B(\infty) = 6.723 \times 10^{-3}$ where "∞" is understood to mean the total trapping period, which ended when no more flies were captured. Note that $A(t) < 1$ means that flies released at the trap were not necessarily captured. There is even anecdotal evidence that fruit flies may enter certain kinds of traps, fly around inside them and then exit! An interesting open question is what kind of movement models lead to functional forms such as Eq. (31)? I will briefly discuss a model for these experiments at the end of the next section.

The "Classical" Trapping Problem and Its Extensions

As a first model of a trapping consider an approach based on partial differential equations or difference equations (Jansen and Metz, 1979; Mangel, 1986). Although this method usually turns out to be computationally impracticable, it is a good starting point. The experimental region is represented by a square of length L and the trap is located at the center of the square. Associated with the trap is a "trap radius" r_t with the property that if distance between the pest and the center of the square is less than r_t, then the pest is trapped. We can define a probability density for an untrapped pest as follows:

$$f(x, y, t)dx\,dy = \text{Pr}\{\text{at time } t \text{ the pest is in the small area } dx\,dy \text{ around}$$
$$\text{the point } (x, y) \text{ and is not trapped}\}. \tag{32}$$

The equation that $f(x, y, t)$ satisfies is determined by the movement model. Two models are the diffusion model and the large deviation model of motion. The biological assumption behind the diffusion model is that in a small interval of time the pests are likely to move, but only a short distance. The distribution of displacement in a short interval of time Δt is assumed to be Gaussian with mean displacement $0 + o(\Delta t)$ and variance $D\Delta t + o(\Delta t)$. In this case, the the equation that $f(x, y, t)$ satisfies is

$$\partial_t f = (D/2)\{\partial_x^2 f + \partial_y^2 f\} \tag{33}$$

where $\partial_i f$ denotes the partial derivative of $f(x, y, t)$ with respect to the ith variable, $\partial_i^2 f$ denotes the second partial derivative and D is the diffusion coefficient.

An alternative to the diffusion model (in which the movement is implicitly small scale) is a large deviation model. The assumption here is that it is possible for the pest to make a large movement in a short interval of time. We thus explicitly specify the size and the probability of a displacement. In this case, the appropriate equation for $f(x, y, t)$ is (e.g. Knessl et al., 1984)

$$f(x, y, t) = \sum \sum f(x - \varepsilon_i, y - \varepsilon_j, t - \Delta t) \rho(\varepsilon_i, \varepsilon_j | x - \varepsilon_i, y - \varepsilon_j) \tag{34}$$

where $\rho(u, v | x, y)$ is the probability of taking a jump of size (u, v) from the point (x, y), $(\varepsilon_i, \varepsilon_j)$ is the size of the jump taken and the summation extends over all points in the region of interest. Although they may appear formidable, Eq. (33) and (34) are easily derived. Consider, for example, Eq. (34). We ask: what is the probability that a pest is around the point (x, y) at time t? The answer is this: to get to (x, y) at time t, the pest could have been at some point (u, v) at time $t - \Delta t$ and taken a jump $(x - u, y - v)$ in the interval Δt. Summing over all possible u and v gives Eq. (34). Equation (33) can then be derived by a Taylor expansion of Eq. (34), assuming that only small jumps occur. The reader is encouraged to try such an expansion.

In order to solve Eqs. (33) or (34) initial and boundary conditions are needed. If the pest is uniformly distributed in the region when $t = 0$, the appropriate initial condition is $f(x, y, 0) = 1/(L^2 - \pi r_t^2)$. Since the pest is trapped upon entering the trap radius, one boundary condition is $f(x, y, t) = 0$ when $x^2 + y^2 = r_t^2$. The boundary condition at the edges of the square is more difficult. If there is a single trap, then the only other condition that can be reasonably applied is that $f(x, y, t)$ is bounded for $x, y \to \infty$. Alternately, we might assume that if the pest leaves the square under consideration, it enters another square of size L with a trap at the center. In such a case, a reflecting condition is appropriate; that is, the normal derivative of $f(x, y, t)$ vanishes on the boundary of the square.

The geometry of this problem, a square region with a circular internal boundary, makes it extremely difficult to solve. As an approximation, one can replace the square by a circle of radius R. There are at least three good choices for R: one can inscribe the square of side L in a circle of radius R, circumscribe the square in a circle of radius R, or choose R so that areas are equal. Each of these has an operational interpretation which is left to the reader as an exercise.

If we choose to use an exterior boundary that is a circle, it is natural to switch to polar coordinates. Then, for example, $f(x, y, t) = f(r, t)$ (assuming radial symmetry) and the full problem associated with Eq. (33) is

$$\partial_t f = D(1/r)\partial_r(r\partial_r f)$$
$$f(r_t, t) = 0; \quad \partial_r f|_{r=R} = 0$$
$$f(r, 0) = \begin{cases} 1/(\pi(R^2 - r_t^2)) & \text{if } r > r_t \\ 0 & \text{otherwise} \end{cases} \tag{35}$$

The solution of this problem, and numerous variants of it, can be looked up in Chapter 13 of Carslaw and Jaeger (1959).

As appealing as this approach may be, there are a number of difficulties with the classical trapping problem. Some of the most important ones are:

- *Imperfect trapping*: In most operational situations, it is likely that pests are not always trapped when they enter a trap, but the classical procedure assumes perfect trapping.
- *Nonconstant coefficients*: The classical model assumes that the diffusion coefficient (and drift coefficient, if there is one) are constant over space and time. This is likely not to be the case.
- *Intercell movement*: The classical model assumes that a pest in cell i can not be trapped by a trap in cell j. This assumption will often be violated in real trapping programs because pests can move freely across boundaries of cells and may be attracted from one cell to another by the presence of distant traps.

There are ways to incorporate these ideas into the classical trapping problem. For example, define

$$\Psi(x, y, \mathbf{z})dt = \Pr\{\text{pest is trapped by one of } N \text{ traps during the interval}$$
$$(t, t + dt)|\text{at time } t \text{ the pest is located at } (x, y)$$
$$\text{and the vector of trap locations is } \mathbf{z}\} . \tag{36}$$

The vector of trap locations is understood in the following way:

the vector $\mathbf{z} = (z_{x1}, z_{y1}; z_{x2}, z_{y2}; \ldots z_{xN}, z_{yN})$ where (z_{xi}, z_{yi})

is the center of the ith trap. Two models for the trapping function are the following:

$$\Psi(x, y, \mathbf{z}) = \sum \delta(x - z_{xi})\delta(y - z_{yi})a_i \tag{37a}$$

where $a_i \leq 1$ is a trapping coefficient (with equality corresponding to perfect trapping) and $\delta(u)$ is the Dirac delta function. That is, trapping occurs with probability a_i if the pest enters the trap centered at (z_{xi}, z_{yi}). A model similar to this one has been analyzed by Szabo et al. (1984) for problems related to partial trapping of random walks in chemical physics.

Another model for the trapping function is

$$\Psi(x, y, \mathbf{z}) = \sum a_i[q_i + (x - z_{xi})^2 + (y - z_{yi})^2]^{-v} \tag{37b}$$

where a_i, q_i, and v are parameters. This model is chosen in analogy to certain detection formulas from search theory (Koopman, 1980) but has not been applied to problems in agricultural pest control.

With these trapping models, it can be shown (Mangel, 1981) that the probability density for the location of an untrapped pest now satisfies

$$\partial_t f = D(\partial_x^2 f + \partial_y^2 f) - \Psi(x, y, \mathbf{z})f \tag{38}$$

with the same initial condition. This equation can sometimes be solved by analytically, especially if we are willing to accept approximate techniques such as asymptotic methods (Mangel, 1981). Otherwise, numerical methods are needed.

Rosenstock (1980), in a paper on problems of trapping chemical physics, gives formulae for the mean time to trapping for a random distribution of traps on a lattice. In particular, suppose that the pests execute some kind of random walk to nearest neighbors on a lattice and assume that a fraction q_t of the lattice points are traps. Rosenstock shows that the mean time until trapping is well approximated by the value $(1/\pi q_t)(-\log(\pi q_t) + 1 - C)$ where $C = 0.577216$ is Euler's constant (see Abramowitz and Stegun, 1965).

Any of these models might be employed to analyze trapping data, such as the data developed by Cunningham and Couey. There are two main questions. First, how do such data arise from particular motion models? Second, given such trapping data, what should be done with the data? My colleague Richard Plant is currently working on an extension of a diffusion model, as a means of modeling the Cunningham and Couey data from "first principles". The extension works as follows. A pest located at the point (x, y) at time t may do one of two things between t and $t + \Delta t$: i) It may make a displacement that is normally distributed with mean 0 and variance proportional to Δt. If it moves, there is a certain probability of being trapped. ii) It may settle at the point (x, y) and stop moving. Plant assigns a probability to each of the choices and then derives an extension of Eq. (33). This model leads to a trapping curve similar to the one found by Cunningham and Couey. The model, however, does not explain the origin of the "move/stay" decision. Determining the behavioral origin of this decision is an interface of behavioral ecology (Krebs and Davies (1984)) and applied ecology.

Next consider what is done with trapping data, once it is obtained. For example, suppose that the probability that a pest is ultimately trapped given that initially it is r units away from the trap is $q(r) = Ae^{-Br}$ and that the trap center-to-center distance is L miles, so that each trap is at the center of a square that is L miles

Table 1. Values of q_L for varying R_M

R_M (miles)	q_L for	
	One day	Entire trapping period
0.01	0.48	0.55
0.02	0.36	0.44
0.03	0.28	0.35
0.04	0.21	0.29
0.05	0.17	0.23
0.06	0.13	0.19
0.07	0.11	0.16
0.08	0.086	0.13
0.09	0.071	0.11
0.10	0.059	0.096
0.20	0.016	0.027
0.30	0.007	0.012
0.40	0.0037	0.0066
0.50	0.0024	0.0043
0.60	0.0017	0.0030
0.70	0.0012	0.0022
0.80	0.00096	0.0017
0.90	0.00076	0.0013
1.00	0.00062	0.0011

on a side. The probability q_L that a pest located anywhere in this cell is trapped is given by the average of $q(r)$ over the cell, so that

$$q_L = \iint A \exp(-B(x^2 + y^2)^{1/2}) dx\, dy/L^2$$

where the region of integration is $-L \leq x \leq L$, $-L \leq y \leq L$ and the factor L^2 in the denominator comes from $dx\, dy/L^2$ being the probability that a pest is located in the small area $dx\, dy$. Although q_L can be found numerically, an exact formula can be computed by replacing the square by a circumscribed circle, and inscribed circle, or a circle of equivalent area. Let R_M denote the radius of the circle of interest. Switching to polar coordinates and integrating by parts once gives the result

$$q_L = (2A/R_M^2)[(1/B^2)\{1 - e^{-BR_M}\} - R_M e^{-BR_M}/B] .$$

Table 1 shows values of q_L computed in this manner. This table gives an idea of the efficacy of different trap spacings. It is useful for the analysis of information in trapping.

Analysis of Information in Trapping

This section is concerned with how to best analyze trap catch data to obtain as much information as possible (also see Plant and Wilson, 1985; Wilson et al., 1985). Questions about the analysis of information naturally lead to a Bayesian framework in which two related questions can be asked:

i) What can be said about population levels if there is no trap catch?

ii) What can be said about population levels if there is a positive trap catch? Although a minor modification of question ii) includes question i), it is operationally useful to separate the two. The answers to these questions give information about the level of the pest population. A third question relates the population level or trap catch and the damage caused by the pest (e.g. Prokopy et al., 1982), but that will not be treated here. My opinion is that the third question is still essentially an experimental one.

Once again, consider a large region divided into cells in which the pest population in the ith cell has a negative binomial distribution with parameters m and k. In the analysis that follows, it is assumed that the overdispersion parameter is known (for example by analogy with other trapping situations) and, for pedagogic ease and simplicity, constant but that the mean m of the negative binomial distribution is not known. The objective of the analysis is thus to use the trap data to make statements about the possible values of m. (The analysis given here can be used with minor modification for the case in which the overdispersion parameter $k = k(m)$ as well.) Assume that

$$q_i = \Pr\{\text{trapping a pest in the } i\text{th cell during the period}$$

$$\text{of interest} | \text{a pest is present}\} . \tag{39}$$

The value of q_i depends upon the number of traps in the cell, the efficacy of a single trap, and the length of the trapping interval. In principle, the q_i can be computed from formulas such as Eq. (31). If the mean of the NB distribution takes the value $\mathbf{m} = m$, the trap distribution itself is negative binomial with parameters $q_i m$ and k. That is, the overdispersion parameter is the same, but the mean of the trapping distribution is the mean of the pest distribution times the probability of trapping. (This result can be verified most easily by noting that the trap distribution is obtained by compounding a binomial distribution with parameters N, q_i with a negative binomial distribution with parameters m, k. Use of generating functions, or simply slogging through the resulting summations, leads to the stated result.)

The probability of no trap catch in the ith cell during the interval of interest is thus

$$\Pr\{\text{no catch in the } i\text{th cell}\} = \{k/(k + q_i m)\}^k . \tag{40}$$

If the region of interest is divided into a total of C cells and we treat the trap catches in different cells as independent random variables, the likelihood of no trap catch in any of the cells is

$$\mathscr{L}_C(m) = \prod_{i=1}^{C} \{k/(k + q_i m)\}^k . \tag{41}$$

If all cells have the same trapping probability $q_i = q$ for all i, the likelihood becomes

$$\mathscr{L}_C(m) = \{k/(k + qm)\}^{Ck} . \tag{42}$$

The objective of a Bayesian analysis of the trapping information is the computation of the posterior density $f_p(m|0)$ for the mean of the negative binomial distribution, given no trap catch. Using Bayes's theorem gives

$$
\begin{aligned}
f_p(m|0)dm &= \Pr\{m \leqq \mathbf{m} \leqq m + dm | \text{no catch}\} \\
&= \frac{\Pr\{m \leqq \mathbf{m} \leqq m + dm, \text{ no catch}\}}{\Pr\{\text{no catch}\}} \\
&= f_0(m)dm\,\mathscr{L}_C(m) / \int f_0(m)\mathscr{L}_C(m)dm
\end{aligned}
\tag{43}
$$

where $f_0(m)$ is the prior density for the value of $\mathbf{m}$. It is used to summarize prior information about the value of the mean. For a situation in which there is little prior information, two reasonable choices are the *uniform prior* in which

$$f_0(m) = 1 \text{ for all values of } 0$$

and the *noninformative prior* in which

$$f_0(m) = [m(k + m)]^{-1/2} .$$

The uniform prior attributes equal prior weight to all values of $\mathbf{m}$. The noninformative prior (DeGroot 1970, Martz and Waller 1982) gives more prior weight to small values of $\mathbf{m}$. The non-informative prior is chosen, roughly, so that the data change only the position but not the shape of the posterior distribution. Neither of these is integrable on the interval $(0, \infty]$ and are thus called improper prior densities. It will be seen, however, that the posterior density given by Eq. (43) will be integrable.

Often, the posterior density itself is not of interest. Instead the interesting quantity is the probability that the mean $\mathbf{m}$ is less than a threshold for action m_T

(which is, at this point, assumed to be given exogenously). The appropriate posterior probability is

$$\Pr\{\mathbf{m} \leq m_T | \text{no trap catch}\} = \int_0^{m_T} f_0(m)\mathscr{L}_c(m)dm \bigg/ \int_0^{\infty} f_0(m)\mathscr{L}_c(m)dm \ . \tag{44}$$

When the uniform prior is used, the integrals can be done exactly, giving

$$\Pr\{\mathbf{m} \leq m_T | \text{no trap catch}\} = 1 - [k/(k + qm_T)]^{Ck-1} \ . \tag{45}$$

When the non-informative prior is used, the resulting integrals can be easily done numerically; a trigonometric substitution to convert the denominator to a finite domain integral. Table 2 shows the results of computations using this approach, for the uniform prior. A table such as this one allows us to interpret the trap catch.

The Bayesian approach is especially well suited for sequential decision problems in which traps are periodically inspected. For such sequential problems, the posterior density from period t becomes the prior density in period $t + 1$.

An alternative approach is based on likelihood arguments (Edwards 1972). In the absence of trap catch, the likelihood $\mathscr{L}_c(m)$ takes its maximum value when $m = 0$ and is a monotonically decreasing function of m. Although the maximum likelihood value of $\mathbf{m}$ is 0, we can construct confidence intervals directly from the likelihood function. Hudson (1971) provides an approximate method for doing this. The method consists of considering an interval of the form

Table 2. Probability that $\mathbf{m} < 1$ for the uniform prior with $k = 2$ and no trap catch

Number of traps	Prob $\{\mathbf{m} < 1\}$ for a trap spacing of		
	1 mile	0.5 mile	0.3 mile
10	0.010	0.040	0.107
20	0.023	0.080	0.208
30	0.032	0.119	0.297
40	0.043	0.156	0.377
50	0.053	0.192	0.447
60	0.063	0.226	0.510
70	0.074	0.258	0.545
80	0.084	0.289	0.614
90	0.094	0.319	0.657
100	0.104	0.348	0.696
200	0.197	0.576	0.908
300	0.281	0.724	0.972
400	0.356	0.820	0.992
500	0.423	0.883	0.997
600	0.483	0.924	0.999
700	0.537	0.950	~1*
800	0.585	0.968	~1
900	0.628	0.979	~1
1000	0.667	0.986	~1

*Here ~1 denotes values that exceed 0.9995

Table 3. Likelihood $\mathscr{L}_C(1)$ for different values of C (number of cells)

Trap Spacing	$\mathscr{L}_C(1)$ for		
(miles)	$C = 10$	$C = 100$	$C = 1000$
1	0.995	0.946	0.577
0.5	0.979	0.807	0.117
0.1	0.626	0.009	~ 0

$\{m : \mathscr{L}_C(m) \geq e^{-\beta}\mathscr{L}_C(\hat{m})\}$ where $\hat{m}$ is the MLE and β is a parameter. Under relatively general conditions, Hudson shows that the choice $\beta = 2$ leads to likelihood intervals that are approximate 95% confidence intervals. For the situation of no trap catch, the MLE is $\hat{m} = 0$ with likelihood 1 so that the confidence interval is simply $\{m : \mathscr{L}_C(m) \geq e^{-\beta}\}$ and Hudson's method is easily applied. Table 3 shows likelihoods for various values of C and trap spacing.

Analysis of Information When There Is Trap Catch

Next consider the situation in which there is trap catch. The trap data can take two forms:

i) *Presence–absence data.* The trap information in this case is that C_p of the traps had pests (positive counts) and $C_n = C - C_p$ of the traps had no pests (negative counts). Kuno (1969) and Plant and Wilson (1985) discuss methods for presence–absence sampling that differ from the ones discussed here.

ii) *Actual counts.* In this case, the data consist of the actual trap counts, denoted by X_i which is the number of pests trapped in the ith cell.

The kinds of questions that we want to answer concern the information provided by the trapping (e.g., what can be said about the value of the mean **m** of the NB distribution) and what kind of action should be taken, given the information provided by the trapping.

The answers to these questions can be built up in a manner analogous to the methods used in the previous section. For simplicity of presentation, the case in which all $q_i = q$ will be the only one considered here. Extensions to differing q_i are relatively straightforward, but also are problem dependent.

For the case of presence–absence sampling, the likelihood function depends upon the value of $q, \mathbf{m}, C_p$ and C_n. It is given by

$$\mathscr{L}(m, C_p, C_n) = [k/(k + qm)]^{C_n k}[1 - (k/(k + qm))^k]^{C_p}. \tag{46}$$

The maximum likelihood estimate (MLE) for **m** is found by differentiating Eq. (46) with respect to m, setting the derivative equal to 0 and solving. This gives the MLE

$$\hat{m} = (k/q)[(1 + (C_p/C_n))^{(q/k)} - 1]. \tag{47}$$

Although appealing for its simplicity, the MLE given in Eq. (47) can be highly biased, in the sense that $E(\hat{m})$ may deviate considerably from the true value of the m that generates the data. Mangel and Smith (1989) describe ways of eliminating the bias in the MLE $\hat{m}$.

A second approach is to consider the likelihood ratio $\mathscr{R}(m_T, \hat{m})$ defined as the ratio of the likelihood of the threshold value of **m** to the MLE value of **m**. This ratio is easily computed. It suffers the same drawback as the MLE procedure itself in that using the MLE $\hat{m}$ may be very misleading.

A third approach is to use a Bayesian procedure and derive a posterior distribution analagous to Eq. (43) using the likelihood function (46). This approach does not suffer from the biased nature of the MLE and also provides a very natural way for incorporating additional information.

For the situation in which actual trap counts are used, consider the question of estimating the population mean in a single cell. This is justified if the cells are assumed to be relatively large and thus trap counts in cells may be viewed as independent variables. The objective is then to estimate m_i (i.e., the value of the NB mean in the ith cell) and from this mean compute the distribution of the population vector $\{N_i\}$. Again for pedagogic ease, only one cell is considered, so that the subscripts can be dropped. Then the datum is that X pests were trapped in the cell, and we want to find the mean of the NB distribution that generates the trap catch in the cell. The NB distribution (21) can be reinterpreted as the likelihood $\mathscr{L}(m|x, q)$ that the mean **m** takes the value m, conditioned on the data that x pests were trapped when the trapping probability for a single pest is q.

Table 4. Likelihood of values of **m** relative to the MLE value when 6 flies are trapped with trap spacing 1 mile

m	Relative likelihood
500	5.56×10^{-3}
1500	0.230
2500	0.600
3500	0.853
4500	0.972
5500	0.999
6500	0.978
7500	0.931
8500	0.872
9500	0.811
10500	0.752
11500	0.694
12500	0.642
13500	0.593
14500	0.549
15500	0.510
16500	0.473
17500	0.440
18500	0.410
19500	0.383

Ignoring all of the terms in (21) that are independent of the value of m gives

$$\mathscr{L}(m|x, q) \propto (k + qm)^{-k}[qm/(k + qm)]^x .\tag{48}$$

The MLE for **m** is $\hat{m} = x/q$, a very natural result. Since the data X have a NB distribution, the MLE $\hat{m}$ will also have a NB distribution. Bayesian methods using the likelihood (48) or general likelihood argument can now be applied to find the distribution of $\hat{m}$.

The likelihood function (48) is nearly flat for operational values of the parameters. For example, if $q = 0.0011$ (corresponding roughly the 1 trap/2.6 km^2 from the Cunningham and Couey curve), $k = 2$, and $x = 6$ flies trapped, then the MLE for **m** is $\hat{m} = 5455$, but values of m in the range [2000, 16,000] are roughly half as likely as the MLE value (Table 4 shows the likelihood ratio over a wide range of values of **m**.) This suggests that we should be wary about reporting point estimates to decision makers without confidence that the point estimate is very accurate. It is almost always better to report reasonable ranges for the values of the parameters of interest and let the decision makers determine how to use the information. To do otherwise is often recipe for double disaster. First, the decision maker may end up with a tremendous error because he or she did not consider the range of eventualities that could arise from the decision. (For example, the point estimate may indicate that the pest population is low and the decision is made accordingly when in fact the population is high—only a few pests were detected—with obvious consequences.) Second, the analyst loses credibility with the decision maker; this is often an irreplaceable loss.

Mangel et al. (1984) bring all of these ideas together in a study of the delimiting of pest infestations, with particular application to the medfly problem in California in the early 1980s.

Area Effects in Resistance Management and Sterile Insect Methods

In this section, two problems associated with area wide control of pests are considered. The first problem is the simultaneous management of resistance to pesticide and optimization of crop yield. In this case, the question is whether to treat only the region containing the crop with pesticide, or the entire region. The second problem concerns the effects of either incorrectly assessing the extent of a pest infestation or of external sources of pests when trying to control through the sterile insect method.

Managing Pesticide Resistance

The general difficulty in the management of pesticide resistance is the mixture of a renewable and non-renewable resource system. The crops are a renewable resource, however, susceptibility to pesticide is fundamentally an exhaustible resource since each time a crop is sprayed and insects are killed, some genetic selection for resistance occurs and thus some of the susceptibility to the pesticide is lost. The literature associated with management of pesticide resistance is considerable (e.g., Comins, 1977; Feder, 1979; Feder and Regev, 1975; Hall and Norgaard, 1973; Heuth and Regev, 1974; Kable and Jeffrey, 1980; Knipling, 1984; Lewis, 1981; May and Dobson, 1986; Moffitt and Farnsworth, 1981; Omer et al., 1980; Plapp et al., 1979 and Sawicki et al., 1978, 1980) and will not be reviewed here. Instead the basic scientific information needed for formulation of a simple model will be discussed and then the model developed.

In order to deal with the management of pesticide resistance, we must be consider some aspects of population genetics. In many cases we can use single locus, two allele models to describe resistance (e.g. Omer et al., 1980). The two alleles are the resistance allele (R) and the susceptible allele (r); three types of pests are homozygous resistant (RR) individuals, heterozygous (Rr) individuals, and homozygous susceptible (rr) individuals. Often, individuals with the susceptibility allele will be more fit in the absence of spraying. For example, growth rates and fecundity of Rr or rr individuals may be higher than those of RR individuals. On the other hand, the RR individuals will have a higher fitness when pesticide is applied, in the sense that for the same dose of pesticide fewer RR individuals are killed than Rr or rr individuals. If $k_i(d)$ denotes the fraction of type i individuals killed at pesticide level d (usually measured in parts per million, ppm, of pesticide) then $k_{RR}(d) < k_{Rr}(d) < k_{rr}(d)$ for virtually all doses. In general, for i fixed the $k_i(d)$ curves are sigmoidal functions of d, rising from 0% towards 100% as d increases. The percent kill is usually measured in a scale called *probits* in which 5 probits are 50% kill and $5 \pm x$ probits is $50\% \pm x$-standard deviations of a normal $(0, 1)$ random variable. For example, 6 probits is about 85% kill and 7 probits is about 97.5% kill.

In a population that is well mixed and at equilibrium, the evolution of the frequency of the alleles can be described by the Hardy–Weinberg formula (e.g. Emlen, 1984 or Roughgarden, 1979) which states that $RR:Rr:rr = p^2:2p(1-p):(1-p)^2$ where p is the frequency of the R allele. If $p(n)$ is the frequency of the R allele in generation n, then $p(n)$ satisfies the difference equation (Comins, 1977)

$$p(n + 1) = [S_{RR}p(n)^2 + S_{Rr}p(n)(1 - p(n))]/\mathscr{S} \tag{49}$$

where

$$\mathscr{S} = S_{RR}p(n)^2 + 2S_{Rr}p(n)(1 - p(n)) + S_{rr}(1 - p(n))^2 \tag{50}$$

and S_{ij} is the survivorship probability of an individual of genotype ij in generation n. Perhaps the most important aspect of the dynamics in Eq. (49) is that if $p(n)$ is small, then

$$p(n + 1) \approx (S_{Rr}/S_{rr})p(n) + o(p(n)) \tag{51}$$

so that there is exponential growth or decay of resistance for small levels of resistance and the rate of growth or decay just depends upon relative survivorships. Omer et al. (1980) give firm evidence that the growth of resistance to a pesticide is a phenomenon that can be very well described by the simple dynamics just described. May and Dobson (1986) analyze the rate of evolution of pesticide resistance and use an equation analogous to (51) to explain the relative constancy of the number of generations taken for a significant development of resistance.

The specific question of interest here is the role of refugia. That is, suppose that there are two regions. One of them (called the field) has crops in it and the other (called the pool) has natural vegetation that supports the pest. Suppose that the pest immigrates from the pool to the field. The question is: Do we treat both regions (area wide control) or just the region containing the crop. The answer must be determined by the pay-off in terms of crop yield:

Approach	Crop yield	Resistance
Area wide	Increase	Presumably increases
Crop only	May decrease	?

The role of analysis is to help develop methods that can be used to assess this trade-off. The following model is a variant of work done by Mangel and Plant (1983), Plant et al. (1985) and Stefanou et al. (1985). Imagine that the season is broken into descrete time periods, with $t = 1$ denoting the start of the season and $t = T$ denoting the end of the season. The crop dynamics are assumed to be

$$C(t+1,j) = C(t,j)r_c(1 - \Delta(t))$$
$$C(1,j) = C_0 \tag{52}$$

where C_0 is the initial crop biomass, $C(t,j)$ is the biomass of the crop at the start of period t in year j, r_c is the intrinsic growth rate of the crop of $\Delta(t)$ is the damage to the crop in period t caused by the pest. A simple model of this damage function might be

$$\Delta(t) = 1 - \exp(-\gamma(t)[X_{RR}(t) + X_{Rr}(t) + X_{rr}(t)]) \tag{53}$$

where $\gamma(t)$ is a measure of the damage that the individual pest can cause during period t and $X_{ij}(t)$ is the number of pests of genotype ij at the start of period t.

The astute reader will recognize that a number of implicit assumptions have been made in Eqs. (51)–(53). For example:

- *No density dependence for the crop*: In the absence of pests, the crop is assumed to grow expontentially throughout the season. This assumption is easily modified.
- *All pests do the same damage.* The damage function in Eq. (53) ignores all age structure (pests of different ages might have different levels of damage) and other complications of the pest population dynamics.

In a model of a particular pest-crop system, these assumptions would need careful study and validation. For a qualitative understanding of the problem, however, another assumption is helpful. Assume that there are only two types of pests: resistants (R) and susceptibles (S) (Mangel and Plant, 1983). Let $R_f(t,j)$ and $S_f(t,j)$ denote the population level of resistant and susceptible pests in the field at the start of period t in year j and let $R_p(t,j)$ and $S_p(t,j)$ denote the population levels in the pool. Assume that, as long as there is no spraying, the field provides a better habitat for growth of the pest so that pests migrate from the pool to the field. Let $\mu(j)$ denote the fraction of resistant pests in the pool at the start of year j and let $I(t)$ denote the immigration to the field during period t. Assuming that the migration of the pests is independent of their resistance type, the number of pests moving from pool to field during period t is $I(t) = i(t)(R_p(t,j) + S_p(t,j))$ where $i(t)$ is the fraction of pests moving in period t. The dynamics within a year, in the absence of spraying thus become

$$R_f(t+1,j) = \lambda_R R_f(t,j) + \mu(j)I(t)$$
$$S_f(t+1,j) = \lambda_S S_f(t,j) + (1 - \mu(j))I(t)$$
$$R_p(t+1,j) = \max\left[\gamma\lambda_R R_p(t,j) - \mu(j)I(t), 0\right]$$
$$S_p(t+1,j) = \max\left[\gamma\lambda_S S_p(t,j) - (1 - \mu(j))I(t), 0\right] . \tag{54}$$

In these equations, it is understood that $\lambda_R < \lambda_S$ so that in the absence of spraying susceptible pests grow faster than resistant pests and that $\gamma < 1$, so that the field is a superior habitat for the pests. The initial condition when solving Eq. (54) connects one year to the next. Assume that at the end of the growing season (period T) the crop is harvested and the pests in the field and pool mix. Let $R(T,j) = R_f(T,j) + R_p(T,j)$ denote the total population of resistant pests in period T in year j and let $S(T,j)$ denote the total population of susceptible pests in period T in year j. Assuming that the pests can only overwinter and breed in the pool, the initial conditions for Eq. (54) become

$$R_f(1,j) = S_f(1,j) = 0$$
$$R_p(1,j) = r_0(R(T,j-1))$$
$$S_p(1,j) = s_0(S(T,j-1)) \tag{55}$$

where $r_0(z)$ and $s_0(z)$ are fecundity functions. The other connection between years is through the fraction of resistant pests, $\mu(j)$. That is

$$\mu(j+1) = r_0(R_p(T,j))/[r_0(R_p(T,j)) + s_0(S_p(T,j))] . \tag{56}$$

A spray schedule corresponds, in this model, to a sequence $U(j) = \{U_1, U_2, \ldots, U_{T-1};j\}$ where U_i is the spray level applied in period i in year j. From it, we compute the fraction of resistant and susceptible pests killed in period i. Thus, for example, if $S_f(t)$ were the susceptible population at the start of period t in the absence of spraying, after spray is applied the population level will be $S_f(t)k_S(U_t)$. The crop dynamics are still given by Eq. (52) and the damage function can still be modelled by Eq. (53), with an appropriate modification. All of these equations can now be combined to study the efficacy of different spraying strategies over years. To do this, define a crop-value function

$$V = \sum_{j=1}^{H} \delta^{j-1}C(T,j) \tag{57}$$

where H is the time horizon over which planning is done and $\delta \leq 1$ is a "discounting" factor that weights future crop yields. Note that if $\delta = 1$, then the crop yield in year H is as important as the yield in year 1.

The value function V depends upon many parameters in the model and upon the control strategy U. A good way to view it, however, is that the underlying state variable in the model is $\mu(j)$. That is, the relatively complicated dynamics in Eqs. (54) and following are used to relate $\mu(j)$ and $\mu(j+1)$. Holding all parameters constant, including the spray strategy U, allows us to formally write $\mu(j+1) = F(\mu(j))$, where $F(\cdot)$ is determined through the dynamics of the pest. The appropriate initial condition is now that $\mu(1) = \mu_0$, assumed to be given. The value function can be determined in an iterative fashion (this is essentially the method of deterministic dynamic programming, without the optimization step). Let $V(x, y; U)$ denote the crop value function from year y to year H, given that $\mu(y) = x$ and that spraying strategy U is applied. First note that $V(x, y; U)$ satisfies the end condition that

$$V(x, H; U) = \delta^{H-1} C(T, H) .\tag{58}$$

Second, note that for values of $y < H$, we can determine $V(x, y; U)$ iteratively. That is,

$$V(x, y; U) = \delta^{y-1} C(T, y) + V(F(x), y + 1; U) .\tag{59}$$

By iterating these equations backwards, we can study the value associated with a given spraying strategy.

The final step, which will not be taken here, would be to optimize over the spraying strategies (e.g. Plant, Mangel and Flynn, 1985). It is often better, however, to report the results of a broad range of strategies, than to simply report the "optimal" strategy. This is particularly true in a problem such as the one just described, since so many assumptions are used in the model. The importance of reporting ranges of options, and not just single "best" strategies, is as important here as it is in the estimation of pest populations.

Area-Wide Effects in the Sterile Insect Method

The basic idea of the sterile insect method (SIM) can be traced to a paper by E.F. Knipling (1955). In this section, area-wide effects associated with the SIM will be discussed; there are many other interesting topics that will not be discussed (e.g. Barclay (1980, 1982, 1987a, b), Barclay and MacKauer (1980), Berryman (1967), International Atomic Energy Agency (1984), Ito (1977), Ito and Kawamoto (1979), Ito and Koyama (1982), and Plant (1986)).

The basic idea of the SIM is to "dilute" the reproductive potential of a population by adding sterile insects, typically males, to the population. The method is most effective for insects which mate only once in their lives; if a sufficient number of sterile males is used, many females will mate with steriles and thus produce no offspring. If the population has a 1:1 sex ratio, then it is sufficient to track females only. Let $F(n)$ denote the number of females in generation n. If the population has non-overlapping generations and grows exponentially, in the absence of sterile males the dynamics of $F(n)$ are

$$F(n + 1) = rF(n)\tag{60}$$

where r is the intrinsic growth rate of the population. The idea is to add sterile males to the population, until $r < 1$. In particular, if $S(n)$ denotes the number of sterile males released in generation n, the dynamics in Eq. (60) are replaced by

$$F(n + 1) = r[F(n)/(F(n) + S(n))]F(n) \tag{61}$$

where the first two terms on the right hand side now represent the effective growth rate in the presence of sterile insects. Plant (1986) discusses more realistic models for the pest dynamics and SIM; the models used here, such as Eq. (61) are chosen mainly for pedagogic purposes.

The dynamical properties of Eq. (61) can be studied through its fixed points. Consider the case in which a constant number of sterile insects, $S(n) = S$, is released in each generation. The fixed point F of Eq. (61) satisfies

$$F = rF^2/(F + S) \tag{62}$$

so that there is a fixed point at $F = 0$ (not very surprising) and another fixed point at

$$F_u = S/(r - 1) . \tag{63}$$

It is easy to demonstrate that the fixed point F_u is unstable in the sense that if the initial population $F(0) < F_u$ then $F(n)$ decreases towards 0 whereas if the initial population $F(0) > F_u$, the population grows without bound. (Recall the original dynamics are exponential growth and that a constant number of sterile insects are being released in each generation.) We now flip Eq. (63) around, to determine the minimum number of steriles that must be released S_t to have the population decrease from one generation to the next. This is called the threshold release level and is given by

$$S_t = F(n)(r - 1) \tag{64}$$

with the property that

$$F(n + 1) = \begin{array}{ll} < F(n) & \text{if } S(n) > S_t \\ = F(n) & \text{if } S(n) = S_t \\ > F(n) & \text{if } S(n) < S_t . \end{array} \tag{65}$$

Another way of interpreting Eq. (65) is that $F(n) \to 0$ if $S(n)$ exceeds the threshold S_t in each generation, $F(n)$ stays constant if $S(n)$ equals the threshold S_t and $F(n)$ grows if $S(n)$ is less than the threshold. Typical dynamics for the case of a constant, sufficiently large number of releases are that $F(n)$ decreases very slowly at first, but then the decrease in $F(n)$ accelerates with n (see Plant and Mangel 1987 for more details).

The value of $F(n)$ is thus extremely important in actual implementation of the SIM. In general, $F(n)$ will be determined by trapping, so that the first section of this chapter and this section are tied together through the operations of detection and control.

If the SIM is applied in a large region, it may again be worthwhile to divide the region into cells. Then let $F_i(n)$ denote the number of females in the ith cell at the start of generation n. Assume that the areas are sufficiently large to be treated independently and that the insects reproduce and then disperse. Let $G_i(n)$ denote the population level in the ith cell after reproduction but before dispersal. If $S_i(n)$ steriles are placed in the ith cell during generation n, then

$$G_i(n) = rF_i(n)^2/[F_i(n) + S_i(n)] .\tag{66}$$

The population in the ith cell at the start of generation $n + 1$ is then

$$F_i(n + 1) = \mathcal{F}(\underline{G}(n))\tag{67}$$

where $\underline{G}(n)$ is the vector $(G_1(n), G_2(n), \ldots, G_C(n))$ and $\mathcal{F}(\underline{z})$ is the model for movement. The total population at the start of generation n is then $F_T(n) = \sum F_i(n)$, with the summation extending over all cells. In most SIM control programs, the total number of sterile insects available for use is limited. This adds one additional constraint that $\sum S_i(n) \leq S_T(n)$, where $S_T(n)$ is the total number of sterile insects available in generation n. A relatively complex stochastic optimization problem arises in a natural fashiong: Minimize $E\{F_T(H)\}$ through choices of $\{S_i(n)\}$ where H is the time horizon. The problem is stochastic because of imperfect information obtained through trapping. In the real application of SIM, producing sterile insects may also be a major difficulty (see, e.g. Plant, 1986) especially for pests with low individual fecundity such as the tsetse fly.

Prout (1978) developed a number of elaborations of the SIM. Two of the most important are i) the SIM in populations with a carrying capacity and ii) the SIM when there is migration into the region being treated (that is, non-area wide treatment). In Prout's model with a carrying capacity, Eq. (60) is replaced by

$$F(n + 1) = F(n)\{rK/(K + (r - 1)F(n))\} .\tag{68}$$

The carrying capacity in this model is $F(n) = K$. If a fixed number of steriles S are released in each generation (Prout calls this "hard" release), then the dynamics are replaced by

$$\begin{aligned}F(n + 1) &= F(n)\{F(n)/(F(n) + S)\}[rK/\{K + (r - 1)F(n)(F(n)/(F(n) + S))\}] \\ &= rKF(n)^2/[K(F(n) + S) + (r - 1)F(n)^2] .\end{aligned}\tag{69}$$

The fixed points F of this equation now satisfy the cubic equation

$$(r - 1)F^3 - K(r - 1)F^2 + KSF = 0 .\tag{70}$$

Note that $F = 0$ is always a solution of this equation. It is the only real solution as long as $K(r - 1) < 4S$. If $K(r - 1) > 4S$, there are two additional, real solutions of Eq. (70), with the middle root unstable.

Next, consider the effect of migration on the population with a carrying capacity. That is, assume that pests move into the region being treated with the SIM. Let M denote the number of migrants into the region in each generation. The dynamics for $F(n)$ then become

$$\begin{aligned}F(n + 1) &= rK[F(n)^2 + M(F(n) + S)]/[K(F(n) + S) \\ &\quad + (r - 1)(F(n)^2 + M(F(n) + S)]\end{aligned}\tag{71}$$

and the fixed points of this equation satisfy

$$(r - 1)F^3 + (r - 1)(M - K)F^2 + \{MS(r - 1) + K(S - Mr)\}F = rKMS .\tag{72}$$

Note that $F = 0$ is no longer a fixed point of the system. The operational interpretation of this result is that when migration occurs, eradication can not be achieved solely through use of the sterile insect method.

In order to analyze Eq. (72), it helps to introduce scaled variables, in which the carrying capacity is used to nondimensionalize population levels. Set

$$x = F/K$$
$$m = M/K$$
$$s = S/K \tag{73}$$

so that Eq. (72) becomes

$$f(x; m, s) = (r - 1)x^3 + (r - 1)(m - 1)x^2 + \{m(sr - s - r) + s\}x = rms . \tag{74}$$

The equilibria of the pest dynamics can now be determined by study of the bifurcations of the cubic equation (74). The following properties are determined (see Plant and Mangel, 1987, for more details):

i) When $m = 0$ and $s < s_c = (r - 1)/4$, the equation $f(x; 0, s) = 0$ has three solutions. One of them is the origin.

ii) When $m = 0$ and $s > s_c$, the only solution of $f(x; 0, s) = 0$ is $x = 0$ so that eradication is possible.

iii) If $s > s_c$ and m is slightly positive, the solution of $f(x; m, s) = 0$ shifts from the origin to a value of $x > 0$, so that eradication is not possible. As m increases, two additional real roots of the equation appear, so that there is a region of multiple steady states of the population. As m increases further, the only root of $f(x; m, s) = 0$ is a large one. The reader is encouraged to work out the details and to see how the bifurcations provide information about the population structure.

Conclusions

Hopefully, the reader has seen the wealth of interesting and challenging problems that arise in agricultural pest control. In recent years, a number of books on the subject of pest management have appeared (e.g., Huffaker and Rabb, 1984; Kogan, 1986; Conway, 1985; Curry and Feldman, 1987). The book of Curry and Feldman is closest to this paper in spirit and approach; it provides a natural departure point for continuing with the material presented here.

Different agricultural problems will require different, and often new, types of mathematics. Even so, it is possible to provide a few general guidelines which will help make the analysis and modelling as good as possible:

- Don't be tied down by what's been done before; think broadly and widely about the problem.
- Be prepared for the unexpected (see Holling, 1987), since a model is always a caricature of reality. It is easy to miss—especially on the first attempt—crucial driving factors.
- Get as close to the problem as possible by spending considerable amounts of

time with the biologists who know the pest and crop and the decision maker who has the responsibility of choosing the action. Understand the biology of pest and crop and the sociology of policy making and try to integrate these in the analysis, as much as is possible. (Barrett, 1984).

- Build confidence. Often this is most easily done by listening and asking good questions, rather than acting as if you've got all the answers. Another way is to solve a "trivial" problem that is of interest to the people you're working with.
- Be problem, not technique, oriented (Barrett, 1985). The objective of bringing analysis to agricultural problems is to be able to solve agricultural problems, not to find problems which fit a particular mathematical technique. Trying to force a problem into the form so that a favorite technique can be used is often recipe for disaster.

Acknowledgements. Many of the ideas described here were developed over a number of years of enjoyable collaboration with Richard E. Plant and Paul Smith. Conversations with Peter Kareiva, and his extensive comments on the first draft, are greatly appreciated. This work has been partially supported by the Agricultural Experiment Station of the University of California, the California Department of Food and Agriculture, and by the National Science Foundation.

Bibliography

Note: Not all of the references listed in this bibliography were cited in the text. They are included with the idea that they might be helpful to individuals who are trying to develop libraries associated with pest management

Abramowitz, M., Stegun, I. (1964) Handbook of Mathematical Functions. Dover, New York

Ackoff, R.L. (1979) The future of operational research is past. Journal of the Operational Research Society *30*, 93

Ackoff, R.L. (1979) Resurrecting the future of operational research. Journal of the Operational Research Society *30*, 189

Aluja, M., Liedo, P.F. (1986) Perspectives on future integrated management of fruit flies in Mexico. In: Mangel, M. (eds.) Pest Control: Operations and Systems Analysis in Fruit Fly Management. Springer, New York, pp. 9–42

Anderson, J.R., Dillon, J.L., Hardaker, B. (1977) Agricultural Decision Analysis. Iowa State University Press, Ames, Iowa

Ascher, W. (1977) An Appraisal for Policy Makers and Planners. Johns Hopkins University Press, Baltimore, Maryland

Austin, J.E. (1981) Agroindustrial Project Analysis. Johns Hopkins University Press, Baltimore, Maryland

Baker, P. (1985) Medfly dispersal, attraction and trapping. Final Report MOSCAMED Program, Tapachula, Mexico

Barclay, H. (1980) Models for the sterile insect release method with concurrent release of pesticides. Ecological Modelling *11*, 167

Barclay, H. (1982) Pest population stability under sterile releases. Research Population Ecology *24*, 405

Barclay, H. (1987a) Models for pest control: complementary effects of period releases of steriles pests and parasitoids. Theoretical Population Biology (to appear)

Barclay, H. (1987b) Combining methods of pest control: complementarity of methods and a guiding principles. Natural Resources Modelling (to appear)

Barclay, H., Mackauer, M. (1980) The sterile insect release methods for pest control: a density dependent model. Ecological Entomology *9*, 81

Barfiled, C.S., O'Neil, R.J. (1984) Is an ecological understanding a prerequisite for pest management? Florida Entomologist 67, 42

Barrett, G.W. (1984) Applied ecology: An integrative paradigm for the 1980s. Environmental Conservation 11, 319–322

Barrett, G.W. (1985) A problem-solving approach to resource management. Bioscience 35, 423–427

Beddington, J.R., Free, C.A., Lawton, J.H. (1978) Characteristics of successful natural enemies in models of biological control of insect pests. Nature 273, 513

Berger, J.O. (1983) Statistical Decision Theory. Springer, Berlin Heidelberg New York

Berryman, A. (1967) Mathematical description of the sterile male principle. Canadian Entomologist 99, 859

Broadbent, S.R., Kendall, D.G. (1953) The random walk of Tricho-trongylus Retortaeformis. Biometrics, December, pp. 460–467

Carlson, G.A. (1970) A decision-theoretic approach to crop disease prediction and control. American Journal of Agricultural Economics 52, 216

Carlson, G.A. (1977) Long-run productivity of insecticides. American Journal of Agricultural Economics 59, 543

Carslaw, H.S., Jaeger, J.C. (1959) Conduction of Heat in Solids. Oxford University Press, Oxford, England

Cavalloro, R. (ed.) (1983) Fruit Flies of Economic Importance. Balkema, Rotterdam, The Netherlands

Cavalloro, R. (ed.) (1984) Statistical and Mathematical Methods in Population Dynamics and Pest Control. Balkema, Rotterdam, The Netherlands

Cavalloro, R., Crovetti, A. (eds.) (1985) Integrated Pest Control in Olive Groves. Balkema, Rotterdam, The Netherlands

Comins, H. (1977) The management of pesticide resistance. Journal of Theoretical Biology 65, 399

Cunningham, R., Couey, H.M. (1985) Mediterranean fruit fly (Diptera: Tephritidae): Distance response curves to trimedlure to measure trapping efficiency. Environmental Entomology 15, 71–74

Curry, G.L., Feldman, R.M. (1987) Mathematical Foundations of Poplation Dynamics. Texas A&M University Press, College Station, Texas

Day, R.H. (1965) Probability distributions of field crop yields. Journal of Farm Economics 47, 713

Debouzie, D., Thioulouse, J. (1986) Statistics to find spatial and temporal structures in populations. In: Mangel, M. et al. (eds.) Pest Control: Operations and Systems Analysis in Fruit Fly Management. Springer, Berlin Heidelberg New York, pp. 263–282

DeGroot, M.H. (1970) Optimal Statistical Decisions. Mc-Graw Hill, New York

Dowell, R.V., Wange, L.K. (1986) Process analysis and failure avoidance in fruit fly programs. In: Mangel, M. et al. (eds.) Pest Control Operations and Systems Analysis in Fruit Fly Management. Springer, Berlin Heidelberg New York, pp. 43–65

Edwards, A.W.F. (1972) Likelihood. Cambridge University Press, Cambridge, England

Feder, G. (1979) Pesticides, information and pest management under uncertainty. American Journal of Agricultural Economics 61, 97

Feder, G., Regev, U. (1975) Biological interactions and environmental effects in the economics of pest control. Journal of Environment Economics and Management 2, 75

Feller, W. (1968) An Introduction of Probability Theory and Its Applications, vol. 1. Wiley, New York

Feller, W. (1971) An Introduction to Probability Theory and Its Applications, vol. 2. Wiley, New York

Friedman, B. (1948) A simple urn model. Communications in Pure and Applied Mathematics 2, 59

Georghiou, G.P., Lagunes, A., Baker, J.D. (1983) Effect of insecticide rotations on the evolution of resistance. In: Proceedings of the International Congress on Pesticide Chemicals, Kyoto, Japan. Pergamon Press, London

Georghiou, G.P., Taylor, C.E. (1977) Operational influences in the evolution of insecticide resistance. Journal of Economic Entomology 70, 653

Gillis, J. (1956) Centrally biased discrete random walk. Quarterly Journal of Mathematics (Oxford) 2, 144–152

Hall, D.C., Norgaard, R.B. (1973) On the timing and application of pesticides. American Journal of Agricultural Economics 55, 198

Hargrove (1981) Tsetse dispersal reconsidered. Journal of Animal Ecology 50, 351

Hassell, M.P. (1978) The Dynamics of Arthropod Predator Prey Systems. Princeton University Press, Princeton, New Jersey

Hassell, M.P. (1980) Foraging stategies, population models and biological control: a case study. Journal of Animal Ecology 49, 603

Hawkes, C. (1972) The estimation of the dispersal rate of the adult cabbage root fly (Erioschia Brassicae (Bouche)) in the presence of brassica crop. Journal of Applied Ecology 9, 617

Holling, C.S. (1986) The resilience of terrestrial ecosystems: local surprise and global change. Chapter 10 in: Clark, W.C., Munn, R.E. (eds.) Sustainable Development of the Biosphere. Cambridge University Press, Cambridge, England, pp. 292–317

Hudson, D.J. (1971) Interval estimation from the likelihood function. Journal of the Royal Statistical Society 33, 256–262

Hueth, D., Regev, U. (1974) Optimal agricultural pest management with increasing pesticide resistance. American Journal of Agricultural Economics 56, 543

Huffaker, C., Rabb (1984) Ecological Entomology

International Atomic Energy Agency (Wagrammerstrasse 5, POB 100, A-1400, Vienna, Austria) (1984) The Sterile Insect Technique for Tsetse Control or Eradication in Developing Countries in Africa.

International Institute for Applied Systems Analysis (Laxenburg Austria). (1980) Expect the Unexpected: An Adaptive Approach to Environmental Management.

Ito, Y. (1977) A model of sterile insect release for eradication of the melon fly. Dacus cucurbitae Coquillett. Appl. Ent. Zool. 12, 303

Ito, Y., Kawamoto, J. (1979) Number of generations necessary to attain eradication of an insect pest with sterile insect release method: a model study. Researches in Population Ecology 20, 216

Ito, Y., Koyama, J. (1982) Eradication of the melon fly: role of population ecology in the successful implementation of the sterile insect method. Protection Ecology 4, 1

Jansen, M.J.W., Metz, J.A.J. (1979) How many victims will a pitfall trap make? Acta Biotheoretica 28, 98

Johnson, N.L., Kotz, S. (1969) Discrete Distributions. Wiley Interscience, New York

Kable, P.F., Jeffrey, H. (1980) Selection for tolerance in organisms exposed to sprays of biocide mixtures: a theoretical model. Phytopathology 70, 8

Kendall, M., Stuart, A. (1973) The Advanced Theory of Statistics, vol. 2. Griffin and Company, London, England

Knessl, C., Mangel, M., Matkowksy, B.J., Schuss, Z., Tier, C. (1984) Asymptotic solution of the Kramers-Moyal equation and first passage times for Markov jump processes Physical Review A29, 3359

Knipling, E.F. (1955) Possibilities of insect control or eradication through the use of sexually sterile males. Journal of Economic Entomology 45, 459

Knipling, E.F. (1984) Influence of insecticide use patterns on the development of resistance to insecticides: a theoretical study. Southwestern Entomologist 9, 351

Kogan, M. (1986) Ecological Theory and IPM.

Koopman, B.O. (1980) Search and Screening. Pergamon Press, New York Krebs, J.R., Davies, N.B. (1984) Behavioural Ecology. Blackwell Scientific Press, Oxford

Kuno, E. (1969) A new method of sequential sampling to obtain the population estimates with a fixed level of precision. Researches in Population Ecology 11, 127

Levins, R., Wilson, M. (1980) Ecological theory and pest management. Annual Review of Entomology 25, 287

Lewis, T. (1981) Pest monitoring to aid insecticide use. Philosophical Transactions of the Royal Society of London B295, 153

Luttrell, C.B. Gilbert, R.A. (1976) Crop yields: random, cyclical or bunchy? American Journal of Agricultural Economics 58, 522

Mangel, M. (1981) Search for a randomly moving object. SIAM Journal on Applied Mathematics 40, 327

Mangel, M. (1982) Applied mathematicians and naval operators. SIAM Review 24; 289

Mangel, M. (1985) Decision and Control in Uncertain Resource Systems. Academic Press, New York

Mangel, M. (1986) Trapping and fruit fly management. In: Mangel, M. et al. (eds.) Pest Control: Operations and Systems Analysis in Fruit Fly Management. Springer, Berlin Heidelberg New York

Mangel, M., Clark, C.W. (1983) Uncertainty, search and information in fisheries. Journal of the International Council for the Exploration of the Seas 41, 93

Mangel, M., Carey, J.R., Plant, R.E. (1986) Pest Control: Operations and Systems Analysis in Fruit Fly Management. Springer, Berlin Heidelberg New York

Mangel, M., Plant, R.E. (1983) Multiseasonal management of an agricultural pest. I: Development of the theory. Ecological Modelling 20, 1

Mangel, M., Plant, R.E., Carey (1984) Rapid delimiting of pest infestations: a case study of the mediterranean fruit fly. Journal of Applied Ecology 21, 563

Mangel, M., Smith, P.E. (1989) Presence–absence plankton sampling in fisheries management. Canadian Journal of Fisheries and Aquatic Sciences (to appear)

Mangel, M., Stefanou, S., Wilen, J.E. (1985) Modelling Lygus hesperus injury to cotton yields. Journal of Economic Entomology 78, 1009

Martz, H.F., Waller, R.A. (1982) Bayesian Reliability Analysis. Wiley Interscience, New York

May, R.M., Dobson, A.P. (1986) Population dynamics and the rate of evolution of pesticide resistance. In: Pesticide Resistance: Strategies and Tactics to Management. National Academy Press, Washington, DC, pp. 170–193

McClendon, R.W., Mitchell, E.B., Jones, J.W., McKinion, J.M., Hardee, D.D. (1976) Computer simulation of pheromone trapping systems as applied to boll weevil population suppression: a theoretical example. Environmental Entomology 5, 799

McPhee, W.J., Nestmann, E.R. (1983) Predicting potential fungicide resistance in fungal populations using continuous culturing technique. Phytopathology 73, 1230

Mellor, J.W., Gavian, S. (1987) Famine: Causes, prevention, and relief. Science 235, 539–545

Metz, J.A.J., Wedel, M., Anguio, A.F. (1985) Discovering an epidemic before it has reached a certain level of prevalence. Biometrica 39, 765

Minogue, K.P., Fry, W.E. (1983a) Models for the spread of disease: model description. Phytopathology 73, 1168

Minogue, K.P., Fry, W.E. (1983b) Models for the spread of disease: some experimental results. Phytopathology 73, 1173

Moffitt, L.J., Farnsworth, R.L. (1981) Bioeconomic analysis of pesticide demand. Agricultural Economics Research 33, 12

Morse, P.M. (1977) In at the Beginnings: A Physicist's Life. MIT Press, Cambridge, Mass

Murdoch, W.W., Chesson, J., Chesson, P.L. (1985) Biological control in theory and practice. American Naturalist 125, 344

Omer, S.M., Georghiou, G.P., Irving, S.N. (1980) DDT/pyrethroid resistance inter-relations in Anopheles stephensi. Mosquito News 40, 200

Perkins, J.H. (1982) Insects, Experts, and the Insecticide Crisis. Plenum Press, New York

Perry, J.N. (1981) Taylor's power law for dependence of variance on mean in animal populations. Applied Statistics 30, 254–263

Perry, J.N., Taylor, L.R. (1986) Stability of real interacting populations in space and time: implications, alternatives and the negative binomial k_c. Journal of Animal Ecology 55, 1053–10681

Plant, R.E. (1986) The sterile insect technique: a theoreticalperspective. In: Mangel, M. et al. (eds.) Pest Control: Operation and Systems Analysis in Fruit Fly Management. Springer Berlin Heidelberg New York, pp. 361–386

Plant, R.E. (1986) Uncertainty and the economic threshold. Journal of Economic Entomology 79, 1–6

Plant, R.E., Wilson, L.T. (1985) A Bayesian method for sequential sampling and forecasting in agricultural pest management. Biometrics 41, 203

Plant, R.E., Wilson (1986) A computer based pest management aid for San Joaquin valley cotton. In: 1986 Beltwide Cotton Production Research Conference Proceedings

Plant, R.E., Mangel, M. (1987) Modeling and simulation in agricultural pest management. SIAM Review 29, 235–262

Plant, R.E., Mangel, M., Flynn, L. (1985) Multiseasonal management of an agricultural pest. II: The economic optimization problem. Journal of Environmental Economics and Management 12, 45

Plapp, F.W., Browning, C.R., Sharpe, P.J. (1979) Analysis of rate of development of insecticide resistance based on simulation of a genetic model. Environmental Entomology 8, 494

Pouliquen, L.Y. (1970) Risk Analysis in Project Appraisal. Johns Hopkins University Press, Baltimore, Maryland

Prokopy, R., Hubbell, G.L., Adams, R.G., Hauschild, K.I. (1982) Visual monitoring trap for tarnished plant bugs on apple. Environmental Entomology *11*, 200

Prout, T. (1978) The joint effects of the release of sterile males and the immigration of fertilized females on a density regulated population. Theoretical Population Biology *13*, 40

Reutlinger, S. (1970) Techniques for Project Appraisal Under Uncertainty. Johns Hopkins University Press, Baltimore, Maryland

Rodgers, D. (1972) Random search and insect population models. Journal of Animal Ecology *41*, 369

Rodgers, D. (1979) Tsetse population dynamics and distribution: a new analytical approach. Journal of Animal Ecology *48*, 825

Rosenstock, H.B. (1980) Absorption time by a random trap distribution. Journal of Mathematical Physics *21*, 1643

Roughgarden, J. (1979) Theory of Population Genetics and Evolutionary Ecology: An Introduction. MacMillan Press, New York

Sage, A (1977) Systems Engineering: Methodology and Applications. IEEE Press, New York

Sawicki, R.M., Devonshire, A.L., Rice, A.D., Moores, G.D., Petzing, S.M., Cameron, A. (1978) The detection and distribution of organophosphorous and carbamate insecticide resistant Myzus persicae (Sulx.) in Britain in 1976. Pesticide Science *9*, 189

Sawicki, R.M., Devonshire, A.L., Payne, R.W., Petzing, S.M. (1980) Stability of resistance in the peachpotato aphid Myzus persicae (Sulxer). Pesticide Science *11*, 33

Sawyer, A.J., Haynes, D.L. (1985) Simulating the spatiotemporal dynamics of the cereal leaf beetle in a regional crop system. Ecological Modelling *30*, 83

Sawyer, A.J., Haynes, D.L. (1986) Cereal leaf beetle spatial dynamics: simulation with a random diffusion model. Ecological Modelling *33*, 89

Shoemaker, C.A. (1981) Applications of dynamic programing and other optimization methods in pest management. IEEE Transactions on Automatic Control AC *26*, 1125

Stenseth, N.C. (1981) How to control pest species: application of models from the theory of island biogeography in formulating pest control strategies. Journal of Applied Ecology *18*, 773

Stern, V.M., Smith, R.F., van den Bosch, R., Hagen, K.S. (1959) The integrated control concept. Hilgardia *29*, 81

Stokey, E., Zeckhauser, R. (1978) A Primer for Policy Analysis. Norton, New York

Squire, L., van der Tak, H.G. (1975) Economic Analysis of Projects. Johns Hopkins University Press, Baltimore, Maryland

Szabo, A., Lamm, G., Weiss, G.H. (1984) Localized partial traps in diffusion processes and random walks. Journal of Statistical Physics *34*, 225–238

Taylor, L.R. (1971) Aggregation as a species characteristic. In Patil, G.P. et al. (eds.) Statistical Ecology. Pennsylvania State University Press, State College, Pennsylvania, pp. 357–377

Taylor, L.R. (1986) Synoptic dynamics: migration and the Rothamsted insect survey. Journal of Animal Ecology *55*, 1

Taylor, L.R., Woiwod, I.P., Perry, J.N. (1979) The negative binomial model as a dynamic ecological model for aggregation, and the density dependence of k. Journal of Animal Ecology *48*, 289

Watson, T.F., Moore, L., Ware, G.W. (1975) Practical Insect Pest Management. Freeman, W.H., San Francisco, Calif

Wellington, W.G. (1977) Returning the insect to insect ecology: some consequences for pest management. Environmental Entomology *6*, 1

Wickwire, K. (1977) Mathematical models for the control of pests and disease: a survey. Theoretical Population Biology *11*, 182

Wilson, L.T., Gonzalez, D., Plant, R. (1985) Predicting sampling frequency and economic status of spider mites on cotton. In: 1985 Beltwide Cotton Production Research Conference Proceedings

Wolf, W.W., Kishaba, A.N., Toba, H.H. (1971) Proposed method for determining density of traps required to reduce an insect population. Journal of Economic Entomology *64*, 872

Wood, R. (1986) Control strategies designed to reduce the chance of resistance with special reference to tephritid fruit flies. In: Mangel, M. et al. (eds.) Pest Control: Operations and Systems Analysis in Fruit Fly Management. Springer, Berlin Heidelberg New York, pp. 399–438

Yekutiel, P. (1981) Lessons from the big eradication campaigns. World Health Forum *2*, 465

Part III. Epidemiology

Three Basic Epidemiological Models

Herbert W. Hethcote

Contents

1. Introduction

There are three basic types of deterministic models for infectious diseases which are spread by direct person-to-person contact in a population. Here these simplest models are formulated as initial value problems for systems of ordinary differential equations and are analysed mathematically. Theorems are stated regarding the asymptotic stability regions for the equilibrium points and phase plane portraits of solution paths are presented. Parameters are estimated for various diseases and are used to compare the vaccination levels necessary for herd immunity for these diseases. Although the three models presented are simple and their mathematical analyses are elementary, these models provide notation, concepts, intuition and foundation for considering more refined models. Some possible refinements are disease-related factors such as the infectious agent, mode of transmission, latent period, infectious period, susceptibility and resistance, but also social, cultural, Ecology by providing a sound intuitive understanding and complete proofs for the three most basic epidemiological models for microparasitic infections.

The study of disease occurrence is called epidemiology. An epidemic is an unusually large, short term outbreak of a disease. A disease is called endemic if it persists in a population. The spread of an infectious disease involves not only disease-related factors such as the infectious agent, mode of transmission, latent period, infectious period, susceptibility and resistance, but also social, cultural, demographic, economic and geographic factors. The three models considered here are the simplest prototypes of three different types of epidemiological models. It

Table 1. Classification of infectious diseases by agent and mode of transmission

	Mode of transmission			
Agent	Person → person	Person → environment environment → person	Reservoir → vector vector → person	Reservoir → person
Virus	Measles Chickenpox Mumps Rubella Smallpox Influenza Poliomyelitis Herpes HIV (AIDS virus)		Arboviruses: yellow fever dengue fever encephalitis tick fever sandfly fever	Rabies
Bacteria	Gonorrhea Tuberculosis Pneumonia Meningitis Strep throat	Typhoid fever Cholera	Plague	Brucellosis Tuleramia Anthrax
Protozoa	Syphillis	Amebiasis	Malaria Trypanosomiasis	
Helminths			Schistosomiasis Filariasis Onchocerciasis	Trichinosis

is important to understand their behaviour before considering general models incorporating more of the factors above.

Table 1 classifies diseases by agent and method of transmission. This useful classification scheme is similar to one presented by K. Dietz in 1974. The models considered here are suitable for diseases which are transmitted directly from person to person. More complicated models must be used when there is transmission by insects called vectors or a reservoir of nonhuman infectives. Epidemiological models are now widely used as more epidemiologists realize the role that modeling can play in basic understanding and policy development.

Justifications of mathematical modeling of the transmission of infectious diseases are given in the next section. The essential assumptions and terminology are given in Section 3. The SIS model analysed in Section 4 is for diseases for which infection does not confer immunity. SIR models for diseases where infection does confer immunity are considered for epidemics in Section 5 and for endemic situations in Section 6. Section 7 is devoted to herd immunity and its implication for vaccination for specific diseases. The discussion in Section 8 summarizes and refers to more complicated models.

2. Why Do Epidemiologic Modeling?

Even though vaccines are available for many infectious diseases, these diseases still cause suffering and mortality in the world, especially in developing countries. In developed countries chronic diseases such as cancer and heart disease have received more attention than infectious diseases, but infectious diseases are still a more common cause of death in the world. Recently, the human immunodeficiency virus (HIV) which can lead to acquired immunodeficiency syndrome (AIDS) has become an important infectious disease in both developing and developed countries.

The transmission mechanism from an infective to susceptibles is understood for nearly all infectious diseases and the spread of diseases through a chain of infections is known. However, the transmission interactions in a population are very complex so that it is difficult to comprehend the large scale dynamics of disease spread without the formal structure of a mathematical model. An epidemiological model uses a microscopic description (the role of an infectious individual) to predict the macroscopic behavior of disease spread through a population.

In many sciences it is possible to conduct experiments to obtain information and test hypotheses. Experiments with infectious disease spread in human populations are often impossible, unethical or expensive. Data is sometimes available from naturally occurring epidemics or from the natural incidence of endemic diseases; however, the data is often incomplete due to underreporting. This lack of reliable data makes accurate parameter estimation difficult so that it may only be possible to estimate a range of values for some parameters. Since repeatable experiments and accurate data are usually not available in epidemiology, mathematical models and computer simulations can be used to perform needed theoretical experiments. Calculations can easily be done for variety of parameter values and data sets.

Mathematical models have both limitations and capabilities that must recognized. Sometimes questions cannot be answered by using epidemiological models, but sometimes the modeler is able to find the right combination of available data, an interesting question and a mathematical model which can lead to the answer.

Comparisons can lead to a better understanding of the processes of disease spread. Modeling can often be used to compare different diseases in the same population, the same disease in different populations, or the same disease at different times. Comparisons of diseases such as measles, rubella, mumps, chickenpox, whooping cough, poliomyelitis and others are made in London and Yorke (1973), Yorke and London (1973), Yorke et al. (1979), Hethcote (1983), Anderson and May (1982) and in the article on rubella in this volume by Hethcote (1989).

Epidemiological models are useful in comparing the effects of prevention or control procedures. Hethcote and Yorke (1984) use models to compare gonorrhea control procedures such as screening, rescreening, tracing infectors, tracing infectees, post-treatment vaccination and general vaccination. Communicable disease models are often the only practical approach to answering questions about which prevention or control procedure is most effective. Quantitative predictions of epidemiological models are always subject to some uncertainty since the models are idealized and the parameter values can only be estimated. However, predictions of the relative merits of several control methods are often robust in the sense that the same conclusions hold over a broad range of prameter values and a variety of models. Strategies for rubella vaccination are compared using a cost benefit analyses in the article on rubella by Hethcote (1989) in this volume.

Optimal strategies for vaccination can be found theoretically by using modeling. Longini, Ackerman and Elveback (1978) use a epidemic model to decide which age groups should be vaccinated first to minimize cost or deaths in an influenza epidemic. Hethcote (1988) uses a modeling approach to estimate the optimal age of vaccination for measles. A primary conclusion of this paper is that better data is needed on vaccine efficacy as a function of age in order to better estimate the optimal age of vaccination. Thus epidemiological modeling can be used to identify crucial data that needs to be collected.

An underrecognized value of epidemiological modeling is that it leads to a clear statement of the assumptions about the biological and sociological mechanisms which influence disease spread. The parameters used in an epidemiological model must have a clear interpretation such as a contact rate or a duration of infection. Models can be used to assess many quantitative conjectures. For example, one could check a conjecture that AIDS incidence would decrease if 90% of the sexually active heterosexual population started using condoms consistently. Epidemiological models can sometimes be used to predict the spread or incidence of a disease. For example, Hethcote (1983) predicted that rubella and Congenital Rubella Syndrome will eventually disappear in the United States because the current vaccination levels using the combined measles-mumps-rubella vaccine are significantly above the threshold required for herd immunity for rubella. An epidemiological model can also be used to determine the sensitivity of predictions to changes in parameter values. After the parameters are identified which have the greatest influence on the predictions, it may be possible to design studies to obtain better estimates of these parameters.

3. Assumptions and Notation

The population under consideration is divided into disjoint classes which change with time t. The susceptible class consists of those individuals who can incur the disease but are not yet infective. The infective class consists of those who are transmitting the disease to others. The removed class consists of those who are removed from the susceptible-infective interaction by recovery with immunity, isolation, or death. The fractions of the total population in these classes are denoted by $S(t)$, $I(t)$ and $R(t)$, respectively.

In the epidemiological models here, the following assumptions are made:

1. The population considered has constant size N which is sufficiently large so that the sizes of each class can be considered as continuous variables. If the model is to include vital dynamics, then it is assumed that births and natural deaths

occur at equal rates and that all newborns are susceptible. Individuals are removed by death from each class at a rate proportional to the class size with proportionality constant μ which is called the daily death removal rate. This corresponds to a negative exponential age structure with an average lifetime of $1/\mu$.

2. The population is homogenously mixing. The daily contact rate λ is the average number of adequate contacts per infective per day. An adequate contact of an infective is an interaction which results in infection of the other individual if he is susceptible. Thus the average number of susceptibles infected by an infective per day is λS, and the average number of susceptibles infected by the infective class with size NI per day in λSNI. The daily contact rate λ is fixed and does not vary seasonally. The type of direct or indirect contact adequate for transmission depends on the specific disease. The number of cases per day λSNI, which is called the incidence, is a mass action law since it involves the product of S and I.

3. Individuals recover and are removed from the infective class at a rate proportional to the number of infectives with proportionality constant γ, called the daily recovery removal rate. The latent period is zero (it is defined as the period between the time of exposure and the time when infectiousness begins). Thus the proportion of individuals exposed (and immediately infective) at time t_0 who are still infective at time $t_0 + t$ is $\exp(-\gamma t)$, and the average period of infectivity is $1/\gamma$ (Hethcote, Stech and van den Driessche, 1981c).

The removal rate from the infective class by both recovery and death is $\gamma + \mu$ so that the death-adjusted average period of infectivity is $1/(\gamma + \mu)$. Thus the average number of adequate contacts (with both susceptibles and others) of an infective during the infectious period is $\sigma = \lambda/(\gamma + \mu)$, which is called the contact number. This quantity is also called the basic reproductive rate (Anderson and May, 1981, 1982; May, 1986) even though it is a number and not a rate. Since the average number of susceptibles infected by an infective during the infectious period is σS, the quantity σS is called the replacement number.

If recovery does not give immunity, then the model is called an SIS model, since individuals move from the susceptible class to the infective class and then back to the susceptible class upon recovery. If individuals recover with permanent immunity, then the model is an SIR model. If individuals recover with temporary immunity so that they eventually become susceptible again, then the model is an SIRS model as considered in Hethcote (1976) and Hethcote, Stech and van den Driessche (1981a). If individuals do not recover, then the model is an SI model. In general, SIR models are appropriate for viral agent diseases such as measles, mumps, and smallpox, while SIS models are appropriate for some bacterial agent diseases such as meningitis, plague, and venereal diseases, and for protozoan agent diseases such as malaria and sleeping sickness (see Table 1).

A basic concept in epidemiology is the existence of thresholds; these are critical values for quantities such as the contact number, population size or vector density that must be exceeded in order for an epidemic to occur or for a disease to remain endemic. The formulations used here are somewhat different from the more classical formulations of Hamer (1906), Ross (1911), Kermack and McKendrick (1927) and others, as given in Bailey (1975). Here they involve the fractions of the populations in the classes instead of the numbers in the classes because these formulations have much more intutive threshold conditions involving the contact number instead of the population sizes. See Hethcote (1976, p. 339) or Hethcote and Van Ark (1987) for further comparisons of formulations in terms of proportions and numbers in the classes.

4. The SIS Model

The first model is for diseases for which infection does not confer immunity. It is called an SIS model since individuals return to the susceptible class when they recover from the infection. Using the notation in Section 3, the compartmental diagram for an SIS model is given in Fig. 1. Naturally occurring births and deaths (vital dynamics) are included, but the behavior of solutions is similar when vital dynamics are not included.

The initial value problem (IVP) for this SIS model formulated in terms of class sizes is

$$(NS(t))' = -\lambda SNI + \gamma NI + \mu N - \mu NS$$
$$(NI(t))' = \lambda SNI - \gamma NI - \mu NI \tag{4.1}$$
$$NS(0) = NS_0 > 0, \quad NI(0) = NI_0 > 0, \quad NS(t) + NI(t) = N$$

where λ is a positive constant and primes denote derivatives with respect to time t. If each equation above is divided by the constant population size N, then the IVP in terms of the fractions in the classes is

$$S'(t) = -\lambda IS + \gamma I + \mu - \mu S$$
$$I'(t) = \lambda IS - \gamma I - \mu I \tag{4.2}$$
$$S(0) = S_0 > 0, \quad I(0) = I_0 > 0, \quad S(t) + I(t) = 1.$$

Note that the IVP (4.2) involves the daily contact and removal rates, but not the population size N. This model is appropriate for some bacterial agent diseases such as gonorrhea, meningitis and streptococcal sore throat. Here all parameters in (4.2) are nonnegative and only nonnegative solutions are considered since negative solutions have no epidemiological significance.

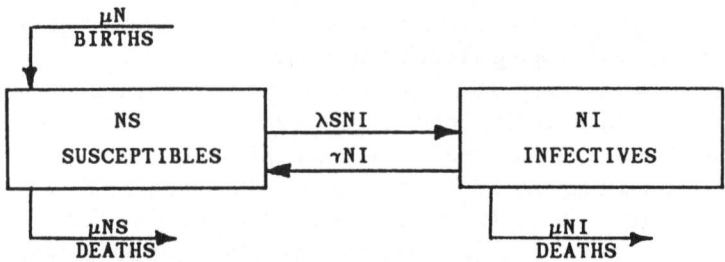

Fig. 1. The compartmental diagram for the SIS model.

Since $S(t)$ can be found from $I(t)$ by using $S(t) = 1 - I(t)$, it is sufficient to consider

$$I'(t) = [\lambda - (\gamma + \mu)]I - \lambda I^2$$
$$I(0) = I_0 > 0. \tag{4.3}$$

Since this is a Bernoulli differential equation, the substitution $y = I^{-1}$ converts (4.3) into a linear differential equation from which the unique solution of (4.3) is found to be

$$I(t) = \begin{cases} \dfrac{e^{(\gamma+\mu)(\sigma-1)t}}{\sigma[e^{(\gamma+\mu)(\sigma-1)t}-1]/(\sigma-1) + 1/I_0} & \text{for} \quad \sigma \neq 1 \\[2ex] \dfrac{1}{\lambda t + 1/I_0} & \text{for} \quad \sigma = 1 \end{cases} \tag{4.4}$$

where σ is the contact number $\lambda/(\gamma + \mu)$ defined in Section 3. The theorem below follows from the explicit solution (4.4).

Theorem 4.1. *The solution $I(t)$ of (4.3) approaches $1 - 1/\sigma$ as $t \to \infty$ if $\sigma > 1$ and approaches 0 as $t \to \infty$ if $\sigma \leq 1$.*

This theorem means that for a disease without immunity with any positive initial infective fraction, the infective fraction approaches a constant endemic value if the contact number exceeds 1; otherwise, the disease dies out. Although the model (4.2) reduces to a one dimensional IVP (4.3), we show SI phase diagrams for this model in Fig. 2 so that they can be compared with the phase diagrams for the other models.

Here the threshold quantity is the contact number σ and the critical threshold value is 1. Note that the replacement number σS is 1 at the endemic equilibrium point. A threshold result for an SI model is obtained from Theorem 4.1 by taking the removal rate γ to be zero in the model. If both the removal rate γ and the birth and death rate μ are zero, then $\sigma = \infty$ so that there is no threshold and eventually everyone is infected. This model with $\gamma = \mu = 0$ is the "simple epidemic

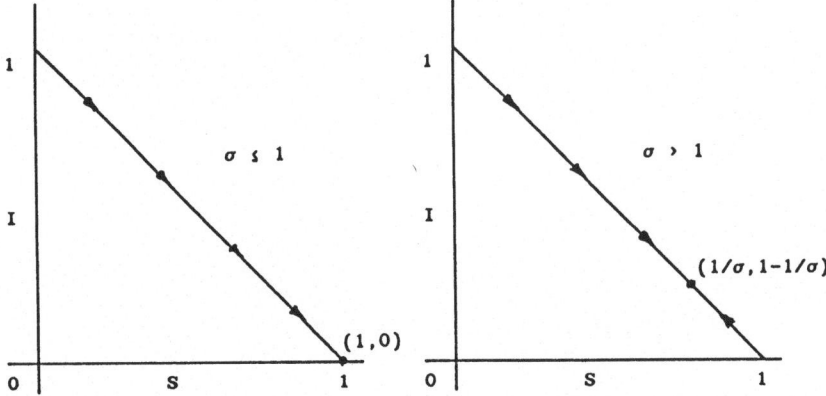

Fig. 2. Phase diagrams for the SIS model. Note that the paths are on the line $S + I = 1$.

model" considered in Bailey (1975, p. 20). Some authors such as May (1986) use other terminology; he uses the basic reproductive rate R_0 is place of the contact number σ and effective reproductive rate $R_0 S$ instead of the replacement number σS.

The prevalence is defined as the number of cases of a disease at a given time so that it corresponds to NI. Since the incidence is defined to be the number of new cases per unit time, it corresponds to the λSNI term in model (4.1). At an endemic equilibrium the prevalence is equal to the incidence times the average duration of infection $1/(\gamma + \mu)$ since the right side of the second equation in (4.1) is zero at an equilibrium.

The incidence and the prevalence of some diseases oscillate seasonally in a population. This oscillation seems to be caused by seasonal oscillation in the contact rate λ. For example, the incidence of childhood diseases such as measles and rubella increase each year in the winter when children aggregate in schools (London and Yorke, 1973; Yorke and London, 1973; Dietz, 1976; Schenzle, 1984).

If the contact rate λ changes with time t, then the λ in models (4.1)–(4.3) are replaced by $\lambda(t)$. If $\lambda(t)$ is periodic with period p, then Hethcote (1973) has found the asymptotic behavior of solutions $I(t)$ of (4.3). If the average contact number $\bar{\sigma} = \bar{\lambda}/(\gamma + \mu)$ satisfies $\bar{\sigma} \leq 1$, then $I(t)$ damps in an oscillatory manner to 0 for large t. However, if $\bar{\sigma} > 1$, then $I(t)$ approaches an explicit periodic solution for large t. These behaviors are shown in Figs. 3 and 4.

Gonorrhea is an example of a disease for which infection does not confer immunity. Fig. 5 shows the actual seasonal oscillation of reported cases of gonorrhea from 1946 to 1984. Numerous models for gonorrhea transmission dynamics and control including a seasonal oscillation model are presented in Hethcote and Yorke (1984).

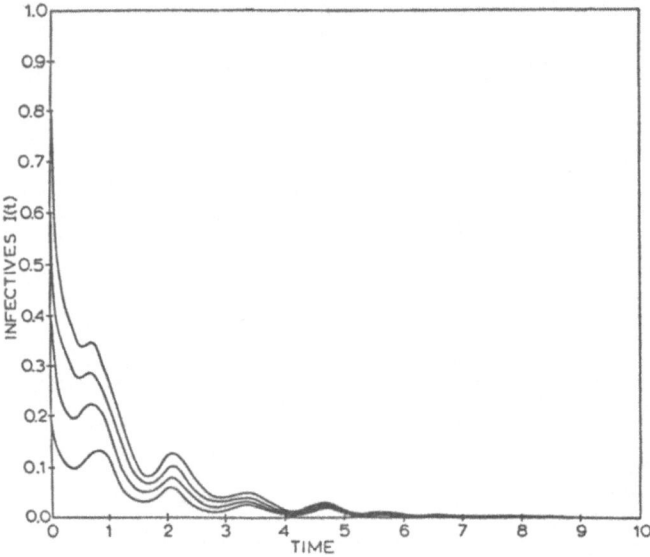

Fig. 3. Solutions of initial value problem (4.3) with periodic $\lambda(t)$ and various values of I_0. Here $\lambda(t) = 2 - 1.8 \cos 5t$ and $\gamma = 4$ so that the average contact number is $\bar{\sigma} = 0.5 < 1$. From Hethcote (1973).

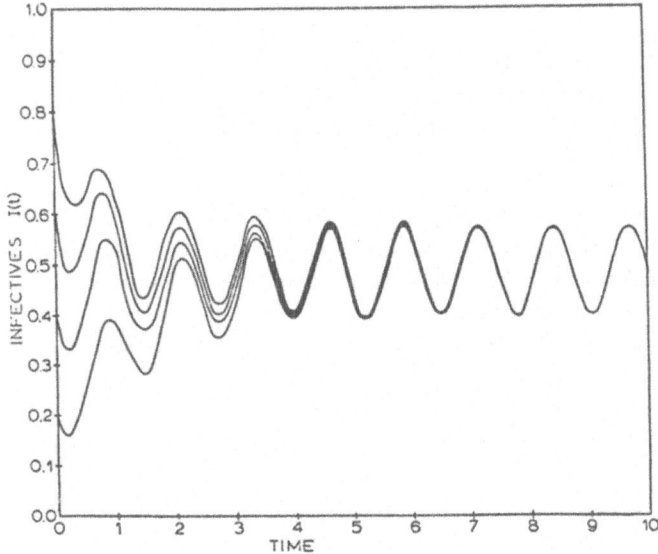

Fig. 4. Solutions as in Fig. 3 except that here $\gamma = 1$ so that $\bar{\sigma} = 2 > 1$. From Hethcote (1973).

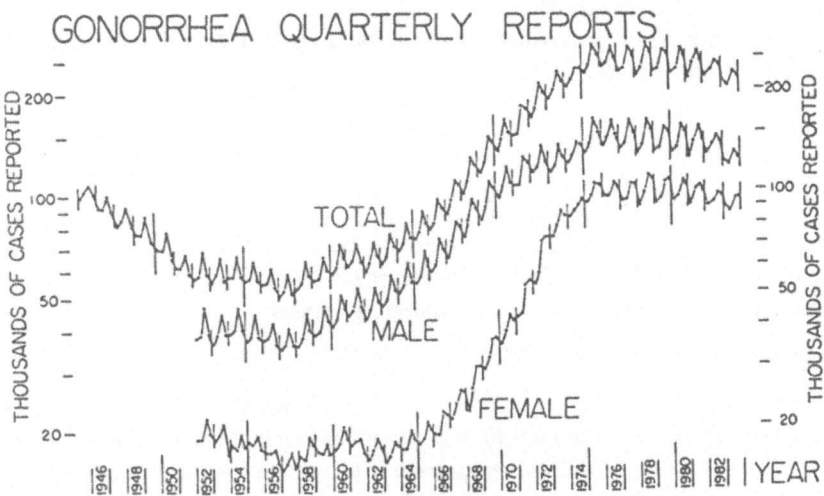

Fig. 5. Reported cases of gonorrhea in women and men in the United States. From Hethcote and Yorke (1984).

5. The SIR Model Without Vital Dynamics

Here and in Sect. 6 we consider diseases for which infection confers permanent immunity. When such an SIR disease goes through a population in a relatively short time (less than one year), then this disease outbreak is called an epidemic.

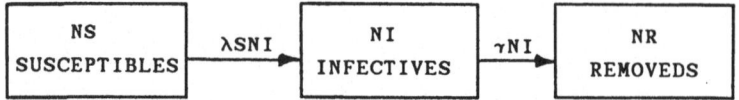

Fig. 6. The compartmental diagram for the SIR model without vital dynamics.

Since an epidemic occurs relatively quickly, the model does not include births and deaths (vital dynamics). Epidemics are common for diseases such as influenza, measles, rubella and chickenpox. Using the notation in Sect. 3, the compartmental diagram for this model is given in Fig. 6.

The initial value problem (IVP) for the SIR model without vital dynamics given in Fig. 6 is

$$(NS(t))' = -\lambda SNI$$
$$(NI(t))' = \lambda SNI - \gamma NI$$
$$(NR(t))' = \gamma NI \tag{5.1}$$
$$NS(0) = NS_0 > 0, \quad NI(0) = NI_0 > 0, \quad NR(0) = NR_0 \geq 0$$
$$NS(t) + NI(t) + NR(t) = N$$

where λ and γ are positive constants.

If each equation in (5.1) in divided by the constant population size N, then the IVP for the fractions $S(t)$ and $I(t)$ is

$$S'(t) = -\lambda SI$$
$$I'(t) = \lambda SI - \gamma I \tag{5.2}$$
$$S(0) = S_0 > 0, \quad I(0) = I_0 > 0.$$

Since $R(t)$ can always be found from $S(t)$ and $I(t)$ by using $R(t) = 1 - S(t) - I(t)$, it is sufficient to consider the IVP (5.2) in the SI phase plane. The epidemiologically reasonable region in the SI plane is the triangle given by

$$T = \{(S, I) | S \geq 0, I \geq 0, S + I \leq 1\}. \tag{5.3}$$

Theorem 5.1. *Let $(S(t), I(t))$ be the solutions of (5.2). If $\sigma S_0 \leq 1$, then $I(t)$ decreases to zero as $t \to \infty$. If $\sigma S_0 > 1$, then $I(t)$ first increases up to a maximum value* Im *equal to $1 - R_0 - 1/\sigma - [\ln(\sigma S_0)]/\sigma$ and then decreases to zero as $t \to \infty$. The susceptible fraction $S(t)$ is a decreasing function and the limiting value $S(\infty)$ is the unique root in $(0, 1/\sigma)$ of the equation*

$$1 - R_0 - S(\infty) + [\ln(S(\infty)/S_0)]/\sigma = 0. \tag{5.4}$$

The threshold quantity in Theorem 5.1 is the initial replacement number σS_0 where $\sigma = \lambda/\gamma$ is the contact number. The natural logarithm is denoted by ln. This theorem states that if the initial replacement number is greater than one, then an epidemic occurs since the prevalence (the infective fraction) increases to a peak and then decreases to zero. Otherwise, there is no epidemic since the prevalence decreases to zero. The infection spread stops during an epidemic because the replacement number $\sigma S(t)$ becomes less than one when $S(t)$ becomes small; however,

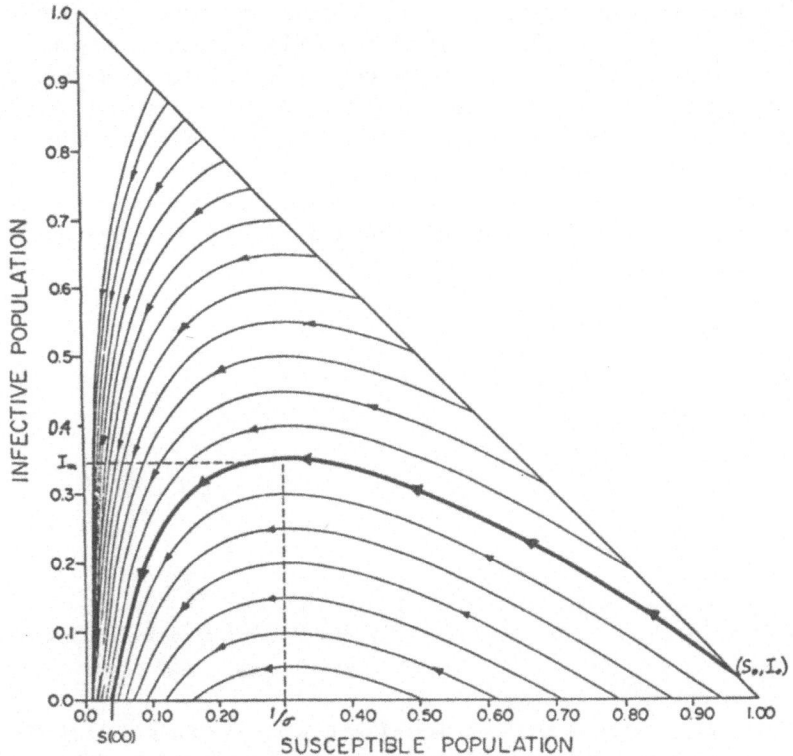

Fig. 7. Phase diagram for the SIR model without vital dynamics with $1/\sigma = 0.30$.

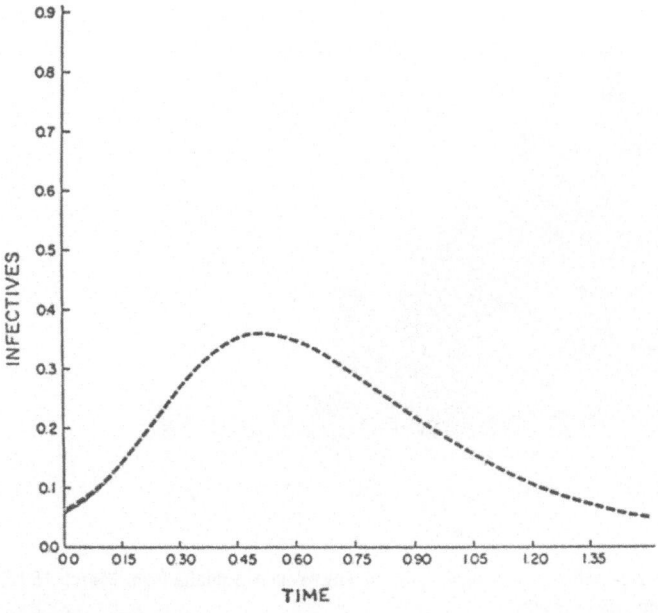

Fig. 8. An epidemic curve starting at $I_0 = 0.06$ with $\lambda = 10$ and $\gamma = 3$ so that $\sigma = 10/3$.

the final susceptible population $S(\infty)$ is not zero. A phase portrait corresponding to system (5.2) is given in Fig. 7. The proof of Theorem 5.1 is given in the Appendix.

Figure 8 shows an epidemic curve which is the prevalence $I(t)$ as a function of time; the incidence or number of new cases per day would also increase to a peak and then decrease. Incidences for examples of epidemics are given in Figs. 9 to 13.

If an epidemic occurs in a homogeneous population and there is no vaccination during the epidemic, then it is possible to estimate the contact number for the disease in that population from epidemic data (Hethcote and Van Ark, 1987). Since

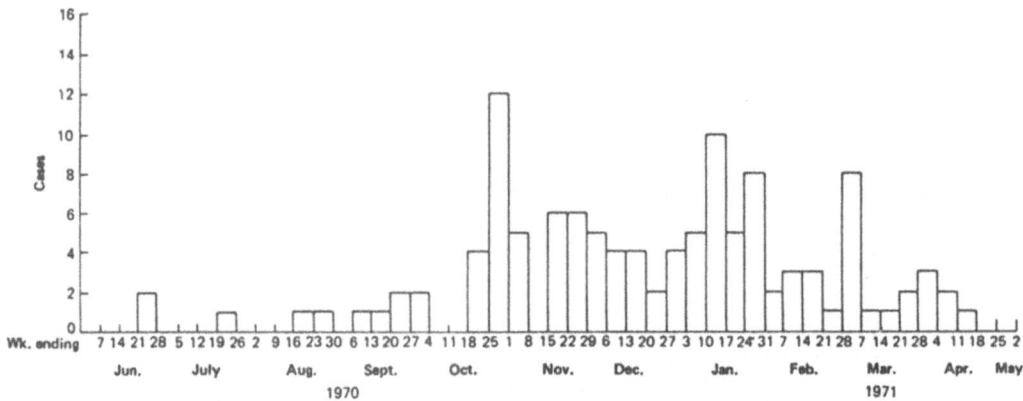

Fig. 9. An epidemic curve for infectious hepatitis in Barren County, Kentucky, USA in 1970 and 1971. The data are irregular, but the general shape is consistent with Fig. 8. Figure from CDS (1971a).

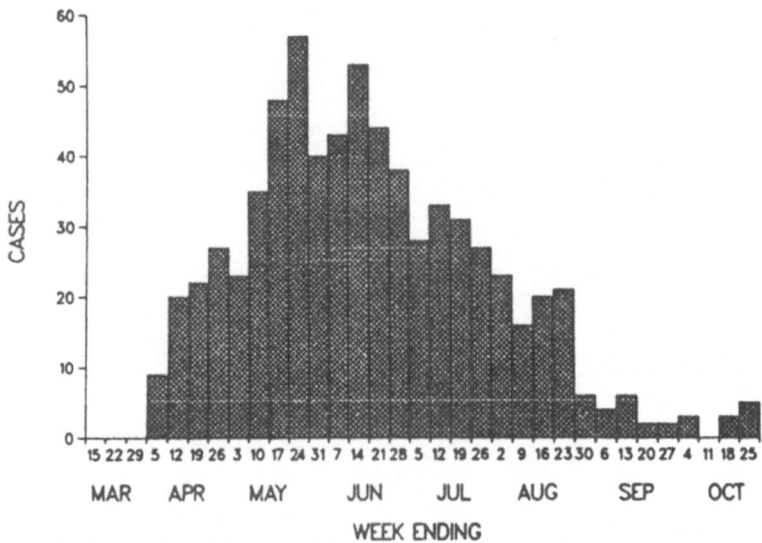

Fig. 10. Cases of non-A, non-B hepatitis in a refugee camp in Tug Wajale, Somalia from March 15 to October 25, 1986. Figure from CDS (1987b).

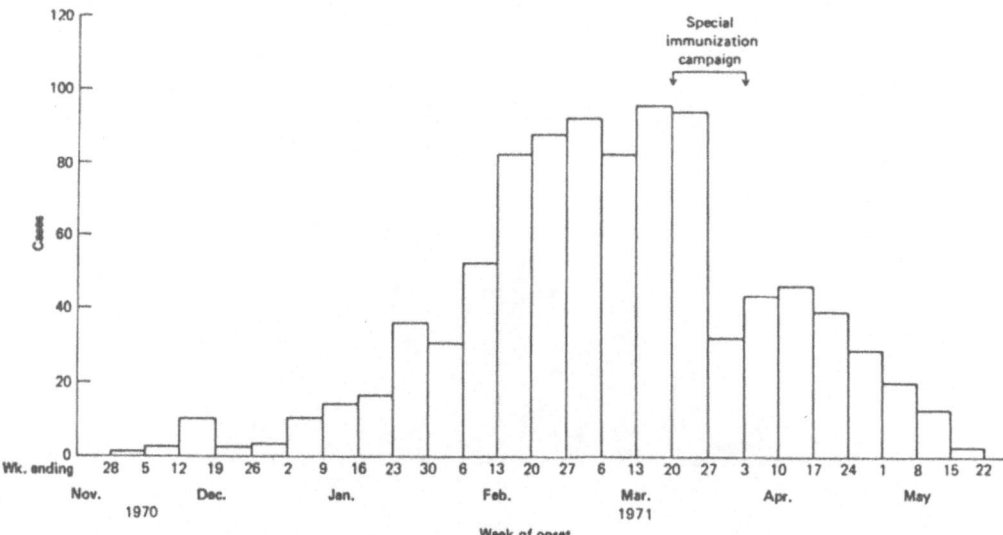

Fig. 11. An epidemic curve for measles cases in Dallas, Texas, USA in 1970 and 1971. Note the interruption due to the special immunization campaign. Figure from CDS (1971b).

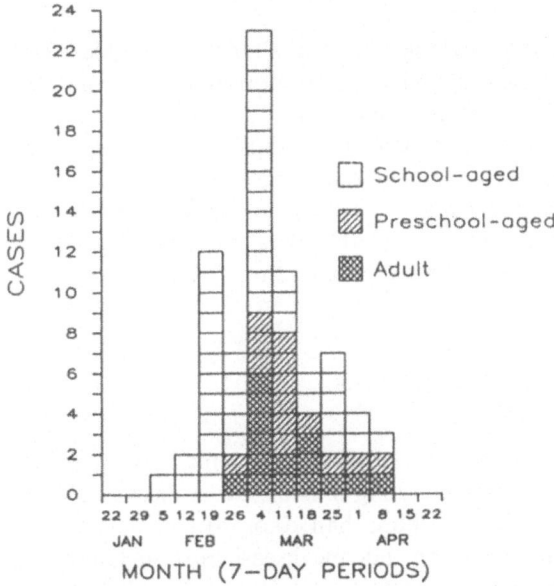

Fig. 12. Measles cases by data of onset in Hobbs, New Mexico from January 22 to April 22, 1984. Figure from CDC (1984).

an epidemic enters a population as one or very few cases, I_0 is negligibly small so that $S_0 = 1 - R_0$. Then (5.4) can be solved for σ to obtain

$$\sigma = \frac{\ln(S_0/S(\infty))}{S_0 - S(\infty)}. \tag{5.5}$$

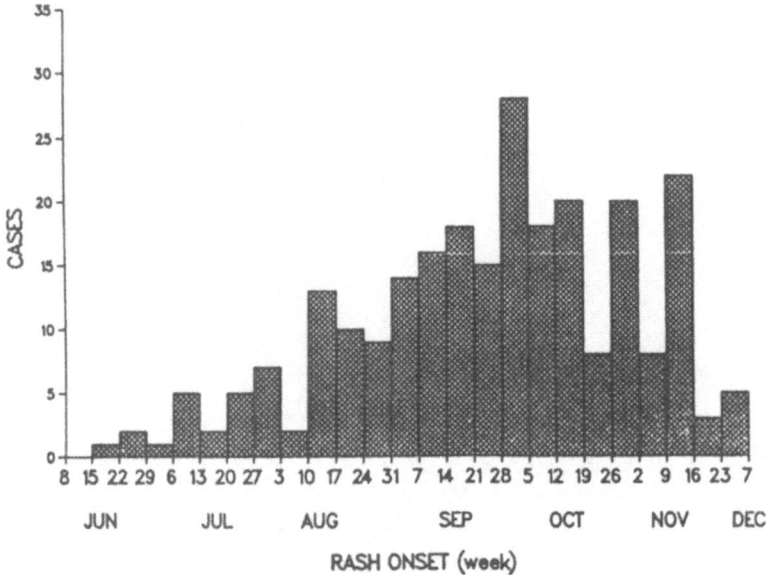

Fig. 13. Reported measles cases by data of rash onset in Dade County, Florida in 1986. The data are somewhat irregular since the latent period for measles is about 2 weeks. Figure from CDC (1987a).

If the susceptible fractions before the epidemic (S_0) and after the epidemic $(S(\infty))$ are measured by serologic studies (i.e., testing immune responses in blood samples), then the contact number σ can be estimated using (5.5).

Evans (1982) reports on serosurveys conducted on freshman at Yale University. The fractions susceptible to rubella at the beginning and end of their freshman year were 0.25 and 0.0965 so that (5.5) leads to the estimate $\sigma = 6.2$. For influenza, the fraction susceptible at the start and end of their freshman year were 0.911 and 0.5138, which leads to a contact number estimate of $\sigma = 1.44$.

6. The SIR Model with Vital Dynamics

An SIR epidemiological model is considered as in Sect. 5, but here we model the disease behavior in the population over a long time period. A disease is called endemic if it is present in a population for more than 10 or 20 years. Because of the long time period involved, a model for an endemic disease must include births as a source of new susceptibles and natural deaths in each class. Using the notation and assumptions in Sect. 3, the compartmental diagram for the SIR model with vital dynamics is given in Fig. 14.

The initial value problem (IVP) for the SIR model with vital dynamics is

$$(NS(t))' = -\lambda SNI + \mu N - \mu NS$$
$$(NI(t))' = \lambda SNI - \gamma NI - \mu NI$$
$$(NR(t))' = \gamma NI - \mu NR \tag{6.1}$$

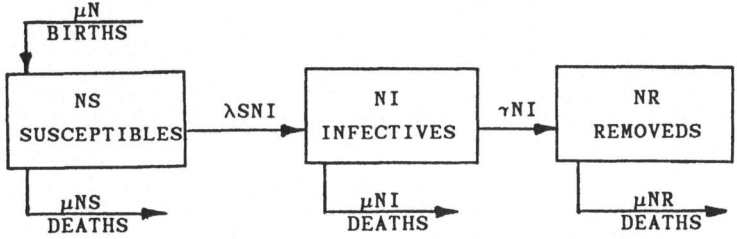

Fig. 14. The compartmental diagram for the SIR model with vital dynamics.

$$NS(0) = NS_0 > 0, \quad NI(0) = NI_0 \geq 0, \quad NR(0) = NR_0 \geq 0$$
$$NS(t) + NI(t) + NR(t) = N$$

where the contact rate λ, the removal rate constant γ and the death rate constant μ are positive constants.

If each equation in (6.1) is divided by N, then the IVP in terms of $S(t)$ and $I(t)$ is

$$S'(t) = -\lambda SI + \mu - \mu S$$
$$I'(t) = \lambda SI - \gamma I - \mu I \tag{6.2}$$
$$S(0) = S_0 > 0, \quad I(0) = I_0 \geq 0.$$

As in Section 5, it is sufficient to consider the IVP (6.2) since $R(t)$ is given by $R(t) = 1 - S(t) - I(t)$. The asymptotic behaviors of solution paths in the SI phase plane are described in the following theorem.

Theorem 6.1. *If $\sigma \leq 1$, then the triangle T defined by (5.3) is an asymptotic stability region for the equilibrium point $(1,0)$. If $\sigma > 1$, then $T - \{(S,0)|0 \leq S \leq 1\}$ is an asymptotic stability region for the equilibrium point*

$$(1/\sigma, \mu(\sigma - 1)/\lambda). \tag{6.3}$$

Figures 15 and 16 are phase plane portraits for the two possibilities described in the theorem. The theorem above can be explained intuitively in terms of the contact number $\sigma = \lambda/(\gamma + \mu)$, which is the threshold quantity. If the contact number is less than one so that an infective replaces itself with less than one new infective, then the disease dies out. Moreover, the susceptible fraction eventually approaches one since everyone is susceptible when the disease has disappeared and all of the removed people who are immune have died.

If the contact number is greater than one, the initial infective fraction I_0 is small, and the initial susceptible fraction S_0 is large so that $\sigma S_0 > 1$, then S decreases and I first increases to a peak and then decreases just as it would for an epidemic (compare Figs. 16 and 8). However, after the infective fraction has decreased to a low level, the susceptible fraction slowly starts to increase due to the births of new susceptibles. When the susceptible fraction gets large enough, there is a second smaller epidemic and so on as the path spirals into the equilibrium point (6.3). At this endemic equilibrium point, the replacement number σS is 1 since if the replacement number were greater or less than 1, then the infective fraction would

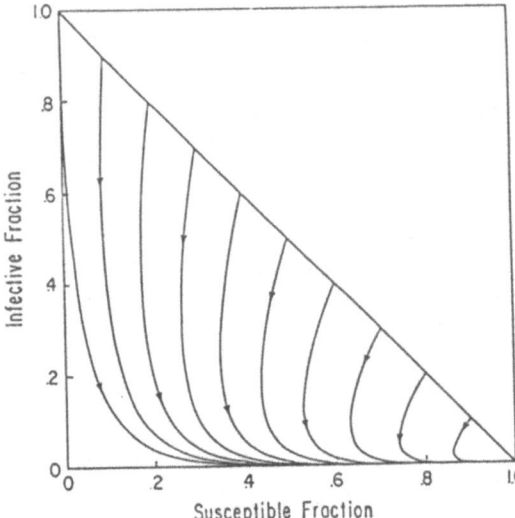

Fig. 15. Phase plane portrait for SIR model with vital dynamics when the contact number is $\sigma = 0.5 < 1$. From Hethcote (1976).

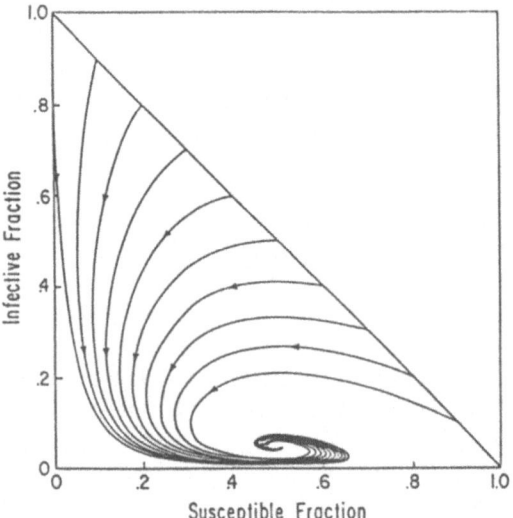

Fig. 16. Phase plane portrait for SIR model with vital dynamics when the contact number is $\sigma = 2 > 1$. From Hethcote (1976).

be increasing or decreasing, respectively. One advantage of precise threshold results such as Theorem 6.1 is that the effects of changes in the parameter values on the asymptotic behavior can be determined directly. The proof of Theorem 6.1 is given in the Appendix.

A method was presented at the end of Section 5 for estimating the contact number from epidemic data. There are two ways to estimate contact numbers from data for SIR diseases which are endemic. The first method involves estimating the susceptible fraction S_e from a serological survey (i.e., testing immune responses in blood samples). It is assumed that the sample is randomly chosen from a homogeneously mixing population and that the disease has reached an endemic equilibrium in the population. Since $S_e = 1/\sigma$ in (6.3), the contact number σ can be estimated using

$$\sigma = 1/S_e. \tag{6.4}$$

This estimation is also valid for the SIS model considered in Section 4 and for endemic SIR disease when part of the population has been vaccinated (Hethcote and Van Ark, 1987).

The second method is useful for childhood diseases in which the susceptible fraction is a decreasing function of age. Dietz (1975) has used an age-structured model in an unvaccinated homogeneously mixing population to derive the formula

$$\sigma = 1 + L/A \tag{6.5}$$

where L is the average lifetime and A is the average age of attack for an endemic disease. Thus if L and A are estimated for an endemic disease in a population, then (6.9) can be used to estimate the contact number σ.

The formula (6.5) can be derived heuristically from the model (6.2). The incidence rate at the endemic equilibrium is $\lambda I_e S_e$ so that λI_e is the incidence proportionality constant. The waiting time to get the infection is distributed as a negative exponential so that the average age at first infection is

$$A = 1/\lambda I_e = 1/[\mu(\sigma - 1)]. \tag{6.6}$$

Solving this equation for σ yields (6.5) since the average lifetime L is $1/\mu$.

7. Herd Immunity and Vaccination

A population is said to have herd immunity for a disease if enough people are immune so that the disease would not spread if it were suddenly introduced somewhere in the population. If the population is homogeneously mixing and the immune people are distributed uniformly in the population, then herd immunity will be obtained if a large enough uniformly distributed fraction is immune. The contact number σ gives the average number of adequate contacts (i.e., those which are sufficient for transmission if all contacted people were susceptible) of an infective during the infectious period. In order to prevent the spread of infection from an infective, enough people must be immune so that the replacement number satisfies $\sigma S < 1$. That is, the susceptible fraction must be small enough so that the average infective infects less than one person during the infectious period.

Herd immunity in a population is achieved by vaccination of susceptibles in the population. If R is the fraction of the population which is immune due to vaccination, then since $S = 1 - R$ when $I = 0$, herd immunity is achieved if $\sigma(1 - R) < 1$ or

$$R > 1 - 1/\sigma. \tag{7.1}$$

For example, if the contact number is 5, at least 80% must be immune to have

Table 2. Estimates of contact numbers and herd immunity fractions from data (Anderson, 1982) on average ages of attack and average lifetimes

Disease	Location	A	L	$\sigma = 1 + A/L$	Minimum R for herd immunity
Measles	England and Wales, 1956–1959	4.8	70	15.6	0.94
	USA, 1912–1928	5.3	60	12.3	0.92
	Nigeria 1960–1968	2.5	40	17.0	0.94
Whooping cough	Maryland, USA, 1943	4.3	70	17.3	0.94
	England and Wales, 1944–1978	4.5	70	16.5	0.94
Chickenpox	Maryland, USA, 1943	6.8	70	11.3	0.91
Diphtheria	Virginia and New York, USA 1934–1947	11.0	70	7.4	0.86
Scarlet fever	Maryland, USA, 1908–1917	8.0	60	8.5	0.88
Mumps	Maryland, USA, 1943	9.9	70	8.1	0.88
Rubella	England and Wales, 1979	11.6	70	7.0	0.86
	West Germany, 1972	10.5	70	7.7	0.87
Poliomyelitis	USA, 1955	17.9	70	4.9	0.80
	Netherlands, 1960	11.2	70	4.3·	0.86
Smallpox	India	12	50	5.2	0.81

herd immunity. If σ is 10, then 90% must be immune for herd immunity. If σ is 20, then 95% immunity corresponds to herd immunity. These results are intuitively reasonable since a higher contact number corresponds to a more easily spread disease so that a larger percentage must be immune to achieve herd immunity.

Table 2 contains data (Anderson, 1982) on the average age A of attack and the average lifetime L for various diseases. The estimates of contact numbers σ in Table 2 are calculated using (6.5) and the minimum immune fraction R is estimated using (7.1). Although the estimates of contact numbers σ in Table 2 are based on many simplifying assumptions, they do lead to crude comparisions of the approximate immunity levels necessary for herd immunity for these diseases.

Attainment of herd immunity for a disease can be quite difficult. Although smallpox was eliminated by vaccination from most developed countries by 1958, it remained endemic in some developing countries. The World Health Organization started a program in 1958 to eradicate smallpox throughout the world (WHO, 1980). Even though high vaccination percentages were achieved in some countries, the disease persisted, primarily because the vaccinations were not uniformly distributed in the population. Eventually the disease was eliminated from more and more countries until the last case occurred in Somalia in 1977. The eradication of smallpox was partly due to herd immunity and partly due to containment efforts such as surveillance, patient isolation and vaccination of all possible contacts when a case occurred (Fenner, 1983). The contact number for smallpox is estimated (see Table 2) to be 5 from data in India. Since eradication in the world of smallpox, which has a low contact number, was difficult, it seems that eradication in the world of diseases with higher contact numbers would be even more difficult.

For various reasons a small fraction of those who are vaccinated do not become immune. This fraction of primary vaccine failures is usually about 0.05 or 0.10, but it can be 0.2 or 0.4 for some influenza vaccines. Vaccine efficacy (VE) is defined as the fraction of those vaccinated who become immune. For example, for measles or rubella vaccination at age 15 months, the vaccine efficacy is approximately 0.95 (Hethcote, 1983). Since the immune fraction R satisfies $R = (V)$ (VE) where V is the vaccinated fraction in the population, inequality (7.1) implies that herd immunity is achieved if the vaccinated fraction V satisfies

$$V > (1 - 1/\sigma)/\text{VE}. \tag{7.2}$$

Measles and rubella have some similarities so it is interesting to compare them. The contact number for rubella is approximately 7 so that herd immunity is obtained if the immune fraction R satisfies $R > 0.86$. If a vaccine efficacy of 0.95 is used, then herd immunity occurs in a homogeneously mixing population if the vaccinated fraction V satisfies $V > 0.91$. The contact number for measles is approximately 15 in a modern developed country so that herd immunity occurs if the immune fraction R satisfies $R > 0.94$. If the vaccine efficiency is 0.95, then herd immunity is achieved if the vaccinated fraction V satisfies $V > 0.99$.

It initially appears that measles may be about twice as difficult to eradicate by herd immunity as rubella since for herd immunity, the *unimmune percentage* must theoretically be less than 14% for rubella and less than 6% for measles. However, it is actually much harder to achieve herd immunity for measles since the *unvaccinated percentage* must be less than 9% for rubella and less than 1% for measles. Indeed, in the USA measles has persisted despite major elimination efforts, while rubella incidence seems to be decreasing (CDC, 1981; CDC, 1986b). Although measles is very difficult or impossible to eradicate with a one dose program, it is easier to achieve herd immunity with a two dose program (Hethcote, 1983).

Hence, although the endemic SIR model is very simple, it has been possible to estimate parameters from it and to use these estimates to get a rough comparision between the immune fractions necessary for herd immunity for various diseases. For further discussion of models with vaccination and applications, see Hethcote (1978), Anderson (1982), Anderson and May (1982, 1983, 1985), Hethcote (1983), May (1986) and Hethcote and Van Ark (1987).

8. Discussion

The SIS model in Sect. 4 and the SIR model with vital dynamics in Sect. 6 have two intuitively appealing features. The first is that the disease dies out if the contact number σ satisfies $\sigma \leq 1$ and the disease remains endemic if $\sigma > 1$. The second is that at an endemic equilibrium, the replacement number is 1; i.e., the average infective replaces itself with one new infective during the infectious period. Although the contact number threshold criterion is the same for diseases without and with immunity, the infective fraction approached asymptotically for large time is higher for diseases without immunity than for diseases with immunity (compare Figs. 2 and 16). In Hethcote, Stech and van den Driessche (1981c) the Soper (1929) model for an SIR disease with vital dynamics is shown to be ill-posed since some solution paths leave the triangle T and R becomes negative.

By comparing Theorems 5.1 and 6.1 it is clear that the asymptotic behaviors for SIR models without and with vital dynamics are very different. The SIR model without vital dynamics might be appropriate for describing an epidemic outbreak during a short time period, whereas the SIR model with vital dynamics would be appropriate over a longer time period. Viral agent diseases such as measles, chickenpox, mumps, and influenza may have occasional large outbreaks in certain communities and yet be endemic at a low level in larger population groups. The threshold quantity for the SIR model without vital dynamics is the initial replacement number σS_0. In this model, no epidemic occurs if $\sigma S_0 < 1$ and an epidemic occurs if $\sigma S_0 > 1$.

The latent period is the time in which an individual is infected but is not yet infectious. The latent period is approximately 15 days for chickenpox, 10 days for measles, and 2 days for influenza. The latent period has been ignored in the three basic models considered here because the thresholds and asymptotic behaviors are essentially the same for the models which include latent periods. The fraction of the population that is in the latent period is often called $E(t)$ or the exposed fraction. Various SEIS models are analysed in Hethcote, Stech and van den Driessche (1981b). Some SEIR models with vital dynamics are considered in Hethcote and Tudor (1980). Longini (1986) shows that the formula (5.5) also holds for an SEIR model without vital dynamics.

Instead of assumption 2 in Section 3, it is sometimes assumed that susceptibles become infectious at a rate proportional to the product of the number of susceptibles NS and the number of infectives NI with proportionality constant β. By comparing the resulting initial value problem with (4.1), (5.1) or (6.1), we see that $\beta = \lambda/N$ and thus the assumption that β is constant implies that the daily contact rate λ is proportional to the population size N. The daily contact rate would probably increase if the population within a fixed region increased (i.e., the population density increased). However, it seems more likely that the daily contact rate λ is independent of population size since λ might be the same for a large population in a large region and a small population in a small region. Hethcote and Van Ark (1987) consider model formulation for heterogeneous populations and discuss a "city and villages" model where confusion between β and λ has led to misleading results. Consequently, it seems best to carefully separate the daily contact rate λ and the population size N as we have done in assumption 2. Moreover, threshold statements involving contact numbers are more appealing intuitively than the population size threshold statements as given in Bailey (1975).

Although the models discussed here do provide some insights and useful comparisons, most models now being applied to specific diseases are more complicated. Hethcote, Stech and van den Driessche (1981c) have surveyed the mathematical epidemiology literature using the classifications introduced in this article. More recent references are given below for some more refined models. Many more complicated models are considered in other articles in this volume.

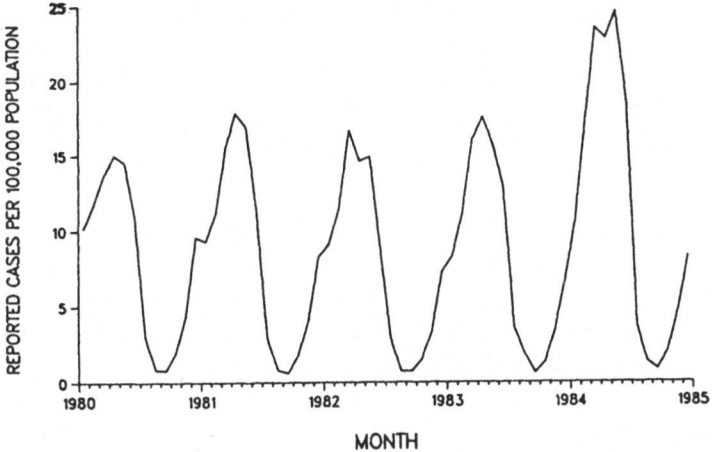

Fig. 17. Seasonal oscillation in the incidence of chickenpox (varicella) in the United States Between 1980 and 1984. Figure from CDC (1986a).

The prevalence for many diseases varies periodically because of seasonal changes in the daily contact rates. For example, the seasonal oscillation in the incidence of chickenpox is shown in Fig. 17. An SIS model with a periodic contact rate has been considered briefly in Sect. 4. Other epidemiological models with periodic contact rates are described in an article in this volume by Hethcote and Levin (1989). Other models without periodic contact rates can also have periodic solutions and are also described in the article mentioned above. These other models leading to periodic solutions have features such as a delay corresponding to temporary immunity, nonlinear incidence, variable population size or cross immunity with age structure.

The three basic epidemiological models in this article have assumed that the population being considered is uniform and homogeneously mixing; however, most infectious diseases actually spread in a diverse or dispersed population. Hence it is desirable to consider a population divided into different subpopulations. Mathematical aspects of models for heterogeneous populations are described in the survey of Hethcote, Stech and van den Driessche (1981c) and, more recently, in Hethcote and Thieme (1985) and in Hethcote and Van Ark (1987). Since gonorrhea transmission occurs in a very heterogeneous population, the models in Hethcote and Yorke (1984) for gonorrhea involve from 2 to 8 subpopulations. A spatially heterogeneous "city and villages" example is considered in May and Anderson (1984a, 1984b) and again in Hethcote and Van Ark (1987). Parameter estimation methods similar to those presented in Sections 5 and 6 are developed for heterogeneous population models in Hethcote and Van Ark (1987).

Models for populations where the disease causes enough deaths to influence the population size are considered by Anderson and May (1979) and May and Anderson (1979). Since contact rates between age groups vary greatly, it is often important to consider models with age structure. These models are considered in papers such as Kermack and McKendrick (1927), Dietz (1975), Hoppensteadt

(1975), Longini et al. (1978), Hethcote (1983), and Anderson and May (1983, 1985). Age-structured models have been used to compare the UK and USA strategies for rubella vaccination as described in the article by Hethcote (1989) in this volume. Models for measles are considered in Fine and Clarkson (1982), Hethcote (1983) and Anderson and May (1983). Epidemiological models for influenza with age structure and cross immunity are presented in the article by Castillo-Chavez et al. (1989) in this volume.

Epidemiological models with spatial spread are surveyed by Mollison (1977) and more recently by Mollison and Kuulasmaa (1985). The spatial spread of fox rabies has been considered by Anderson et al. (1981). See Radcliffe and Rass (1986) and the references cited therein for thresholds, final sizes, pandemic theorems and asymptotic speeds of propagation of travelling epidemic waves. The spread of influenza throughout the world has recently been modeled and is described in Rvachev and Longini (1985).

As indicated in Sect. 2, infectious disease models are useful in comparing control procedures. See Wickwire (1977) for a survey of models for the control of infectious diseases. The optimal uses of vaccination for influenza are considered in Longini, Ackerman and Elveback (1978). Control strategies for rubella and comparisons using cost benefit analyses are described in the article by Hethcote (1989) on rubella in this volume. Gonorrhea control procedures are compared in Hethcote and Yorke (1984).

The purpose of this article has been to introduce the most basic ideas, assumptions, notation and formulations for epidemiological models in order to prepare the reader for the study of more refined models and their applications to specific diseases. There is a great need for individuals to understand and analyse specific diseases through modeling and to use modeling to investigate and compare methods for decreasing their incidence.

References

Anderson, R.M. (1982) Directly transmitted viral and bacterial infections of man. In: Anderson, R.M. (ed.) Population Dynamics of Infectious Diseases. Theory and Applications. Chapman and Hall, NewYork, pp. 1–37.

Anderson, R.M., Jackson, H.C., May, R.M., Smith, A.D.M. (1981) Populations dynamics of fox rabies in Europe. Nature 289, 765–777.

Anderson, R.M., May, R.M. (1979) Population biology of infectious diseases I. Nature 280, 361–367.

Anderson, R.M., May, R.M. (1981) The population dynamics of microparasites and their invertebrate hosts. Phil. Trans. Roy. Soc. London B291, 451–524.

Anderson, R.M., May, R.M. (1982) Directly transmitted infectious diseases: control by vaccination. Science 215, 1053–1060.

Anderson, R.M., May, R.M. (1983) Vaccination against rubella and measles: quantitative investigations of different policies. J. Hyg. Camb. 90, 259–325.

Anderson, R.M., May, R.M. (1985) Vaccination and herd immunity to infectious diseases. Nature 318, 323–329.

Bailey, N.T.J. (1975) The Mathematical Theory of Infectious Diseases, 2nd edn. Hafner, New York.

Castillo-Chavez, C., Hethcote, H.W., Andreasen, V., Levin, S.A., Liu, W.M. (1988) Cross-immunity in the dynamics of homogeneous and heterogeneous populations, In: T.G. Hallam, L. Gross, and S.A. Levin (eds.) Mathematical Ecology, World Scientific Publishing, Singapore, 303–316.

Centers for Disease Control (1971a) Infectious hepatitis—Kentucky. Morbidity and Mortality Weekly Report 20, 136–137.

Centers for Disease Control (1971b) Measles—Dallas, Texas, Morbidity and Mortality Weekly Report 20, 191–192.

Centers for Disease Control (1981) Rubella—United States, 1978–1981. Morbidity and Mortality Weekly Report 30, 513–515.

Centers for Disease Control (1984) Measles in an immunized school-aged population—New Mexico. Morbidity and Mortality Weekly Report 34, 52–59.

Centers for Disease Control (1986a) Annual summary 1984: reported morbidity and mortality in the United States. Morbidity and Mortality Weekly Report 33(54).

Centers for Disease Control (1986b) Rubella and congenital rubella syndrome—United States 1984–1985. Morbidity and Mortality Weekly Report 35, 129–135.

Centers for Disease Control (1987a) Measles—Dade County, Florida. Morbidity and Mortality Weekly Report 36, 45–48.

Centers for Disease Control (1987b) Enterically transmitted non-A, non-B hepatitis—East Africa, Morbidity and Mortality Weekly Report 36, 241–244.

Coddington, E.A., Levinson, N. (1955) Theory of Ordinary Differential Equations. McGraw-Hill, New York.

Coleman, C.S. (1978) Biological cycles and the fivefold way. In: Braun, M., Coleman, C.S., Drew, D.A. (eds.) Differential Equation Models. Springer, New York, pp. 251–278.

Dietz, K. (1975) Transmission and control of arbovirus diseases. In: Ludwig D. and Cooke, K.L (eds.) Epidemiology. SIMS 1974 Utah Conference Proceedings, SIAM, Philadelphia, pp. 104–121.

Dietz, K. (1976) The incidence of infectious diseases under the influence of season fluctuations. In: Berger, J., Buhler, R., Repges, R., Tantu, P. (eds.) Mathematical Models in Medicine. Lecture Notes in Biomathematics, vol. 11. Springer, New York, pp. 1–15.

Evans, A.S. (1982) Viral Infections of Humans 2nd edn. Plenum Medical Book Company, New York.

Fenner, F. (1983) Biological control, as exemplified by smallpox eradication and myxomatosis. Proc. Roy. Soc. London B218, 259–285.

Fine, P.E.M., Clarkson, J.A. (1982) Measles in England and Wales I: An analysis of factors underlying seasonal patterns, and II: The impact of the measles vaccination programme on the distribution of immunity in the population. Int. J. Epid. 11, 5–14 and 15–24.

Guckenheimer, J., Holmes, P. (1983) Nonlinear Oscillations, Dynamical Systems and Bifurcations of Vector Fields. Springer, New York.

Hamer, W.H. (1906) Epidemic disease in England. Lancet 1, 733–739.

Hethcote, H.W. (1973) Asymptotic behavior in a deterministic epidemic model. Bull. Math. Biology 35, 607–614.

Hethcote, H.W. (1974) Asymptotic behavior and stability in epidemic models. In: van den Driessche P. (ed.) Mathematical Problems in Biology. Lecture Notes in Biomathematics, vol. 2. Springer, Berlin Heidelberg New York, pp. 83–92.

Hethcote, H.W. (1976) Qualitative analysis for communicable disease models. Math. Biosci. 28, 335–356.

Hethcote, H.W. (1978) An immunization model for a heterogeneous population. Theor. Prop. Biol. 14, 338–349.

Hethcote, H.W. (1983) Measles and rubella in the United States. Am. J. Epidemiol. 117, 2–13.

Hethcote, H.W. (1988) Optimal ages of vaccination for measles. Math. Biosci. 89, 29–52.

Hethcote, H.W. (1989) Rubella. In: Levin, S.A., Hallam, T.G., Gross, L. (eds.) Applied Mathematical Ecology. Biomathematics, vol. 18. Springer, Berlin, Heidelberg, New York.

Hethcote, H.W., Levin, S.A. (1988) Periodicity in epidemiological models. In: Levin, S.A., Hallam, T.G. Gross, L. (eds.) Applied Mathematical Ecology. Biomathematics, vol. 18. Springer, Berlin, Heidelberg, New York, 193–211.

Hethcote, H.W., Stech, H.W., van den Driessche, P. (1981a) Nonlinear oscillations in epidemic models. SIAM J. Appl. Math. 40, 1–9.

Hethcote, H.W. Stech, H.W., van den Driessche, P. (1981b) Stability analysis for models of diseases without immunity. J. Math. Biology 13, 185–198.

Hethcote, H.W., Stech, H.W., van den Driessche, P. (1981c) Periodicity and stability in epidemic models: a survey. In: Busenberg, S. and Cooke, K.L. (eds.) Differential Equations and Applications in Ecology, Epidemics and Populations Problems. Academic Press, New York, pp. 65–82.

Hethcote, H.W., Tudor, D.W. (1980) Integral equation models for endemic infectious diseases. J. Math. Biol. *9*, 37–47.

Hethcote, H.W., Van Ark, J.W. (1987) Epidemiological models for heterogeneous populations: proportionate mixing, parameter estimation and immunization programs. Math. Biosci. *84*, 85–118.

Hethcote, H.W., Yorke, J.A. (1984) Gonorrhea Transmission Dynamics and Control. Lecture Notes in Biomathematics, vol. 56, Springer, Berlin Heidelberg New York.

Hoppensteadt, F. (1975) Mathematical Theories of Populations. Demographics, Genetics and Epidemics. SIAM, Philadelphia.

Jordan, D.W., Smith, P. (1977) Nonlinear Ordinary Differential Equations. Oxford University Press, Oxford.

Kermack, W.O., McKendrick, A.G. (1927) A contribution to the mathematical theory of epidemics. Proc. Roy. Soc. *A115*, 700–721.

London, W.A., Yorke, J.A. (1973) Recurrent outbreaks of measles, chickenpox and mumps. I. Am. J. Epid. *98*, 453–468.

Longini, I.M., Jr. (1986) The generalized discrete-time epidemic model with immunity: a synthesis. Math. Biosci. *82*, 19–41.

Longini, I.M., Jr., Ackerman, E., Elveback, L.R. (1978) An optimization model for influenza A epidemics. Math. Biosci. *38*, 141–157.

May, R.N. (1986) Population biology of microparasitic infections, In: Hallam T.G. and Levin, S.A. (eds.) Mathematical Ecology. Biomathematics, vol. 17. Springer, Berlin, Heidelberg, New York, pp. 405–442.

May, R.M., Anderson, R.M. (1979) Population biology of infectious diseases II. Nature *280*, 455–461.

May, R.M., Anderson, R.M. (1984a) Spatial heterogeneity and the design of immunization programs. Math. Biosci. *72*, 83–111.

May, R.M., Anderson, R.M. (1984b) Spatial, temporal, and genetic heterogeneity in host populations and the design of immunization programmes. IMA J. of Math. App. Med. Biol. *1*, 233–266.

Miller, R.K., Michel, A.N. (1982) Ordinary Differential Equations. Academic Press, New York.

Mollison, D. (1977) Spatial contact models for ecological and epidemic spread. J.R. Statist. Soc. Ser. *B39*, 283–326.

Mollison, D., Kuulasmaa, K. (1985) Spatial epidemic models: theory and simulations. In: Bacon, P.J. (ed.) Population Dynamics of Rabies in Wildlife. Academic Press, London, pp. 291–309.

Radcliffe, J., Rass, L. (1986) The asymptotic speed of propagation of the deterministic nonreducible *n*-type epidemic. J. Math. Biol. *23*, 341–359.

Ross, R. (1911). The Prevention of Malaria, 2nd edn. Murray, London.

Rvachev, L.A., Longini, I.M. Jr. (1985) A mathematical model for the global spread of influenza. Math. Biosci. *75*, 3–22.

Schenzle, D. (1984) An age structured model of pre and post-vaccination measles transmission. IMA J. Math. Appl. Biol. Med. *1*, 169–191.

Soper, H.E. (1929) Interpretation of periodicity in disease prevalence. J.R. Statist. Soc. *92*, 34—73.

World Health Organization (1980) The Global Eradication of Smallpox. Final report, WHO, Geneva.

Yorke, J.A., London, W.P. (1973) Recurrent outbreaks of measles, chickenpox and mumps II. Am. J. Epid. *98*, 469–482.

Yorke, J.A., Nathanson, N. Pianigiani, G. Martin, J. (1979) Seasonality and the requirements for prepetuation and eradication of viruses in populations. Am. J. Epidemiol. *109*, 103–123.

Appendix

Proof of Theorem 5.1. The triangle T given by (5.3) is positively invariant since no direction vectors at the boundary of T are outward. More precisely, $S = 0$ implies $S' = 0$, $I = 0$ implies $I' = 0$, and $S + I = 1$ implies $(S + I)' = -\gamma I \leqq 0$. Moreover, every point on the S axis where $I = 0$ is an equilibrium point (EP). The EP for $S < 1/\sigma$ are neutrally stable and the EP for $S > 1/\sigma$ are neutrally unstable.

From (5.2) and the positive invariance of T, it follows that $S(t)$ is nondecreasing and $S(t) \geq 0$ so that a unique limit $S(\infty)$ exists. Since $R'(t) = \gamma I \geq 0$ and $R(t)$ is bounded above by 1, the limit $R(\infty)$ exists. Since $I(t) = 1 - S(t) - R(t)$, the limit $I(\infty)$ exists. Moreover, $I(\infty) = 0$ since otherwise $R'(t) > \gamma I(\infty)/2$ for t sufficiently large so that $R(\infty) = \infty$ which contradicts $R(\infty) \leq 1$. The solution paths

$$I = 1 - R_0 - S + [\ln(S(t)/S_0)]/\sigma \qquad (A.1)$$

are found from $dI/dS = -1 + 1/\sigma S$. The conclusions stated follow directly from (A.1) and the observation that a solution path has a maximum infective fraction Im when $\sigma S = 1$.   □

Proof of Theorem 6.1. This proof uses standard phase plane methods found in differential equation books such as Coddington and Levinson (1955), Jordan and Smith (1977) and Miller and Michel (1982). This proof was first presented in Hethcote (1976). The equilibrium points (EP) in the SI phase plane are $(1,0)$ and the EP given by (6.3). The characteristic roots of the linearization around the EP $(1,0)$ are $-\mu$ and $(\gamma + \mu)(\sigma - 1)$ so that this EP is a stable node if $\sigma < 1$ and a saddle if $\sigma > 1$. The triangle defined by (5.4) is positively invariant since no path leaves through a boundary. More precisely, $S = 0$ implies $S'(t) = \mu > 0$, $I = 0$ implies $I'(t) = 0$, and $S + I = 1$ implies $(S + I)' = -\gamma I \leq 0$. Moreover, there is a path along the S axis approaching the EP $(1,0)$.

The Poincarè-Bendixson theorem (Coddington and Levinson, 1955; Miller and Michel, 1982) implies that bounded paths in the phase plane approach either an EP, a limit cycle or a cycle graph (Coleman, 1978). If $\sigma \leq 1$, then $(1,0)$ is the only EP in the triangle T. There is no limit cycle contained in T since limit cycles must contain at least one EP in their interior. There is no cycle graph in T since a homoclinic loop is not possible from a stable EP (Guckenheimer and Holmes, 1983). Thus all paths in T approach the EP $(1,0)$ if $\sigma \leq 1$.

If $\sigma > 1$, then the EP $(1,0)$ is a saddle with an attractive path along the S axis and a repulsive path into T with slope $-1 + \gamma/(\gamma + \mu)$. If $\sigma > 1$, then the EP (6.3) is in the interior of T and it is locally asymptotically stable since the characteristic roots of the linearization around it have negative real parts. The Bendixson–Dulac test (Jordan and Smith, 1977, p. 91; Hethcote, 1976) with multiplying factor $1/I$ leads to

$$\frac{\partial}{\partial S}\left(-\lambda S + \frac{\mu}{I} - \frac{\mu S}{I}\right) + \frac{\partial}{\partial I}(\lambda S - \gamma - \mu) = -\lambda - \frac{\mu}{I} < 0$$

so that there are no limit cycles or cycle graphs in T. The only path in T approaching the EP $(1,0)$ is the S axis. Thus all paths in T except the S axis approach the EP given by (6.3).

This stability result for $\sigma > 1$ can also be proved by using a Liapunov function. The Liapunov function used here also works for the SIRS model in Hethcote (1976), but the Liapunov function given in Hethcote (1974) does not. If $S = S_e(1 + U)$ and $I = I_e(1 + V)$ where (S_e, I_e) is the EP (6.3), then

$$U'(t) = -\lambda I_e U(1 + V) - \lambda I_e V - \mu U$$
$$V'(t) = (\gamma + \mu)U(1 + V)$$

and the positively invariant triangle T given by (5.3) becomes

$$T^* = \{(U, V): U \geq -1, V \geq -1, S_e U + I_e V \leq 1 - S_e - I_e\}.$$

The Liapunov function

$$L = U^2/2 + \sigma I_e[V - \ln(1 + V)]$$

is positive definite for $V > -1$ and the Liapunov derivative is

$$L' = -\lambda I_e U^2(1 + V) - \mu U^2 \leq 0.$$

The set $E = \{(U, V): L' = 0\}$ is the V axis where $U = 0$. Since $U = 0$ implies $U' = -\lambda I_e V$, the only positively invariant subset of E is the origin. By the Liapunov–Lasalle theorem (Miller and Michel, 1982, p. 226), the EP $(0,0)$ in UV coordinates is locally asymptotically stable. Since the Liapunov curves $L(U, V) = C$ fill the upper half plane above $V = -1$ as C approaches infinity, this half plane is an asymptotic stability region for the EP $(0,0)$. This implies the result stated for $\sigma > 1$ in Theorem 6.1. $\square$

The Population Biology of Parasitic Helminths in Animal Populations

A.P. Dobson

From the 2nd Workshop on Mathematical Biology, Trieste, 1986

Contents

Introduction

Most of the other chapters on infectious diseases in this book have concentrated on *microparasites*, the viruses, bacteria and protozoa, in human populations. This chapter will concentrate on mathematical models for *macroparasites*, the parasitic helminths, in wild, domestic and laboratory populations of animals. Instead of describing any group of models in specific detail, the development of the models from a common source is illustrated. Similarly, no formal proofs of the models' properties are given, emphasis is instead placed upon the applications of the models to the control of diseases caused by parasitic helminths. Citations to the relevant papers form an introduction to the reader wishing to either examine the mathematical properties of the models or the more specific details of their application to a specific epidemiological problem.

Macroparasites

Anderson and May (1979a) define the macroparasites to include the parasitic helminths (the nematodes, trematodes, cestodes and acanthocephala) and the parasitic arthropods (ticks, fleas and some copepods). The majority of recently developed mathematical models in this area have been concerned with the parasitic helminths and in particular those species that are important as pathogens of either domestic livestock or species harvested by man for recreational or food purposes.

Rather than discuss models for different taxonomic classes of hosts, an artificial mathematical taxonomy has been imposed upon recently studies. Table 1 attempts to illustrate how the different major taxonomic classes of helminths fit into this framework of life-cycles whose dynamics require models of differing levels of complexity. The table also indicates that all fundamental types of macroparasite life cycle have now been modeled with the possible exception of nematode species whose life cycles involve more than one sequential species of intermediate host. Initially, models for *monoxenic* parasites will be considered, then, in less detail, models for *heteroxenic* life cycles. The former group of parasites tend to utilize only one species of definitive host and transmission proceeds either directly between hosts or via free-living infective stages. Heteroxenic parasites, in contrast, utilize more than one species of host in an obligatory sequence, with one or more species of intermediate host serving as transmission agents between any two definitive hosts. Definitive hosts may be defined as the host where the parasite reproduces sexually, while the intermediate hosts may serve either as sites for asexual reproduction or may simply function as *paratenic* or transport hosts.

Basic Macroparasite Model

Crofton (1971b) was probably the first person to develop a simple mathematical model for host-parasite dynamics which captured all the essential features of these associations. His difference equation model is essentially a limiting case of the more general host-parasite models that require a differential equation formulation (May, 1977a). Anderson and May (1978) and May and Anderson (1978) proposed the following three equation model for the dynamics of simple direct life cycle macroparasites

$$dH/dt = (a - b)H - \alpha \sum_{i=1}^{\infty} ip(i) \tag{1}$$

Table 1. Life cycle characteristics of different parasitic helminths and a guide to models of these life cycles.*

Taxonomic Group	Life cycle features						Species studied and refs.
	lc	h	d	i	a	pp	
Platyhelminthes Class:							
Monogenea	M	L	+	0	(+)	0	Gyrodactylus bullatarudis (1, 2)
Cestoda	H	F	+	+	0	+	Caryophyllaeus laticeps (3)
	H	L	+	+	0	+	Hymenolepis diminuta (4, 5, 6)
	H	F	+	+	+	+	Echinococcus granulosus (7)
Digenea	H	B	+	+	+	0	Schistosoma mansoni (8–15)
	H	F	+	+	+	0	Fasciola hepatica (16, 17)
Acanthocephala	H	B	+	+	0	+	General models (18)
Nematoda	M	F	+	0	0	0	Ostertagia ostertagi (19, 20)
	M	F	+	0	0	0	Haemonchus contortus (21)
	M	L	+	0	0	0	Heligmosomoides polygyrus (22)
	M	F	+	0	0	0	Trichostrongyle tenuis (23, 24)
	M	L	+	0	0	0	Romanomermis culicivorax (25)
Ciliates	M	L	+	0	(+)	0	Ichythyophthirius multifillis (26, 27)

*The abbreviations used in the central column of this table signify the following: H or M—heteroxenic or monoxenic life cycle; F or L—field or laboratory based data (B = both); the 'd' and 'i' columns indicate whether the models have concentrated on the parasite in its definitive or intermediate host, while the 'a' column indicates whether asexual reproduction occurs in the intermediate host (in the monogeneans and ciliates, this occurs in the definitive host). The final column, 'pp', indicates whether or not transmission to the definitive host is via a predator prey relationship. In all case presence of a trait is indicated by a '+', while its absence is denoted by a '0'. The numbers in the final column correspond to the numbered references in the following list: (1) Scott, 1982; (2) Scott and Anderson, 1984; (3) Anderson, 1976; (4) Keymer, 1980; (5) Keymer, 1982b; (6) Keymer and Anderson, 1979; (7) Roberts, Lawson and Gemmell, 1986; (8) Macdonald, 1965; (9) Nasell, 1976a; (10) May, 1977b; (11) Bradley and May, 1978; (12) Cohen, 1976, 1977; (13) Barbour, 1978, 1982; (14) Anderson and Crombie 1984, 1985; (15) Anderson and May (1979b); (16) Wilson, Smith and Thomas, 1982; (17) Smith, 1982, 1984a–d; (18) Dobson and Keymer (1986); (19) Smith, Grenfell and Anderson, 1987a, b, c; (20) Grenfell, Smith and Anderson, 1986, 1987a, b; (21) Smith, 1988; (22) Berding, Keymer, Murray and Slater, 1987a, b; (23) Hudson, Dobson and Newborn, 1985; (24) Hudson and Dobson, 1988; (25) Tingley and Anderson, 1987; (26) McCallum 1982, 1985, 1986; (27) McCallum and Anderson, 1984.

$$dP/dt = \beta WH - (b+u)P - \alpha H \sum_{i=1}^{\infty} i^2 p(i) \tag{2}$$

$$dW/dt = \lambda P - wW - \beta WH. \tag{3}$$

Here a and b are the birth and death rates of the host population, H. The parasite population, P, causes host mortality at a per capita rate, α, while parasite fecundity is denoted by λ, u is the mortality rate of the adult parasites, while w is the mortality rate of the free living stages, W, and β is the rate at which these free-living stages locate definitive hosts. The two summation terms in equations 1 and 2 allow the statistical distribution of the parasites to be taken into consideration.

Aggregation, Causes and Consequences

Parasitic helminths are invariably aggregated in their distributions in both their definitive and intermediate hosts. A large proportion of the host population only harbors a few parasite individuals, while the majority of the parasite population are concentrated in only a few host individuals (Crofton, 1975a; Anderson and May, 1978). Laboratory studies of infection dynamics (Anderson, Whitfield and Dobson, 1978) and field studies of hosts in different age classes (Halvorsen and Anderson, 1984), indicate the levels of aggregation tend to vary dynamically through time or with host age (Fig. 1). Increasing evidence suggests that small genetic differences in susceptibility of hosts to infection may be a prime component in producing these distributions (Schad and Anderson, 1985; Wakelin 1984, 1985). Similarly, differences in feeding rates, body size or differential use of different areas of habitat also tend to increase levels of parasite aggregation. In contrast, increased parasite pathogenicity or other forms of density-dependent increases in parasite mortality will tend to reduce levels of aggregation (Anderson & Gordon, 1983). The negative binomial distribution has been widely used to describe the distributions of parasites in their host populations in empirical studies, and the appropriate terms for this function may be substituted for the summation terms in equations 1 and 2 (Anderson, 1978a).

Although detailed models of parasite-host dynamics require consideration of the whole life cycle of the parasite, a lot of the fundamental dynamics of parasitic helminths may be understood by initially assuming that transmission stages operate on a much faster time scale than the processes which occur in the definitive host. If we assume that the free-living stage of the parasite is relatively short-lived ($w \gg 1$), with respect to the dynamics of hosts and adult parasites, then we may set equation 3 at equilibrium, with $W^* = \lambda P/(w + \beta H)$. Defining $H_0 = w/\beta$ gives a single transmission parameter whose magnitude varies inversely with the intensity of transmission. Substitution of the summation terms by the appropriate expressions for the negative binomial distribution and appropriate rearrangements give the basic two equation macroparasite-host model (Anderson and May, 1978)

$$dH/dt = (a-b) - \alpha P \tag{4}$$

$$dP/dt = \lambda PH/(H + H_0) - (u+b+\alpha) - \alpha(P/H)((k+1)/k). \tag{5}$$

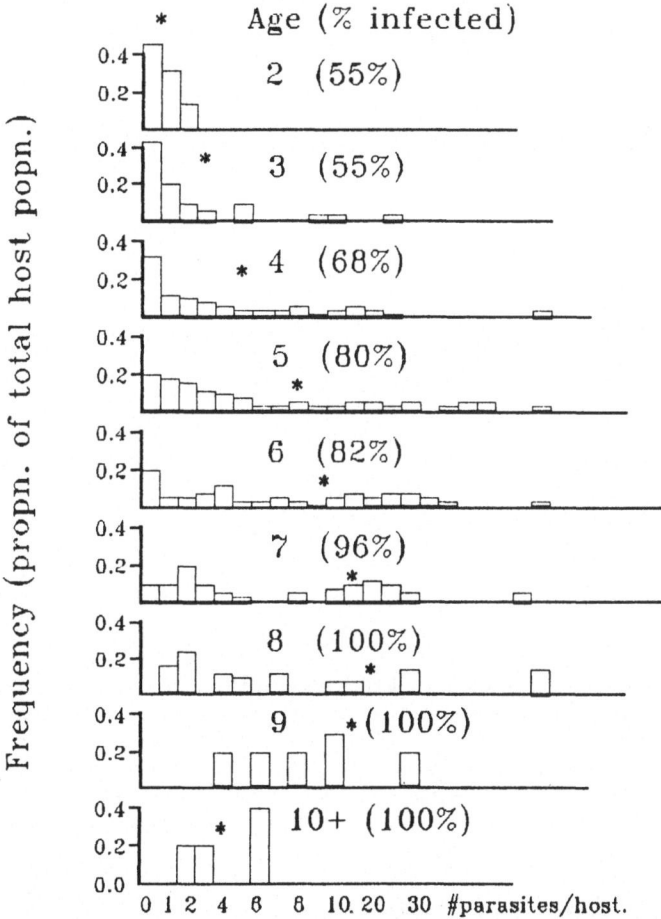

Fig. 1. The distribution of plerocercoids of the cestode *Diphyllobothrium ditremum* in arctic charr *Salvelinus alpinus* of different ages. The frequency distributions show the proportion in each frequency class, and the dots show the mean parasite burden at each age (after Halvorsen & Andersen, 1984). A more detailed analysis of this data appears in Pacala and Dobson (1987).

Although this 'fast transmission' assumption will be relaxed later, it allows transmission to be modelled in a phenomonological manner while concentrating upon the population dynamics interactions between the adult parasites and their definitive host. Thus, the more fundamental interactions at the core of all macroparasite-host relationships may be examined before the complications of different modes of transmission are explored in any detail. Most of the models discussed in the rest of this chapter are variations and extensions of this basic model for direct life cycle macroparasites whose dynamic properties are described in Anderson and May (1978) and May and Anderson (1978). In its basic form it exhibits patterns of dynamic behavior that are stable, providing that the parasites are aggregated in their distribution ($k < 1$) and levels of parasite induced host mortality are sufficiently high for the parasite to regulate the host (Fig. 2).

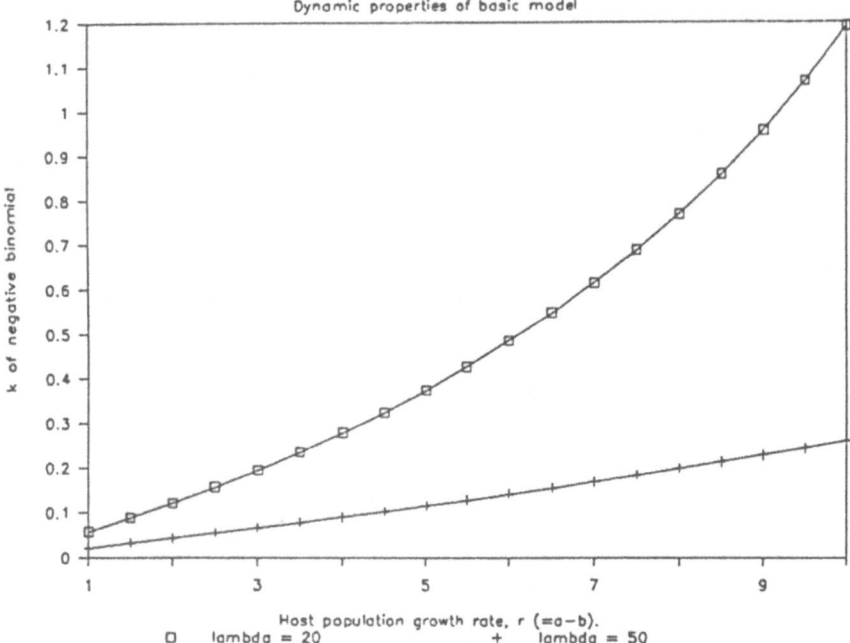

Fig. 2. Stability properties of the basic macroparasite model (Anderson and May, 1978). The boundary between the region where the parasite is able to regulate host population growth and the region where the host population grows exponentially (at a rate $a - b - \alpha$), is plotted for two values of λ. The parasite is unable to regulate the host for combinations of r and k beneath the line. Other parameters are $\alpha = 0.1$, $b = 0.5$, and $u = 1.0$. In the region above the line the parasites and hosts return asymptotically to equilibrium when perturbed. Increasing values of k lead to oscillations in the abundance of parasite and host.

The Basic Reproductive Rate of the Parasite, R_0

The basic reproductive rate of a parasite, R_0, may be defined as the total number of offspring produced by each parasite that survive to infect a host individual in the absence of density dependent constraints on fecundity and survival (Anderson, 1985). An expression for R_0 may be obtained by considering the growth rate of the parasite when first introduced into a population of uninfected hosts,

$$R_0 = \frac{\lambda H}{(H + H_0)(u + b + \alpha)} = T_1/M_1 M_2. \tag{6}$$

Setting the total transmission rate, $T_1 = \lambda \beta H$, and the net mortality rates of the adult and free-living stages of the parasite as $M_1 = (u + b + \alpha)$ and $M_2 = \beta H + w$, respectively, allows us to see that R_0 essentially equals the product of the transmission rate and the life expectancies of the two stages in the life cycle. As the parasite is only able to establish when R_0 exceeds unity, we can use equation (6) to determine the threshold number of hosts required to sustain a parasite population,

$$H_T = w M_1/\beta(\lambda - M_1). \tag{7}$$

When parasite induced rates of host mortality are low, or when egg production rates and transmission are high, the parasites are able to establish in relatively small host populations.

The rest of this chapter will illustrate how the addition of more detail allows the basic model to be adopted for a variety of different types of macroparasites. In each case more information is added by either reinstating equation 3 or using other additional equations which describe the dynamics of the parasite's free-living stages, or of the parasite in its intermediate hosts.

Case Study *Trichostrongylus tenuis* in Red Grouse

The basic model has been used in its most general form in a study of the influence of the parasitic nematode *Trichostrongylus tenuis* on the red grouse *Lagopus lagopus scoticus* (Hudson, Dobson and Newborn, 1985; Hudson and Dobson, 1988). This bird is an important game species in the North of England and Scotland (Hudson and Watson, 1985). The parasite lives in the caecae of infected hosts; the birds are infected within the first few weeks of life, and parasite burdens build up rapidly. Manipulative field experiments indicate that the parasite's influence on host fecundity is more significant than its impact on host survival (Hudson, 1986a, b). This is an important destabilizing mechanism in the life cycle and tends to produce oscillations in the numbers of both parasites and hosts (May and Anderson, 1978; Hudson et al., 1985).

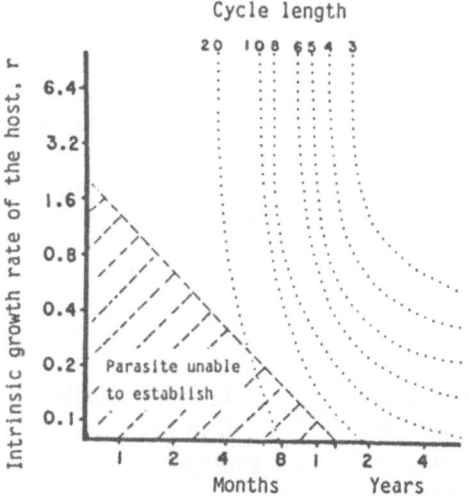

Fig. 3. Dynamic behaviour for a model of the grouse nematode system at different values of host population growth rate and life expectancy of the parasite free-living stage. In the shaded area at the bootom left portion of the graph, the parasite is unable to establish within the host population. Everywhere else on this plane the populations will exhibit stable limit cycles of abundance. The dotted contour lines give the expected cycle lengths for any combination of parameter values (after Hudson and Dobson, 1988).

Inclusion of a density-dependent term that regulates the grouse in the absence of the parasites and consideration of the dynamics of the free living stages (which in this system *are* potentially as long-lived as either the adult parasites or the hosts) gives rise to a model which exhibits stable limit cycles of host and parasite abundance for a wide range of parameter values (Fig. 3). The close correspondence between these cycles and those observed in field studies of the parasite and its host suggest that the parasite may be responsible for the patterns of oscillatory abundance exhibited by many grouse populations in the North of England (Potts, Tapper & Hudson, 1984).

The Dynamics of the Infection Process

More detailed models have been developed for the dynamics of the infection process in a laboratory study of an external parasite of fish, *Ichthyophthirius multifiliis* (McCallum 1982, 1986; McCallum and Anderson, 1984). The dynamics of this protozoan species, 'white spot' disease, are sufficiently similar to those of many parasitic helminths that it bears 'honorary' macroparasite status. In an experimental study of the dynamics of infection of the fish hosts by the free-swimming 'tomites' (infective stages), McCallum (1982) illustrated a linear relationship between the number of tomites the host was exposed to and the number of parasites that successfully establish. There was however a high degree of variability in the number of parasites established on hosts subjected to the same number of infective stages. The variance of parasite burden per host increased approximately linearly with the square of mean parasite burden. McCallum (1982) suggests that this is most probably a consequence of differences in the proportion of tomites infecting a host that survive to form trophozoites (adults). Assuming that the number of tomites encountering a host is a Poisson variate, it can be shown that, if a mean number, T, of tomites encounter each host and if the proportion, s, of these that survive to form trophozoites varies between hosts, the expected number of parasites establishing per host, E(P). and the variance in number of parasites, var(P), are given by (McCallum, 1982):

$$E(P) = T\bar{s}, \tag{8}$$

$$\text{var}(P) = T^2 \text{var}(s) + T\bar{s}. \tag{9}$$

Here $\bar{s}$ is the mean proportion of tomites surviving on the host population, and var(s) is the variance of this proportion between hosts. If this distribution is to be negative binomial in form, the parameter, k, of the probability model that inversely measures the degree of aggregation is:

$$k = (\bar{s})^2/[\text{var}(s)]. \tag{10}$$

The parameter k should therefore be independent of the mean parasite burden, which provides some justification for the assumption made in the basic model (eqns. 4 and 5) that the parasite distribution may be described by a negative binomial with constant k. Although it should be borne in mind that in free-running infections, the distribution will become distorted by both the deaths of heavily infected individuals and heterogeneities in the distribution of infective stages (Anderson and Gordon, 1982; Keymer and Anderson, 1979).

Regulation of Direct Life Cycle Macroparasites

A variety of density dependent mechanisms can act to regulate parasite abundance once the parasite has successfully located the definitive host (Keymer, 1982a). Models that consider both the dynamics of infection and density dependent reductions in establishment and survival in further detail have been developed to examine ways of controlling the parasitic nematodes that live in the alimentary tracts of domestic cattle (Smith and Grenfell, 1985; Smith, Grenfell and Anderson, 1987a, b, c; Grenfell, Smith and Anderson, 1986, 1987a, b). The majority of these papers have concentrated upon *Ostertagia ostertagi*, one of the most prevalent and harmful parasites of cattle in the temperature world.

Case Study *Ostertagia ostertagia* in Cattle

The models used in this study consist of a set of first-order differential equations which model the change in abundance of the parasite from the time when calves are first released into a pasture containing infective larvae. A reanalysis of the very detailed trickle infection experiments carried out by Michel (1970) suggested that the parasite is regulated by the host's immune response in three different ways:

(a) A reduction in the *per capita* fecundity of the female worms.
(b) A reduction in the rate of establishment of the ingested L_3 larvae.
(c) An increase in the death rate of mature worms.

The population dynamic consequences of density-dependent reductions in the fecundity of adult female helminths have been discussed in detail by Anderson (1982). Berding (1986) shows that the presence of such a function is a sufficient condition to provide a unqiue stable equilibrium in any model for a direct life-cycle helminth. The two other forms of regulation may be more significant in the dynamics of *Ostertagia*, where long term experiments suggest that the calves' abomasal mucosa (stomachlining) may eventually become refractory to the infective larvae. This leads to an exponential decline in the rate of establishment for the duration of exposure. Complementarily, the mortality rate of the mature and immature 5th stage worms ($\simeq$ adults) is an increasing function of initial infection density (Anderson and Michel, 1977; Grenfell, Smith and Anderson, 1987a).

The detailed nature of the model for the entire *Ostertagia* life cycle requires that it be solved numerically. However, it has been developed in sufficiently general a way to permit its use as a management tool for cattle stocked at different densities under a wide range of different climatological conditions (Smith and Grenfell, 1985). The model has proved particularly useful in determining the response of the parasite to a variety of drugs administered under various regimes of treatment (Fig. 4). Further developments of these models may include the economic costs of

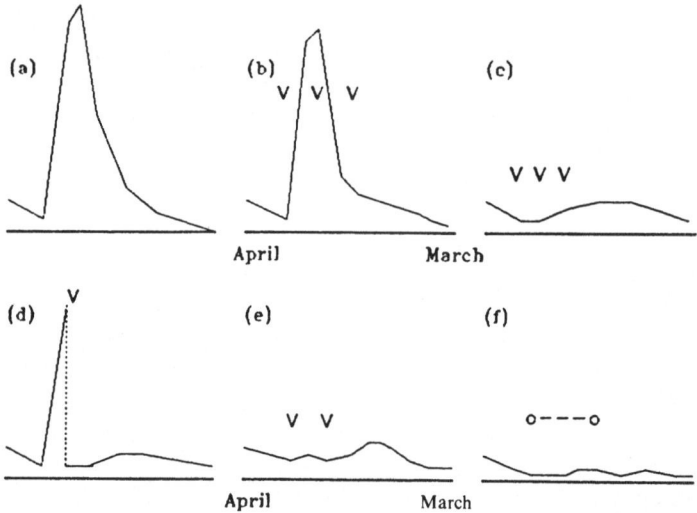

Fig. 4. Examples of the use of the model for Ostertagia to differentiate between different methods of controlling the parasite using anthelmintic drugs. The examples given here compare the impact of three hypothetical anthelmintics on the midsummer rise in $L3$ larval abundance.

Anthelmintic A—half-life less than 24 h, effective against all parasite stages.
Anthelmintic B—half-life around 4 days, effective against all parasite stages.
Anthelmintic C—half-life less than 24 h, only effective against fifth stage worms.

The period shown in each graph runs from April of one year to the end of March in the next; calves are turned out in May and housed in November, under climatic conditions typical of SE England (from Smith and Grenfell, 1985). The protocols tested are:

(a) Control (no anthelmintic therapy).
(b) Therapeutic administration of A at the time of the mid-summer rise.
(c) Prophylactic administration of A, three, six and nine weeks after turnout.
(d) Movement of the calves to a second pasture immediately following a prophylatctic does of A.
(e) Prophylactic administration of B three and eight weeks after turnout.
(f) Prophylactic administration of C via an intraruminal sustained release device. Effective quantities of anthelmintic are released for 80–100 days.

different treatments and potential effects of the development of resistance to the chemotherapeutic agent by the parasite.

The regulatory mechanisms adduced for *O. ostertagia* in cattle are found to provide a satisfactory model for the regulation of *Haemonchus contortus* populations (Barger et al., 1984; Smith (1988)) and may, with some qualifications, apply to most of the important trichostrongylid nematode parasites of sheep (Smith and Galligan 1988). However, neither the *Ostertagia* model, nor the sheep nematode models entirely account for the dynamical behaviour of the parasite populations they portray. For example, none of these models deals in a very satisfactory way with the phenomenon of arrested development (hypobiosis), a mechanism by which the parasite appears to avoid the adverse consequences of the host's immune response and the inimical environmental conditions that usually prevail during those months when most arrested larvae are found. Nor, in the case of *H. contortus*, is it yet possible to represent the dramatic manifestations of 'self-curve' during which the entire established worm burden is expelled over a matter of days.

Population Dynamic Consequences of the Immune Response

In the majority of parasitic helminths, the immune responses that produce density dependent reductions in parasite establishment and survival tend to be characterized by a marked dependence on the degree of antigenic stimulation (the host's accumulated experience of infection over time), an ability to provide only partial protection against reinfection, and a transient efficacy (Anderson and May 1985a; Mitchell, 1979). The population dynamic consequences of this have been explored in a number of studies of macroparasites in laboratory mice.

Immune Responses and Parasite Dynamics Within Individual Hosts

Studies on *Schistosoma mansoni* in laboratory mice indicate that the level of immune response produced is proportional to the rate at which the host acquires parasites (Anderson and Crombie 1985; Crombie and Anderson, 1985). In the absence of acquired resistance to infection and under conditions of constant exposure to infective stages, average parasite burdens per host would be expected to rise monotonically as the duration of exposure increases. If the mean worm burden per host at time t is defined as $M(t)$ and u and Λ are the *per capita* parasite death rate and host infection rate, then changes in $M(t)$ with time (or equivalently host age) may be defined by the simple differential equation

$$dM(t)/dt = \Lambda - uM(t). \tag{11}$$

If the hosts are uninfected at time (age) zero ($M(0) = 0$), then the age prevalence curve will be given by

$$M(t) = (\Lambda/u)[1 - e^{-ut}]. \tag{12}$$

The mean parasite burden will thus equilibrate at Λ/u as the hosts get older. However, current understanding of acquired immunity to *S. mansoni* infection in mice suggests that the host's immune response primarily acts to decrease the rate of establishment of adult parasites (Dean, 1983). Anderson and May (1985a) suggest that this effect may be modeled by assuming that parasite establishment decays linearly as the accumulated past experience of infection increases. The per capita average input of parasites $\bar{\Lambda}$ now takes the form

$$\bar{\Lambda}(M(a)) = \Lambda[1 - \varepsilon \int_0^a M(a')e^{-(a-a')}da']. \tag{13}$$

Here, Λ denotes the initial per capita rate of infection in naive hosts. This rate is decreased in a linear manner as the accumulated sum of past experience of infection

($\int_0^a M(a')da'$). Acquired immunity is assumed to have a duration of $1/\sigma$ units of time such that the summed experience of infection at age a' is decreased by a factor $\exp[-\sigma(a-a')]$ by age a. The parameter ε records the strength of acquisition of immunity. This model assumes that acquired immunity acts to decrease establishment as a function of past worm burdens, but may be readily modified to consider factors such as past exposure to infection (Anderson and Crombie, 1985). Note that the linear nature of this assumption is only valid when ε is small. If the rate of acquisition of immunity is high, this assumption would ultimately result in a negative input of parasites! However, it greatly facilitates analytical investigations of the properties of the acquired immunity model.

The full model now takes the form (Anderson and May, 1985a)

$$dM(a)/dt = \bar{\Lambda}[M(a)] - uM(a) \tag{14}$$

$$M(a) = D + \frac{\Lambda}{\lambda}\left(\frac{(p_1+\sigma)}{p_i}e^{p_1 a} - \frac{(p_2+\sigma)}{p^2}e^{p_2 a}\right). \tag{15}$$

Here $D = \Lambda\sigma/(\sigma u + \varepsilon\Lambda)$, $p_1 = -(u+\sigma-\lambda)/2$, $p_2 = -(u+\sigma+\lambda)/2$, and $\lambda = [(u-\sigma)^2 - 4\varepsilon\Lambda]^{1/2}$. The models may produce a wide range of dynamic behaviour (Fig. 5), ranging from monotonic growth to a stable average worm burden in older age classes (if Λ and ε are small and σ large), from convex patterns with average worm burden declining in older age classes (Λ large and σ small), to damped oscillations

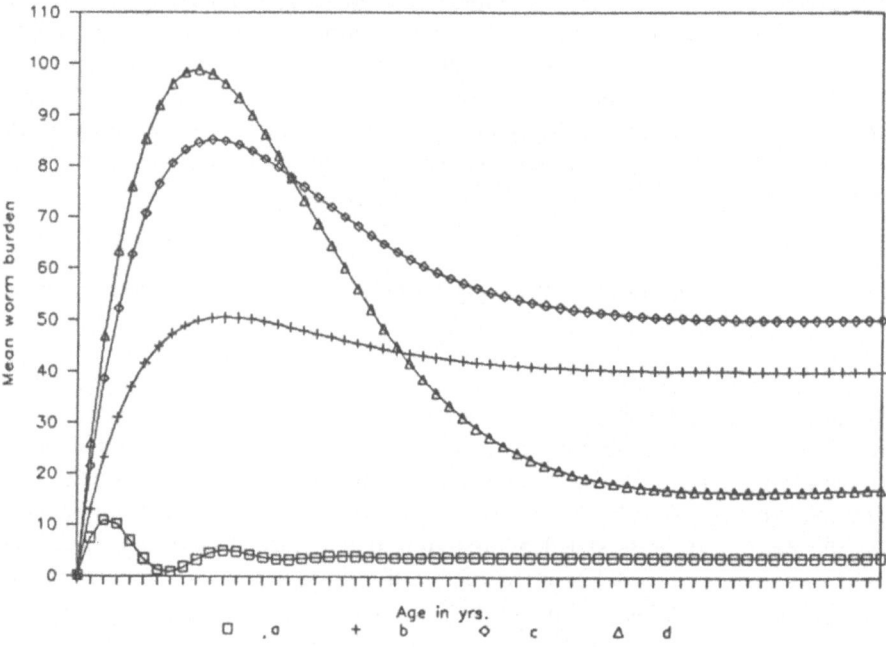

Fig. 5. Illustration of the various patterns of change in average worm burden with age generated by the acquired immunity model defined by Eqn. 15 in the main text. Parameter values: bottom line at age 10 years, $\Lambda = 10$, $u = 0.2$, $\varepsilon = 0.05$, $\sigma = 0.2$; second bottom line at age 10 years $\Lambda = 10$, $u = 0.2$, $\varepsilon = 0.0005$, $\sigma = 0.1$; second top line at age 10 years $\Lambda = 20$, $u = 0.2$, $\varepsilon = 0.0005$, $\sigma = 0.05$; top line at age 10 years, $\Lambda = 30$, $u = 0.2$, $\varepsilon = 0.0005$, $\sigma = 0.01$ (all per year) [after Anderson and Crombie, 1985].

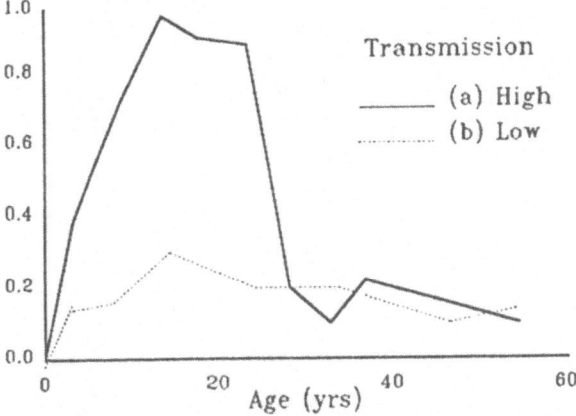

Fig. 6. Observed age-average intensity of infection curves for *S. mansoni* in areas of high (top line at age 20 years, data from Abdel–Waheb *et al*, 1980) and low (bottom line at age 20 years, data from Siongok *et al*, 1976) transmission intensity. The y-axis is scaled in units of mean intensity of infection (eggs per gram of faeces × 1000). Redrawn from Anderson and Crombie (1985).

in intensity of infection as age increases (Λ, σ and ε large). Of particular importance is the ability of the model to produce asymptotic growth to a stable plateau and convex curves with variation in only one parameter, the initial force of infection Λ. These patterns seem to correspond to the paterns observed in sets of laboratory and field data (Fig. 6), suggesting that acquired immunity may be of major importance in restricting the size of worm burdens in older age classes. The public health significance of this type of models is discussed further in Anderson and May (1985a, b) and Anderson and Crombie (1985).

Nutrition and Density Dependence

The nutritional state of the host is a further important component of its ability to mount an efficient immune response (Keymer, Crompton and Walters, 1983; Keusch 1985). In studies of laboratory rats maintained under different dietary regimes, Slater and Keymer (1986, 1987) demonstrated significant impairment in the ability of malnourished hosts to control the rates of establishment of the parasitic nematode *Heligmosomoides polygyrus* (Fig. 7a, b). Considerable variation was also observed in the fecundity of the worms at different parasite burdens (Fig. 7c).

The effects have been included into models of parasite host dynamics by Berding, Keymer, Murray and Slater (1987a, b). These models are similar to those described by Anderson and May (1985a) except that the host's immune response affects the survival of adult parasites rather than larval establishment. Interestingly, the hosts on a low nutrition diet are best described by a model of a relatively simple immigration-death process (similar to Eqs. 11 and 12), while the data for well-nourished hosts require consideration of a function for the host's acquired immunity to the parasite (similar in form to Eqs. 13, 14 and 15). Differences in the ability of well nourished and poorly nourished hosts to mount an immune response will obviously

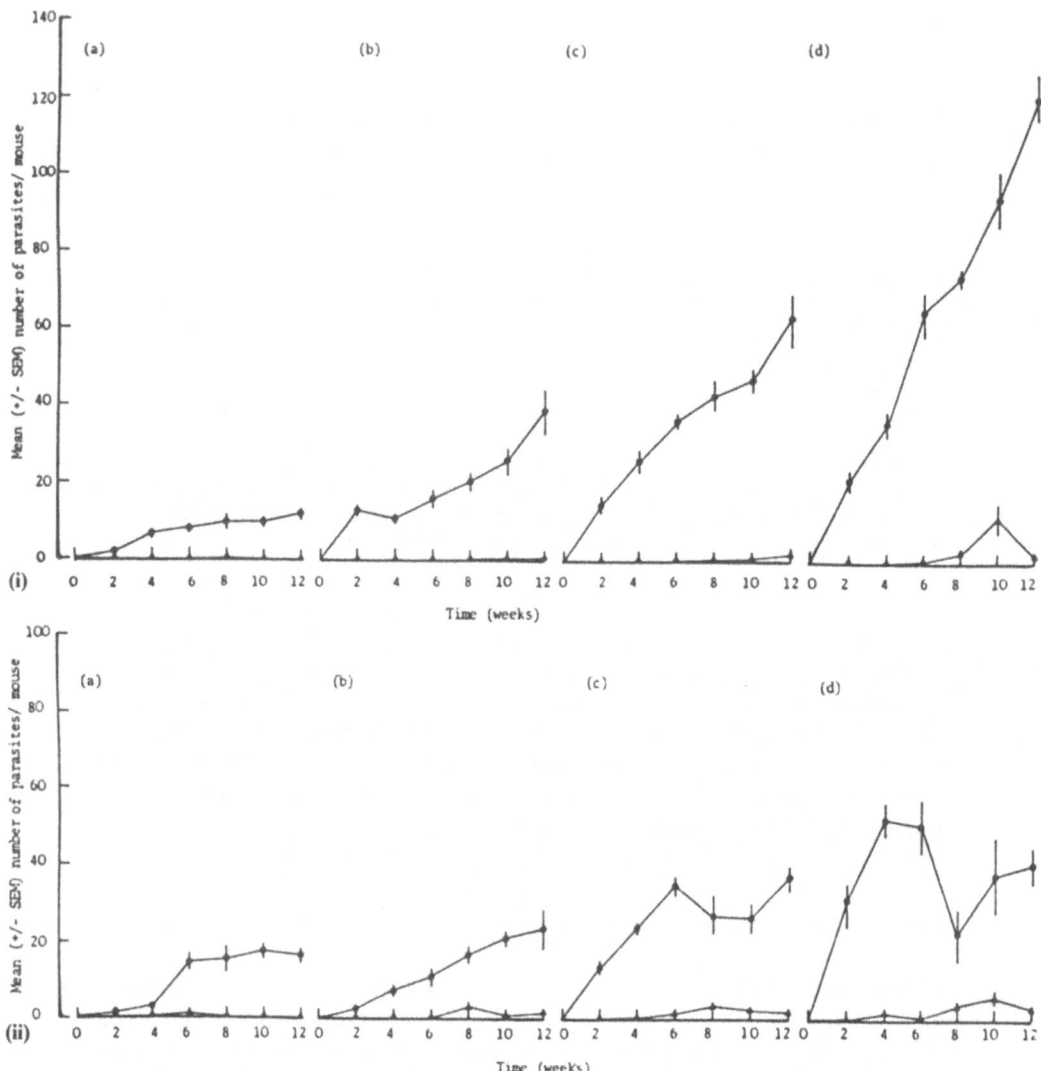

Fig. 7(i), (ii)

be important in attempts to control parasitic diseases. Further examination of the relative benefits of food, chemotherapy and vaccination is an area where models of this type can be invaluable in determining control strategies for these diseases in human populations.

Immune Responses and Fish Populations

The consequences of an immune response by the host have also been explored in two studies of external macroparasites on laboratory populations of fish (Scott and

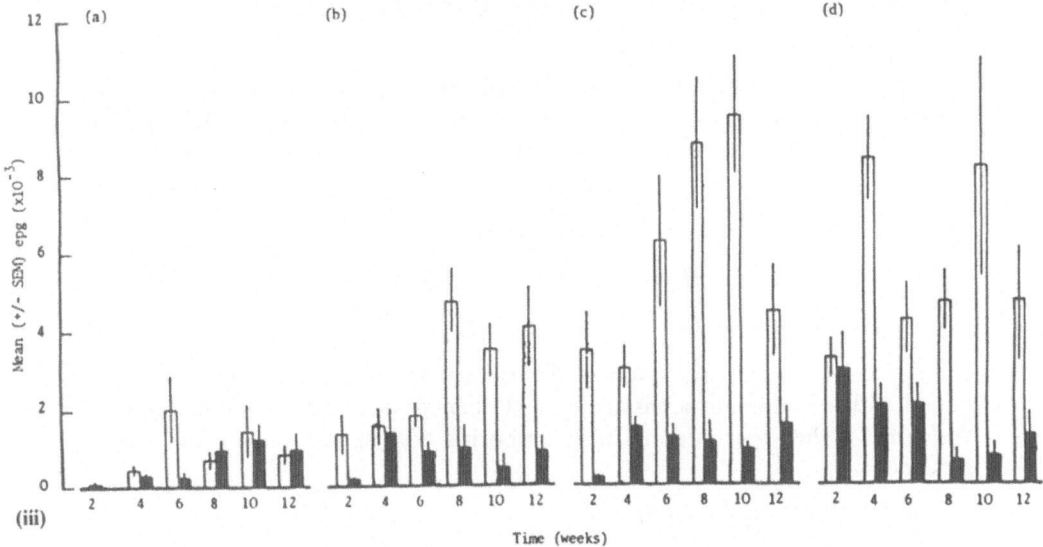

Fig. 7. Comparison of the effects of host nutrition on the establishment and fecundity of *Heligmosimoides polygyrus* in laboratory mice. **(i)** The recovery rate of larval (triangle) and adult (circle) worms from mice feeding on a 22% (low) protein diet. Mice were repeatedly infected with 5(a), 10(b), 20(c) and 40(d) larvae/2 weeks. **(ii)** Recovery of larval and adult worms over time from mice feeding on an 8% (high) protein diet (infection rates as for a). **(iii)** Change in egg output over time during the repeated infection experiments (open bars—2% protein diet; shaded bars—8% protein diet). Mice were repeatedly infected with 5(a), 10(b), 20(c) and 40(d) larvae/2 weeks. Figures redrawn from Slater and Keymer (1987).

Anderson, 1984; McCallum, and Anderson 1984; McCallum, 1986). Here the transient patterns of immunity produce more complex patterns of dynamic behavior in both the experimental data and the proposed models.

Genetic Basis for Heterogeneity in the Host's Response

Both studies of parasitic helminths in laboratory mice discussed above also consider variation in the responses of different individuals of the host populations to infection by the parasite (Anderson and Crombie 1985; Anderson and May 1985a; Berding, Keymer, Murray and Slater, 1987a, b). Considerable evidence is accumulating to suggest that these differences have a large genetic component. This variation is likely to be important in producing the aggregated distributions observed in most host parasite systems (Wakelin, 1984, 1985). Further exploration of the dynamic consequences of genetic variability between hosts will also be highly important in determining how to design 'targeted' or 'selective' chemotherapy programs in areas where parasitic helminths are endemic in both humans and domestic livestock.

Models for Competition Between Parasite Species

The basic model can be extended to consider the dynamics of a system where two different species of parasite utilize the same host species (Dobson, 1985). Here we need to include additional terms which take into consideration how the statistical distribution of the different parasite species effects their ability to interact.

$$dH/dt = (a - b)H - \alpha_1 H.E(i) - \alpha_2 H.E(j) \tag{16}$$

$$dP_1/dt = \lambda P_1 H/(H + H_0) - P_1(u + b) - H[\alpha_1 E(i^2) + \alpha_2 E(i.j)] \tag{17}$$

$$dP_2/dt = \lambda P_2 H/(H + H_0) - P_2(u + b) - H[\alpha_2 E(i^2) + \alpha_1 E(i.j)]. \tag{18}$$

The expectation terms will again be determined by the nature of the underlying statistical distribution, which is again most likely to be negative binomial in form. The term for the expectation of hosts harboring both species of parasites may be expanded:

$$E(i.j) = E(i).E(j) + \text{cov}(i.j). \tag{19}$$

In the simplest case we may assume that the covariance term equals zero; the joint expectation is thus equal to the product of the two individual expectations. This case corresponds to exploitation competition, or joint use of the same resources by the hosts (Miller, 1967). Negative antagonistic interactions between the parasites would correspond to 'interference' competition and would necessitate consideration of the covariance term, which may be expanded to give

$$\text{cov}(i.j) = r_{i.j}.s_i.s_j. \tag{20}$$

Interactions between the two parasite species may be positive or negative depending on the biological details of the transmission process and the strength of the interaction between the two species in concomitantly infected hosts. Where both species are heteroxenic and are transmitted to the definitive host by ingestion of the same species of intermediate host, then $r_{i.j}$ may be positive. When interactions between the two species are strong, $r_{i.j}$ may be negative (Dobson, 1985).

Population Dynamics of Competition Between Parasites

The equations may be simplified by setting the host population size constant and examining the dynamics of the two parasites. The phase planes for the interaction then resemble those for competititon between two free-living species, although here the slopes and intersects of the zero-growth isoclines are determined by the pathogenicity and degree of aggregation of the two parasite species (Fig. 8). Essentially, the more aggregated the two species, the more likely they are to coexist in the same host population. Similar results have been obtained from models of insects that utilize plants (Atkinson and Shorrocks, 1981; Ives and May 1985). When the dynamics of the host population are considered in more detail, persistance of the two parasites and their host requires one of the following inequalities to

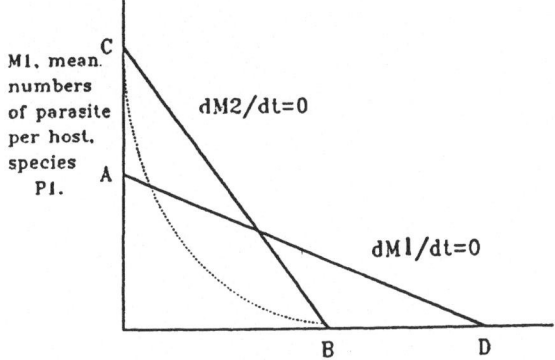

M2, mean numbers of parasite/host, species P2

Fig. 8. Phase plane analysis illustrating the stability properties of the basic model for competition between two parasite species in a host population of constant size. The solid line illustrates the case for exploitation competition, while the dashed line illustrates the case for asymmetric interference competition (e.g. Species 1 has an antagonistic effect on species 2, but this is not reciprocated). The slopes of the two zero growth isoclines are determined by the ratios of the rates of resource utilization (α_1 & α_2) and the degree of aggregation of each species. The intersections of each isocline with its own axis, A and B, corresponds to the equilibrium mean parasite burden of each species in the absence of a competitor. These intersections have to occur inside the other intersections, C and D, if the equilibrium at their joint intersections is to be stable. This only occurs when both parasites are aggregated in their distributions (after Dobson, 1985).

be met:

$$\text{if } \alpha_2 > \alpha_1 \quad \text{then} \quad r/k_1 > (\alpha_2 - \alpha_1) \tag{21}$$

$$\text{if } \alpha_1 > \alpha_2 \quad \text{then} \quad r/k_2 > (\alpha_1 - \alpha_2). \tag{22}$$

Thus a three species equilibrium is dependent upon the the host's intrinsic growth rate, r, the relative pathogenicities, α_1, and the degrees of aggregation of the two parasite species, k_1 (Dobson 1985). Further increases in host population growth rate may allow further species of parasite to establish (Fig. 9), while increasing degrees of aggregation in the distribution of each parasite species reduces to the actual level of interspecific competition between different parasite species.

These models can be used to quantitatively assess the potential for using competing parasite species as biological control agents for pathogens of economic or medical importance. In most cases potential competitors would be introduced into the intermediate hosts of parasites with heteroxenic life cycles (Lim and Heyneman, 1972). The most important criterion for identifying a successful control agent is an ability to infect a high proportion of the host population. When combined with intermediate levels of pathogenicity and high rates of transmission, such a competitor is likely to be quite effective (Dobson, 1985). In contrast, antagonistic interactions between the parasites species contribute only secondarily to the success of the control, as only a small proportion of the hosts will be infected with both species of parasite.

dM1/dt = 0

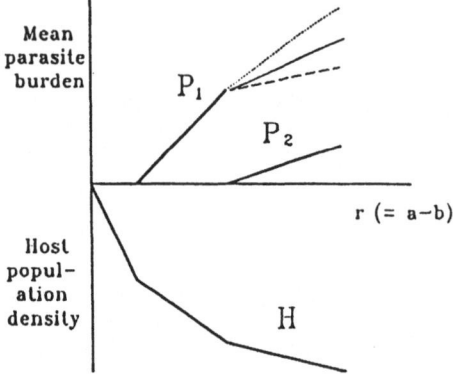

Fig. 9. The effect of the growth rate of the host population on the population densities of two potentially competing species of macroparasite. In the absence of the parasites the hosts are regulated by other density dependent factors to a carrying capacity r/Δ (note that the inverse of host population density is plotted on the lower axis). The two parasite species are assumed to have different pathogenicities with $\alpha_1 < \alpha_2$. Species P_1 cannot establish until $r (= a - b) > \alpha_1$; species P_2 cannot establish until $r > \alpha_2$, or, in the presence of species P_1, until $r > \alpha_1 + \alpha_2$. Both parasites cause a reduction in the host population density. The impact of the two parasites species on each other depends on both their pathogenicity and the sign and magnitude of the covariance term. When the covariance term is positive (dotted line), the second species causes less of a reduction in the mean burden of species P_1, then when the covariance equals zero (thin solid line) or is negative (dashed line). (after Dobson, in prep).

Models for Heteroxenic Macroparasites

Many species of parasitic helminths have life cycles which utilize one or more species of intermediate host. In general, the regulatory processes in these life cycles are similar to those in direct life cycle macroparasites. Both the density dependent reduction in the establishment, survival and fecundity of adult worms and their statistical distribution in the definitive host remain of prime importance in determining recruitment to the parasite population. However, when transmission is more complex, the dynamics of the intermediate host population have to be considered, and developmental time delays may occur between successive 'generations' of adult worms. Two types of heteroxenic life cycle will be discussed here: initially the case is considered where transmission between successive hosts occurs via free-living stages; we then examine cases where transmission to the definitive host is mediated via a predator-prey relationship, with the intermediate host a prey item in the diet of the definitive host.

Reproductive Rates in Parasites with Complex Life Cycles

As always, in order to examine possible strategies for controlling the parasite or to quantitatively compare potential mechanisms by which heteroxenic life cycles might have evolved, it is useful to derive an expression for the basic reproductive rate of the pathogen, R_0 (Anderson 1985). If it is assumed that the definitive hosts are very long-lived, then the dynamics of the life cycle may be modeled using the following set of coupled differential equations:

$$\frac{dP_2}{dt} = \eta_2 H_2 C - \pi P_2 - \delta H_2 \left[\frac{P_2}{H_2} + \left\{ \frac{P_2}{H_2} \right\}^2 \frac{(k+1)}{k} \right] \tag{23}$$

$$dW/dt = \lambda P_2 - \gamma_1 W - \eta_1 W H_1 \tag{24}$$

$$\frac{dP_1}{dt} = \eta_1 W H_1 - (u + \varepsilon b) P_1 - \Delta H_1 \left[\frac{P_1}{dt} + \left(\frac{P_1}{H_1} \right)^2 \right] \tag{25}$$

$$dC/dt = \theta P_1 - \gamma_2 C - \eta_2 C H_2. \tag{26}$$

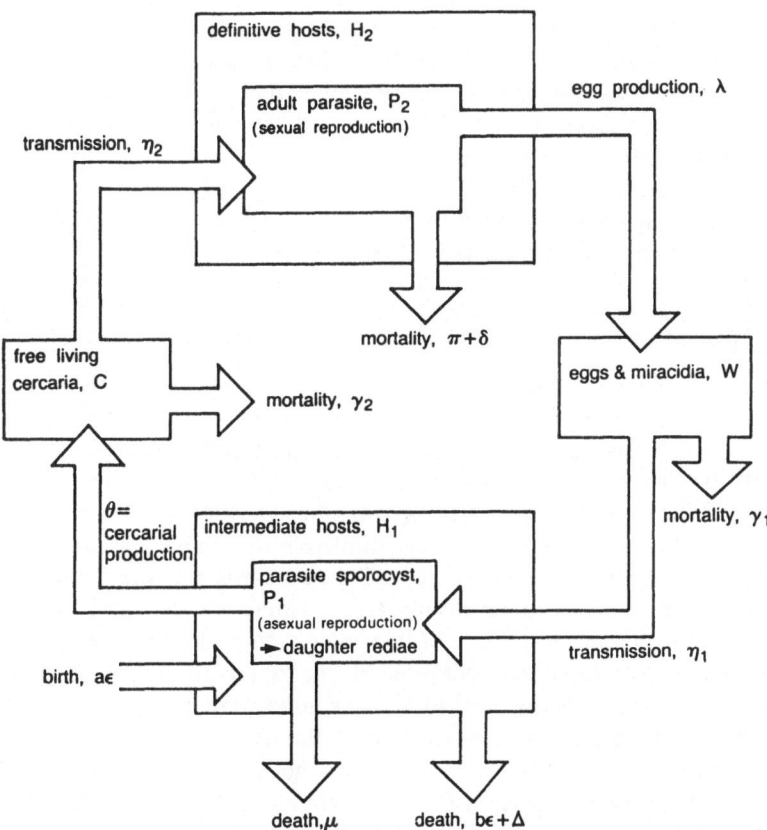

Fig. 10. Flow chart of the life cycle of a heteroxenic 'schistosome-like' parasite which utilizes two species free-living infective stages and reproduces asexually in its intermediate host. Additional parameters used in this model are Δ, the per capita rate of parasite induced mortality of the snail intermediate hosts and θ, the rate of asexual reproduction of the parasites in the snails. The two free-living stages, eggs/miracidia, W, and cercaria, C, have average life expectancies of $1/\gamma_1$ and $1/\gamma_2$ and are able to locate and infect the next host stage in the life-cycle at rates η_1 and η_2 respectively. The definitive host population, H_2, is assumed to remain constant with respect to all other stages in the life cycle.

The most characteristic trematode parasite with a life cycle of this type would be schistosomiasis, the parasite which causes bilharzia. The basic life cycle is illustrated as a flow chart in Fig. 10, where the parameters used in Eqs. 23–27 are also defined. Essentially, the adult parasites, P_2, in the definitive hosts, H_2, release eggs which develop into free-living stages (micracidia), W, that actively seek out the snail intermediate hosts, H_1. Here it is assumed that the parasites are randomly distributed in their snail hosts and that only a small proportion of these hosts are infected; more aggregated distributions are likely to be the case in field populations (Anderson and May, 1979b). If we further assume that in the initial stages of an outbreak the majority of infected snails are infected by only one parasite, then the mean burden in the intermediate hosts, n, is likely to be less than unity, and the terms in $1 - e^{-n}$ may be approximated by n. The parasites reproduce asexually in their intermediate hosts at a rate, θ; this leads to the production of a second type of free-living stage (cercaria), C, which complete transmission by actively searching for definitive hosts to infect. As discussed further below the asexual reproduction of the parasite within the snail intermediate host is often damaging and may result in reduced fecundity and survival of this host. In some cases initial reductions in intermediate host fecundity may be compensated for by increased longevity, thus ε causes decreases in both 'a' and 'b'.

The equations can again be collapsed to obtain an expression for R_0, the basic reproductive rate of the parasite (Anderson, 1985; Dobson, 1988):

$$R_0 = \frac{T_1 T_2}{M_1 M_2 M_3 M_4}. \tag{28}$$

Here $T_1 = \lambda\eta_1 H_1$, $T_2 = \theta\eta_2 H_2$, $M_1 = (\varepsilon b + u + \Delta)$, $M_2 = (\pi + \delta)$, $M_3 = (\eta_1 H_1 + \gamma_1)$ and $M_4 = (\eta_2 H_2 + \gamma_2)$. The expression is again the product of the transmission terms and the life expectancies of the parasite at each stage of the life cycle. When compared with the appropriate expression for a monoxenic life cycle (Eq. 6), it seems that indirect life cycles will be advantageous when the life expectancy of the parasite in the definitive host is short or when definitive host populations are small and intermediate hosts plentiful. Asexual reproduction of the parasite within its intermediate host is importance in increasing R_0, providing this does not lead to significant increases in the mortality rate of infected intermediate hosts, M_1.

The most obvious way to control parasites of this type is to reduce either of the two transmission rates. In many cases this can be done by reducing the contact between potential hosts and the aquatic or damp habitats where transmission is likely to take place. Increased mortality of either the adult parasites in their definitive hosts or the larval stages and the intermediate hosts by chemotherapy will also be important in reducing R_0. Although it should be borne in mind that unless control continues until complete eradicatation is achieved, the parasite may rapidly recover once chemotherapy is terminated.

Expressions may also be obtained for the threshold numbers of definitive and intermediate hosts required for a population of parasites of this type to become established.

$$H_{1T} = \frac{\gamma_1 M_1 M_2 M_4}{\eta_1 (\lambda T_2 - M_1 M_2 M_4)}, \tag{29}$$

$$H_{2T} = \frac{\gamma_2 M_1 M_2 M_3}{\theta \eta_2 (\lambda T_2 - M_1 M_2 M_3)}. \tag{30}$$

Here H_{1T} is the threshold number of snail intermediate hosts, and H_{2T} is the threshold number of definitive hosts. In both cases reductions in parasite pathogenicity are important in reducing the thresholds for establishment. Asexual reproduction by the parasite in its intermediate host is also important in reducing the numbers of definitive hosts required to sustain an infection.

Infection Dynamics of Cercaria and Miracidia

Although reducing rates of transmission may seem an efficient method of controlling parasites with this type of life cycle, these parameters are always the hardest to estimate for any pathogen. In heteroxenic species with several different transmission stages these problems increase significantly. Nevertheless, a combination of detailed laboratory and field studies have been used to develop a number of models which deal specifically with the dynamics of transmission (Anderson, 1978b; Anderson, Whitfield, Dobson and Keymer, 1978; Anderson, Mercer, Wilson and Carter, 1982; Anderson and Crombie, 1984; Carter, Anderson and Wilson, 1982; Smith and Grenfell, 1984). These have considerably improved our ability to determine the relative importance of the different factors which determine transmission. The majority of work on the population dynamics of parasites with two free-living infective stages has concentrated upon schistosomiasis and fascioliasis (Anderson and May, 1979b; Barbour, 1978; Bradley and May, 1978; Cohen, 1976, 1977; May, 1977, Nasell, 1976a, b; Nasell and Hirsch, 1973). Macdonald (1965) was the first to use such models to study the transmission dynamics of schistosomiasis.

Proportion of Hosts Infected for Different Values of k

Laboratory studies of the dynamics of infection of snails by free-living miracidia stages in digenean life cycles again emphasize the importance of heterogeneity in the hosts susceptibility in determining observed levels of prevalence (Anderson, 1978b). Heterogeneity in the susceptibility of snails to infection may be generated by differences in size or age, rates of mucus production or genetic differences between hosts. All of this variation will tend to produce aggregated distributions of the parasites in the intermediate host population. The percentage of hosts infected through time when hosts are exposed to different densities of miracidia has been examined in simple laboratory experiments designed to examine the dynamics of transmission. Anderson (1978b) shows that if we assume that distribution of infections to be random (or Poisson) then $I(t)$, the prevalence of infection is given by

$$I(t) = 100[1 - \exp(-M_0(1 - \exp(-\alpha t)))]. \tag{31}$$

Here M_0 is the number of miracidial infective stages to which the hosts are exposed, and t is the duration of exposure. In contrast, if the parasites are aggregated in

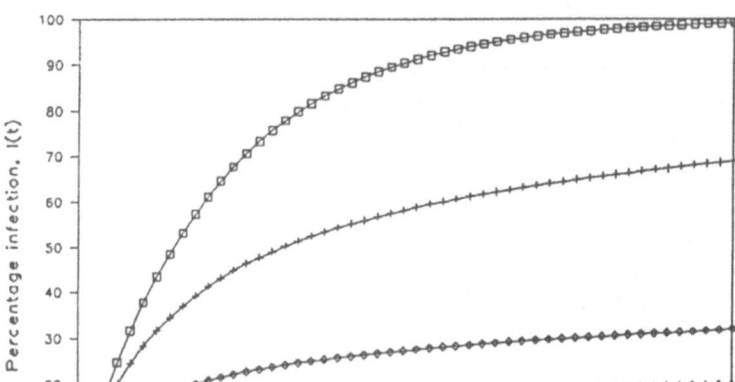

Fig. 11. The relationship between the percentage infection $I(t)$, and miracidial exposure density, M_0, for various categories of dispersion of miracidial infections per snail (after Anderson, 1978). (a) Poisson model (Eq. 31); (b) negative binomial (Eq. 32; $k = 0.5$); (c) negative binomial ($k = 0.1$); (d) negative binomial ($k = 0.05$) ($\alpha = 0.05$, $t = 2$).

their distribution, then the proportion of hosts infected is given by

$$I(t) = 100[1 - (k/[M_0(1 - \exp(-\alpha t))])^k].\tag{32}$$

Here k is again a parameter of the negative binomial distribution which varies inversely with the degree of aggregation. As the parasites become more aggregated in their distribution, or the snails become more heterogeneous in their susceptibility to infection, a lower proportion of hosts are infected (Fig. 11). Experimental corroboration of this model was provided by an analysis of the transmission dynamics of *F. hepatica* miracidia (Smith, 1987).

Effect of Host Death Due to Parasite Pathogenicity

In a survey of empirical data on snail infection by schistosome parasites, Anderson and May (1979) develop some models to illustrate the relative importance of the duration of the latent period of infection and the mortality rates of infected and uninfected snails in determining the patterns of prevalence of infection in snail populations. In many cases the life expectancy of infected snails declines due to the influence of the parasite and may be less than the latent period of infection. This leads to still lower levels of parasite prevalence in the snail population. Anderson and Crombie (1984) develop this study further and illustrate how variation in the rates at which snails are exposed to infection and age dependent variation in susceptibility tend to produce different patterns of age-prevalence in snail populations.

Temperature Dependence of Transmission

Studies on both schistosomiasis (Anderson et al., 1982; Carter et al., 1982) and fascioliasis (Smith and Grenfell, 1984; Wilson, Smith and Thomas, 1982) have examined the dependence of the survival rate of parasite transmission stages on temperature. In all cases some optimum combination of temperature and humidity conditions maximizes transmission and hence R_0. In the fascioliasis case this data has been used to determine how meteorological conditions determine outbreaks and to develop models for predicting when to treat hosts in the event of an imminent outbreak (Wilson et al., 1982).

Control of Fascioliasis by Chemotherapy

In a series of papers on *Fasciola hepatica*, Smith (1982, 1984a, b, c, d) illustrates how the model developed by Wilson et al. (1982), a more specific case of Eqs. 23–27, may be used to predict both the age-structure of the adult parasites in the sheep hosts and to examine the relative merits of different types of control strategies for the parasite. This approach was necessary since most flukicides (drugs to treat the parasite) are not equally effective over the whole range of age-classes of the parasite. It follows that the density-dependent and density-independent factors which govern the age distribution curve of the parasite population within individual hosts ultimately determine the efficiency of chemotherapy.

Parasites and Predator Prey Relationships

Similar approaches to those discussed above to calculate R_0, may be adopted when considering life cycles where transmission of the parasites to the definitive host occurs via a predator-prey relationship (Anderson, 1985). In many life cycles of this type the parasite alters the intermediate hosts behavior and increases its susceptibility to predation (Dobson and Keymer, 1986; Dobson, 1988). This type of life cycle may be modeled by a modified set of equations similar to Eqs. 23–27 (Fig. 12).

$$\frac{dP_2}{dt} = \alpha\rho H_2 H_1\left[\frac{P_1}{H_1} + \left(\frac{P_1}{H_1}\right)^2\right] - \pi P_2 - \delta H_2\left[\frac{P_2}{H_2} + \left(\frac{P_2}{H_2}\right)^2\left\{\frac{k+1}{k}\right\}\right] \tag{33}$$

$$dW/dt = \lambda P_2 - \gamma W - \eta W H_1 \tag{34}$$

$$\frac{dP_1}{dt} = \eta W H_1 - (u+b)P_1 - \alpha\rho H_2 H_1\left[\frac{P_1}{H_1} + \left(\frac{P_1}{H_1}\right)^2\right] \tag{35}$$

$$dH_1/dt = (a-b)H_1 - \rho H_1 H_2 - \alpha\rho P_1 H_2. \tag{36}$$

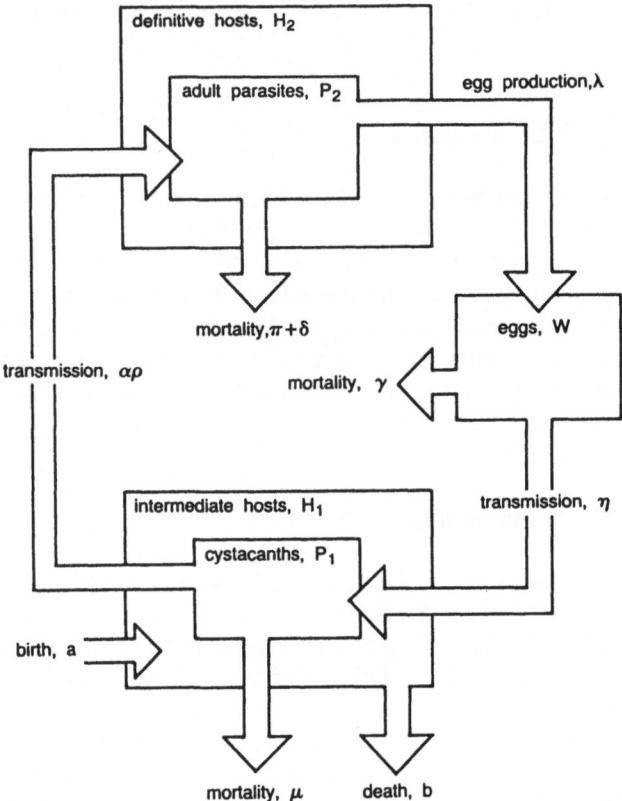

Fig. 12. Flow chart of the life cycle of a heteroxenic parasite that uses one free-living infective stage and is then transmitted to its definitive host via a predator-prey relationship.

Here we have assumed a simple linear functional response for the rate at which the definitive hosts (predators) attack the intermediate hosts (prey). Uninfected prey are attacked at a rate ρ, while the presence of a parasite increases this rate by a factor α, per parasite. The equations can again be used to calculate an expression for the basic reproductive rate of the parasite, R_0

$$R_0 = \frac{T_1 T_2}{M_1 M_2 M_3}. \tag{37}$$

Where $T_1 = \lambda \eta H_1$, $T_2 = \alpha \rho H_2$, $M_1 = (\pi + \delta)$, $M_2 = (b + u + \alpha \rho H_2)$ and $M_3 = (\eta H_1 + \gamma)$. Thus, R_0 is again the product of the net rates of transmission and the life expectancies of the three different life history stages of the parasite. Parasite induced changes in the host's susceptibility to predation tend to increase R_0. However, they are more effective in reducing the threshold numbers of hosts required to sustain a population of parasites:

$$H_{1T} = \frac{\gamma M_1 M_2}{\eta(\lambda T_2 - M_1 M_2)} = \frac{\gamma M_1 (b + u + \alpha \rho H_2)}{\eta(\lambda \alpha \rho H_2 - M_1 (b + u + \alpha \rho H_2))}, \tag{38}$$

$$H_{2T} = \frac{M_1 M_3 (b + u)}{\alpha \rho (T_1 - M_1 M_3)}. \tag{39}$$

This suggests that the mechanisms that parasite's use to change host behavior may have evolved as an adaptation to ephemeral definitive hosts that are either present at low population densities or only occasionally visit the habitat occupied by the intermediate hosts (Dobson, 1988). Control of parasites with this type of life cycle is considerably exasperated by the low numbers of definitive hosts required to sustain the parasite. Although it might be possible to treat infected definitive hosts with chemotherapy, parasites of this type are fairly non-specific towards their definitive hosts and often a relatively inconspicuous species of definitive host may act as a reservoir of infections for the host of medical or economic importance.

Cod Worm and Seal Culling

A particular example of this may be pertinent to the dynamics of the cod worm *Ascaris simplex*, an important parasite of commercial fish stocks in the North Atlantic. The parasite has a life cycle fairly similar to the one described by Eqs. 33–36, except that a second species of intermediate host is included in the life cycle. The parasite therefore uses marine mammals as definitive hosts, crustaceans as primary intermediate hosts and gadoid fish such as cod as second intermediate hosts. Transmission to both the fish and mammal hosts is via a predator prey relationship. If we assume the transmission through the crustacean stage to occur very rapidly, then we can concentrate on the dynamics of the parasite in the fish and mammal hosts. An expression may be readily obtained for the mean parasite burden in the fish hosts

$$P_1/H_1 = (a - b - \rho H_2)/\alpha \rho H_2. \tag{40}$$

This expression is interesting in that it suggests that reduction in the numbers of mammal definitive hosts, H_2, may lead to increases in the mean parasite burden in the fish host, P_1/H_1. Essentially, this occurs because heavily parasitized fish are more susceptible to predation and reducing the numbers of predators increases the life expectancy of infected fish in the absence of other density dependent constraints. Obviously the result is dependent upon the fairly gross simplifications of the model; however, it does suggest that in some cases culling of seals may not be a productive way of reducing the incidence of codworm in commerical fish stocks (Dobson and May, 1986).

Case Study Control of Echinococcosis and Cystercercosis

The most detailed study of a economically important parasites of this type has been undertaken for echinococcosis and cystercercosis (Gemmell, Lawson and Robert, 1986, 1987; Gemmell et al., 1986; Roberts, Lawson and Gemmell, 1986, 1987). These cestode parasites utilize vertebrates as both definitive and intermediate hosts and may occasionally infect man. Transmission of the parasites is maintained through a predator-prey relationship, where the carnivore (dogs, foxes) ingests the larval form when eating its prey (sheep-goats and cattle in *Echinococcus granulosus* and rodents in the case of *E. multilocularis*.) Human infections arise when man accidentally ingests the infective egg stages, and colossal *hydatid* cysts form due to asexual reproduction of the parasite. The major form of regulation in the life cycle appears to be the development of an immune response in the intermediate host. Examination of the dynamics of these two parasites suggests that *E. granulosus* is relatively easy to control as it is usually present as an endemic infection with a reproductive rate close to unity. The development of safe livestock slaughtering facilities and more rigorous quarantine procedures for infected sheep have helped reduce transmission. In contrast, the transmission of *E. multilocularis* is dependent on the density of small mammals and fluctuations in the numbers of these hosts may lead to values of R_0 that give rise to occasional outbreaks (Gemmell, Lawson and Roberts, 1987). Control of the wild sylvatic hosts of *E. multilocularis* is thus much harder to achieve than for *E. granulosis*.

Dynamics of Complex Life Cycles

The dynamics of complex life cycle parasites may be explored further by making assumptions about the rates at which the various life cycle stages operate and collapsing the total sets of equations accordingly. References for the appropriate models for schistosomiasis are given above. Keymer (1982) describes the dynamics of cestode life cycles, where the densities of both the predator (definitive) host and prey (intermediate) host are held constant. These models always give rise to a single asymptotically stable equilibrium. If only the dynamics of the definitive host are held constant, then models of heteroxenic life cycles tend to have more complex

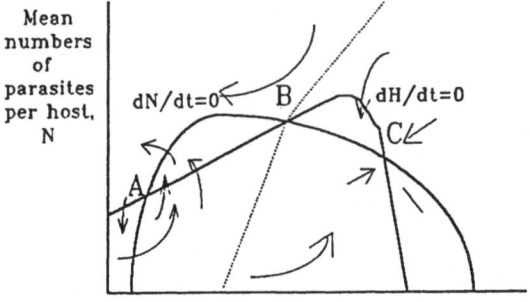

Intermediate host population size, H1

Fig. 13. Phase plane analysis of the stability properties of a model for parasites with a life cycle as depicted in Figure 12. This diagram assumes that the intermediate hosts are regulated to some upper level of abundance in the absence of the parasite. The axes intersect in three places; *A* is stable but oscillatory, *B* is an unstable equilibrium and *C* corresponds to an asymptotically stable upper equilibrium. The upper stable equilibrium is attained when the intermediate host population size exceeds some breakpoint (indicated by the broken line). The arrows illustrate the movement of the two populations when perturbed from equilibrium. Redrawn from Dobson and Keymer (1986).

dynamics. In a comparative study of population dynamics in the acanthocephala, Dobson and Keymer (1986), illustrate that models with two potential stable states are possible. One of these has oscillatory tendencies similar to those of Lotka–Volterra predator prey relationships; the other is more stable and similar in properties of those of the basic macroparasite model (Fig. 13).

Conclusion

The basic models for macroparasites and their hosts developed by Anderson and May (1977) and May and Anderson (1977) have been extended in a number of ways to examine both the control of specific parasitic helminths and the population dynamics and evolution of the different classes of parasitic helminth in more general detail. Although this review has attempted to concentrate on macroparasites in non-human populations, most of the parasites discussed above have human equivalents which are the cause of considerable suffering and debilitation throughout the world. An introduction to the vast literature on mathematical models for these parasites is given in the papers by Anderson and May (1985a, 1988), Dietz (1982), Cohen (1976, 1977) and Barbour (1982).

In concluding, it seems important to stress that the successful development of models for parasite-host relationships has been critically dependent upon the constant investigation of the large body of field and laboratory data that exist for parasitic helminths. The area has been particularly productive in that hypotheses generated by the models have been used by management to design experimental tests of the models and to provide independent ways of estimating parameter values. However, there is still much to be done in the modeling of macroparasites and their hosts. More models are required of the evolution of heteroxenic life cycles from monoxenic ones, while the full dynamic behavior of different types of heteroxenic macroparasites requires considerable further exploration. Similarly, no models have been developed for the coevolution of macroparasites and their hosts nor the evolution of resistance of macroparasites to chemotherapy. Although these models may await further empirical studies of the interactions between pathogenicity, aggregation and parasite fecundity, sensible extrapolation from present studies may help suggest future areas for experimental development.

Acknowledgements. I am most grateful to Si Levin and Tom Hallam for inviting me to the workshop. The contents of the chapter benefited from conversations over many years with Roy Anderson, Bryan Grenfell, Peter Hudson, Anne Keymer, Bob May, Gary Smith, Marilyn Scott, Phil Whitfield and various members of the biology departments at Imperial College and Kings College, London University and the Population Biology Group at Princeton University. I am particularly grateful to Gary Smith for his comments on this manuscript.

References

Abdel-Wahab, M.F., Strickland, G.T., El-Sahly, A., Ahmed, L., Zakaria, S., El Kady, N., Mahmoud, S. (1980) *Schistosomiasis mansoni* in an Egyptian village in the Nile data. American Journal of Tropical Medicine and Hygiene *29*, 868–874.

Anderson, R.M. (1976) Seasonal variation in the population dynamics of *Caryophyllaeus laticeps*. Parasitology *72*, 281–305.

Anderson, R.M. (1978a) The regulation on host population growth by parasitic species. Parasitology *76*, 119–157.

Anderson, R.M. (1978b) Population dynamics of snail infection by miracidia. Parasitology *77*, 201–224.

Anderson, R.M. (1982) The population dynamics and control of hookworm and roundworm infections. In: Anderson, R.M. (ed.) The Population Dynamics of Infectious Diseases: Theory and Applications. Chapman & Hall, London.

Anderson, R.M. (1985) Reproductive strategies of trematodes. In: Adiyodi, K.G., Adiyodi, R.G. (eds.) Reproductive Biology of Invertebrates. John Wiley, New York.

Anderson, R.M., Crombie, J. (1984) Experimental studies of age-prevalence curves for *Schistosoma mansoni* infections in populations of Biomphalaria glabrata. Parasitology *89*, 79–104.

Anderson, R.M., Crombie, J.A. (1985) Experimental studies of age-intensity profiles of infection: *Schistosoma mansoni* in snails and mice. pp. 111–145. In: Rollinson, D., Anderson, R.M. (eds.) Ecology and Genetics of Host-Parasite Interactions. Linnean Society, Academic Press, London.

Anderson, R.M., Gordon, D.M. (1982) Processes influencing the distributions of parasite numbers within host populations with special emphasis on parasite-induced host mortalities. Parasitology *85*, 373–398.

Anderson, R.M., May, R.M. (1978) Regulation and stability of host-parasite population interactions: I Regulatory processes. Journal of Animal Ecology *47*, 219–249.

Anderson, R.M., May, R.M. (1979a) Population biology of infectious disease: Part I. Nature *280*, 361–367.

Anderson, R.M., May, R.M. (1979b) Prevalence of schistosome infections within molluscan populations: obserced patterns and theoretical predictions. Parasitology *79*, 63–94.

Anderson, R.M., May, R.M. (1985a) Herd immunity to helminth infection and implications for control. Nature *315*, 493–496.

Anderson, R.M., May, R.M. (1985b) Helminth infections of humans: mathematical models, population dynamics, and control. Advances in Parasitology *24*, 1–101.

Anderson, R.M., May, R.M. (1989) The population biology of human infectious diseases. Monographs in Population Biology, Oxford University Press (in press).

Anderson, R.M., Mercer, J.G., Wilson, R.A., Carter, N.A. (1982) Transmission of *Schistosoma mansoni* from man to snail: experimental studies of miracidial survival and infectivity in relation to larval age, water temperature, host size and host age. Parasitology *85*, 339–360.

Anderson, R.M., Michel, J.F. (1977) Density dependent survival in populations of *Ostertagia ostertagi*. International Journal for Parasitology *7*, 321–329.

Anderson, R.M., Whitfield, P.J., Dobson, A.P. (1978) Experimental studies of infection dynamics: infection of the definitive host by the cercariae of *Transversotrema patialense*. Parasitology *77*, 189–200.

Anderson, R.M., Whitfield, P.J., Dobson, A.P., Keymer, A.E. (1978) Concomitant predation and infection processes: an experimental study. Journal of Animal Ecology *47*, 891–911.

Atkinson, W.D., Shorrocks, B. (1981). Competition on a divided and ephemeral resource: a simulation model. Journal of Animal Ecology *50*, 461–71.

Barger, A., LeJambre, L.F., Georgi, J.R., Davies, H.I. (1985) Regulation of *Haemonchus contortus* populations in sheep exposed to continuous infection. International Journal for Parasitology *15*, 529–533.

Barbour, A.D. (1978) Macdonald's model and the transmission of bilharzia. Transactions of the Royal Society of Tropical Medicine and Hygiene *72*, 6–15.

Barbour, A.D. (1982) Schistosomiasis. In: Anderson, R.M. (ed.) Population Dynamics of Infectious Diseases. Chapman and Hall, London, pp. 180–208.

Berding, C. (1986) A note on the role of density-dependent fecundity in the population dynamics of direct life-cycle helminths. Mathematical Biosciences *81*, 1–12.

Berding, C., Keymer, A.E., Murray, J.D., Slater, A.F.G. (1987a) The population dynamics of acquired immunity of helminth infection. Journal of theoretical Biology *122*, 459–471.

Berding, C., Keymer, A.E., Murray, J.D., Slater, A.F.G. (1987b) The population dynamics of acquired immunity to infection: experimental and natural transmission. Journal of theoretical Biology *126*, 167–182.

Bradley, D.J., May, R.M. (1978) Consequences of helminth aggregation for the dynamics of schistosomiasis. Transactions of the Royal Society for Tropical Medicine and Hygiene *72*, 262–73.

Carter, N.P., Anderson, R.M., Wilson, R.A. (1982) Transmission of *Schistosoma mansoni* from man to snail: laboratory studies on the influence of snail and miracidial densities on transmission success. Parasitology *85*, 361–72.

Cohen, J.E. (1976) Schistosomiasis, a human-parasite system. In: May, R.M. (ed.) Theoretical Ecology: Principles and Applications Blackwell Scientific, Oxford.

Cohen, J.E. (1977) Mathematical models of schistosomiasis. Annual Review of Ecology and Systematics *8*, 209–33.

Crofton, H.D. (1971a) A quantitative approach to parasitism. Parasitology *63*, 179–93.

Crofton, H.D. (1971b) A model of host-parasite relationships. Parasitology *63*, 343–64.

Crombie, J.A., Anderson, R.M. (1985) Population dynamics of *Schistosoma mansoni* in mice repeatedly exposed to infection. Nature *315*, 491–493.

Dietz, K. (1982) The population dynamics of onchocerciasis. In: Anderson, R.M. (ed.) Population Dynamics of Infectious Diseases. Chapman & Hall, London, pp. 209–241.

Dobson, A.P. (1985) The population dynamics of competition between parasites. Parasitology *91*, 317–347.

Dobson, A.P. (1988) Parasite induced changes in host behaviour: population dynamic consequences and evolutionary considerations. Quarterly Review of Biology *63*, 139–165.

Dobson, A.P., Keymer, A.E. (1986) Life history models. In: Crompton, D.W.T. and Nickol, B.B. (eds.) Biology of the Acanthocephala. Cambridge University Press, pp. 347–384.

Dobson, A.P., May, R.M. (1986) The effects of parasites on fish populations—theoretical aspects. In: Howell, M.J. (eds.) Parasitology—Quo Vadit? Proceedings of the Sixth International Congress of Parasitology. Australian Academy of Science, Canberra, pp. 363–370.

Dean, D.A. (1983) A review. Schistosoma and related genera: acquired resistance in mice. Experimental Parasitology *4*, 1–104.

Gemmell, M.A., Lawson, J.R., Roberts, M.G. (1986) Population dynamics in echinococcosis and cysticercosis: biological parameters of *Echinococcus granulosus* in dogs and sheep. Parasitology *92*, 599–620.

Gemmell, M.A., Lawson, J.R., Roberts, M.G. (1987) Towards global control of cystic and alveolar hydatid diseases. Parasitology Today *3*, 144–151.

Gemmell, M.A., Lawson, J.R., Roberts, M.G., Kerin, B.R., Mason, C.J. (1986) Population dynamics in echinococcosis and cysticercosis: comparison of the response of *Echinococcus granulosus, Taenia hydatigena* and *T. ovis* to control. Parasitology *93*, 357–369.

Grenfell, B.T., Smith, G., Anderson, R.M. (1986) Maximum-likelihood estimates of the mortality and migration rates of the infective larvae of *Ostertagia ostertagi* and *Cooperia oncophora*. Parasitology *92*, 643–652.

Grenfell, B.T., Smith, G., Anderson, R.M. (1987a) The regulation of *Ostertagia ostertagi* populations in calves: the effect of past and current experience of infection on proportional establishment and parasite survival. Parasitology *95*, 363–372.

Grenfell, B.T., Smith, G., Anderson, R.M. (1987b) A mathematical model of the population biology of *Ostertagia ostertagi* in calves and yearlings. Parasitology 389–406.

Halvorsen, O., Anderson, K. (1984) The ecological interactions between arctic char, *Salvelinus alpinus* (L.), and the plerocercoid stage of *Diphyllobothrium ditremum*. Journal of Fish Biology *25*, 305–316.

Hudson, P.J. (1986a) The effect of a parasitic nematode on the breeding performance of red grouse. Journal of Animal Ecology *55*, 85–92.

Hudson, P.J. (1986b) Red Grouse. The Biology and Management of a Wild Gamebird. Bourne Press, Bournemouth.

Hudson, P.J., Dobson, A.P. (1989) The population dynamics of *Trichostrongylus tenuis* in Red grouse. Parasitology (in press).

Hudson, P.J., Watson, A. (1985). Exploited animals: Red grouse. Biologist *32*, 13–18.

Hudson, P.J., Dobson, A.P., Newborn, D. (1985). Cyclic and non-cyclic populations of red grouse: a role for parasitism? In: Rollinson, D., Anderson, R.M. (eds.) Ecology and Genetics of Host-Parasite Interactions. Academic Press, London, pp. 77–90.

Keusch, G.T. (1985) Nutrition and immune funtion. In: Warren, K.S., Mahmoud, A.A.F. (eds.) Tropical and Geographical Medicine. McGRaw Hill, New York, pp. 212–218.

Keymer, A.E. (1980) The influence of *Hymenolepis diminuta* on the survival and fecundity of the intermediate host, *Tribolium confusum*. Parasitology *81*, 405–421.

Keymer, A.E. (1982a) Density-dependent mechanisms in the regulation of intestinal helminth populations. Parasitology *84*, 573–587.

Keymer, A.E. (1982b) Tapeworm Infections. In: Anderson, R.M. (eds.) Population Dynamics of Infectious Diseases. Chapman and Hall, London, pp. 107–138.

Keymer, A.E., Anderson, R.M. (1979) The dynamics of infection of *Tribolium confusum* by *Hymenolepis diminuta*: the influence of infective-stage density and spatial distribution. Parasitology *79*, 195–207.

Keymer, A.E., Crompton, D.W.T., Walters, D.E. (1983) Parasite population biology and host nutrition: dietary fructose and *Moniliformis* (Acanthocephala). Parasitology *87*, 265–278.

Lim, H.K., Heyneman, D. (1972) Intramolluscan intertrematode antagonism: a review of factors influencing the host-parasite system and its possible role in biological control. Advances in Parasitology *10*, 191–268.

Macdonald, G. (1965) The dynamics of helminth infections with special reference to schistosomes. Transactions of the Royal Society of Tropical Medicine and Hygiene *59*, 489–506.

May, R.M. (1977a) Dynamical aspects of host-parasite associations: Crofton's model revisited. Parasitology *75*, 259–276.

May, R.M. (1977b) Togetherness among schistosomes: its effects on the dynamics of the infection. Mathematical Biosciences *35*, 301–43.

May, R.M., Anderson, R.M. (1978) Regulation and stability of host-parasite population interactions. II Destabilizing processes. Journal of Animal Ecology *47*, 249–268.

May, R.M., Anderson, R.M. (1979) Population biology of infectious diseases: Part II. Nature *280*, 455–461.

McCallum, H.I. (1982) Infection dynamics of *Ichthyophthirius multifillis*. Parasitology *85*, 475–488.

McCallum, H.I. (1985) Population effects of parasite survival of host death: experimental studies of the interaction of *Ichthyophthirius multifillis* and its fish host. Parasitology *90*, 529–547.

McCallum, H.I. (1986) Acquired resistance of black mollies Poecilia latipinna to infection by *Ichthyophthirius multifillis*. Parasitology *93*, 251–261.

McCallum, H.I., Anderson, R.M. (1984) Systematic temporal changes in host susceptibility to infection: demographic mechanisms. Parasitology *89*, 195–208.

Michel, J.F. (1970) The regulation of populations of *Ostertagia ostertagi* in calves. Parasitology *79*, 157–68.

Miller, R.S. (1967) Pattern and process in competition. Advances in Ecological Research *4*, 1–74.

Mitchell, G.F. (1979) Responses to infection with metazoan and protozoan parasites in mice. Advances in Immunology *28*, 451–511.

Nasell, I. (1976a) A hybrid model of schistosomiasis. Theoretical Population Biology *10*, 47–69.

Nasell, I. (1976b) On eradication of schistosomiasis. Theoretical Population Biology *10*, 133–44.

Nasell, I., Hirsch, W.M. (1973) The transmission dynamics of schistosomiasis. Communications in Pure and Applied Mathematics *26*, 395–453.

Pacala, S.W., Dobson, A.P. (1988) The relation between the number of parasites per host and host age: population dynamic causes and maximum-likelihood estimation. Parasitology *96*, 197–210.

Potts, G.R., Tapper, S.C., Hudson, P.J. (1984) Population fluctuations in red grouse: analysis of bag records and a simulation model. Journal of Animal Ecology *53*, 21–36.

Roberts, M.G., Lawson, J.R., Gemmell, M.A. (1986) Population dynamics in echinococcosis and cystercosis: mathematical model of the life-cycle of *Echinococcus granulosus*. Parasitology *92*, 621–641.

Roberts, M.G., Lawson, J.R., Gemmell, M.A. (1987) Population dynamics in echinococcosis and cystercosis: mathematical model of the life-cycles of *Taenia hydatigena* and *T. ovis*. Parasitology *94*, 181–197.

Schad, G.A., Anderson, R.M. (1985) Predisposition to hookworm infection in man. Science *228*, 1537–40.

Scott, M.E. (1982) Reproductive potential of *Gyrodactylus bullatarudis* (Monogenea) on guppies (Poecilia reticulata). Parasitology *85*, 217–236.

Scott, M.E., Anderson, R.M. (1984) The population dynamics of *Gyrodactylus bullatarudis* (Monogenea) within laboratory populations of the fish host Poecilia reticulata. Parasitology *89*, 159–194.

Siongok, T.K.A., Mahmoud, A.A.F., Ouma, J.H., Warren, K.S., Muller, A.S., Handy, A.K., Houser, H.B.

(1976) Morbidity in *Schistosomiasis mansoni* in relation to intensity of infection: study of a community in Machakos, Kenya. American Journal of Tropical Medicine and Hygiene 25, 273–284.

Smith, G. (1982) An analysis of variations in the age structure of *Fasciola hepatica* populations in sheep. Parasitology 84, 49–61.

Smith, G. (1984a) Density-dependent mechanisms in the regulation of *Fasciola hepatica* populations of sheep. Parasitology 88, 449–461.

Smith, G. (1984b) Analysis of anthelmintic trial protocols using sheep experimentally or naturally infected with *Fasciola hepatica*. Veterinary Parasitology 16, 83–94.

Smith, G. (1984c) Chemotherapy of ovine fascioliasis: use of an analytical model to assess the impact of a series of discrete dose of anthelmintic on the prevalence and intensity of infection. Veterinary Parasitology 16, 95–106.

Smith, G. (1984d) The impact of repeated doses of anthelmintic on the intensity of infection and age structure of Fasciola hepatica populations in sheep. Veterinary Parasitology 16, 107–115.

Smith, G. (1987) The relationship between the density of *Fasciola hepatica* miracidia and the net rate of miracidial infections in *Lymnaea trunculata*. Parasitology (in press).

Smith, G. (1988) The population biology of the parasitic stages of *Haemonchus contortus*. Parasitology 96, 185–195.

Smith, G., Grenfell, B.T. (1984) The influence of water temperature and pH on the survival of *Fasciola hepatica* miracidia. Parasitology 88, 97–104.

Smith, G., Grenfell, B.T. (1985) The population biology of *Ostertagia ostertagia*. Parasitology Today 1, 76–81.

Smith, G., Grenfell, B.T., Anderson, R.M. (1987a) The development and mortality of the non-infective free-living stages of *Ostertagia ostertagi* in the field and in laboratory culture. Parasitology 92, 471–482.

Smith, G., Grenfell, B.T., Anderson, R.M. (1987b) The regulation of *Ostertagia ostertagi* in calves: density dependent control of fecundity. Parasitology 95, 373–388.

Smith, G., Grenfell, B.T., Anderson, R.M. (1987c) Population biology of *Ostertagia ostertagi* and anthelmintic strategies against ostertagiasis in calves. Parasitology 95, 407–420.

Smith, G., Galligan, D. (1988) Mathematical models of the population biology of *Ostertagia ostertagia* and Teladorsagia circumcincta and the economic evaluation of control strategies. Veterinary Parasitology 27, 73–83.

Slater, A.F.G., Keymer, A.E. (1986) Epidemiology of *Heligmosomoides polygyrus* in mice: experiments on natural transmission. Parasitology 93, 177–187.

Slater, A.F.G., Keymer, A.F. (1987) *Heligmosomoides polygyrus* (Nematoda): the influence of dietary protein on the dynamics of repeated infection. Proceedings of the Royal Society of London B 229, 69–83.

Tingley, G., Anderson, R.M. (1987) Experimental sex determination and density-dependent population regulation in the entomophagous nematode *Romanomermis culicivorax*. Parasitology.

Wakelin, D. (1984) Evasion of the immune response: survival within low responder individuals of the host population. Parasitology 88, 639–657.

Wakelin, D. (1985) Genetics, immunity and parasite survival. In: Rollinson, D., Anderson, R.M. (eds.) Ecology and Genetics of Host-Parasite Interactions Linnean Society, Academic Press, London.

Wilson, R.A., Smith, G., Thomas, M.R. (1982) Fascioliasis. In: Anderson, R.M. (ed.) Population Dynamics of Infectious Diseases. Chapman and Hall, London, pp. 262–319.

Simple Versus Complex Epidemiological Models

Joan L. Aron

Lecture Notes for Second Autumn Course on Mathematical Ecology

Contents

Abstract

Mathematical models of the population dynamics of disease can contribute to a better understanding of epidemiological patterns and disease control. Simple models with few assumptions lead to general conclusions of a qualitative nature. More detailed models are useful for quantitative conclusions. However, the comparative merits of simple versus complex epidemiological models are not always readily apparent. The problem of determining an appropriate degree of complexity is discussed in reference to age structure in a model of transmission which has been applied to the study of measles immunization.

Introduction

Mathematical models of the population dynamics of disease can contribute to a better understanding of epidemiological patterns and disease control. However, the process of developing useful mathematical models depends on a collaborative relationship between mathematicians and epidemiologists. Much of the process involves debate about the appropriate level of structural detail. The desire of mathematicians to characterize completely a model tends to push towards simplicity, yet the demand for epidemiologists to analyze real problems tends to push towards complexity.

The purpose of this chapter is to clarify the usefulness of simplicity and complexity by presenting a simple model and considering the merits of several extensions. The discussion complements two other closely-related chapters on models of diseases spread by person-to-person contact. In this volume, Hethcote (1989a) delves into the mathematical properties of the simple models. In the companion volume, May (1986) surveys the complicating features which have been incorporated into the simple models. This chapter places more emphasis on the debates about the role of complexity, with special attention paid to age structure and measles immunization.

Simple Model of Epidemic Spread

The SIR model is a simple model of epidemic spread in which disease is transmitted by direct contact between hosts who become immune after a single infection. The SIR model is most appropriate for measles but it can be applied to many diseases, especially the traditional childhood diseases. As in any model of transmission, the basic feature is that susceptibles pick up infection from infective hosts and then become infective themselves. Implicit in the formulation as a system of differential equations is the assumption that disease transmission events are sufficiently regular so that chance occurrences (i.e., stochastic effects) do not play a major role.

The SIR model has three epidemiological classes—susceptibles (S), infectives (I) and recovereds/immunes (R). The variables X, Y and Z are used to denote the number of individuals in each of these classes. Susceptibles become infectives at a rate determined by a mass-action contact term for susceptibles and infectives ($\beta X Y$). Infectives become permanently immune at a characteristic rate (γY). The quantity $1/\gamma$ is the average duration of infectivity. Without births, deaths or migration, the total population size, N, is fixed. (Some applications use the "R" class to refer to removals, i.e., deaths from the disease. The total population size is fixed only if its definition is extended to include these deaths.) The system is presented in (1) below.

$$\frac{dX}{dt} = -\beta X Y$$

$$\frac{dY}{dt} = \beta X Y - \gamma Y$$

$$\frac{dZ}{dt} = \gamma Y$$

$$X + Y + Z = N$$

(1)

In such a closed population, the chain of infection must terminate. In order to consider diseases which persist in the population, the model must allow an influx of susceptibles. The simplest way is to introduce the susceptibles via births. That is, neglecting the short period during which infants are protected by maternal antibodies, all births are assumed to enter the susceptible class. At the same time, mortality is introduced into the model to maintain a fixed population size. The resulting system is shown in (2) below.

$$\frac{dX}{dt} = -\beta X Y + B - \mu X$$

$$\frac{dY}{dt} = \beta X Y - \gamma Y - \mu Y$$

$$\frac{dZ}{dt} = \gamma Y - \mu Z \tag{2}$$

$$B = \mu N, \quad X + Y + Z = N$$

Here the total population density, N, is fixed because births (B) exactly balance deaths ($\mu X + \mu Y + \mu Z$). However, the introduction of vital dynamics into the system means that the population is no longer closed.

The very simple SIR model clearly demonstrates the role of influx of susceptibles in sustaining transmission of an infection which causes permanent immunity. If an infection persists in a small, remote population for which births and immigrants are few, one can easily rule out the hypothesis that every infective person quickly becomes immune. As an illustration, the persistence of hepatitis B in remote Eskimo populations contrasts with the episodic appearance and disappearance of polio (Yorke et al., 1979). The explanation for the difference is that polio is an SIR-type disease while hepatitis B virus can induce a chronic carrier state which facilitates persistence in even extremely isolated populations.

The SIR model can also give some insight into how to measure the conditions for spread of disease. The behavior of the epidemic model in (1) depends upon the dimensionless quantity, $\beta N/\gamma$, which is called the basic reproductive rate, R. R represents the number of new infectives generated directly from a single infective introduced into a susceptible population. It is the product of βN, the rate at which new contacts are made by an infective in a susceptible population, and $1/\gamma$, the duration of infectivity. A single infective does not generate an epidemic if R is less than unity. An epidemic occurs, i.e., the number of infectives increases, if R exceeds unity. When the epidemic dies out, everyone is either susceptible or immune. The proportion of the population which was never infected, X_∞/N, varies inversely with the basic reproductive rate,

$$R = \frac{1}{1 - \dfrac{X_\infty}{N}} \ln \frac{1}{\dfrac{X_\infty}{N}} \tag{3}$$

Thus, the fraction remaining susceptible after an epidemic can be used to estimate R.

The behavior of the SIR model with vital dynamics in (2) depends upon the dimensionless quantity, $\beta N/(\gamma + \mu)$. This quantity too is the basic reproductive rate and it differs from the previous definition only because the duration of infectivity must account for possible death. However, if (as is typical) γ is much larger than

μ, then the duration of infectivity, $1/(\gamma + \mu)$, is approximately $1/\gamma$. For most practical purposes, the basic reproductive rate is the same as above, $\beta N/\gamma$. As in the closed-population model, if R is less than unity, an infective introduced into the population will not generate an increasing number of infectives. However, now if R exceeds unity, there is a stable equilibrium for which disease is endemic. Moreover, the fraction susceptible at equilibrium, X^*/N, varies inversely with the basic reproductive rate

$$R = N/X^* \tag{4}$$

Thus, Eqs. (3) and (4) show that fewer susceptibles remaining at the end are associated with greater reproductive rates of infection.

Such relationships can be used to check epidemiological presumptions concerning mode of transmission and contact rate. For example, an increase in population density should increase the basic reproductive rate. For an endemic disease described by (2), there is a corresponding decrease in the fraction susceptible, X^*/N. The reduced fraction susceptible means that, on average, people are infected (i.e., lose their susceptibility) at younger ages. Consequently, one would expect the average age at infection to be lower in urban (high-density) areas than in rural (low-density) areas. This relationship is corroborated for many respiratory infections (Anderson, 1982). Equally important, however, is the reversal observed for hepatitis A (Lobel and McCollum, 1965). The low ages at infection in rural areas imply that an alternative factor such as sanitation is more important than population density in determining contact rates.

The SIR model with vital dynamics may be extended to investigate the effect of an immunization program. The simplest assumption is that a random fraction, p, of infants can be permanently immunized. The dynamics in the unimmunized fraction, $N(1 - p)$, are as follows:

$$\frac{dX}{dt} = -\beta XY + B(1 - p) - \mu X$$

$$\frac{dY}{dt} = \beta XY - \gamma Y - \mu Y \tag{5}$$

$$\frac{dZ}{dt} = \gamma Y - \mu Z$$

$$B = \mu N, \quad X + Y + Z = N(1 - p)$$

The system in (5) is exactly the same as the system in (2) except that the density of the population is reduced by the percentage immunized. Thus, the average age at infection should rise. The net result is fewer but older cases after immunization. Despite this shift, the susceptible fraction at equilibrium, X^*/N, is still the inverse of the original basic reproductive rate, R (see Eq. (4)). An intuitive explanation depends on the notion of an equilibrium in which one infective, on average, replaces one infective. Since, in a totally susceptible population, an infective could generate R new infectives, the existence of the equilibrium means that only $1/R$ of the population is susceptible. Thus, effective immunization does not reduce the size of the at-risk population if transmission persists. However, a sufficiently high immunization fraction, p,

$$R(1 - p) < 1\left(\text{or } p > 1 - \frac{1}{R}\right) \tag{6}$$

means that the average infective will generate less than one (i.e., $R(1 - p)$) new infective. The inequality in (6) represents the principle of herd immunity because the susceptible population may be protected from epidemics if enough people are immunized. In principle, the disease should be eradicated.

Thus, the SIR model provides important epidemiological insights because of its simplicity. The role of the influx of susceptibles in maintaining disease transmission is clear in the absence of extraneous detail. Analysis of the simple model reveals semi-quantitative behavior such as the tendency for the average age at infection to be higher in areas of low population density or high levels of immunization. The principle of herd immunity demonstrates the possibility of eradication by immunization. It is also easy to identify logical inconsistencies such as the incorrect notion that an immunization program must reduce the size of the susceptible population in the long run.

Age Structure in Simple Model

In more detailed studies, one of the limitations of the simple SIR model (and its variants) is the unrealistic characterization of age. Infection rates, recovery rates and mortality rates are all assumed to be independent of the age of the host. However, before introducing refinements, it is necessary to formulate the role of age structure implicit in the simple models.

An age and time dependent set of equations corresponding to (2) is written below in terms of $\tilde{X}(a, t)$, $\tilde{Y}(a, t)$, and $\tilde{Z}(a, t)$, which are, respectively, the numbers of susceptibles, infectives and immunes, of age a and time t.

$$\partial \tilde{X}(a, t)/\partial t + \partial \tilde{X}(a, t)/\partial a = -[\lambda(t) + \mu]\tilde{X}(a, t)$$
$$\partial \tilde{Y}(a, t)/\partial t + \partial \tilde{Y}(a, t)/\partial a = \lambda(t)\tilde{X}(a, t) - [\gamma + \mu]\tilde{Y}(a, t)$$
$$\partial \tilde{Z}(a, t)/\partial t + \partial \tilde{Z}(a, t)/\partial a = \gamma \tilde{Y}(a, t) - \mu \tilde{Z}(a, t) \tag{7}$$
$$\lambda(t) = \beta \int_0^\infty \tilde{Y}(a, t)\, da$$
$$\tilde{X}(0, t) = B \quad \tilde{Y}(0, t) = 0 \quad \tilde{Z}(0, t) = 0$$
$$\tilde{X}(a, 0) = X_0(a) \quad \tilde{Y}(a, 0) = Y_0(a) \quad \tilde{Z}(a, 0) = Z_0(a).$$

In order that the total population size N be fixed, the total births, B, must be chosen to match the total deaths, i.e.,

$$B = \int_0^\infty \mu(X_0(a) + Y_0(a) + Z_0(a))da \tag{8}$$

At equilibrium, the rate of exposure in (7) depends on the total number of infectives in the population which can be calculated directly from (2). The total numbers of susceptibles, infectives and immunes (respectively, X^*, Y^* and Z^*)

satisfy

$$\frac{X^*}{N} = \frac{1}{R}, \quad \frac{Y^*}{N} = \left(1 - \frac{1}{R}\right)\frac{\mu}{\gamma + \mu}, \quad \frac{Z^*}{N} = \left(1 - \frac{1}{R}\right)\frac{\gamma}{\gamma + \mu} \tag{9}$$

if $R > 1$ where $R = \dfrac{\beta N}{\gamma + \mu}$

Thus, λ, the rate of exposure of susceptibles at equilibrium, is fixed at βY^* where Y^* is taken from (9). The numbers of susceptibles, infectives and immunes at each age ($\bar{X}(a)$, $\bar{Y}(a)$ and $\bar{Z}(a)$, respectively) are then found by solving the time-invariant system below.

$$\frac{d\bar{X}(a)}{da} = -\lambda\bar{X}(a) - \mu\bar{X}(a)$$

$$\frac{d\bar{Y}(a)}{da} = \lambda\bar{X}(a) - \gamma\bar{Y}(a) - \mu\bar{Y}(a)$$

$$\frac{d\bar{Z}(a)}{da} = \gamma\bar{Y}(a) - \mu\bar{Z}(a) \tag{10}$$

$$\lambda = \beta Y^*$$

$$\bar{X}(0) = B \quad \bar{Y}(0) = 0 \quad \bar{Z}(0) = 0$$

The total number at each age, $\bar{N}(a)$, is the sum of $\bar{X}(a)$, $\bar{Y}(a)$ and $\bar{Z}(a)$. From (10), it is easily seen that

$$\frac{d\bar{N}(a)}{da} = -\mu\bar{N}(a)$$
$$\bar{N}(0) = B \tag{11}$$

Upon solving the equations in (10) and (11), we have

$$\bar{X}(a) = Be^{-(\lambda+\mu)a}$$
$$\bar{N}(a) = Be^{-\mu a} \tag{12}$$

The relationship between the total fraction susceptible at equilibrium and the average age at infection (mentioned in the previous section on the simple model) can now be made explicit.

$$\frac{X^*}{N} = \frac{\int_0^\infty \bar{X}(a)da}{\int_0^\infty \bar{N}(a)da} = \frac{A}{L} \tag{13}$$

$$A = 1/(\lambda + \mu)$$

$$L = 1/\mu$$

Here, A is the average age of the susceptible population (or, equivalently, the average age at infection) while L is the life expectancy of the population as a whole.

From (9) and (13), it follows that

$$R = \frac{L}{A} \tag{14}$$

Eqn. (14) is the explicit formulation of the earlier statement that a lower age at infection is associated with a larger basic reproductive rate.

A similar analysis may be used for the simple model of immunization in (5). Instead of (9), the endemic equilibrium for the total number of (unimmunized) susceptibles, infectives and immunes (X^*, Y^* and Z^*, respectively) satisfies

$$\frac{X^*}{N} = \frac{1}{R}, \quad \frac{Y^*}{N} = \left(1 - \frac{1}{R'}\right)(1 - p)\frac{\mu}{\gamma + \mu}, \quad \frac{Z^*}{N} = \left(1 - \frac{1}{R'}\right)(1 - p)\frac{\gamma}{\gamma + \mu}$$

$$\text{if } R' > 1 \quad \text{where} \quad R' = R(1 - p), \quad R = \frac{\beta N}{\gamma + \mu}. \tag{15}$$

The numbers of (unimmunized) susceptibles, infectives and immunes at each age ($\bar{X}(a)$, $\bar{Y}(a)$ and $\bar{Z}(a)$, respectively) may be found by solving the system in (10) with the births of susceptibles adjusted by the immunized fraction, i.e., $\bar{X}(0) = B(1 - p)$. Since Y^* is smaller under immunization, it follows that λ, the rate of infection, is smaller and there are fewer cases. However, it also follows that A, the average age at infection, is larger. Eq. (13) must be modified to account for the reduced number of susceptible births

$$\frac{X^*}{N} = \frac{A_I(1 - p)}{L} \tag{16}$$

where A_I is the average age at infection after immunization. A comparison of Eq. (13) and Eq. (16) shows that because the susceptible fraction is unchanged, the increase in the average age at infection must exactly compensate the reduced number of susceptible births. This is a more formal presentation of the result that immunization produces fewer older cases while maintaining the size of the susceptible population.

Adding Age Dependence

The age structure inherent in the simple models is due to the cumulative effect of infection and death in each birth cohort. Even if the rate parameters do not depend on age, there are fewer older people and they are more likely to be immune. Nevertheless, processes of infection and mortality often depend strongly on age in ways not captured in the simple model. The problem is not to decide if age dependence exists (it obviously does) but to determine the circumstances under which it affects conclusions drawn from the model. This section will examine the incremental effect of several forms of age dependence.

The model assumption that infants are born susceptible is a simplification which ignores passive immunity. For some diseases, infants are protected for several months as maternally-derived antibodies gradually decay. If passive immunity is incorporated into the model, "age" becomes duration of susceptible life. One way to construct "age" is by subtracting the average duration of passive immunity from the real age (Anderson and May, 1983; Hethcote, 1983). Since the effect of subtracting some months is greatest at low ages, the adjustment is most important where childhood infections typically strike at very young ages (as in many developing countries). However, the magnitude of the adjustment may be somewhat offset by the shorter duration of maternally-derived immunity in developing countries, at least for measles (Black et al., 1986). It may also be necessary to alter the duration of passive immunity after a vaccination program has been in effect for a generation if children of vaccinated mothers have a shorter period of passive immunity (Lennon and Black, 1986).

The model assumption that mortality is independent of age results in an unrealistic exponential age distribution. Since measures of disease burden are often weighted by the age distribution, the evaluation of disease control policies depends upon this assumption. The effect of incorporating more realistic age distributions is examined here with an age-specific measure of two hypothetical policies, $s_1(a)$ and $s_2(a)$, and their age-weighted comparison of $\int_0^\infty s_1(a)\,c(a)\,da$ and $\int_0^\infty s_2(a)\,c(a)\,da$. An important evaluation issue is if policy preference can be reversed by an alternative choice of age-specific weights $c(a)$. (Without loss of generality, we assume that the aggregate measure for the first policy is greater than the aggregate measure for the second policy.) Under certain conditions, it is possible to rule out such reversals. One obvious condition is that $s_1(a)$ be consistently greater than $s_2(a)$. A less obvious pair of conditions is

$$\int_0^\infty s_1(a)\,da \geqq \int_0^\infty s_2(a)\,da \quad \text{for all } a$$

and   (17)

$c(a)$ does not increase with a

(See Appendix for Proof). The first condition in (17) deals with epidemiological properties of the model. For example, if age-specific prevalence of infection is the measure, then a policy of reducing transmission in the model (10) satisfies that condition. However, in other models, a policy of reducing transmission may not satisfy that condition (e.g., see Aron (1983)). The second condition in (17) deals with demographic properties of the model. That condition is satisfied by any stationary age distribution (Frauenthal, 1986). Thus, in an analysis of the equilibrium, weighting ages according to their numbers in the population does not introduce reversals. However, policy concerns may require weights that violate that condition. An excellent example is the analysis of rubella vaccination programs because the primary objective is to protect pregnant women from delivering infants with congenital rubella syndrome.

The exclusion of childhood rubella cases introduces the problem that childhood immunization may increase disease in the ages of interest. When protection of immunized individuals reduces transmission, unimmunized girls are more likely to be exposed as women. Alternatives to routine childhood rubella immunization include immunization of pre-pubertal girls only. The evaluation of optimal policies has varied somewhat among investigators (Hethcote, 1989b). Hethcote accounts for differences by the choice of the cost function which can include the immunized fraction, the number of vaccinations and cases of congenital rubella syndrome. However, this problem is also sensitive to the choice of the age distribution. In examining the incidence of rubella in women between 16 and 40 years of age, Anderson and May (1983) note that an exponential age distribution and a realistic age distribution produce markedly different results.

The interaction between epidemiology and demography is even more complicated if infection may be fatal. Case fatalities act to reduce the observed rate of increase of the proportion immune at each age (McLean, 1986). Moreover, infections associated with relatively high mortality often occur in areas where the population is growing rapidly. May and Anderson (1985) and McLean (1986) show that the birth rate of the population appears in the criteria for eradication. However, more work need to be done to extend these developments. The dependence of measles mortality on other health problems is ignored. Both models use asymptotically stable age distributions which do not reflect the changing vital rates in developing countries. Extreme assumptions are used to link population density and rates of infection. On the one hand, growth may be accompanied by steadily increasing rates of infection due to greater density (May and Anderson, 1985). On the other hand, the rate of infection may be stable if exposure depends not on the density of infectives but on the proportion of infectives in the population (McLean, 1986). Both models are clearly simplifications which do not take into account social and environmental consequences of population growth.

Age-dependent rates of infection and mortality also affect the interpretation of the average age at infection in a stationary population. The concept of average age at infection involves subtle distinctions which can be best appreciated by considering the system in (10) which is re-written below with age-dependent parameters.

$$\frac{d\bar{X}(a)}{da} = -\lambda(a)\bar{X}(a) - \mu(a)\bar{X}(a)$$

$$\frac{d\bar{Y}(a)}{da} = \lambda(a)\bar{X}(a) - \gamma(a)\bar{Y}(a) - \mu(a)\bar{Y}(a)$$

$$\frac{d\bar{Z}(a)}{da} = \gamma(a)\bar{Y}(a) - \mu(a)\bar{Z}(a) \tag{18}$$

$$\bar{X}(0) = B, \quad \bar{Y}(0) = 0, \quad \bar{Z}(0) = 0$$

The link between the rate of infection and the parameter of transmission is discussed in more detail elsewhere (May, 1986, p. 410). From (18), the generalization of (12) is

$$\bar{X}(a) = Bl_x(a)$$

$$\bar{N}(a) = Bl_n(a)$$

$$l_x(a) = e^{-\int_0^a (\lambda(w) + \mu(w))dw} \tag{19}$$

$$l_n(a) = e^{-\int_0^a \mu(w)dw}$$

The susceptible fraction at equilibrium may then be expressed as

$$\frac{X^*}{N} = \frac{A}{L}$$

$$A = \int_0^\infty l_x(a)da \tag{20}$$

$$L = \int_0^\infty l_n(a)da.$$

The quantities A and L are in the form of an average waiting time for a birth cohort. L is the expectation of life at birth (the average waiting time until death) and A is the expectation of susceptible life at birth (the average waiting time until infection or death). However, A, a cohort measure, is not the only possible definition of average age at infection. An alternative is A_c, a cross-sectional measure which represents the average age of the cases

$$A_c = \frac{\int_0^\infty a\lambda(a)l_x(a)da}{\int_0^\infty \lambda(a)l_x(a)da}. \tag{21}$$

The intuitive use of the average age of the cases as an estimator for A is not generally valid.

The distinction between A and A_c has not been stressed because most applications to date fall into two areas where the measures are identical. Under the common assumption that infection and mortality are independent of age, A and A_c are identical. Even when age-dependence is introduced, the assumption that (virtually) all infection occurs before any mortality insures that A and A_c are (virtually) identical.

The neglect of the effect of age-dependent mortality has been reinforced by the common re-expression of the system in (18) in terms of the fractions in each age group $x(a) = \bar{X}(a)/\bar{N}(a)$, $y(a) = \bar{Y}(a)/\bar{N}(a)$ and $z(a) = \bar{Z}(a)/\bar{N}(a)$,

$$\frac{dx(a)}{da} = -\lambda(a)x(a)$$

$$\frac{dy(a)}{da} = \lambda(a)x(a) - \gamma(a)y(a) \tag{22}$$

$$\frac{dz(a)}{da} = \gamma(a)y(a)$$

$$x(0) = 1, \quad y(0) = 0, \quad z(0) = 0$$

Since mortality is assumed to affect all epidemiological classes equally, there is no effect of mortality on the proportion in each age group. On the basis of the system in (22), the most commonly used definition of average age at infection is not A or A_c but A', which is a cohort measure dependent only on the infection rates

$$A' = \int_0^\infty \exp\left(-\int_0^a \lambda(w)dw \right) da \tag{23}$$

The distinction between A and A' too has been blurred because they are virtually identical when mortality is low. The distinction is significant because the susceptible fraction in Eq. (20) is expressed more easily in terms of A.

The differences among the definitions of the average age at infection are illustrated by an example in which the rate of infection, λ, is independent of age and everyone lives exactly L years. We see that

$$A = \frac{1}{\lambda}(1 - e^{-\lambda L})$$

$$A_c = \frac{1}{\lambda} - \frac{Le^{-\lambda L}}{1 - e^{-\lambda L}} \tag{24}$$

$$A' = \frac{1}{\lambda}$$

When many people die before becoming infected (λL is small), the three measures are distinct. When infection is widespread before death at age L (λL is large), A, A_c and A' are virtually identical.

For completeness, another cross-sectional measure is the average age of the susceptible population, A_x,

$$A_x = \frac{\int_0^\infty a l_x(a)da}{\int_0^\infty l_x(a)da} \tag{25}$$

The quantity A_x is useful in characterizing the entire population that would be at risk if exposed. Clearly, the difference between A_x in Eq. (25) and A_c in Eq. (21) exists only if the rate of infection is age-dependent. A, A_x and A_c are identical if λ and μ are independent of age.

Age-dependent infection rates are particularly important in the analysis of immunization. The simple model can predict that the average age at infection will increase but not the amount by which it will increase. Moreover, the number of susceptibles may change after immunization (Schenzle, 1984). The fraction of the population, p, which must be effectively immunized to interrupt transmission can no longer be calculated using Eqs. (4) and (6) from the simple SIR model. Fortunately, however, the simple SIR model may still be useful in setting bounds on the "true" value of p. If infection rates consistently fall with age, then the critical fraction to be immunized is less than would be calculated with the simple model (Anderson and May, 1985). Intuitively, as immunization increases the average age at infection, the lower infection rates for older people should make it easier to eradicate the disease. Conversely, if infection rates consistently rise with age, then the critical fraction to be immunized is greater than would be calculated with the simple model. For example, if the simple model predicted the critical fraction to be 90% but infection rates rose consistently with age, then the more realistic model would predict the critical fraction to be between 90% and 100%. If it were known that only 60% of the population would accept vaccination, then the impossibility of eradication would be immediately obvious.

Application to Measles in England and Wales

Another way to understand the significance of age structure is to focus on one practical application. The problem chosen is the evaluation of the impact of measles immunization in England and Wales. The British example is a good one because the measles immunization program which began in 1968 has had a major impact but has not interrupted transmission. Consequently, this problem has attracted many to consider the SIR model of mass transmission and its extensions. The British government also has very good population and health records which Fine and Clarkson (1982a, 1982b) have used to reconstruct the measles experience in England and Wales from 1950 to 1979. This discussion is restricted to the dynamics of transmission. A complete evaluation of measles immunization needs to account for the serious sequelae of measles (such as encephalitis) whose prevention is the reason for immunization (e.g., see Hinman et al. (1983) and Miller (1983)).

The importance of detailed age structure in studies of measles immunization has been the subject of ongoing debate. A key observation has been Fine and Clarkson's (1982b) estimate that the number of susceptibles has remained fixed at around 4 to 4.5 million throughout the period 1950 to 1979. Since the simple SIR model of vaccination predicts a constant number of susceptibles in the absence of eradication, it may be that age structure is not important in determining the overall impact of the immunization program (e.g., May (1986)). Schenzle (1984) challenges this explanation because the number of susceptibles in an age-structured model with very little transmission among adults may appear to be constant for years before rising. The predicted slow rise in the number of susceptibles coupled with imprecise estimates of this number makes it difficult to draw conclusions. The model differences also affect the theoretical vaccination coverage required for eradication. In a model without age structure, Anderson and May (1982) estimate that 96% of the population would have to be effectively immunized. (Note that the fraction vaccinated has to be even higher because vaccination is not 100% effective.) The assumption of very little transmission among adults leads Schenzle to a very low estimate of 76%. Anderson and May (1985) also produced an age-structured model which predicts that 89% effective immunization would be sufficient. The differences among the three predictions of 76%, 89% and 96% are essentially due to the assumption about disease contact rates for adults—low for 76%, intermediate for 89% and high for 96%. The difficulty in interrupting measles transmission in the United States (e.g., see Davis et al. (1987)) suggests that the high figure of 96% is probably closest to the truth. Further refinement depends on more data on the epidemiology of measles (especially for adults) in the era of immunization.

Despite the equivocal effects of age structure on overall transmission, there is little debate about its importance in epidemiological observations of children. The enhanced contact rates among school-aged children definitely influence the average age at infection. The level of measles vaccine uptake in England from 1970 to 1980 was about 50%, but was accompanied only by a modest rise in the average age of the cases from 4.5 to 5.5 years (Fine and Clarkson, 1983). With an uptake of 50%, the simple model predicts a doubling of the average age at infection (see Eq. (16)). For children in developed countries, it is also generally agreed that a model with little or no mortality is appropriate. In this setting, the distinction between the cohort and cross-sectional measures of the average age at infection is not necessary.

The discussion up to now has ignored time-dependent epidemiological features. Time dependence enters into an examination of characteristic fluctuations in incidence about the average and into an analysis of the response to immunization programs before an equilibrium is reached. These issues have practical significance in evaluating immunization programs and validating models of transmission.

The latent period between exposure and the development of infection is a major factor in time-dependent phenomena. Latency, however, does not affect the basic reproductive rate (the number of new infections generated) unless there is a probability of dying during the period of latency. Since this probability is usually negligible, the basic reproductive rate is essentially independent of latency. The SIR approach is sufficient for the steady state equilibrium, but the SEIR approach is needed for time-dependence. The SEIR model incorporates an exposed but latent class (the "E") into the transition from susceptible to infective. The time for one infection to generate subsequent infections ("generation time") is the sum of the period of latency and the period of infectivity. However, the partition of the generation time into its two components or the choice between fixed and variable intervals are of relatively minor importance (Schenzle, 1984; May, 1986). For measles, the average generation time is around two weeks.

Before the immunization program began, the most striking feature of the epidemiology of measles was the alternation between years of major and minor epidemics (Fine and Clarkson, 1982a). One approach to understanding fluctuations in incidence examines the propensity for these model systems to produce weakly-damped oscillations while responding to perturbations (Castillo–Chavez et al., 1989). The frequency and the rate of damping of the oscillations depends on the pattern of age-structured contacts (Anderson and May, 1985). Anderson and May (1985) point out that a pattern of age-structured contacts with much transmission among school-aged children generates a two-year cycle. An alternative approach emphasizes the strong seasonality of transmission which is coupled to the schedule of the school year (Fine and Clarkson, 1982a). Seasonal contact in the SEIR model generates cycles of major and minor epidemics (London and Yorke, 1973; Aron and Schwartz, 1984). Moreover, age-structured contacts in a seasonal model can account for asymmetry between major and minor epidemics; the apparent contact rate declines during the school year during major epidemics but not during minor epidemics (Schenzle, 1984). However, the asymmetry between major and minor epidemics might be a statistical artefact due to clustering of cases in schools and neighborhoods (Fine and Clarkson, 1982a).

Unfortunately, it is difficult to resolve issues involving the detailed temporal fluctuations of measles incidence. The sources of data—birth records, measles notifications and vaccination reports—are subject to reporting and sampling errors to which models of temporal change are especially sensitive. One example is the identification of parameters of the simple SEIR model for comparison with age-structured models. Schenzle (1984) deduces that an estimate of 13 for the basic reproductive rate is consistent with observations on infection rates and the number of susceptibles. However, other estimates range from 13.7 to 18.0 (Anderson, 1982, p. 12). Aron and Schwartz (1984) show that the model's ability to generate the pattern of major and minor epidemics is sensitive to the value of the basic reproductive rate. The problems are even more serious with transient responses to immunizations. Model predictions are sensitive to the exact starting point of the immunization program during the two-year cycle of major and minor epidemics (Anderson and May, 1985; Schenzle, 1984). Non-random distribution of vaccine, which is difficult to assess, also affects the epidemiological response (Fine and Clarkson, 1983). Even where data are available, interpretations differ. The distinction between signal and noise in epidemiological time series is often in the eye of the beholder. Where Anderson and May (1985) observe a lengthening period of fluctuations in response to immunizations, Fine and Clarkson (1983) find "little evidence of an increase in duration between periodic epidemics in these national statistics". Instead, Fine and Clarkson (1983) see a repeated pattern reminiscent of the minor epidemic years before the immunization program. Studies of temporal fluctuations will probably require going beyond highly aggregated national statistics.

Conclusion

Models must simplify to be useful. The key design problem is deciding what to ignore. This decision must be placed firmly in the context of a specific question. If interests are qualitative or semi-quantitative, then a basic model, such as the simple SIR model, is sufficient. A limited number of assumptions is indeed a virtue in developing a conceptual understanding of epidemiological patterns. If quantitative results are desired, basic models need to be re-evaluated.

The topic of interest drives the selection of model features. Models of age-specific infection rates require age structure but models of the steady-state do not need a latent period between exposure and infectivity. The appropriate selection may also vary with the epidemiological setting. The United States has a much higher measles immunization rate than Britain and most of its cases are clustered in outbreaks. There is a greater role for chance events as well as outbreak control so that the mass-action measles model would be inappropriate. The most pressing epidemiological question in the United States with regard to measles is its ability to generate outbreaks in heavily immunized populations (Davis et al., 1987). In Africa, it may be necessary to take into account the ability of measles vaccine to reduce the severity of illness. Aaby et al. (1986) present evidence from Guinea-Bissau that measles vaccination not only reduces susceptibility to infection but also reduces the rate of mortality and the degree of infectiousness among those vaccinees who become infected.

Progress depends on collaboration between mathematicians and epidemiologists. Available data do not always permit complete evaluation of competing hypotheses. Models can identify important gaps. Careful comparison between epidemiological observations and theoretical models leads to a better understanding of the dynamics of infection.

References

Aaby, P., Bukh, J., Leerhoy, J., Lisse, I.M., Mordhorst, C.H., Pedersen, I.R. (1986) Vaccinated children get milder measles infection: a community study from Guinea-Bissau. J. Inf. Dis. *154*, 858–863

Anderson, R.M. (1982) Directly transmitted viral and bacterial infections of man. In: Anderson, R.M. (ed.) Population Dynamics of Infectious Diseases. Chapman and Hall, London, Chap. 1

Anderson, R.M., May, R.M. (1982) Directly transmitted infectious diseases: control by vaccination. Science *215*, 1053–1060

Anderson, R.M., May, R.M. (1983) Vaccination against rubella and measles: quantitative investigations of different policies. J. Hyg. (Camb.) *90*, 259–325

Anderson, R.M., May, R.M. (1985) Age-related changes in the rate of disease transmission: implications for the design of vaccination programmes. J. Hyg. (Camb.) *94*, 365–436

Aron, J.L. (1983) Dynamics of acquired immunity boosted by exposure to infection. Math. Biosci. *64*, 249–259

Aron, J.L., Schwartz, I.B. (1984) Seasonality and period-doubling bifurcations in an epidemic model. J. Theor. Biol. *110*, 665–679

Black, F.L., Berman, L.L., Borgono, J.M., Capper, R.A., Carvalho, A.A., Collins, C., Glover, O., Hijazi, Z., Jacobson, D.L., Lee, Y.-L., Libel, M., Linhares, A.C., Mendizabal-Morris, C.A., Simoes, E., Siqueira-Campos, E., Stevenson, J., Vecchi, N. (1986) Geographic variation in infant loss of maternal measles antibody and in prevalence of rubella antibody. Am. J. Epidem. *124*, 442–452

Castillo-Chavez, C., Hethcote, H.W., Andreasen, V., Levin, S.A., Liu, W. (1989) Epidemiological models with age structure, proportionate mixing and cross-immunity. J. Math. Biol. *27*, 233–258

Davis, R.M., Whitman, E.D., Orenstein, W.A., Preblud, S.R., Markowitz, L.E., Hinman, A.R. (1987) A persistent outbreak of measles despite appropriate prevention and control measures. Am. J. Epidem. *126*, 438–449

Fine, P.E.M., Clarkson, J.A. (1982a) Measles in England and Wales—I: An analysis of factors underlying seasonal patterns. Int. J. Epidem. *11*, 5–14

Fine, P.E.M., Clarkson, J.A. (1982b) Measles in England and Wales—II. The impact of the measles vaccination programme on the distribution of immunity in the population. Int. J. Epidem. *11*, 15–25

Fine, P.E.M., Clarkson, J.A. (1983) Measles in England and Wales III. Assessing published predictions of the impact of vaccination on incidence. Int. J. Epidem. *12*, 332–339

Frauenthal, J.C. (1986) Analysis of age-structure models. In: Hallam, T.G., Levin, S.A. (eds.) Mathematical Ecology. Biomathematics, vol. 17. Springer, Berlin Heidelberg New York, pp. 117–147

Hethcote, H.W. (1983) Measles and rubella in the United States. Amer. J. Epidem. *117*, 2–13

Hethcote, H.W. (1989a) Three basic epidemiological models. In: Levin, S.A., Hallam, T.G., Gross, L. (eds.) Applied Mathematical Ecology. Biomathematics, vol. 18. Springer, Berlin Heidelberg New York, pp. 119–144

Hethcote, H.W. (1989b) Rubella. In: Levin, S.A., Hallam, T.G., Gross, L. (eds.) Applied Mathematical Ecology. Biomathematics, vol. 18. Springer, Berlin Heidelberg New York, pp. 212–234

Hinman, A.R., Orenstein, W.A., Bloch, A.B., Bart, K.J., Eddins, D.L., Amler, R.W., Kirby, C.D. (1983) Impact of measles in the United States. Rev. Infect. Dis. *5*, 439–444

Lennon, J.L., Black, F.L. (1986) Maternally derived measles immunity in era of vaccine-protected mothers. J. Pediatrics. *108*, 671–676

Lobel, H.O., McCollum, R.W. (1965) Some observations on the ecology of infectious hepatitis. Bull. WHO *32*, 675–682

London, W.P., Yorke, J.A. (1973) Recurrent outbreaks of measles, chickenpox and mumps. I: Seasonal variation in contact rates. Amer. J. Epidem. *98*, 453–468

May, R.M. (1986) Population biology of microparasitic infections. In: Hallam, T.G., Levin, S.A. (eds.) Mathematical Ecology. Biomathematics, vol. 17. Springer, Berlin Heidelberg New York, pp. 405–442

May, R.M., Anderson, R.M. (1985) Endemic infections in growing populations. Math. Biosci. *77*, 141–156

McLean, A. (1986) Dynamics of childhood infections in high birthrate countries. In: Hoffman, G.W. Hraba, T. (eds.) Immunology and Epidemiology. Lecture Notes in Biomathematics, vol. 65. Springer, Berlin Heidelberg New York, pp. 171–197.

Miller, C.L. (1983) Current impact of measles in the United Kingdom. Rev. Infect. Dis. *5*, 427–32

Schenzle, D. (1984) An age-structured model of pre- and post-vaccination measles transmission. IMA J. Math. Appl. Med. Biol. *1*, 169–191

Yorke, J.A., Nathanson, N., Pianigiani, G., Martin, J. (1979) Seasonality and the requirements for perpetuation and eradication of viruses in populations. Am. J. Epidem. *109*, 103–123

Mathematical Appendix

Lemma 1. *If $\{c_i\}_{i=1,m}$ and $\{G_i\}_{i=1,m}$ have the properties that $c_i \geq 0$, $c_i \geq c_{i+1}$ and $\sum_{i=1}^{k} G_i \geq 0 \forall k\varepsilon[1,m]$, then $\sum_{i=1}^{m} c_i G_i \geq 0$.*

Proof of Lemma 1. The proof follows by induction on $\{G_i\}_{i=1,k}$. The sum is obviously nonnegative for $k = 1$.

$$\sum_{i=1}^{k+1} c_i G_i = c_k \left[\sum_{1}^{k} \left[\frac{c_i}{c_k} - 1 \right] G_i \right] + c_{k+1} \left[\frac{c_k}{c_{k+1}} \left[\sum_{1}^{k} G_i \right] + G_{k+1} \right]$$

The left term on the right hand side is nonnegative if we assume that the property holds for the sequence $\{G_i\}_{i=1,k}$. This follows because the sequence $\{c_i/c_k - 1\}_{i=1,k}$ is non-negative and non-increasing. The right term on the right hand side is non-negative because it is greater than $c_{k+1} \sum_{1}^{k+1} G_i$. Therefore, $\sum_{1}^{k+1} c_i G_i \geq 0$, Q.E.D.

Lemma 2. *If $\int_{0}^{a} g(w)dw \geq 0 \forall a \geq 0$ and if $c(w)$ does not increase with w, then $\int_{0}^{a} g(w)c(w)dw \geq 0$.*

Proof of Lemma 2. If we partition $[0,a]$ into n intervals $[0,a_1], [a_1,a_2], \ldots, [a_{n-1},a_n]$ such that $g(w)$ does not change sign in any interval, then

$$\int_{0}^{a} g(w)c(w)dw = \sum_{i=1}^{n} \int_{a_{i-1}}^{a_i} g(w)c(w)dw = \sum_{i=1}^{n} \hat{c}_i G_i$$

where $G_i = \int_{a_{i-1}}^{a_i} g(w)dw$ and $\hat{c}_i$ is a mean value. Since $c(w)$ does not increase with w, $\{\hat{c}_i\}_{i=1,n}$ is a non-increasing sequence. The result now follows from Lemma 1.

Lemma 2 applies to the condition in (17) if $g(w)$ is replaced by the difference $(s_1(w) - s_2(w))$.

Note: If $c(a)$ is differentiable, the proof of Lemma 2 can be simplified using integration by parts.

$$\int_0^a g(w)c(w)dw = c(a)\int_0^a g(w)dw - \int_0^a\int_0^w g(v)dv\,c'(w)dw \geq 0$$

Periodicity in Epidemiological Models

Herbert W. Hethcote and Simon A. Levin

Contents

Various epidemiological mechanisms have been shown to lead to periodic solutions. The most direct way in which periodicity arises is through extrinsic forcing by a parameter such as the contact rate, but periodicity can also arise autonomously. Cyclic models of SIRS or SEIRS type can have periodic solutions if there is a large time delay in the removed class. Epidemiological models with nonlinear incidence of certain general forms can have periodic solutions. Some models with variable population size and disease-related deaths have periodic solutions; most of these are host-parasite models where the parasite lifetime is much shorter than that of the host. Recently, periodic solutions have been found numerically in age structured models with cross immunity between two viral strains.

1. Introduction

Periodicity and other oscillatory behaviors have been observed in the incidence of many infectious diseases, including measles, mumps, rubella, chickenpox, poliomyelitis, diphtheria, pertussis and influenza. In some locations the incidence of some diseases such as chickenpox, mumps and poliomyelitis goes up and down every year. For example, there were yearly outbreaks of chickenpox and mumps from 1929 to 1970 in New York City (London and Yorke, 1973). The reported incidence of gonorrhea in the United States has oscillated seasonally for at least the last forty years (Hethcote and Yorke, 1984). This period of one year appears to be due to a seasonal variation in some factor such as the contact rate. Contact rates may vary seasonally due to weather changes and the periodic aggregation of children in schools.

The observed interepidemic periods are longer for some diseases. For example, measles outbreaks occurred every two to three years in New York City and Baltimore from 1929 to 1970 (London and Yorke, 1973). Fine and Clarkson (1982, 1983) found that, although measles incidence oscillates biennially in England and Wales, the contact rate varies annually with increases at the openings of school terms and decreases at the closings. In many countries the interepidemic period is 2 to 6 years for infectious diseases such as whooping cough, poliomyelitis, chickenpox, rubella, mumps, diphtheria and scarlet fever (Anderson, 1982).

Although the periodic outbreaks may look like epidemics, the disease often remains endemic in the country or continent by low-level transmission between the major outbreaks. The susceptible fraction may be low after an outbreak, but usually increases as new susceptibles are born or as recovered people lose their immunity. When the susceptible fraction is sufficiently large, a new outbreak may be triggered by the low-level background transmission or by a case from outside the country. A related phenomenon (Liu, 1987) occurs when the disease has a refuge in a second host population, which can serve as a source to trigger an epidemic when the susceptible fraction of the original host population is sufficiently large.

Because of the observed periodicity in the incidence of many diseases, there has been great interest in the ways in which periodic solutions can arise in epidemiological models. In a survey of epidemic models, Hethcote et al. (1981c) found that the mathematical epidemiology literature up to 1981 supported the conjecture that without external forcing or time-dependent coefficients, a single-population epidemiological model with bilinear incidence rates, constant population size and constant parameter values can have periodic solutions if and only if the model is cyclic of SIRS or SEIRS type and individuals can be "significantly delayed" in the removed class by a mechanism such as a large constant period of temporary immunity.

This was a useful way to organize and unify the conclusions from the mathematical epidemiology literature up to 1981. However, the conjecture was based on models with bilinear incidence λIS; recent work (Liu, Levin and Iwasa, 1986; Liu, Hethcote and Levin, 1987) described in Sect. 4 shows that nonlinear incidence can lead to periodic solutions even in models that are not cyclic. Furthermore, other modifications of the basic assumptions also can lead to periodicity and other sustained oscillations. Therefore, rather than trying to patch the 1981 conjecture with additional qualifications, we have chosen to survey the known ways in which periodicity can arise. Undoubtedly, this list will be enlarged in the future. The benefit of having such a catalogue available goes beyond simply demonstrating that similar phenomena in nature can have distinct causes; it also provides a systematic way to organize investigations of the mechanisms underlying observed patterns.

Models with periodic coefficients are described in Sect. 2 and models with delays in the removed class are considered in Sect. 3. Section 4 is devoted to models with nonlinear incidence and Sect. 5 considers models with variable population sizes. Models with age structure are presented in Sect. 6. It is useful to classify the various mechanisms as follows. Periodicity can arise from causes related to the structure of the epidemiological model itself, including delays or periodicities in the parameters; such influences are considered in Sects. 2 to 4. Alternatively, periodicity may have little to do with epidemiology and may arise from demographic or other causes. For example, any model with variable population size or age structure, if based on an underlying demographic model that has an oscillatory character, can be expected to demonstrate the same character when the population is subdivided into epidemiological classes. Thus, as discussed to some extent in Sects. 5 and 6, epidemiological models with variable population size or age structure can exhibit oscillations that have nothing to do with the epidemiology. Perhaps most intriguing are the class of models, only recently investigated, that exhibit oscillations due to an interplay between epidemiological and demographic features. Examples include the interplay between age specific mortality and the existence of multiple strains in models of fixed population size; the influence of alternative host species in influenza and other diseases and the oscillatory and chaotic behavior that can arise indirectly when epidemiological models are forced by periodic parameter variation.

The paper complements the 1981 survey by focusing exclusively on periodicity and related sustained oscillations in epidemiological models and by including recent results. It must be recognized that recurrent epidemics in nature often are damped in magnitude in successive outbreaks. Many models such as those with age-dependent contact rates do exhibit slowly damped oscillations, and cannot be eliminated as explanations of observed natural variations.

In the models that follow, let $S(t)$ be the fraction of the population that is susceptible at time t, $E(t)$ be the fraction that is latent (i.e., infected but not yet infectious), $I(t)$ be the fraction that is infectious, and $R(t)$ be the fraction that is removed by immunity (i.e., removed from the susceptible-infective interactions, but not removed from circulation in the population). Models are usually named according to the flow between classes in the model; for example, in an SEIRS model, an individual is susceptible, then exposed (latent), then infectious, then immune upon recovery and finally susceptible again when the temporary immunity disappears. See Hethcote (1976, 1989) for the three basic epidemiological models and Hethcote, Stech and van den Driessche (1981c) for the formulation of functional differential and integral equation models. Pitfalls in the formulation of models for heterogeneous populations are explained in Hethcote and Van Ark (1987).

2. Models with Periodic Coefficients

In a chapter on recurrent epidemics and endemicity, Bailey (1975) described the deterministic models used in 1929 by Soper and the similar models used in the early 1940's by Wilson and Worcester (1945a, b). These models have damped oscillations; however, a periodic infection rate leads to periodic solutions with the same period as the infection rate. Using stochastic models, Bartlett (1956, 1960) presented the concept of a critical community size above which oscillations in disease incidence can be maintained in a community.

Many authors have found that periodic coefficients in deterministic epidemiological models lead to periodic solutions. For an SIS model, Hethcote (1973) found explicit periodic solutions for a differential equation with a periodic contact rate. Cooke and Kaplan (1976) proved that periodic solutions exist for a scalar delay integral equation describing an SIS model with periodic coefficients. Their proof used fixed points on cones in Banach spaces. Similar methods were used by Nussbaum (1977, 1978), Smith (1977, 1978, 1979) and Busenberg and Cooke (1978a, b) to prove the existence of periodic solutions for a variety of models with periodic coefficients including an SEIRS model with three delays.

For *n* subgroups in an SIS model, Lajmanovich and Yorke (1976) proved global stability for the endemic equilibrium above the threshold. Aronsson and Mellander (1980) extended this result by proving that when the contact rates are periodic, the periodic solution is globally asymptotically stable. Hirsch (1984) generalized the Lajmanovich and Yorke result to a more general SIS model. Smith (1986) proved that when the Hirsch model is periodic, the periodic solution is globally stable above the threshold. These results of Hirsch and Smith are important since they imply that the Lajmanovich and Yorke, and Aronsson and Mellander results have a certain robustness; namely, the dynamical behavior is the same as long as the incidence and removal terms in the differential equations have the same general behavior. The Hirsch result suggests that any observed fluctuations in incidence are not due to the intrinsic dynamics of the disease so they must be due to fluctuations in epidemiological or environmental factors or in reporting.

Hethcote and Yorke (1984) analyzed an SIS gonorrhea model that subdivided the population into groups of men and women, and with contact rates that have a small periodic component. Using a perturbation analysis they found that a 5% oscillation in the contact rates leads to results consistent with the observed 6% and 10% oscillations in incidence in women and men, respectively. Moreover, the observed peak incidences in August to October are due to an earlier peak in the contact rate in the summer months.

Biennial Oscillations in Measles

Measles incidence has an approximately biennial oscillation in which the peak in the high year may be 80 times the peak in the low year. Numerous epidemiological models have been proposed as explanations of this two year period. London and Yorke (1973) used a delay-differential SEIR model with a contact rate that varied yearly by a factor of two. They found that the contact number and latent period for measles lead to periodic solutions with a two year period, but the parameter values for mumps and chickenpox yield one year periods. Stirzaker (1975) used perturbation methods to analyze a periodic SIR model with all births joining the susceptible class and all deaths in the removed class, but this model is not well posed (Hethcote et al. 1981c).

Dietz (1976) studied two well-posed models: an SIR model with births into the susceptible class and deaths in each class, and the SEIR model

$$S'(t) = -\lambda(t)IS + \mu - \mu S$$
$$E'(t) = \lambda(t)IS - (\varepsilon + \mu)E$$
$$I'(t) = \varepsilon E - (\gamma + \mu)I$$
$$R'(r) = \gamma I - \mu R$$
$$S + E + I + R = 1$$

where $\lambda(t)$ is periodic. The behaviors of the two models are essentially the same. For parameter values consistent with measles, Dietz found that the natural damping period with a constant contact rate is about 2 years; his numerical calculations with periodic contact rates yielded biennial oscillations, provided the contact rate oscillations percentage was not too small. Longer periods of oscillations of 3, 4 or 6 years were found numerically for other parameter ranges. Since the period of the damped oscillations is an integral multiple of the one year periodicity of the contact rate, the oscillations become undamped. This phenomenon, called subharmonic resonance, is a plausible explanation of the biennial oscillation in measles incidence; but it is probably too simplistic in view of the work of Schenzle (1984) described in Sect. 6.

Grossman, Gumowski and Dietz (1977) used formal perturbation methods on the model above to obtain biennial periodic solutions. Grossman (1980) extended this perturbation analysis to models with delays. Smith (1982a, 1982b) proved the existence of two-year periodic solutions of the SIR model with vital dynamics and showed that for many values of n, n-year periodic solutions coexist and are simultaneously stable. Schwartz (1983) found numerically that an SEIR measles model has coexisting solutions with periods of 1, 2 and 3 years. Schwartz and Smith (1983) showed both analytically and numerically that an SEIR model has coexisting stable, large-amplitude subharmonic solutions of period n for many values of n, including $n = 2$. They suggested that random environmental changes could perturb the system from the domain of attraction of one sub-harmonic to another so that the incidence would look aperiodic. As a possible example, they cited measles, which often exhibits either two or three year intervals between major outbreaks.

Aron and Schwartz (1984a, b) showed numerically for the SEIR model that small amplitude periodic solutions exhibit a sequence of period-doubling bifurcations as the amplitude of the seasonal variation in transmission increases: this predicts a transition to chaotic behavior. The period-doubled solutions all appear as alternating high- and low-incidence years; effectively only biennial epidemics can be explained by this model since the longer periods change only the low years, in ways that cannot be distinguished statistically. The appearance of period-doubling occurs for higher basic reproduction rates corresponding to measles, but not for lower rates corresponding to rubella. Thus they suggested that the seasonal mechanism generating biennial epidemics may not be able to account for small-amplitude recurrent epidemics of arbitrary periodicity. Schwartz (1985) numerically determined the basins of attraction for the coexisting small and large amplitude periodic solutions. The basins of attraction for two coexisting

stable periodic solutions are intertwined in a complicated (fractal) manner, so that
it may not be possible to predict the outbreak type due to the uncertainty in the
initial data.

A seasonally-forced SEIR model can have a sequence of period doubling
subharmonic bifurcations such that the dynamical system can exhibit chaos (Aron
and Schwartz, 1984a). Schaffer (1985) and Schaffer and Kot (1985) have studied
epidemiological data and models related to periodic childhood diseases by using
time series analysis and recent dynamical systems methods. They have sought to
determine if the epidemiological dynamics are 1) damped oscillations maintained
by noise, or 2) stable cycles with noise where the period is some multiple of the
seasonal (yearly) forcing, or 3) inherently chaotic fluctuations.

Nonlinear systems of differential equations can exhibit a variety of dynamic
behaviors including point attractors, stable limit cycles, toroidal flow (including
quasiperiodic and phase-locked trajectories) and various sorts of chaotic attractors.
Chaotic systems are deterministic, yet their solutions can look like stochastic
systems. Some chaotic systems have maximal algorithmic complexity in the sense
that the solutions are indistinguishable from a set of random numbers and the
most economical description is the time series of the solution. Other chaotic systems
have some structure and methods have been developed to detect this structure.

Plotting the orbit in phase space can reveal whether the system corresponds
to a known kind of chaos. Takens (1981) showed how to construct the phase
portrait for a system with n state variables from a univariate time series. Orbits
of fractal or Hausdorff dimension almost 2 or between 2 and 3 often occur
for ecology and epidemiological models. Lyapunov exponents are generalized
eigenvalues that average the motion over an attractor. A fractal dimension
(computed from the Lyapunov exponents) near 2 means the flow is essentially
two-dimensional so that the Poincaré section (a cross section perpendicular to the
flow) is thin. A difference equation corresponding to the Poincaré or time-1 map
may turn out to be a standard type about which much is known.

Schaffer and Kot (1985) using the methods above analyzed measles data in
New York City and Baltimore and found essentially two-dimensional chaos in the
presence of noise (the fractal dimension is about 2.5). Chickenpox data from New
York City did not yield a one dimensional map so they suggested that there is a
simple yearly cycle with noise. Schaffer (1985) not only analyzed actual data for
diseases such as measles and chickenpox, but also analyzed simulated data. For
example, data with noise were generated using a seasonally forced SEIR model
with parameter values in the chaotic region. When this simulated data set was
analyzed, the phase space reconstruction, Poincaré section and time-1 map were
similar to those obtained from actual data. Thus the chaotic theory is not
inconsistent with the commonly accepted SEIR model. Unfortunately, there does
not seem to be any precise statistical measure of goodness of fit, so the similarities
observed are only visual. Schaffer also showed that Bartlett's (1956, 1960) hypothesis
that recurrent epidemics are sustained by chance perturbations did not lead to
data that yield phase space reconstructions, Poincaré sections and time-1 maps
that were similar to those for measles data.

Schwarz (1988) provides further numerical results on the seasonally forced

SEIR model. As the amplitude of the yearly periodic forcing increases from zero, the following behavior with parameter values for measles is found numerically: 1) a stable period-2 orbit bifurcates off the period-1 orbit which becomes unstable, 2) then a stable period-3 branch of orbits and an unstable orbit branch appear at a saddle-node bifurcation, 3) a stable period-4 branch of orbits and an unstable orbit branch appear at a saddle-node bifurcation, 4) the period-3 branch of attractive orbits undergo period doubling bifurcations leading to a chaotic attractor, 5) the period-4 saddle orbit branch appears to be the cause of the disappearance of the period-3 branch chaotic attractor and to lead to a period-4 branch chaotic attractor. No chaos is observed along the period-1 branch for fluctuations of the seasonal forcing that are less than 20%.

Myxomatosis

In an analysis of the interaction between European rabbits and the viral disease myxomatosis, Dwyer, Levin and Buttel (preprint, 1988) considered a very detailed SIR model involving age structure, demographic seasonality, and detailed accounting of the etiology of disease. With demographic seasonality suppressed, they examined the sensitivity of the model to parameter variation. At low natural mortalities, the typical behavior of the model is rapidly damped oscillation to a stable equilibrium. However, as natural mortality is increased, there is a bifurcation to periodic behavior determined by the maturation delay that is incorporated in the population's life table. As the natural mortality is increased further, the period of the oscillation increases; the amplitude first increases and then decreases, and ultimately the population crashes.

When demographic seasonality is introduced into the model above, the annual period used to force the demographic parameters interacts in fascinating ways with the natural oscillatory dynamics represented in the analysis above. When the natural period of oscillation is much less than a year, as it is at low natural mortalities, the intrinsically and extrinsically induced fluctuations remain separable: there is a dominant annual period, with several shorter and less pronounced oscillations superimposed upon it. At the other extreme, that of a long natural period, a complementary situation exists: annual fluctuations are superimposed on the broader trends that are exhibited by the unforced model. In the middle regions, in which the natural fluctuations appear with a period not too different from 12 months, a complicated interference pattern results in apparently chaotic dynamics. In general, computer simulations provide us a powerful tool to explore the interplay between multiple sources of fluctuation and should lead to new insights in years to come.

3. Models with Delays in the Removed Class

Numerical simulations by Hoppensteadt and Waltman (1971), Mosevich (1975), and Boland and Powers (1977) of an integral equation SEIRS model with three time delays suggested the existence of nontrivial periodic solutions. Periodic solutions do not exist for the ordinary differential equation SIRS model since solutions always approach equilibrium points (Hethcote, 1976). Green (1978) considered the SIRS model with constant time delays in the infectious and removed classes and showed, using a combination of numerical and Hopf bifurcation techniques, that the associated system of delay-differential equations has nontrivial periodic solutions. Busenberg and Cooke (1980) subsequently pointed out that this approach could be used to prove the existence of a periodic bifurcation from the endemic equilibrium for this SIRS model.

After all of the numerical studies above, it seemed that some epidemiological models with delays could have periodic solutions; but the essential structure causing the periodicity was unknown. In particular, it was not known if any models with one delay or no delays could have periodic solutions. Since it is important for those who use models for specific diseases to known which models lead to periodic solutions, answers to the questions above became a goal of much research.

Hethcote, Stech and van den Driessche (1981a) considered an SIRS model with an infective removal rate proportional to the infectious fraction and a constant length of immunity w. For $t > w$ the system is:

$$I'(t) = \lambda I S - \gamma I$$

$$R(t) = \int_{t-w}^{t} \gamma I(u)\,du \qquad\qquad (3.1)$$

$$S(t) + I(t) + R(t) = 1$$

They showed that as the ratio of the period of immunity w to the average infectious period $1/\gamma$ increases the endemic equilibrium for (3.1) loses its asymptotic stability and there is a Hopf bifurcation to a locally asymptotically stable periodic solution. Similarly, they showed that for particular parameter values, the ordinary differential equation $SIR_1 R_2 \cdots R_n S$ model has periodic solutions whenever $n \geq 3$. This model corresponds to an SIRS model with a gamma-distributed time delay in the removed class. They also showed that the endemic equilibrium for the SIRS model with an arbitrarily distributed time delay in the infectious class and an exponentially distributed delay in the removed class is always locally asymptotically stable; hence, periodic solutions cannot arise by Hopf bifurcation.

Thus, a large constant length of immunity or a delay induced by a sequence of at least three removed classes is sufficient for the existence of periodic solutions. Stech and Williams (1981) considered the SIRS model with an arbitrarily distributed time delay in the removed class and obtained qualitative conditions implying stability of the endemic equilibrium, as well as qualitative conditions implying its instability. Their results suggest that at a change in stability there is generically a Hopf bifurcation from the endemic equilibrium. Van den Driessche (1981) incorporated the effects of vital dynamics into (3.1) and showed that this tends to have a stabilizing effect on the endemic equilibrium.

At about the same time Gripenberg (1980) used a bifurcation method to show that a class-age SIRS model has periodic solutions. Also, Diekmann and Montijn (1982) showed that periodic solutions arise by Hopf bifurcation for a class-age model equivalent to an SEIRS model. Smith (1986) also showed that periodic solutions can occur in an SEIRS model. For SEIS models with distributed time delays in the exposed and infectious class, Hethcote, Stech and van den Driessche (1981b) proved that above the threshold the endemic equilibrium is always locally asymptotically stable if both time delays are constant or if any one delay is exponentially distributed. For SIR and SEIR models with permanent immunity, vital dynamics and distributed infectious period, Hethcote and Tudor (1980) showed that periodic solutions do not occur.

The experience gained from the papers above and others led Hethcote, Stech and van den Driessche (1981c) to the conjecture already mentioned regarding when periodic solutions can occur in epidemiological models. This conjecture, however, did not consider that periodic solutions could arise in models with nonlinear incidence as described in the next section.

4. Models with Nonlinear Incidence

The incidence rate is the rate of new infections. In most epidemiological models, the incidence rate is assumed to be bilinear in the infective fraction I and the susceptible fraction S. This bilinear incidence rate of λIS is consistent with the law of mass action in chemistry, in which a chemical reaction occurs during a collision between two active molecules which are moving randomly in the gas or liquid. Epidemiological models assume that individuals are moving around in a community and that the infection is transmitted when a susceptible and an infective come into contact or "collide". The bilinear incidence rate λIS is certainly a reasonable assumption; however, nonlinear incidence rates that include saturation or that are nearly bilinear are also reasonable. Some models with nonlinear incidence can have periodic solutions without seasonal variation in the parameters.

The use of bilinear incidence is so well entrenched in the literature that many researchers feel very strongly that it should be accepted as the null hypothesis unless there is a compelling reason to the contrary. Dietz (personal communication) has argued to us that nonlinear incidence rates do not represent reasonable epidemiological models, and that their consideration should be actively discouraged and replaced by heterogeneous population models built on the assumption of bilinear incidence within and between groups. Our view is different. Certain functional forms other than bilinear incidence can be motivated on first principles; and indeed, at high densities of infectives, saturation effects seem to us as much a logical necessity as are saturation effects in population growth models. Non-linearities of the other form, in which infection rates increase disproportionately as the density of infectives increases, represent more special situations that can only hold over limited ranges of variables. Analyses of such models are nonetheless very suggestive and are likely to be qualitatively correct near bifurcation points.

A variety of nonlinear incidence rates have been used. Wilson and Worcester (1945a, b) used λIS^q, but concluded that $q = 1$ fits the data best. Severo (1969) formulated models with incidence rates kI^pS^q where $q < 1$, but did not analyse them. Yorke and London (1973) simulated measles outbreaks and found that an incidence rate given by $g(I)S = \beta I(1 - CI)S$ with positive C and time dependent β gives good agreement with observation. Capasso and Serio (1978) used $g(I)S$ where $g(I)$ tends to a saturation level when I gets large, e.g., $g(I) = \beta I/(1 + \beta \delta I)$. May and Anderson (1979) derived a similar form using different time scales to approximate complex transmission processes of parasitic infections with indirect life cycles. Gani (1978) argued that similar modifications should be used in a model of myxomatosis in rabbits. General functions to represent contact rates were used by Wang (1978) and Cooke (1979).

Hethcote, Stech and van den Driessche (1981b) used the incidence rate $g(I)S$ in an SEIS model. They assumed that $g(I)$ is $C'[0, 1]$, $g(I)$ is positive on $(0, 1]$, $g(0) = 0$ and $g(I)/I$ is nonincreasing. Clearly, no incidence rate of the form kI^pS can satisfy these assumptions unless $p = 1$; however, more general forms such as those given in the previous paragraph qualify. For incidence rates of the form $g(I)S$ where $g(I)$ satisfies the assumptions, there is still a threshold value such that the disease dies out below the threshold and approaches an endemic equilibrium above the threshold. Thus these nonlinear incidence rates do not lead to periodic solutions arising through Hopf bifurcation.

Cunningham (1979) indicated that incidence rates of the form $k(IS)^p$ with $p > 1$ may lead to periodic solutions. Gabriel, Hanisch and Hirsch (1981) used a nonlinear incidence rate in a helminth disease model where reproduction depends on the probability of pairing worms within the host. In a model with a delay in the intermediate host, they found several positive equilibrium points and periodic solutions. Liu, Levin and Iwasa (1986) discussed possible mechnisms that could lead to nonlinear incidence and showed that an SIRS model with nonlinear incidence kI^pS^q can have periodic solutions for $p > 1$.

. Liu, Hethcote and Levin (1987) considered the following SEIRS model with vital dynamics and nonlinear incidence.

$$S'(t) = -\lambda I^p S^q + \mu - \mu S + \delta R$$
$$E'(t) = \lambda I^p S^q - (\varepsilon + \mu)E$$
$$I'(t) = \varepsilon E - (\gamma + \mu)I$$
$$R'(t) = \gamma I - (\delta + \mu)R$$
$$S + E + I + R = 1$$

They also considered the SIS, SIR, SIRS, SEIS and SEIR models that are limiting cases of the SEIRS model above. The phase space portrait of the system above changes greatly as p changes, but does not change significantly when q changes. For $p = 1$ this SEIRS model has the usual behavior in the sense that the disease disappears below the threshold and approaches an endemic equilibrium above the threshold. Thresholds seem reasonable in infectious disease models; in particular, the notion that a contact number or reproductive number (Hethcote, 1976; Hethcote, 1987; Anderson and May, 1982) must exceed unity fits biological intuition. For $0 < p < 1$ the trivial equilibrium is always unstable and the positive equilibrium is asymptotically stable. An unrealistic aspect when $0 < p < 1$ is that there is no threshold so that the disease always remains endemic.

For $p > 1$, the disease disappears below the threshold. Above the threshold, there are three equilibria for $p > 1$ due to a saddle-node bifurcation of the positive equilibrium at the threshold. The trivial equilibrium is always locally asymptotically stable, and the small positive equilibrium always is a saddle; however, the large positive equilibrium can be stable or unstable depending on the values of p and the contact number $\sigma = \lambda\varepsilon/[(\gamma + \mu)(\varepsilon + \mu)]$. Liu, Hethcote and Levin (1987) showed that both stable and unstable periodic solutions can arise by Hopf bifurcation.

Note that periodic solutions can occur in the SEIR model ($\delta = 0$) with nonlinear incidence if $p > 1$. Thus nonlinear incidence can lead to periodicity, even if the model is not cyclic due to temporary immunity as in Sect. 3. Let us consider a typical childhood disease such as measles, mumps, rubella or chickenpox with average latent period $1/\varepsilon = 12$ days $= 1/30$ year, average (death-adjusted) infectious period $1/(\gamma + \mu) = 6$ days $= 1/60$ year, average lifetime $1/\mu = 75$ years and permanent immunity so that $\delta = 0$. Then Hopf bifurcation can occur for $p > p_1 = 1.00067$. If we choose $p = 1.01$ and $q = 1$, then there are three equilibria for $\sigma > \sigma^* = 1.15053$, and Hopf bifurcation occurs at $\sigma = \sigma^{**} = 20.667$. Note that the threshold value σ^* is approximately 1.

If $\sigma < \sigma^* = 1.15053$ then there is only the trivial equilibrium and the disease disappears. For any $\sigma > 2$ the infectious fraction I at the small positive equilibrium is less than 10^{-30} so that it is practically indistinguishable from $I = 0$ and the basin of attraction for this trivial equilibrium point is very small. The large positive equilibrium has an infectious fraction I that is approximately 0.000222 times $(1 - 1/\sigma)$. The stability of the periodic solutions around this large positive equilibrium for σ near σ^{**} can be determined by calculating the sign of a stability constant A given by (42) in Liu, Hethcote and Levin (1987). If A is positive, then there is a subcritical Hopf bifurcation and an unstable periodic solution surrounds the large positive equilibrium for $\sigma > \sigma^{**}$. If A is negative, then there is a supercritical Hopf bifurcation and a stable periodic solution surrounds the large positive equilibrium for $\sigma < \sigma^{**}$. The radius r of these periodic solutions is approximated for σ near σ^{**} by

$$r = 4\left[\frac{d(\sigma - \sigma^{**})}{-A}\right]^{1/2}$$

where d is the derivative with respect to σ of the real part of the complex eigenvalues evaluated at $\sigma = \sigma^{**}$. Thus periodic solutions can occur for realistic parameter values and for p values very near 1; however, these periodic solutions might be so small that they are indistinguishable from the positive equilibrium.

Hethcote, Lewis and van den Driessche (1989) analyzed an SIRS epidemiological model with a time delay in the removed class and a nonlinear incidence rate. They found the thresholds and number of equilibria and evaluated stability. For some parameter values, periodic solutions arise by Hopf bifurcation from the large nontrivial equilibrium state. For this model periodic solutions occur for $1/2 < p < 1$ while they did not occur for $p \leq 1$ in the nonlinear incidence model of Liu, Hethcote and Levin (1987). Moreover, periodic solutions can occur for any $p \geq 1$, while they only occurred for $p \geq p_1 > 1$ in the model of Liu, Hethcote and Levin (1987).

5. Models with Variable Population Size

Many epidemiological models assume that the total population size is constant. Thus if these models include vital dynamics, the birth rate must be equal to the death rate in the population. This assumption is reasonable in human populations for diseases where death due to the disease is rare. Even if the population is growing a few percent per year, this change is often considered to be insignificant and is ignored. However, if the death rate due to the disease is significant, then this must be incorporated and the total population cannot be considered to be constant. Indeed, in animal populations disease-related deaths may regulate the population size.

Anderson and May (1979) and Ross (1916) considered models with disease-related death and variable population size. For example, if $N(t)$ is the total population size and $X(t)$, $Y(t)$ and $Z(t)$ are the numbers in the susceptible, infectious and removed classes, then one SIRS model is

$$
\begin{aligned}
X'(t) &= -\beta X Y + aN - bX + \gamma Z \\
Y'(t) &= \beta X Y - (v + b + \alpha) Y \\
Z'(t) &= vY - (\gamma + b)Z \\
N'(t) &= (a - b)N - \alpha Y
\end{aligned}
\tag{5.1}
$$

where a is the birth rate, b is the natural death rate, and α is the disease-related death rate. For this model, Anderson and May (1979) found that the disease dies out if $r = a - b < 0$ and it persists if $r > 0$. If $0 < r < \alpha(1 + v/(b + \gamma))$, then an equilibrium is approached since the disease-related death regulates the population size. If $r > \alpha(1 + v/(b + \gamma))$, then the sizes of the infectious class, the removed class and the total population grow exponentially while the susceptible class approaches a constant size as $t \to \infty$. Other models for microparasitic (viruses, bacteria, protozoans) infections are also formulated in Anderson and May (1979). One of these models gives good fits to two diseases in laboratory mice populations. Except for the fox rabies model below, the variable population size models of Anderson and May (1979) with direct transmission do not seem to have periodic solutions.

Anderson et al. (1981) proposed the following model for fox rabies in Europe.

$$
\begin{aligned}
X'(t) &= \beta X Y + aX - (b + \gamma N)X \\
I'(t) &= \beta X Y - (\sigma + b + \gamma N)I \\
Y'(t) &= \sigma I - (\alpha + b + \gamma N)Y \\
N'(t) &= aX - (b + \gamma N)N - \alpha Y
\end{aligned}
\tag{5.2}
$$

Here N is the total population size and X, I and Y are the numbers in the susceptible, exposed and infective classes. For certain parameter values, they found numerically that this model has stable periodic limit cycles. The periodicity may occur because the exposed and infectious foxes do not reproduce in this model.

May and Anderson (1979) developed models for both microparasites and macroparasites (helminths and arthropods) transmitted either directly or indirectly by one or more intermediate hosts. They considered the overall dynamics and the mechanisms that yield cyclic states or multiple stable states in the prevalence of the host population. In most of their models for microparasitic and macroparasitic infections, the host population either is regulated to a stable size by the disease for else it grows exponentially. However, May and Anderson (1978) found numerically that the model 5.1 with nonlinear incidence of the form $\lambda X Y/(H_0 + X)$ can lead to stable limit cycles. This term is the result of a host-parasite model where the free-living infective stage of the parasite "collapses" due to a quasi-steady-state assumption. This "collapse" is due to the differences in the time scales of the host and parasite lifetimes.

Anderson and May (1981) presented a variety of models for microparasites and their invertebrate hosts. One model includes dynamical equations for the populations of free-living infective stages of microparasites:

$$X'(t) = -vWX + a(X + Y) - bX + \gamma Y$$
$$Y'(t) = vWX - (\alpha + b + \gamma)Y$$
$$H'(t) = rH - \alpha Y$$
$$W'(t) = \lambda Y - (\mu + vH)W$$

$$(5.3)$$

where the total host population size H is divided into X susceptibles and Y infectives, and W is the number of free-living infective stages of the microparasites. They found numerically that this model has stable limit cycles for some parameter values.

Hence it seems that a variety of host-parasite models with disease-related deaths and variable population sizes can lead to periodic solutions. These host-parasite models are somewhat like predator-prey models that have periodic solutions.

Mena (1988) has analyzed rigorously various models with disease-related deaths, including the models above of Anderson and May. For these models he has determined thresholds, proved local and global stability, determined neutral stability surfaces (where the real part of conjugate pairs of characteristic roots changes sign) and proved that periodic solutions exist by Hopf bifurcation.

6. Models with Age Structure

For many diseases the susceptible fraction of age a decreases as the age a increases. Moreover, the number of contacts of an infective with susceptibles often depends not only on the age of the infective, but also on the ages of the susceptibles. School children tend to have a high contact rate since they mix a lot in school while preschool children, adults and older people often have lower contact rates. Thus it is important to consider epidemiological models with age structure.

Age-structured models were formulated by some of the pioneers in epidemiological modeling such as Ross and Hudson (1917) and Kermack and McKendrick (1927). Dietz (1975) presented the following SIR model with age structure:

$$\frac{\partial x}{\partial a} + \frac{\partial x}{\partial t} = -\lambda(t)x - \mu x$$

$$\frac{\partial y}{\partial a} + \frac{\partial y}{\partial t} = \lambda(t)x - (\gamma + \mu)y$$

$$\frac{\partial z}{\partial a} + \frac{\partial z}{\partial t} = \gamma y - \mu z \qquad\qquad (6.1)$$

$$\lambda(t) = \beta \int_0^\infty y(s, t)ds$$

with initial and boundary conditions

$$x(a, 0) = x_0(a), \quad y(a, 0) = y_0(a), \quad z(a, 0) = z_0(a)$$
$$x(0, t) = N\mu, \quad y(0, t) = 0, \quad z(0, t) = 0$$

where x, y and z are the susceptibles, infectives and removeds, respectively, N is the total population size, μ is the birth and death rate, $\lambda(t)$ is the total infectivity and γ is the removal rate. Similar models with a chronological age a and a class age c in the infectious class were formulated and analysed by Hoppenstadt (1975). Longini et al. (1978) used an ordinary differential equation SIR model with the population divided into age groups to compare vaccination programs for influenza.

Although age-structured models with a proportionate mixing contact matrix or a symmetric contact matrix do not seem to have periodic solutions, it appears that an asymmetric contact matrix can lead to periodic solutions. Enderle (1980) used a discrete Reed-Frost type model with three age groups for measles. When he used an asymmetric contact matrix between his age groups, he numerically obtained periodic solutions with a period of about eight years.

Age-structured models have been used to compare the USA and UK strategies for rubella vaccination (Knox, 1980; Dietz, 1981; Hethcote, 1983; Anderson and May, 1983). The implications of these modeling studies of rubella vaccination were summarized in Hethcote (1986). Age-structured models were applied to a variety of diseases in Anderson and May (1982). None of these models with age-specific incidence seems to have periodic solutions.

Ordinary differential equation models with age groups have been used to study measles. Schenzle (1984) used 21 age groups and a symmetric contact matrix which changes when school opens and closes. His model with its seasonally varying contact matrix fits the biennial data for measles in England and Wales. Moreover, his model fits data from Hamburg in 1900–1913 when new children started school in the spring and there were two peaks per year. As in Sect. 2, it is the periodic contact matrix in the Schenzle model that leads to periodic solutions. Tudor (1985) used a model with 5 age groups and a proportionate mixing contact matrix to study measles vaccination in the USA.

Other age-structured models are described in Dietz and Schenzle (1985a). Dietz and Schenzle (1985b) derived the threshold for a general SEIR model with proportionate mixing, chronological age, class age, and vaccination. El-Doma (1986) proved global stability of the steady-state age distribution for an SIS model with age structure and both horizontal and vertical transmission. Thus the current status for periodicity in age-structured models of constant population size is as follows: sustained periodic solutions have occurred only when the contact matrix is asymmetric (Enderle, 1980) or the contact matrix is periodic (Schenzle, 1984).

Castillo-Chavez et al. (1988, 1989) used various models to gain some insight into the observed multiyear periodicities in influenza and into the roles of cross-immunity and age-dependence. Their simulations and analyses with cross immunity between two strains, but without any age structure, suggest that periodic solutions do not occur. On the other hand, when age-structure is introduced for a single viral strain, again only slowly damped oscillations are observed. However, when two strains are coupled by cross immunity, periodic solutions are found numerically for some age structures; in effect, the weakly damped oscillations due to age structure and those due to the interaction between strains mutually excite one another. In particular, if the survivorship is a negative exponential function of age (approximately true in a developing country), then their two-strain model does not seem to have periodic solutions; however, if the survivorship is constant until age 75 years when everyone dies (approximately true in a developed country), then numerical simulations suggest the existence of periodic solutions (Andreasen, 1987). Thus the multiyear periodicities that are observed in influenza incidence could be due to cross immunity between related strains of an influenza subtype. However, the sensitivity of the observed period to the time step used leaves open many questions concerning the robustness of this conclusion.

7. Conclusion

The list of ways in which periodic behavior and related sustained oscillatory behavior can arise in epidemiological models is a growing one; but these can be grouped into a number of categories. The principal dichotomy is between periodicities that are forced externally and those that arise autonomously, although interesting phenomena occur when both factors are present in the same model.

Autonomous oscillations can arise in a variety of ways, all of which involve some sort of delay. For a single host population without age structure and without periodic coefficients, explicit delays in the removed class can be introduced in ways that give rise to oscillations. Such models introduce a time delay operating through the internal structure of the population. Similarly, delays introduced through age-specific contact rates can lead to sustained oscillations, but apparently only if contacts are not symmetric in transmission effectiveness. These restrictions may be relaxed if the population size is allowed to vary, in which case various mechanisms (such as age-specific fecundities) can lead to oscillations. Finally, a somewhat surprising deviation from these generalizations occurs if the usual assumption of bilinear incidence rates is relaxed. The important ecological relevance of the particular incidence functions that lead to sustained oscillations can be debated, but it is clear that the investigation of the mathematical properties of models with nonlinear incidence functions has raised our awareness of the importance of the form of this term.

With the consideration for various diseases of the interactions between multiple host populations, multiple disease types and multiple groups, including age classes, a whole new set of possibilities is opened up. Preliminary investigation of age-structured models with multiple disease types suggest that sustained oscillations are possible even with constant population size. It should not be a surprise if further investigations of multiple-type models demonstrate a variety of ways in which oscillations can be maintained, since interactions between types provide another way that delays can be introduced.

8. References

Anderson, R.M. (1982) Directly transmitted viral and bacterial infections of man. In: Anderson, R.M. (ed.) Population Dynamics of Infectious Diseases. Theory and Applications. Chapman and Hall, New York, pp. 1–37

Anderson, R.M., Jackson, H.C., May, R.M., Smith, A.D.M. (1981) Populations dynamics of fox rabies in Europe. Nature 289, 765–777

Anderson, R.M., May, R.M. (1979) Population biology of infectious diseases I. Nature 280, 361–367

Anderson, R.M., May, R.M. (1981) The population dynamics of microparasites and their invertebrate hosts. Phil. Trans. Roy. Soc. London B291, 451–524

Anderson, R.M., May, R.M. (1982) Directly transmitted infectious diseases: control by vaccination. Science 215, 1053–1060

Anderson, R.M., May, R.M. (1983) Vaccination against rubella and measles: quantitative investigations of different policies. J. Hyg. Camb. 90, 259–325

Andreasen, V. (1987) Dynamical behaviour of epidemiological models. Preprint

Aron, J.L., Schwartz, I.B. (1984a) Seasonality and period-doubling bifurcations in an epidemic model. J. Theor. Biol. 110, 665–679

Aron, J.L., Schwartz, I.B. (1984b) Some new directions for research in epidemic models, IMA J. Math. Appl. Med. Biol. 1, 267–276

Aronsson, G., Mellander, I. (1980) A deterministic model in biomathematics: asymptotic behavior and threshold conditions, Math. Biosci. 49, 207–222

Bailey, N.T.J. (1975) The Mathematical Theory of Infectious Diseases, Second Edition, Hafner, New York

Bartlett, M.S. (1956) Deterministic and stochastic models for recurrent epidemics, Proc. Third Berkeley Symp. Math. Stat. Prob. 4, 81–109

Bartlett, M.S. (1960) Stochastic Population Models in Ecology and Epidemiology, Methuen, London

Boland, W.R., Powers, M.W. (1977) A numerical technique for obtaining approximate solutions of certain functional equations arising in the theory of epidemics, Math. Biosci. 33, 297–319

Busenberg, S.N., Cooke, K.L. (1978a) Periodic solutions of delay differential equations arising in some models of epidemics, in Proceedings of the Applied Nonlinear Analysis Conference, Univ. of Texas, Arlington, Academic Press, New York

Busenberg, S.N., Cooke, K.L. (1978b) Periodic solutions of a periodic nonlinear delay differential equation, SIAM J. on Applied Math. 35, 704–721

Busenberg, S.N., Cooke, K.L. (1980) The effect of integral conditions in certain equations modelling epidemics and population growth, J. Math. Biol. 10, 13–32

Capasso, V., Serio, G. (1978) A generalization of the Kermack-McKendrick deterministic epidemic model, Math. Biosci. 42, 43–61

Castillo-Chavez, C., Hethcote, H.W., Andreasen, V., Levin, S.A., Liu, W.M. (1988) Cross-immunity in the dynamics of homogeneous and heterogeneous populations. In Hallam, T.G., Gross, L. and Levin, S.A. (eds.) Mathematical Ecology, World Scientific Publishing Co., Singapore, 303–316

Castillo-Chavez, C., Hethcote, H.W., Andreasen, V., Levin, S.A., Liu, W.M. (1989) Epidemiological models with age-structure and proportionate mixing, J. Math. Biology (to appear)

Cooke, K.L. (1982) Models for endemic infections with asymptomatic cases: one group, Math. Modelling 3, 1–15

Cooke, K., Kaplan, J. (1976) A periodicity threshold theorem for epidemics and population growth, Math. Biosci. *31*, 87–104

Cunningham, J. (1979) A deterministic model for measles, Z. Naturforsch *34c*, 647–648

Diekmann, O., Montijn, R. (1982) Prelude to Hopf bifurcation in an epidemic model: analysis of the characteristic equation associated with a nonlinear Volterra integral equation, J. Math. Biol. *14*, 117–127

Dietz, K. (1975) Transmission and control of arbovirus diseases, in Epidemiology, SIMS 1974 Utah Conference Proceedings, SIAM, Philadelphia, pp. 104–121

Dietz, K. (1976) The incidence of infectious diseases under the influence of season fluctuations, in Mathematical Models in Medicine, Lecture Notes in Biomathematics, No. 11, Springer-Verlag, New York, pp. 1–15

Dietz, K. (1981) The evaluation of rubella vaccination strategies. In: Hiorns, R.W., Cooke. D. (eds.) The Mathematical Theory of the Dynamics of Biological Populations, vol. II. Academic Press, London, pp. 81–87

Dietz, K., Schenzle, D. (1985a) Mathematical models for infectious disease statistics. In: Atikinson, A.C., Fienberg, S.E. (eds.) A Celebration of Statistics. Springer-Verlag, New York, pp. 167–204

Dietz, K., Schenzle, D. (1985b) Proportionate mixing models for age-dependent infection transmission, J. Math. Biol. *22*, 117–120

El-Doma, M. (1987) Analysis of nonlinear integro-differential equations arising in age-dependent epidemic models, Nonlinear Anal. TMA *11*, 913–937

Enderle, J.D. (1980) A stochastic communicable disease model with age-specific states and applications to measles, Ph.D. dissertation, Rensselaer Polytechnic Institute

Fine, P.E.M., Clarkson, J.A. (1982) Measles in England and Wales. I. An analysis of factors underlying seasonal patterns, Int. J. Epidem. *11*, 5–14

Fine, P.E.M., Clarkson, J.A. (1983) Measles in England and Wales. III. Assessing published predictions of the impact of vaccination on incidence, Int. J. Epidem. *12*, 332–339

Gabriel, J.P., Hanisch, H., Hirsch, W.M. (1981) Dynamic equilibria of helminthic infections. In: Chapman, D.G., Gallucci, V.F. (eds.) Quantitative Population Dynamics. Intern. Cooperative Publ. House, Maryland, Stat. Ecology Series *13*, 83–104

Gani, J. (1978) Some problems in epidemic theory, J. Roy. Statist. Soc. Ser. *A140*, 323–347

Green, D. (1978) Self-oscillations for epidemic models, Math. Biosci. *38*, 91–111

Gripenberg, G. (1980) Periodic solutions of an epidemic model, J. Math. Biol. *10*, 271–280

Grossman, Z. (1980) Oscillatory phenomena in a model of infectious diseases, Theor. Pop. Biol. *18*, 204–243

Grossman, Z., Gumowski, I., Dietz, K. (1977) The incidence of infectious diseases under the influence of seasonal fluctuations—analytic approach, in Nonlinear Systems and Applications to Life Sciences, Academic Press, New York, pp. 525–546

Hethcote, H.W. (1973) Asymptotic behavior in a deterministic epidemic model, Bull. Math. Biology *35*, 607–614

Hethcote, H.W. (1976) Qualitative analysis for communicable disease models, Math. Biosci. *28*, 335–356

Hethcote, H.W. (1983) Measles and rubella in the United States, Am. J. Epidemiol. *117*, 2–13

Hethcote, H.W. (1986) Choosing a strategy for rubella vaccination, in Proceedings of the Third International Colloquium on Theoretical Biology and Medicine: Models in Epidemiology, Fontevraud, France, September

Hethcote, H.W. (1989) Three basic epidemiological models. In: Levin, S.A., Hallam, T.G. and Gross, L. (eds.) *Applied Mathematical Ecology*, Biomathematics vol. 18, Springer-Verlag, Berlin, Heidelberg, New York, 119–144

Hethcote, H.W., Lewis, M.A., van den Driessche, P. (1989) An epidemiological model with a delay and a nonlinear incidence rate, J. Math. Biol. *27*, 49–64

Hethcote, H.W., Stech, H.W., van den Driessche, P. (1981a) Nonlinear oscillations in epidemic models. SIAM J. Appl. Math. *40*, 1–9

Hethcote, H.W., Stech, H.W., van den Driessche, P. (1981b) Stability analysis for models of diseases without immunity. J. Math. Biology *13*, 185–198

Hethcote, H.W., Stech, H.W., van den Driessche, P. (1981c) Periodicity and stability in epidemic models:

a survey, In: Busenberg, S., Cooke, K.L. (eds.) Differential Equations and Applications in Ecology, Epidemics and Populations Problems. Academic Press, New York, pp. 65–82

Hethcote, H.W., Tudor, D.W. (1980) Integral equation models for endemic infectious diseases, J. Math. Biol. *9*, 37–47

Hethcote, H.W., Van Ark, J.W. (1987) Epidemiological models for heterogeneous populations: proportionate mixing, parameter estimation and immunization programs, Math. Biosci. *84*, 85–118

Hethcote, H.W., Yorke, J.A. (1984) Gonorrhea Transmission Dynamics and Control. Lecture Notes in Biomathematics, vol. 56. Springer, Berlin Heidelberg New York

Hirsch, M.W. (1984) The differential equations approach to dynamical systems, Bull. Amer. Math. Soc. *11*, 1–64

Hoppensteadt, F. (1975) Mathematical Theories of Populations: Demographics, Genetics and Epidemics, SIAM, Philadelphia

Hoppensteadt, F., Waltman, P. (1971) A problem in the theory of epidemics II, Math. Biosci. *12*, 133–145

Kermack, W.O., Mckendrick, A.G. (1927) A contribution to the mathematical theory of epidemics, Proc. Roy. Soc. *A115*, 700–721

Knox, E.G. (1980) Strategy for rubella vaccination, Int. J. Epidemiol. *9*, 13–23

Lajmanovich, A., Yorke, J.A. (1976) A deterministic model for gonorrhea in a nonhomogeneous population, Math. Biosci. *28*, 221–236

Liu, W.M. (1987) Dynamics of epidemiological models—recurrent outbreaks in autonomous systems, Ph.D. Thesis. Cornell University

Liu, W.M., Levin, S.A., Iwasa, Y. (1986) Influence of nonlinear incidence rates upon the behavior of SIRS epidemiological models, J. Math. Biol. *23*, 187–204

Liu, W.M., Hethcote, H.W., Levin, S.A. (1987) Dynamical behavior of epidemiological models with nonlinear incidence rates. *25*, 359–380

London, W.P., Yorke, J.A. (1973) Recurrent outbreaks of measles, chickenpox and mumps, I, Am. J. Epid. *98*, 453–468

Longini, Jr., I.M., Ackerman, E., Elveback, L.R. (1978) An optimization model for influenza A epidemics, Math. Biosci. *38*, 141–157

May, R.M., Anderson, R.M. (1978) Regulation and stability of hostparasite population interactions. II. Destabilizing processes, J. Animal Ecology *47*, 249–267

May, R.M., Anderson, R.M. (1979) Population biology of infectious diseases II, Nature *280*, 455–461

Mena, J. (1988) Periodicity and stability in epidemiological models with disease-related deaths, Ph.D. thesis in Mathematics, University of Iowa

Mosevich, J. (1975) A numerical method for approximating solutions to the functional equation arising in the epidemic model of Hoppensteadt and Waltman, Math. Biosci. *24*, 333–344

Nussbaum, R. (1977) Periodic solutions of some integral equations from the theory of epidemics, in Nonlinear Systems and Applications to Life Sciences. Academic Press, New York, pp. 235–255

Nussbaum, R. (1978) A periodicity threshold theorem for some nonlinear integral equations, SIAM J. Math. Anal. *9*, 356–376

Ross, R. (1916) An application of the theory of probabilities to the study of a priori pathometry, Part I, Proc. Roy. Soc. *A92*, 204–230

Ross, R., Hudson, H.P. (1917) An application of the theory of probabilities to the study of a priori pathometry—Part III, Proc. Roy. Soc. *A93*, 225–240

Schaffer, W.M. (1985) Can nonlinear dynamics help us infer mechanisms in ecology and epidemiology?, IMA J. Math. Appl. Biol. Med. *2*, 221–252

Schaffer, W.M., Kot, M. (1985) Nearly one dimensional dynamics in an epidemic, J. Theor. Biol. *112*, 403–427

Schenzle, D. (1984) An age structured model of pre and post-vaccination measles transmission, IMA J. Math. Appl. Biol. Med. *1*, 169–191

Schwartz, I.B. (1983) Estimating regions of existence of unstable periodic orbits using computer-based techniques, SIAM J. Num. Anal. *20*, 106–120

Schwartz, I.B. (1985) Multiple recurrent outbreaks and predictability in seasonally forced nonlinear epidemic models, J. Math. Biol. *21*, 347–361

Schwartz, I.B. (1988) Nonlinear dynamics of seasonally driven epidemic models, Preprint

Schwartz, I.B., Smith, H.L. (1983) Infinite subharmonic bifurcations in an SEIR model, J. Math. Biol. *18*, 233–253

Severo, N.C. (1969) Generalizations of some stochastic epidemic models, Math. Biosci. *4*, 395–402

Smith, H.L. (1977) On periodic solutions of a delay integral equation modeling epidemics, J. Math. Biol. *4*, 69–80

Smith, H.L. (1978) Periodic solutions for a class of epidemic equations, J. Math. Anal. and Applic. *64*, 467–479

Smith, H.L. (1979) Periodic solutions for an epidemic model with a threshold, Rocky Mountain J. of Math. *9*, 131–142

Smith, H.L. (1983) Subharmonic bifurcation in an SIR epidemic model, J. Math. Biology *17*, 163–177

Smith, H.L. (1983) Multiple stable subharmonics for a periodic epidemic model, J. Math. Biology *17*, 179–190

Smith, H.L. (1983) Hopf bifurcation in a system of functional equations modeling the spread of an infectious disease, SIAM J. Appl. Math. *43*, 370–385

Smith, H.L. (1986) Cooperative systems of differential equations with concave nonlinearities, J. Nonlin. Anal. T.M.A. *10*, 1037–1052

Stech, H.W., Williams, M. (1981) Stability for a class of cyclic epidemic models with delay, J. Math. Biol. *11*, 95–103

Stirzaker, D.R. (1975) A perturbation method for the stochastic recurrent epidemic, J. Inst. Maths Applics *15*, 135–160

Takens, F. (1981) Detecting strange attractors in turbulence, in Dynamical Systems and Turbulence. In: Rand, D.A., Young, L.S. (eds.) Warwick, 1980, Springer-Verlag, New York, pp. 366–381

Tudor, D.W. (1985) An age dependent epidemic model with application to measles, Math. Biosci. *73*, 131–147

van den Driessche, P. (1981) An SIRS model with constant temporary immunity and constant births and deaths. In: Freedman, H.I., Strobeck, D. (eds.) Population Biology. Lecture Notes in Biomathematics, vol. 52. Springer, Berlin Heidelberg New York, pp. 433–440

Wang, F.J.S. (1978) Asymptotic behavior of some deterministic epidemic models, SIAM J. Math. Anal. *9*, 529–534

Wilson, E.B., Worcester, J. (1945) The law of mass action in epidemiology, Proc. N.A.S. *31*, 24–34

Wilson, E.B., Worcester, J. (1945) The law of mass action in epidemiology, II., Proc. N.A.S. *31*, 109–116

Yorke, J.A., London, W.P. (1973) Recurrent outbreaks of measles, chickenpox and mumps II, Am. J. Epid. *98*, 469–482

Rubella

Herbert W. Hethcote

Contents

1. Introduction

George Maton in 1814 realized that there was a mild illness characterized by rash, adenopathy and no fever that was distinct from scarlatina. This new disease was named rubella by Henry Veale in 1866. In 1942 an Australian ophthalmologist, Norman Gregg, noticed that German measles (rubella) infection in the first trimester of pregnancy caused serious birth defects in the offspring. The rubella virus was isolated in tissue culture in 1962 at two different laboratories. After the severe rubella epidemic in 1964, it was recognized that congenital rubella syndrome (CRS) included not only cardiovascular lesions, cataracts, deafness, mental retardation, central nervous system abnormalities and generalized growth retardation, but also bone lesions, hepatitis, meningoencephalitis, progressive rubella panencephalitis and eventual diabetes mellitus. In 1969, attenuated rubella vaccines became available for use in the United States (Cooper, 1985).

Rubella is a mild illness in most individuals; indeed, many infections occur without evident rash or other symptoms. Congenital rubella syndrome (CRS) occurs in at least 20% of infants born to women who acquired rubella infection during the first trimester of pregnancy. The incidence of CRS is difficult to estimate because of underreporting of cases and the delayed appearance of symptoms in some children. Rubella vaccines are safe, effective and provide permanent immunity (CDC, 1984).

Studies show that in the prevaccine era many people got rubella in childhood. In most studies in developed countries about one-half of the children 6 to 8 years of age had antibody to rubella while 80% to 90% had antibody at 17–22 years of age and the percentage remained constant for older women (Assaad and Ljungars-Esteves, 1985). Rubella epidemics occurred approximately every four to ten years and the susceptibility of the age groups depended on the length of time since the last epidemic. Most studies in European, Asian, African and North American countries indicate that 10% to 25% of women of childbearing age are susceptible to rubella (Assaad and Ljungars-Esteves, 1985; Hinman et al., 1983). These percentages are sometimes lower in less dense rural areas and in island populations with infrequent mainland contacts, partly because the period between rubella epidemics is often longer in these populations.

The last rubella epidemic in 1964–65 in the United States of America (USA) resulted in an estimated 11,000 wasted pregnancies and 20,000 children with congenital defects (Orenstein et al., 1984). There have been no major rubella epidemics in the USA since vaccination was started in 1969 (see Fig. 1 and 2). Rubella vaccines have been extensively used in some North American and European countries, but only sparsely in the developing world. Indeed, a major question is whether rubella vaccination is cost effective in developing countries.

Two distinct approaches to rubella vaccination are used. The USA strategy of universal vaccination of young children seeks to protect pregnant women by interrupting or reducing rubella transmission. Since it started in the United Kingdom (UK), the program of directly protecting women by vaccination of prepubertal girls against rubella is called the UK strategy. After these and other strategies are described in Sect. 2, their long term implications are presented in Sect. 3, cost-benefit comparisons are made in Sect. 4 and short term implications are presented in Sect. 5. The epidemiologist's recommendations in Sect. 6 agree with the theoretical results and lead to guidelines in Sect. 7 for choosing a rubella vaccination strategy. In Sects. 8–12, the details of rubella incidence and vaccination strategies are given for the USA, United Kingdom, France, Netherlands and Africa.

For more information on rubella infection, congenital rubella syndrome, rubella vaccines and rubella epidemiology, see the proceedings of the International Symposium on Prevention of Congenital Rubella Infection held at the Pan American Health Organization in Washington, D.C. in 1984 (Proceedings, 1985).

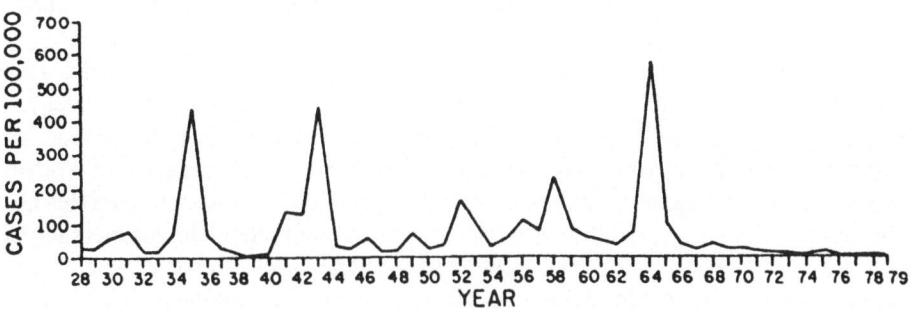

Fig. 1. Reported rubella incidence in ten selected areas of the United States of America from 1928 to 1979. From CDC (1983)

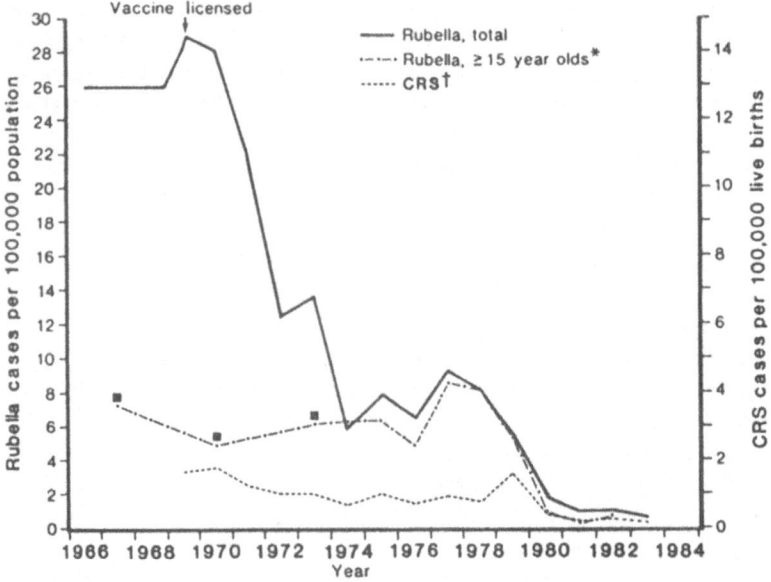

Fig. 2. Incidence rate of reported rubella cases and congenital rubella syndrome cases in the United States of America from 1966 to 1983. From CDC (1983)

* Includes proration of unknown age cases in ≥ 15 year olds. 1983 data are not available.
† Rate per 10^5 births of confirmed and compatible cases of CRS by year of birth. Reporting for recent years is provisional, as cases may not be diagnosed until later in childhood.
■ Average annual United States estimate based on data from Illinois, Massachusetts, and New York City for the 3-year periods 1966–1968, 1969–1971, and 1972–1974. Age specific data were not available for U.S. totals until 1975.

2. Rubella Vaccination Strategies Available

Approximately twenty percent of women infected with rubella during their first trimester of pregnancy have offspring with congenital rubella syndrome (CRS). Consequently, the goal of a rubella vaccination program is to eliminate or reduce rubella infection in pregnant women and the subsequent congenital rubella syndrome in their children.

One simple strategy is to not vaccinate anyone in the population for rubella. This is the most widely-used strategy in the world. The reasoning behind this strategy is that endemic rubella and rubella epidemics will give most of the population disease-acquired immunity at young enough ages so that only a small fraction of women reach childbearing age without having rubella immunity. In some countries this no-vaccination strategy is used thoughtlessly or because of practical limitations such as the lack of money to purchase vaccine and give the shots. However, even if the resources were available for rubella vaccination, the no-vaccination strategy could still be the best strategy in a country. This strategy is used in places such as India, Brazil, People's Republic of China and tropical Africa (Proceedings, 1985).

Another strategy is to vaccinate prepubertal girls between ages 11 and 14. This is a widely-used strategy in European countries and is often called the United Kingdom (UK) strategy. It is also used in Australia, Israel and Japan (Proceedings, 1985). The reasoning behind this strategy is that vaccinated women will be protected and many women who are not vaccinated will be protected by disease-acquired immunity (since the disease will still spread in the susceptible pool of men and young girls).

A third strategy is to vaccinate both boys and girls just after one year of age. This strategy is often called the United States of America (USA) strategy. It is now used in the United States and Canada (Orenstein et al., 1984). The reasoning behind this strategy is that vaccination of many boys and girls will reduce or eliminate the occurrence of rubella so that pregnant women will not be infected. The ultimate goal is to achieve a large enough immune fraction in the population so that herd immunity is obtained and rubella disappears.

Thus the USA strategy is to eliminate rubella in the population while the UK strategy relies on endemic rubella or periodic rubella epidemics to give some girls natural immunity. Clearly it is not possible to compare the effectiveness of these two strategies by comparing the reported cases of rubella. It is reasonable to compare them by comparing the number of cases of CRS or by comparing the number of rubella cases among women of childbearing age.

An additional policy that is often used in conjunction with both the USA and UK strategies is to vaccinate susceptible women whenever they are discovered by private physicians or in family planning clinics. Since approximately forty percent of CRS cases occur among second or later children, vaccination of women just after they have had their first child is a very effective policy (Orenstein et al., 1984).

3. Long-Term Theoretical Implications of Rubella Vaccination Strategies

Mathematical models have been used to study the steady-state or equilibrium prevalence of rubella and the incidence of rubella infection among women of childbearing age. In these models the calculated steady states are reached after an immunization strategy has been applied consistently for 20 to 30 years. Thus these equilibrium models are calculating theoretical long-term rubella incidences. Sect. 5 considers theoretical incidences of rubella infection in the short term ($\leq$ 20 years) after a vaccination strategy is started.

Knox (1980) used a negative-exponential to describe the decrease of the susceptible fraction of the populations as a function of age. Since 25 years is an average childbearing age he assumed that the incidence of CRS would be proportional to the number of women who got rubella between 25 and 25.1 years of age. He found that if less than 70% of the target population can be immunized (i.e., vaccinated with primary vaccine success and permanent immunity), then the UK program of vaccinating prepubertal girls is better than the USA program since the incidence of CRS is less. However, if more than 70% of the target population can be immunized, then he concluded that the USA program of vaccinating all young children is better.

Dietz (1981) used age-structured differential equation models and a cost-benefit approach in which the cost of a vaccination could depend on the age of vaccination. Dietz used the same benefit as Knox of minimizing rubella incidence in women between ages 25 and 25.1. While Knox used the fraction immunized as the cost, Dietz used the number of vaccinations given as the cost. He concluded that for the same fixed cost the UK strategy always results in lower long-term incidence of CRS than the USA strategy.

Because Knox and Dietz obtained different conclusions, Hethcote (1983) also did a cost-benefit analysis. He did not agree with the cost function used by Knox since the cost is not a linear function of the fraction of the population immunized. For example, it is much harder to go from 98% to 99% immunized than from 50% to 51% immunized. The cost function used by Dietz also seemed inadequate since immunizing 49% of all children would be much easier than vaccinating 98% of the prepubertal girls, but the cost in terms of the number of vaccinations would be the same. Hethcote stated that both the fraction immunized and the number of vaccinations given must be considered in determining the cost of a vaccination strategy.

Hethcote (1983) used an age-structured partial differential equations model of Dietz (1981) and the same Knox-Dietz benefit function involving the rubella incidence in women between 25 and 25.1 years of age, but his cost function is different from those of Knox and Dietz. Hethcote used a cost which considered both the fraction immunized and the number of vaccinations given. His conclusion was similar to that of Knox. He found that at the long-term equilibrium, the UK strategy has lower CRS incidence than the USA strategy if the percentage immunized is less than 80% (cf. Knox 70%). When the percentage successfully immunized is above 80%, then the USA strategy is better than the UK strategy since the CRS incidence is enough lower that vaccinating twice as many children as prepubertal girls is worthwhile. Details of the model and the cost-benefit calculations of Hethcote (1983) are given in the next section.

Both Knox (1980) and Dietz (1981) also did calculations with waning vaccine-induced immunity, but we are not presenting these results since the current evidence implies that vaccine immunity is permanent. Knox, Dietz and Hethcote also considered other vaccination strategies such as vaccinating only girls just after 1 year of age, and vaccinating both young girls and prepubertal girls.

Anderson and May (1983) also compared different vaccination policies against rubella. Their model was similar to the age-structured model used previously by Dietz (1981) and Hethcote (1983), but they assumed that all people die at age 75 instead of using a negative-exponential age structure. They used the sum of the products of the fraction vaccinated at each age and the size of the age group as their cost function which is similar to that of Knox. They assumed that CRS incidence was proportional to the sum over the age range 16 to 40 years of the age specific fertility rate times the age specific incidence of rubella infection and used this as their benefit function to be minimized. Their conclusion was similar to that obtained earlier by Knox (1980) and Hethcote (1983) since they found that the UK policy is better when immunization levels are below 84% and the USA policy is better when immunization levels exceed 84%. More recent comparisons by Anderson and Grenfell (1986) of various one-, two- and three-stage vaccination policies in the UK are described in Sect. 9.

For countries where the average ages of infection were low (such as 3 years in The Gambia), Anderson and May (1983) found from a steady-state analysis that vaccination of young children reduced the infectivity in the population, but that more women got rubella between ages 16 and 40. In this case the USA policy of vaccinating young boys and girls seemed to cause more CRS than the no-vaccination policy. The UK strategy of vaccinating prepubertal girls did not cause more rubella infection of women of childbearing age at the endemic equilibrium.

Although Dietz (1981) concluded that the UK strategy was always better, he has acknowledged later in conversations that his conclusion was the result of his cost function. Thus there is no disagreement among the modelers. When the long term equilibrium incidence of CRS after 20 to 30 years is considered, the modeling results of Knox (1980), Hethcote (1983) and Anderson and May (1983) lead to the same conclusion. Namely the UK strategy of vaccinating prepubertal girls is better below a certain immunization level near 80% and the USA strategy of vaccinating young boys and girls is better above this immunization level. It is not very important that their different models led to slightly different estimates of the switchover immunization levels since they all yield the same qualitative result.

4. Cost-Benefit Comparisons of Rubella Vaccination Strategies

The vaccination strategies compared here include those described in Sect. 2. Strategy C2 is vaccination of both boys and girls at age 2 years; G2 is vaccination of only girls at age 2 years; G14 is vaccination of only girls at age 14 years; G2G14 is vaccination of only girls at age two years and at age 14 years; and C2G14 is vaccination of all children at age 2 years and of only girls at age 14 years. Strategies of vaccinating only boys are not considered since they are less effective. Vaccinations are assumed to be given randomly to individuals in the age group. Note that C2 is approximately the USA strategy and G14 is approximately the UK strategy.

The age dependent model used by Hethcote (1983) and described here is similar to the models of Dietz (1981). Consider a homogeneously mixing population of constant size with equal numbers of women and men. The birth and death rates are equal to μ, the birth rate for men and for women is $\mu/2$, and all newborns are susceptible. Let $x_i(a, t)$, $y_i(a, t)$ and $z_i(a, t)$ be the age density functions of the permanent immunity due to infection or vaccination, respectively, where subscript $i = 1$ corresponds to men and subscript $i = 2$ to women. For example, the fraction of the population who are women between the ages a_1 and a_2 and who are infected at time t is

$$\int_{a_1}^{a_2} y_2(a, t) \, da.$$

The contact rate β is the average number of adequate contacts of an infective per day; it is assumed that $\beta/2$ contacts are with men and $\beta/2$ are with women. The infection rate is determined by mass action and the recovery rate is γ times the infective fraction at age a. The vaccinations are given so that a fraction f_1 of the susceptible boys and fraction f_2 of the susceptible girls become immune at age A_1 (2 years) and a fraction g_2 of the susceptible girls become immune at age A_2 (14 years). The system of integrodifferential equations for the model is

$$\frac{\partial x_i}{\partial a} + \frac{\partial x_i}{\partial t} = -(\lambda(t) + \mu) x_i(a, t)$$

$$\frac{\partial y_i}{\partial a} + \frac{\partial y_i}{\partial t} = \lambda(t) x_i(a, t) - (\gamma + \mu) y_i(a, t) \tag{4.1}$$

$$\frac{\partial z_i}{\partial a} + \frac{\partial z_i}{\partial t} = \gamma y_i(a, t) - \mu z_i(a, t)$$

$$\lambda(t) = \beta/2 \int_0^\infty [y_1(s, t) + y_2(s, t)] \, ds$$

for $i = 1, 2$. The initial conditions at $t = 0$ and the matching conditions at ages 0, A_1 and A_2 are

$$x_i(a, 0) = x_{i0}(a), \quad y_i(a, 0) = y_{i0}(a), \quad z_i(a, 0) = z_{i0}(a)$$
$$x_i(0, t) = \mu/2, \quad y_i(0, t) = 0, \quad z_i(0, t) = 0 \tag{4.2}$$
$$x_i(A_1 + 0, t) = (1 - f_i) x_i(A_1 - 0, t), \quad x_2(A_2 + 0, t) = (1 - g_2) x_2(A_2 - 0, t).$$

The conditions on the right and left limits of the susceptible fractions at ages A_1 and A_2 correspond to jump decreases caused by vaccination. Since the death and recovery rates are equivalent to negative exponential distributions, the average lifetime is $1/\mu$ and the average infectious period is $1/(\gamma + \mu)$. The contact number σ is the average number of adequate contacts of an infective during the infectious period. The contact number σ satisfies $\sigma = \beta/(\gamma + \mu)$. The parameter $\lambda(t)$ in (4.1) is called the force of infection.

For large time the solutions of the model approach stable age distributions which are found by setting the time derivatives in Eq. (4.1) to zero. The total population of age a, $K(a) = x_1 + y_1 + z_1 + x_2 + y_2 + z_2$ satisfies $K(a) = \mu e^{-\mu a}$. The fractions of those of age a who are susceptible, infectious and removed are $u_i = x_i/K$, $v_i = y_i/K$ and $w_i = z_i/K = 1 - u_i - v_i$ where $i = 1$ for men and $i = 2$ for women. The initial value problem for the stable age distributions is

$$\frac{du_i}{da} = -\lambda u_i, \quad u_i(0) = 1/2$$

$$\frac{dv_i}{da} = \lambda u_i - \gamma v_i, \quad v_i(0) = 0 \tag{4.3}$$

$$\lambda = \beta/2 \int\limits_0^\infty [v_1(s) + v_2(s)] \mu e^{-\mu s}\, ds$$

$$u_i(A_1 + 0) = (1 - f_i)u_i(A_1 - 0), \quad u_2(A_2 + 0) = (1 - g_2)u_2(A_2 - 0)$$

where $i = 1, 2$.

If the inequality

$$\sigma \left[1 - \frac{f_1 + f_2}{2} e^{-\mu A_1} - \frac{(1 - f_2)g_2}{2} e^{-\mu A_2} \right] \le 1 \tag{4.4}$$

is satisfied, then all solutions approach a stable age distribution given by $u_i(a)$ in (4.5) below with $\lambda = 0$ and $v_i(a) = 0$ so that the disease disappears. The intuitive interpretation is similar to that in the section in this volume on Three Basic Epidemiological Models by Hethcote (1988). If the contact number times the longest possible average susceptible fraction is not greater than one, then the average infective cannot replace itself with at least one new infective during the infectious period so that the disease dies out.

If inequality (4.4) is not satisfied, then (except when there are no initial infectives) the susceptible fraction approaches the stable age distribution

$$u_1(a) = \begin{cases} e^{-\lambda a} & 0 \le a \le A_1 \\ (1 - f_1)e^{-\lambda a} & A_1 < a < \infty \end{cases} \tag{4.5}$$

$$u_2(a) = \begin{cases} e^{-\lambda a} & 0 \le a \le A_1 \\ (1 - f_2)e^{-\lambda a} & A_1 < a \le A_2 \\ (1 - f_2)(1 - g_2)e^{-\lambda a} & A_2 < a < \infty \end{cases}$$

where λ is a positive constant. Susceptible fractions in serosurveys do seem to have negative exponential distributions (Hethcote, 1983). At this endemic stable age distribution the average infective must infect (or reproduce) exactly one new infective so that $\sigma(\bar{u}_1 + \bar{u}_2) = 1$. This implies that the constant λ satisfies

$$\lambda = \mu \left\{ \sigma \left[1 - \frac{f_1 + f_2}{2} e^{-(\lambda + \mu)A_1} - \frac{(1 - f_2)g_2}{2} e^{-(\lambda + \mu)A_2} \right] - 1 \right\}. \tag{4.6}$$

For comparability with the calculations of Knox (1980) and Dietz (1981), we use a contact number σ of 7.62 and an average lifetime $1/\mu$ of 77 years so that the force of infection without vaccination is $\lambda = 0.086$.

A cost function used by Dietz (1981) is

$$C = \frac{f_1 + f_2}{2} e^{-2\mu} + \frac{g_2}{2} e^{-14\mu} \tag{4.7}$$

so that his cost C is the product of the number of vaccinations given and the vaccine efficacy. Vaccine efficacy is the proportion of those vaccinated who actually become immune. Knox (1980) implicitly used $f = f_1 = f_2 = g_2$ as the cost function in his comparisons. As indicated in Sect. 3, both of these cost functions have unrealistic features. A better cost function would consider both the cost of the vaccinations and the cost of a case of congenital rubella syndrome (CRS). The shared cost of a rubella vaccination when it is given with measles and mumps vaccines may be between $5 and \$10. When the rubella vaccine is given alone, the cost is probably \$15 to \$25. The total lifetime cost of a case of CRS has been estimated to be \$220,000 (CDC, 1984). The total cost of each specific immune level and the resulting incidence of CRS could be calculated for each case considered below; however, the results are clear without these details and the simplicity of the model does not justify such precision. Thus the immunized fraction, the number of vaccinations and the cases of CRS are all considered in our cost benefit analysis.

The reason for vaccinating against rubella is to prevent infection in pregnant women and the subsequent congential rubella syndrome (CRS) in their children. The incidence of CRS is assumed in the model to be proportional to the incidence of rubella in women between 25 and 25.1 years of age since 25 years is approximately the average age of conception and 0.1 year is the approximate duration of the period during which the fetus is at risk of CRS. The incidence of CRS in the model is proportional to

$$I = (1 - f_2)(1 - g_2)(e^{-25\lambda} - e^{-25.1\lambda}) \tag{4.8}$$

which depends on f_1, f_2, g_2, σ, μ, A_1 and A_2 since λ is the solution of (4.6). One strategy is better than another if the incidence I of CRS is lower for the former than for the latter. Values of the function C and I are given for various values of f in Table 1.

It can be seen from the I values in Table 1 why Knox (1980) concluded that strategy G14 is better than C2 for the same immunized fraction f at or below 0.7 and C2 is better for fractions above 0.7. For f below 0.7 the higher prevalence when all men are unvaccinated gives more women disease-acquired immunity before they reach child-bearing age. For strategy C2 with f above 0.7, rubella incidence (and hence the incidence of CRS) decreases rapidly as herd immunity at $f = 0.892$ is approached. Below the maximum value of C for G14, strategy G14 is better than C2 for the same C so that Dietz (1981) concluded that G14 is always better than C2.

Consideration of a balance between f and C suggests that when the actual immunized fraction f which can be achieved is at or above 0.80, then the strategy C2 of vaccinating both boys and girls at age 2 years is better than G14, i.e., the reduction in the incidence of CRS by a factor of at least 1.44 seems sufficient to justify the increase in the number of vaccinations by a factor of 2.34. When the actual immunized fraction f is less than 0.80, the strategy G14 of vaccinating only girls at age 14 years is better than C2. The switching point could be determined more precisely by using the dollar estimates given above of rubella vaccinations and of cases of CRS, but such precision is probably not justified in this simplified model.

For values of f above about 0.5, the reduction in the incidence of CRS using strategy G2G14 seems to be sufficient to justify the additional vaccinations over strategy G14. Strategy C2 is better than G2G14 when the immunization fraction is above 0.88; otherwise, G2G14 is better than C2. Strategy C2G14 has the advantage that rubella dies out when the immunization fraction f is above 0.832, but it requires more vaccinations than the other strategies.

Immunization strategies would have to be maintained for at least 20 or 30 years in order to be near the equilibrium situations used to compare the strategies in

Table 1. Incidence of CRS for the vaccination strategies[1]

Immunized fraction of the target population	C2: boys and girls at age 2 years ($f_1 = f_2 = f, g_2 = 0$)		G2: girls at age 2 years ($f_2 = f, f_1 = g_2 = 0$)		G14: girls at age 14 years ($f_1 = f_2 = 0, g_2 = f$)		G2G14: girls at age 2 years and 14 years ($f_1 = 0, f_2 = g_2 = f$)		C2G14: boys and girls at 2 years and girls at 14 years ($f_1 = f_2 = g_2 = f$)	
f	C	$I \times 10^3$	C	$I \times 10^3$	C	$I \times 10^3$	C	$I \times 10^3$	C	$I \times 10^3$
0	0	0.999	0	0.999	0	0.999	0	0.999	0	0.999
0.1	0.097	0.999	0.049	0.948	0.042	0.914	0.091	0.867	0.139	0.913
0.2	0.194	0.979	0.097	0.888	0.083	0.826	0.181	0.731	0.277	0.805
0.3	0.292	0.932	0.146	0.817	0.125	0.736	0.271	0.593	0.417	0.674
0.4	0.390	0.852	0.195	0.734	0.167	0.642	0.362	0.459	0.557	0.524
0.5	0.487	0.734	0.244	0.639	0.208	0.545	0.452	0.333	0.695	0.364
0.6	0.585	0.573	0.292	0.533	0.250	0.445	0.542	0.221	0.835	0.209
0.7	0.682	0.375	0.341	0.414	0.292	0.340	0.633	0.128	0.974	0.084
0.8	0.779	0.160	0.390	0.284	0.334	0.231	0.724	0.058	1.113	0.011
0.85	0.828	0.064	0.414	0.216	0.354	0.175	0.768	0.033	1.182	0
0.90	0.877	0	0.438	0.145	0.375	0.118	0.813	0.015	1.252	0
0.95	0.926	0	0.462	0.073	0.396	0.060	0.858	0.004	1.322	0

[1]The parameters are $\sigma = 7.62$, $1/\mu = 77$ years. Table from Hethcote (1983)

Table 1. Calculations to determine the short term or transient effects of various strategies are considered in the next section.

5. Short-Term Theoretical Implications of Rubella Vaccination Strategies

When a vaccination program is started, the incidence of congenital rubella syndrome (CRS) will change from the preprogram level until eventually it is near the long-term equilibrium level discussed in Sects. 3 and 4. Here we consider the short-term or transient behavior between the time of program initiation and the time when rubella incidence is essentially at the equilibrium level.

Knox (1980) studied the transient behavior by using a dynamic computer model with many complexities of real life such as a heterogeneous population divided by sex and social strata, progressive changes in his attrition parameter, the distribution of age-at-delivery, and the varying time between vaccination of girls and exposure of their fetuses. His simulations with the UK strategy showed that for various f values the incidence of CRS decreased until the steady-states were reached in about 25 years. Half of the ultimate goal was achieved within 10 years.

Knox's simulations with the USA strategy yielded some surprising results. When more than 75% of the young boys and girls were immunized, the incidence of CRS decreased rapidly to the steady-state levels. Steady-states were achieved in 14 to 20 years and the half-way point was reached in about 7 years. These rapid decreases were due to the protection of pregnant women by prevention of exposure to the wild virus. When less than 75% of the young children were immunized and especially in the 40% to 65% range, the early decreases were followed by a severe rebound towards a peak incidence of CRS 20 to 30 years after initiation of the program. These temporary peaks were up to 25% higher than the CRS levels without vaccination. This rebound phenomenon seemed to be due to the movement of girls who were 2 to 10 years old at the initiation. Many had not yet had rubella, were not vaccinated against it and were exposed at much lower levels due to the vaccination program. Thus a large number of susceptibles passed into the reproductive ages when there was still sufficient wild virus to infect a substantial number. For high fractions vaccinated, the wild virus was sufficiently suppressed to prevent this rebound.

Anderson and May (1983) and May (1986) also considered the short-term dynamics after initiation of a rubella vaccination program. Their simulations of the UK strategy showed that the CRS incidence decreased slightly for a few years and then rebounded approximately to the initial level, after which it decreased fairly steadily to the steady-state level. The CRS levels were near the steady-state levels after 20 to 30 years. They claimed that the reported cases of CRS in the United Kingdom were consistent with their simulations.

The Anderson and May (1983) simulations of the USA strategy had more severe rebounds than those found by Knox. They suggested that the differences in rebounds could be due to the unreasonably long iteration interval used by Knox. Their rebounds of the CRS incidence occurred about every 5 years when 20% or 40% of young children were immunized, every 6 or 7 years when 60% were immunized and every 10 or 11 years when 80% were immunized. The peak of the rebound CRS incidence was about twice as high as the no-vaccination level. They claimed that the agreement with the observed incidence of CRS in the United States of America was satisfactory, given the simplicity of their model. They explained the severity of the oscillations in CRS incidence in the simulations as due to dramatic changes in the infectivity level in the population. See Sect. 9 for more recent simulations by Anderson and Grenfell (1986).

When rubella vaccination was started in the USA in 1969, many school children of all ages were vaccinated; moreover, many susceptible women have been vaccinated when they were discovered. Thus the actual rubella policy in the United States has been more complicated than the theoretical USA strategy of vaccinating boys and girls shortly after 1 year of age. Thus the rebounds of CRS incidence predicted by the USA strategy simulations of Knox (1980) and Anderson and May (1983) may not actually occur in the United States of America. However, their simulations did point out the possibility of undesirable rebounds in CRS incidence if a pure USA strategy were adopted without any conjunctive strategies.

6. Recommendations from Epidemiologists

In a paper on rational strategy for rubella vaccination, Hinman et al. (1983) presented two priorities. They stated that the first priority should be to directly protect women of childbearing age by vaccination. Their second priority was to interrupt the transmission of rubella. These priorities were based on experience in the United States and Europe, and on the current knowledge of vaccine characteristics. For example, it is safe to vaccinate women of childbearing age since intrauterine infection with rubella vaccine poses little or no real risk to a fetus. Vaccine virus is not transmissible. The rubella vaccine efficacy is high and primary vaccine success seems to give permanent immunity.

The UK strategy of vaccinating girls before they enter the childbearing period aims to directly protect individual women. The UK strategy does not have much effect on the overall epidemiology of rubella virus transmission. The pure USA strategy of vaccinating young boys and girls just after 1 year of age is intended to interrupt the transmission of rubella and thus indirectly protect pregnant women from rubella infection. Thus the UK strategy is consistent with the first priority of Hinman et al. (1983) while the pure USA strategy is consistent with their second priority.

Adoption of the first priority would involve vaccinating all women 15 to 34 years of age or 15 to 40 years depending on the fertility pattern. Adoption of the second priority would entail initial vaccination of all children aged 1 to 14 years followed by vaccinations of all children at 1 year of age.

Any new rubella vaccination program would probably achieve vaccination levels for which the UK strategy is theoretically better (see Sects. 3 and 4). Vaccination levels high enough to make the USA strategy better would probably require legislation to require immunization before school entry. Thus countries that are not prepared to pass such legislation should adopt the UK strategy when starting a rubella vaccination program.

Although they do not unequivocally state it, we interpret the priorities of Hinman et al. (1983) to also be a recommendation that countries starting a rubella program should strongly consider using the UK strategy instead of the USA strategy. Thus both the theoretical modeling and the epidemiologists' priorities lead to the same conclusions.

7. Choosing a Strategy for Rubella Vaccination

Countries without rubella vaccination programs are beginning to estimate the number of cases of CRS and to consider possible strategies to control rubella. Countries with rubella vaccination programs are evaluating their current programs and considering modifications. The decision regarding new or modified rubella vaccination programs depends on factors in each country such as the frequency of CRS cases, the epidemiology of rubella, fertility patterns, the success of any existing program, the estimation of the vaccination levels attainable under various new or modified strategies, and the costs of vaccination and CRS incidence.

As shown in Sects. 3 and 4 the achievable vaccination levels are crucial in determining the best rubella vaccination strategy. The United States of America and Canada have achieved high enough vaccination levels to make the USA strategy better, but they have done it through laws which require rubella immunization before school entry. Unless a country is prepared to enact similar legislation, it seems unlikely that it can achieve the vaccination levels necessary to make the USA program better. Moreover, it is possible that implementation of the USA strategy at lower levels may lead to a rebound in CRS incidence as predicted by simulations (see Sect. 5). Thus it seems clear that the UK strategy would be better for all countries starting a new rubella vaccination program. This conclusion is consistent with the priorities discussed in Sect. 6. A useful conjunctive strategy to the UK strategy is to vaccinate as many women of childbearing age as possible.

Implementation of a comprehensive vaccination strategy to reduce CRS is a major endeavor which may not be possible in some countries because of cost or other factors, particularly in developing countries. Partial implementation or phased implementation may be necessary. If a phased implementation is necessary, then the priorities in Sect. 6 suggest that the first phase should be to vaccinate as many women of childbearing age as possible. The second phase should be the UK strategy of vaccinating prepubertal girls. The last phase should be to start vaccinating all young children in order to reduce rubella transmission. Vaccination of high enough fractions would eventually lead to elimination of rubella through herd immunity.

8. The United States of America

In the prevaccine era rubella epidemics occurred in the USA every 5 to 10 years as shown in Fig. 1. The United States began a rubella vaccination program in 1969 and reported in 1978 that about two-thirds of the children were vaccinated against rubella (Orenstein et al., 1984). Since then the fractions vaccinated have increased so that 98 percent of all children entering school for the first time in the fall of 1985 were vaccinated against rubella. Thus the immunized fractions are now in the range where the USA strategy is clearly better for the United States according to the theoretical analyses described in Sects. 3 and 4.

The vaccination program adopted in 1969 was directed primarily towards pre-school and school-age children so that over 123 milion doses of rubella vaccine have been used since 1969. The vaccination program in the United States has caused a dramatic decline in reported rubella as shown in Fig. 2 and has eliminated the characteristic 6 to 9 year cycle of rubella epidemics (Orenstein et al., 1984). Reported cases of rubella were 18,269 in 1978, 11,795 in 1979, 3904 in 1980, 2077 in 1981, 2325 in 1982, 970 in 1983, 745 in 1984 and 630 in 1985. Reported cases of CRS were 55 in 1979, 14 in 1980, 10 in 1981, 11 in 1982, 6 in 1983 and 2 in 1984. Since the theoretical analyses in Sect. 4 indicate that herd immunity is possible using the USA strategy and the reported cases are declining steadily, herd immunity for rubella may be achieved soon in the United States.

The occurrence of rubella has declined in all age groups and the incidence pattern has changed. In the years before vaccine licensure, children less than 10 years of age accounted for 60% of the cases while 21% of the total cases were reported among those greater than 15 years of age. In 1975–77 children under 10 years of age accounted for 24% of cases while persons 15 years of age or older made up 62% of cases. In 1982–84 persons 15 years of age or older still accounted for 52% of cases, but this percentage is declining (CDC, 1986a).

However, the possibility of increased rubella activity in older individuals still exists. Rubella susceptibility in adolescents and young adults is still estimated to be 10–20% so that outbreaks can occur in these groups (CDC, 1984). Special efforts must be made to vaccinate all susceptible postpubertal individuals, particularly females who are no longer in school. Although rubella vaccination is not recommended during pregnancy, the risk of CRS if a woman is vaccinated during early pregnancy are minimal so that women should not hesitate to accept the vaccine (CDC, 1984). Underreporting of rubella and CRS is a problem, but the declines in reported cases seem to be due to a decline in actual rubella and CRS cases in the USA.

Local rubella epidemics still occur occasionally in the United States. For example, an outbreak of rubella in a prison in New York City in 1985 involved 49 males and one female ranging in age from 19 to 41 years. The incidence data in Fig. 3 shows that vaccination clinics for inmates and employees seemed to stop the outbreak (CDC, 1985b). This prison outbreak was part of a bigger outbreak in New York City with 45 cases reported in a factory and eighteen cases in five hospitals. Of the 630 cases of rubella reported to CDC in 1985, 184 were cases reported from New

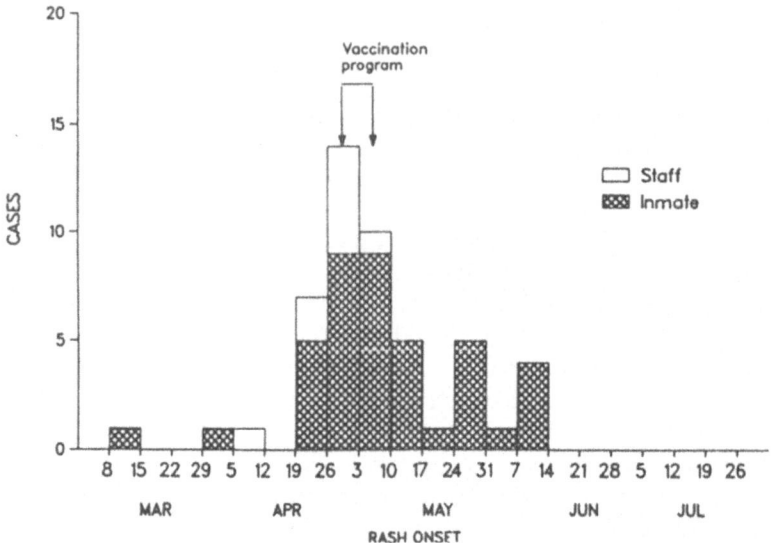

Fig. 3. Rubella incidence among inmates and staff in a New York City prison in 1985. From CDC (1985b)

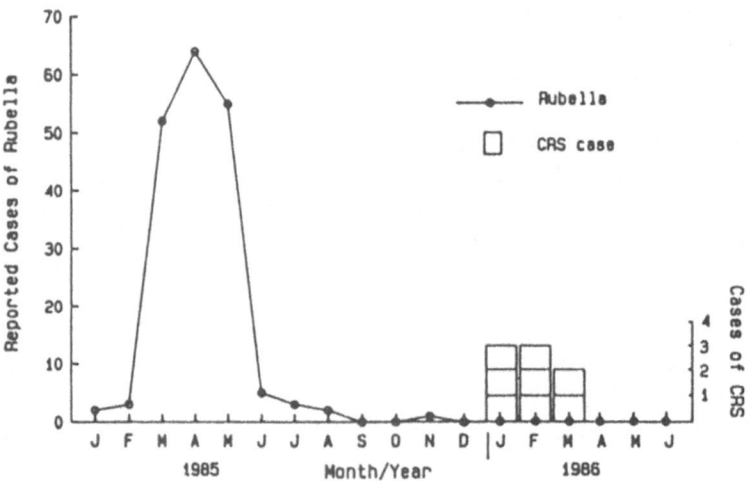

Fig. 4. Reported rubella cases in 1985 and reported congenital rubella syndrome cases in 1986 in New York City. From CDC (1986b)

York City (CDC, 1986b). As shown in Fig. 4, eight suspected cases of CRS were reported during the first 5 months of 1986. These infants were all born between 8 and 10 months after the peak of the 1985 rubella epidemic. The eight mothers were not directly linked to any of the reported cases so there were probably many cases besides those which were reported. None of the eight mothers had been vaccinated even though five had previous children so they could have been vaccinated postpartum. The expense of rubella immunization programs for susceptible women is small compared to the cost of ongoing rubella control and the approximately $220,000 lifetime cost of a case of CRS (CDC, 1986b).

The first priority of Hinman et al. (1983) (Sect. 6) would suggest that efforts to vaccinate susceptible women of childbearing age should be increased in the United States as a supplement to the current vaccination of children. Vaccination should be offered to women of childbearing age excluding those who are pregnant or may become pregnant within 3 months. Vaccination could be offered in family planning clinics, during obstetrical examinations, to postpartum and postabortion women and to women in colleges, schools, military organizations and work places. These vaccinations would provide direct protection to these women and also speed the progress towards herd immunity. Hinman et al. (1983) state that "immediate protection to the population currently at risk of having a pregnancy complicated by rubella... has been the weakest aspect of both the U.S. and the U.K. strategies...".

9. The United Kingdom

As described in Sect. 6, the UK strategy is consistent with the first priority of protecting women of childbearing age. This first priority could be strengthened in the United Kingdom by additional efforts to vaccinate susceptible women of childbearing age. It is estimated that about 90% of prepubertal girls are currently being vaccinated for rubella in the United Kingdom (Tobin et al., 1985) so that it may be appropriate to now try to interrupt rubella transmission. Adoption of the second priority would entail initially vaccinating all children aged 1 to 14 year followed by vaccination of all children at 1 year of age.

There seems to be some doubt about the cost effectiveness of modifying the rubella program in the United Kingdom to include vaccination of all 1 year old children. Dudgeon (1985) believes that this modification would be a disaster because it could lead to lower acceptance rates of a measles-mumps-rubella vaccine. Knox (1985) also recommends that the United Kingdom program not be modified to include vaccination of 1 year old children. His recommendation is based on the rebounds which he calculated using his model for incompletely implemented indirect strategies such as the USA strategy.

It is difficult to estimate what fraction of all children could actually be vaccinated for rubella, but since approximately 57% are now being vaccinated against measles in the United Kingdom (Dudgeon, 1985) and the rubella vaccine would be combined with the measles vaccine, 57% is a reasonable estimate. The analysis of Hethcote (1983) can be used to estimate changes in the equilibrium incidence of CRS. From Table 1 we see that immunization of 90% of 14 year old girls would reduce the equilibrium incidence of CRS to 11.8% of its incidence without any vaccination. If in addition 54% of all young children (57% vaccinated times 95% vaccine efficacy) were successfully immunized against rubella, then the equilibrium incidence of CRS would be reduced to 6.1% of its prevaccine incidence. Recall from Sect. 4 that this value is based on the equilibrium incidence which could be attained about 20 to 30 years after program implementation.

Thus the calculations show that the major additional costs of adding the rubella vaccine to the measles vaccine would eventually reduce CRS incidence by an additional 5.7% (from 11.8% to 6.1%). Table 1 shows that a similar reduction could be achieved by increasing the percentage of 14 year old girls who are vaccinated from 90% to 95%. Hence from one point of view, the benefits of vaccinating many young children for rubella do not seem to be substantial enough to justify the costs involved, particularly since the same reduction could be achieved by a slight improvement in the current United Kingdom program.

Rubella can never be eliminated using the UK strategy with any percentage vaccinated since rubella can always spread among the young girls and men, but as seen in Table 1 it can be eliminated through herd immunity by the combined UK and USA strategies (C2G14). Thus adding rubella vaccine to the measles vaccine could eventually lead to elimination if the percentages of young children vaccinated could be improved. Even at the current 57% vaccination rate for young children, the addition of rubella vaccine would eventually cut CRS incidence in half (from 11.8% to 6.1%). Thus from another point of view, the potential benefits of the modification seem more substantial.

Recently, Anderson and Grenfell (1986) have used a computer simulation model with heterogeneous mixing of 6 age groups to compare various rubella vaccination policies in the UK. The policies are the UK, the combined UK-USA, the UK combined with vaccination of susceptible women of age 25 years and a three-stage policy including the UK and USA strategies combined with vaccination of susceptible women of age 25 years. They use parameter estimates derived from serological data or case notifications given in Nokes, Anderson and Anderson (1986), which reveal that the force of infection is low for young children and adults, but is high for 5 to 15 year olds. Their principal conclusion is that the three-stage policy is best in both the short and long term, provided the vaccination coverage of two year old children exceeds 60% and the vaccination of teenage girls continues at the current high levels.

The decision regarding modification of the rubella program in the United Kingdom to include rubella vaccination of young children is not easy or clear-cut. The comparisons above are useful in evaluating the potential effects of this modication. These comparisons are just one of many factors which must be considered by epidemiological policy makers when making their decision regarding this possible modification.

10. France

Although exact statistics do not seem to be available, the Laboratorie National de la Santé estimates that the incidence in France of congenital rubella syndrome in 1978–79 was about 50–100 cases per year, which is about one case per 7,500 to 15,000 live births (Valleron and Flahault, 1986). This CRS incidence is higher than in the United States where the incidence has decreased from 1.82 cases per 100,000 live births in 1970 to 0.05 cases per 100,000 live births in 1985 (Bart et al., 1985). Thus CRS does seem to be a problem in France.

In 1969 measles (rougeole) vaccination was recommended for infants living in collectivity and for fragile infants. In conjunction with the 1970 European program, rubella (rubéole) vaccination of prepubertal girls and seronegative women was recommended (Bouvet, 1986). Unfortunately, there was not much emphasis on these programs and the vaccination rates were low. In 1983 a true national program for vaccination against measles and rubella was begun. A combination of the USA and UK strategies for rubella vaccination was chosen (Bouvet, 1986). The measles and rubella vaccination recommended to all children between 12 and 15 months of age is paid for by the French government, but the rubella vaccination recommended for the prepubertal girls is not. Bouvet (1986) believes that there are still three obstacles to the program. The first is that the vaccinations are recommended, but are not legally required. The second is the decentralization of the program and the third is the non-free aspect of the vaccination for the prepubertal girls.

It is estimated that approximately 45% of the children in France are vaccinated for measles (Valleron and Flahault, 1986). Thus the fraction of infants currently being vaccinated for rubella is probably also about 45% since the measles and rubella vaccines are now combined. This percentage may increase as the program started in 1983 becomes more influential. It is not known what fraction of prepubertal girls are vaccinated for rubella in France. In the United Kingdom even though the fraction of children vaccinated for measles is about 57%, the fraction of prepubertal girls vaccinated for rubella increased from 35% in 1972 to about 90% currently (Dudgeon, 1985; Tobin et al., 1985). Thus it might be possible to vaccinate more than 45% of the prepubertal girls in France.

The combined USA-UK rubella vaccination strategy now adopted in France is a reasonable policy. The calculations of Hethcote (1983) in Table 1 indicate that when immunization levels are in the neighborhood of 50%, then most of the reduction in CRS is due to the vaccination of prepubertal girls. It may be desirable at this time to place more emphasis on the vaccination of prepubertal girls and seronegative women for rubella since this provides immediate direct protection of women from rubella. Later, when a large fraction of the prepubertal girls are being vaccinated, the vaccination of infants can be emphasized in order to decrease rubella transmission and to achieve elimination through herd immunity. However, without legislation to require rubella immunization of infants, it is unlikely that the immunized fractions in France will increase to the levels necessary for herd immunity.

As described in Hinman et al. (1983), other additional tactics can be used to hasten the reduction in CRS incidence. Specifically a mass vaccination of children aged 1 to 12 years is a reasonable way to start a USA strategy and a mass vaccination of susceptible women of childbearing age is a useful way to start a UK strategy. Moreover, both strategies can be augmented by vaccination of postpartum women and other women as they are encountered in health care centers. These additional tactics would cause CRS incidence to decrease more quickly towards the equilibrium level.

11. The Netherlands

In 1974 the Netherlands adopted the UK strategy of vaccinating girls against rubella at age 11 years. In 1984 the Health Council of the Netherlands advised that a new vaccination strategy against measles, mumps and rubella (MMR) should be initiated. They recommended that a combined MMR vaccine be given to all children at 14 months and 9 years of age. The second MMR vaccination coincides with a DPT shot given at 9 years of age. Rubella vaccine acceptance among 11 year old girls has been estimated to be 92–94% (van Druten, Boo and Plantinga, 1986). Although the Netherlands had rubella epidemics in 1976 and 1979, this is not undesirable since the UK strategy relies on periodic rubella epidemics in the men and young girls to give immunity to girls who are not vaccinated.

There are many reasons why the Health Council of the Netherlands may have recommended the new vaccination program. Combining vaccines has logistical and financial advantages. Two dose programs are more effective than one dose programs (Hethcote, 1983). The high vaccination rate for rubella in prepubertal girls may have encouraged them to now focus on the elimination of rubella. As indicated in the section in this volume on Three Basic Epidemiological Models by Hethcote (1988), the contact numbers for mumps and rubella are similar, but the contact number for measles is higher. Although the recommended program for the Netherlands requires more vaccination than some other possible programs, it should be adequate to achieve herd immunity for all three diseases, provided the vaccination levels are high enough.

Mathematical modeling and computer simulations of rubella and CRS have been done by van Druten, Boo and Plantinga (1986). They are aware of the earlier work on rubella models described in Sects. 3, 4 and 5. Their work has focussed primarily on vaccination programs in the Netherlands. They note that vaccination-induced immunity in part of a population can be a disadvantage for the unvaccinated individual. Vaccination programs which do not eliminate the disease lead to a new steady state (endemic equilibrium) at which the new force of infection is lower than it was prior to vaccination. Thus the average age of infection in unvaccinated individuals increases so that it is possible that more women get rubella infection during pregnancy.

They (van Druten, Boo and Plantinga, 1986) have presented the information on rubella shown in Table 2 using data from Anderson and May (1983) and van Druten et al. (1987). From the average age of infection they calculated the average value of the force of infection as the reciprocal of the average age of infection (corrected by the average length of protection provided by maternal antibodies). The force of infection is a measure of the risk for a susceptible person to contract the infection in a period of one year and is the mean number of adequate contacts with an infective per person per year. The basic reproduction rate or contact number is calculated as one plus the average lifetime divided by the average age of infection. The minimum immunization percentage for herd immunity is one hundred times one minus the reciprocal of the contact number: that is, $100(1 - 1/\sigma)$ where $\sigma = \beta/(\gamma + \mu)$ as in Sect. 4. These equations for the contact number and minimum immunization for

Table 2. Parameter values for rubella in various countries[1]

Location	Time period	Average age of infection	Force of infection	Basic reproduction rate	Minimum immunization for heard immunity
West Germany	1970–77	11–12	0.08–0.09	6.8–7.8	85–87%
USA	1966–68	9–10	0.10–0.11	8.0–9.3	88–89%
England and Wales	1977	9–10	0.10–0.11	8.0–9.3	88–89%
Czechoslovakia	1970–77	8–9	0.11–0.13	8.8–10.4	89–90%
The Netherlands	1964–74	8–9	0.11–0.13	8.8–10.4	89–90%
Scotland	1950–60	6–7	0.14–0.17	11.0–13.5	91–93%
Poland	1970–77	6–7	0.14–0.17	11.0–13.5	91–93%
Gambia[2]	1976	2–3	0.30–0.50	14.2–21.0	93–95%

[1] Data from Anderson and May (1983) except for the Netherlands data from van Druten et al. (1987). Table from van Druten et al. (1987)
[2] The average lifetime is assumed to be 70–75 years in all countries except The Gambia where it is 40 years

herd immunity are explained in the section in this volume on Three Basic Epidemiological Models by Hethcote (1988).

From their results in Table 2, they observe that the force of infection and the basic reproduction rate (contact number) in the Netherlands are higher than those in West Germany, USA and England and Wales, which were used by Knox (1980), Dietz (1981) and Hethcote (1983). Using a force of infection of 0.12 in the Netherlands, they calculate that rubella vaccine acceptance of 92–94% of pre-pubertal girls lowers the incidence of CRS by 90% in the long term. They also calculate that if the USA strategy had been adopted in the Netherlands and the vaccination coverage was below 60%, then the incidence of CRS would be *higher*. Thus in countries with high forces of infection, the USA strategy can lead to more cases of CRS if the vaccination coverage is very low. This result is consistent with the calculations of Anderson and May (1983).

The newly recommended vaccination program in the Netherlands includes MMR vaccination of all children at ages 14 months and 9 years. To achieve the elimination of these diseases in the shortest time, they also propose to vaccinate 4 year old children during the first three years and to continue vaccinating 11 year old girls for rubella for the first two years. They (van Druten, Boo and Plantinga, 1986) calculate that if the new program started in 1987 in the Netherlands with immunity achieved for 90% of the susceptibles of age 1 year and 70–90% of the susceptibles of ages 4 years and 9 years, then elimination of rubella through herd immunity would be achieved before 1991. This rapid elimination seems remarkable; however, one must remember that if the vaccine efficacy is 95%, then this multiple dose program requires vaccinations of 95% of 1 year old children and vaccination of 74–95% of 4 and 9 year old children. These high vaccination rates may be possible to achieve in the Netherlands, but they would be difficult to achieve in many other countries.

12. Africa

The relative magnitude of the rubella problem when compared to other health problems must be assessed in African countries in order to decide whether or not to undertake a rubella vaccination program. Several relevant factors are the age pattern of susceptibility and fertility, the incidence of congenital rubella and the potentially achievable vaccination levels using various programs.

The pattern of rubella incidence in Africa is not clear, but it seems that rubella epidemics may occur approximately every ten years (Assaad and Ljungars-Esteves, 1985). Note that serologic data varies greatly depending on the time since the last rubella epidemic. For example, 100% of children from five to nine years old in Tunisia had rubella antibody in 1969. In a 1966 survey in the Gambia, 83.5% of children had rubella antibody while only 32.8% had rubella antibody in 1976. These and other data show that a rubella epidemic occurred there in 1963–64. A retrospective examination in the late 1970's of clinical records in The Gambia failed to find evidence of CRS cases. However, a rubella outbreak in Zimbabwe in 1978 resulted in a high incidence of CRS (Assaad and Ljungars-Esteves, 1985).

The percentage of susceptibles has varied greatly in various surveys in the 1970's and 1980's (Assaad and Ljungars-Esteves, 1985; Mingle, 1985). In the Gambia and Upper Volta the percentages of 10–14 year old girls with rubella antibody were 93% and 96%. In Ghana, Angola, Ethiopia and Southern Africa, the percentages of 10–14 year old girls with rubella antibody were 70% to 80%. In Nigeria and Togo the percentages with rubella antibody in this group were 49% to 60%.

Since the average age of attack for measles infection is lower in tropical Africa than in prevaccine European countries (Anderson and May, 1983), it seems likely that the average age of attack for rubella is also lower in Africa than the average age of rubella infection of about 10 years in European countries before vaccination started. If the average age of rubella attack really is lower in Africa, then more women would have disease-acquired immunity by the time they reach childbearing age. Thus a no-vaccination policy may be best in Africa. Monjour et al. (1982) concluded from a study that mass vaccination was not justified in Upper Volta.

Certainly, better data on rubella seropositivity over many years and the frequency of congenital rubella would be needed before a decision could be made to start a rubella vaccination program in Africa. The percentages vaccinated for measles in central Africa vary from 50% to 80%. If similar vaccination levels were achievable for rubella and it was decided to start a rubella vaccination program, then the UK strategy would be a good choice for the reasons given in Sects. 3–6. However, a decision to start a rubella vaccination program should not be made hastily, since the currently available data suggest that a no-vaccination policy may be the best rubella strategy for Africa.

References

Anderson, R.M., Grenfell, B.T. (1986) Quantitative investigations of different vaccination policies for the control of congenital rubella syndrome (CRS) in the United Kingdom. J. Hyg. Camb. 96, 305–333

Anderson, R.M., May, R.M. (1983) Vaccination against rubella and measles: quantitative investigation of different policies. J. Hyg. Camb. 90, 259–325

Assaad, F., Ljungars-Esteves, K. (1985) Rubella—World Impact. Rev. Inf. Dis. 7, Suppl. 1: s28-s36

Bart, K.J., Orenstein, W.A., Preblud, S.R. Hinman, A.R. (1985) Universal immunization to interrupt rubella. Rev. Inf. Dis. 7, Suppl. 1: s177-s184

Bouvet, E. (1986) Vaccination contre la rougeole et la rubéole. Le programme francais de vaccination (février 1986). Médecine et Enfance 6, no. 2: 49–52

Cooper, L.Z. (1985) The history and medical consequences of rubella. Rev. Inf. Dis. 7, Suppl. 1: s2-s10

Centers for Disease Control (1984) Rubella and congential rubella surveillance, 1983. In: CDC Surveillance Summaries 33 (no. 4SS): 1SS-10SS

Centers for Disease Control (1985a) Elimination of rubella and congenital rubella syndrome—United States. Morbidity and Mortality Weekly Report 34, 65–66

Centers for Disease Control (1985b) Rubella outbreaks in prisons—New York City, West Virginia, California. Morbidity and Mortality Weekly Report 34, 615–618

Centers for Disease Control (1986a) Rubella and congenital rubella syndrome—United States, 1984–1985. Morbidity and Mortality Weekly Report 35, 129–135

Centers for Disease Control (1986b) Rubella and congenital rubella syndrome—New York City. Morbidity and Mortality Weekly Report 35, 770–779

Dietz, K. (1981) The evaluation of rubella vaccination strategies. In: Hiorns, R.W., Cooke, D. (eds).

The Mathematical Theory of the Dynamics of Biological Populations, vol. II. Academic Press, London, pp. 81–87

Dudgeon, J.A. (1985) Selective immunization: protection of the individual. Rev. Inf. Dis. 7, Suppl. 1: s185-s190

Hethcote, H.W. (1983) Measles and rubella in the United States. American J. of Epidemiology 117, 2–13

Hethcote, H.W. (1989) Three basic epidemiological models. This volume, pp. 119–142.

Hinman, A.R., Bart, K.J., Orenstein, W.A., Preblud, S.R. (1983) Rational strategy for rubella vaccination. Lancet, January 1/8

Hinman, A.R. (1985) Prevention of congenital rubella infection: Symposium summary. Rev. Inf. Dis. 7, Suppl. 1: s212-s215

Knox, E.G. (1980) Strategy for rubella vaccination. Int. J. Epidemiol. 9, 13–23

Knox, E.G. (1985) Theoretical aspects of rubella vaccination strategies. Rev. Inf. Dis. 7, Suppl. 1: s194-s197

May, R.M. (1986) Population biology of microparasitic infections. In: Hallam, T.G., Levin, S.A. (eds). Mathematical Ecology. Biomathematics, vol. 17. Springer, Berlin Heidelberg New York London Paris Tokyo, pp. 405–422

Mingle, J.A.A. (1985) Frequency of rubella antibodies in the population of some tropical African countries. Rev. Inf. Dis. 7, Suppl. 1: s68-s71

Monjour, L., Druilhe, P., Huraux, J.M., Palminteri, R., Froment, A., Kyelem, J.M., Alfred, C., Laplace, J.L., Genfilini, M. (1982) Contribution to the study of rubella epidemiology in rural Upper Volta. Acta. Trop. (Basel) 39, 247–52

Nokes, D.J., Anderson, R.M., Anderson, M.J. (1986) Rubella epidemiology in South East England 1980–1984. J. Hyg. Camb. 96, 281–304

Orenstein, W.A., Bart, K.J., Hinman, A.R., Preblud, S.R., Greaves, W.L., Doster, S.W., Stetler, H.C., Sirotkin, B. (1984) The opportunity and obligation to eliminate rubella from the United States. JAMA 251, 1988–1994

Proceedings of The International Symposium on Prevention of Congenital Rubella Infection (1985). Reviews of Infectious Diseases 7, Supplement 1: s1-s215

Tobin, J.O'H., Sheppard, S., Smithells, R.W., Milton, A., Noah, N., Ried, D. (1985) Rubella in the United Kingdom, 1970–1983. Rev. Inf. Dis. 7, Suppl. 1: s47-s52

Valleron, A.J., Flahault, A. (1986) Personal communication

van Druten, J.A.M., de Boo, Th., Plantinga, A.D. (1986) Measles, mumps and rubella: Control by vaccination. In: Develop. Biol. Standard, vol. 65. Karger, Basel, pp. 53–63

van Druten, J.A.M., de Boo, Th., Doesburg, W.H., Peer, P., Reintjes, A.G.M., Bos, J.M., Plantinga, A.D., Ruitenberg, E.J. (1987) Incidence of the congenital rubella syndrome in the Netherlands. Preprint

Influenza and Some Related Mathematical Models

Wei-min Liu and *Simon A. Levin*

Contents

Despite advances in biology and medical science that have controlled many severe infectious diseases, influenza remains a recurrent problem, initiating new global pandemics because of its ability to change its form. In 1918–1919, an influenza pandemic (Spanish flu) killed about 20 million people and infected perhaps 2 billion. The special feature of this pandemic was a tendency towards bronchopneumonic complications fatal to previously healthy young adults. In Philadelphia, people were dying so quickly that bodies were stacked by the hundreds in temporary morgues, awaiting burial. Such horrible mortality caused tremendous social and economic disruption, and stimulated intensive research into the cause of the disease (Beveridge, 1977).

1. Viruses

As we know today, influenza is caused by highly variable RNA viruses belonging to the orthomyxovirus group. Influenza virus first was isolated from swine in 1931 by Richard Shope, and from man in 1933 by Christopher Andrews and Wilson Smith. Due to the development of modern molecular biology, we know a great deal about

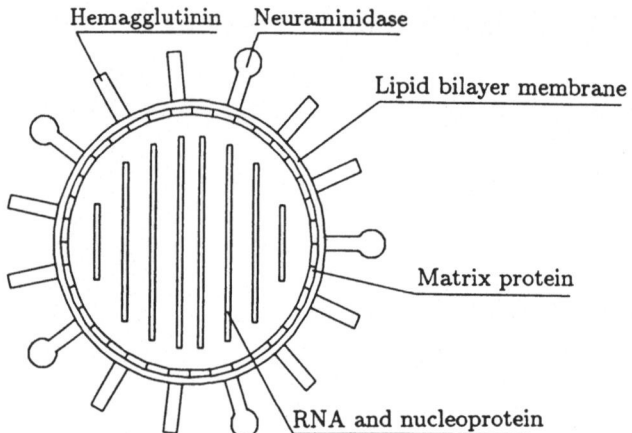

Fig. 1. Schematic diagram of the structure of the influenza virus

the influenza virus, although many points remain unknown. See Kilbourne (1975), Stuart-Harris and Schild (1976), Palese and Young (1982), Webster, Laver, Air and Schild (1982), Selby (1982), Thacker (1986) for details.

The influenza virus is of diameter 80–120 nanometers ($0.8-1.2 \times 10^{-7}$ meters). It has a lipid bilayer membrane with about 500 molecules of hemagglutinin (HA) and about 100 molecules of neuraminidase (NA) on the surface (Fig. 1). Under the membrane there is a shell consisting of about 3000 matrix (M) protein molecules. Inside the matrix protein shell there is a ribonucleoprotein (RNP) complex, which has 7 or 8 single strand RNA segments associated with the nucleoprotein (NP) and 3 different RNA-dependent RNA polymerases (P1, P2 and P3). Recently, at least three non-structural proteins (NS) coded by the virus RNA have been found in infected cells.

Influenza virus is classified into three viral types, A, B and C, according to the nucleoprotein and matrix protein. Influenza A is classified further into subtypes according to the antigenic characteristics of its surface antigens, i.e., HA and NA molecules. So far, 13 distinct HA subtypes and 9 NA subtypes have been discovered. Three subtypes of influenza A are found in man: H1N1, H2N2, and H3N2. The viruses within a subtype can be classified further into different variants according to slight differences among the surface antigens.

Although the core antigens (M and NP) stimulate the production of antibodies, those antibodies do not protect against subsequent infections. Immunity is induced by the surface antigens, particularly by HA. There is no relationship between the antigens of different types; therefore there is no cross-immunity between type A and type B. It also is believed that cross-immunity even between different subtypes of influenza A virus is very weak or does not exist. However, cross-immunity of different variants within a subtype has been observed (Couch and Kasel, 1983).

The influenza virus can change its surface antigenic structure very rapidly to circumvent the host's immune response. There are two kinds of antigenic changes in influenza: drift, the gradual, relatively minor change in antigens, and shift, the sudden, complete change of one or more of the antigens. Drift of surface antigens produces new variants and is the main reason why a subtype of influenza A can survive for several consecutive years in the same host population. Shift produces new subtypes and is the major cause of influenza pandemics.

2. Influenza in Populations

The influenza virus has wide host range, and can infect many domestic and wild vertebrates in addition to humans. Some subtypes of the virus are host specific, while others can infect several host species; the latter behavior is facilitated by antigenic change, since new susceptibility classes can be exploited. The most virulent and the most frequently changing influenza type is influenza A, which causes the pandemics that sweep through the world. The pandemics during the last 100 years have occurred in 1889 (Asiatic), 1918 (Spanish, H1N1), 1957 (Asian, H2N2), 1968 (Hong Kong, H3N2), and 1977 (H1N1). Serological analysis of people born before 1889 suggests that the virus of that year may have had similar antigenic properties to the A/Hong Kong (H3N2) virus of 1968 (Masurel, 1969).

Note that since 1977 two influenza A subtypes H1N1 and H3N2 have cocirculated in the human population, and that the H1N1 subtype is very likely the same as the virus of 1918 Spanish flu. There are basically two hypotheses about the reoccurrence of the H1N1 subtype. The recombination hypothesis assumes that the new H1N1 subtype is produced by recombination of certain other strains. A competing hypothesis is that the H1N1 virus might subsist in a refuge, for example, in some animal population, while awaiting sufficient increase in the number of young susceptible individuals to permit the virus once again to increase within the human population.

3. Influenza in Individuals

Not all influenza infections cause disease. Only 20% to 75% of infections are manifested clinically. The most important symptoms of clinical influenza are headache, fever, aching of the muscles of the back and legs, pains in joints, dry cough and sore throat. The fever temperature may rise to 39°C (101°F) or higher. Sometimes after the body temperature has declined there is a secondary but lower elevation. The latent period of influenza is usually 1 to 3 days; the infectious period averages about 3 days, but may vary from 1 to 6 days.

Usually, the virus only infects the respiratory tract. In exceptional cases, it can be found in blood, heart muscle, and brain. Most influenza deaths are due to the complications of bronchitis and pneumonia. Antibiotics and sulpha drugs can reduce this mortality due to secondary bacterial infection, but they do not affect the course of viral infection. Ultimately, the body's immune system must take over.

In 1957, Alick Issacs discovered that when the influenza virus infects tissue, the cells produce an antiviral substance, which he called interferon. The antiviral action of interferon is not specific to a particular subtype or variant of the virus (Friedman, 1981). However, the amount of interferon in the infected host decreases very quickly after recovery. Therefore, although interferon may prevent an infectious individual from becoming infected by a second strain, it can not provide long lasting widespectral immunity.

4. Transmission

Influenza viruses are usually aerially transmitted between individuals. When an infectious individual coughes or sneezes, many droplets are blown into the air. Some droplets fall to the floor in a few minutes, while some dry out to become droplet nuclei, which are small particles of dried mucus or saliva of $2–4 \times 10^{-6}$ meters in diameter. The droplet nuclei can remain suspended in the air for hours. The influenza virus in the droplet nuclei can also live in the droplet nuclei for many hours if they are not exposed to sunlight. In a crowded enclosed space, there is a very high probability that susceptible people will inhale the droplet nuclei; but the possibility of transmission in open air is relatively small, because of lower concentration and the fact that the sunlight may destroy the virus quickly. However, it is still possible for the virus to be carried by wind over a long distance under suitable conditions.

Generally speaking, influenza has higher incidence in the winter. The common explanation is that seasonal factors involving host physiology, host behavior, and ambient conditions combine to favor the viral transfer during the cold months, when for example, individuals have more contact with each other under enclosed conditions. This is supported by the observation that most outbreaks of equine influenza have been in spring and fall, when there are more races and therefore greater chance of transmission between stables. Furthermore, it is thought that transmission rates are higher among school children in the winter, when they spend most of their time close to one another, in closed spaces, and that incorporation of such factors is essential to the modeling of influenza dynamics.

5. Models

Fine (1982) provides an excellent review of influenza biology and mathematical models. Here, we will discuss briefly those models and others not covered by him. However, there is no pretense of being exhaustive in our coverage.

5.1 Statistical Models of Excess Mortality

For about twenty years, the Centers for Disease Control in Atlanta have used the Serfling model to produce the regression curves and the epidemic threshold curves for weekly pneumonia and influenza mortality, for about 120 US cities. The Serfling model is a descriptive statistical model, where the expected weekly mortality, M_t, in week t is written as a superposition of a linear function, $u + rt$, in t, and the harmonic terms of periods one year (52 weeks) and a half year (26 weeks):

$$M_t = u + rt + A_1 \cos \frac{2\pi t}{52} + B_1 \sin \frac{2\pi t}{52} + A_2 \cos \frac{4\pi t}{52} + B_2 \sin \frac{4\pi t}{52}.$$

The coefficients are obtained by least squares method with the weekly mortality data for the past four or five years, but omitting weeks with epidemic influenza. The curve of 1.64 standard deviations above the regression curve, M_t, is taken to be an epidemic threshold. The probability that an observed mortality in the absence of an epidemic falls above the epidemic threshold is 0.05 (once every twenty weeks). When in twenty weeks two or more weekly data are above the epidemic threshold, it usually indicates an epidemiologically important rise in mortality (Serfling, 1963).

Several improvements have been proposed. Using the proportional mortality (the proportion of all deaths due to pneumonia and influenza) instead of the absolute mortality, one obtains more stable statistics, because this compensates for the inexact reporting period at holidays. Since Serfling's method omits the data of epidemic weeks, it tends to underestimate the level of the epidemics; one can adjust for this by using the estimated mortality for the epidemic weeks.

Choi and Thacker (1981a, b) presented another model for influenza mortality surveillance based on Box-Jenkins-type time series analysis. The method uses an ARIMA (auto-regressive integrated moving average) model. Without assuming independence of successive mortality data, it assigns particular weight to points immediately preceding, and one year preceding, each prediction point.

5.2 Models with Bilinear Incidence Rate

In this and the following sections we discuss some dynamical models of influenza. An often used assumption in dynamical models of infectious disease is that the new incidence is proportional both to the susceptible and infectious subpopulations; in mathematical language, this means that the incidence rate is bilinear in S, the number of susceptibles, and I, the number of infectives. This assumption of bilinearity and homogeneous mixing is a classical one, and pervades the literature (Kermack and McKendrick, 1927; Bailey, 1975; and Fine, 1982). However recent work (Castillo-Chavez, Hethcote, Andreasen, Levin and Liu, 1988a, b) seeks to relax the assumption.

Because we already know the infectious period of influenza, special assumptions about it can be introduced. Spicer (1979) simplified the model of Baroyan and Rvachev et al. (1971, 1977) to fit the data in England and Wales. According to his original notation, the basic model is

$$y_{t+1} = \lambda x_t \sum_{\tau=0}^{t} y_{t-\tau} \psi_\tau, \tag{1}$$

$$x_{t+1} = x_t - y_{t+1}, \tag{2}$$

where x_t is the number of susceptible individuals at time t, y_t is the number of new cases at time t, and ψ_τ is the proportion of cases still infectious τ days after being infected. λ is the transmissibility parameter. He uses the values of ψ_τ suggested by Baroyan et al. (1977):

τ	0	1	2	3	4	5	≥ 6
ψ_τ	1.0	0.9	0.55	0.3	0.15	0.05	0

Since the data available are the weekly deaths registered due to influenza and influenzal pneumonia for England and Wales and separately for Great London, Spicer supplements (1) and (2) with the equation

$$z_{t+1} = k \sum_{\theta=0}^{t+1} y_{t+1-\theta} \phi_\theta, \tag{3}$$

where z_t is the number of registered deaths from time t to $t + 1$, k is the probability that an infected person will die, and ϕ_θ is the probability of dying θ time units after infection. The least squares method is used to find the parameters.

Since 1967, Baroyan, Rvachev and colleagues have published a series of papers on mathematical models for the spread of influenza. Their aim is to forecast the subsequent spatial spread of influenza epidemics. Their model can be written as

$$\frac{dx_i(t)}{dt} = -\frac{\beta_i}{P_i} x_i(t) \int_0^{\bar\tau} y_i(\tau,t) g(\tau) d\tau + \sum_{j=1}^{n} \left(\frac{\sigma_{ji}}{P_j} x_j(t) - \frac{\sigma_{ij}}{P_i} x_i(t) \right), \tag{4}$$

$$\left(\frac{\partial}{\partial t} + \frac{\partial}{\partial \tau} \right) y_i(\tau,t) = \sum_{j=1}^{n} \left(\frac{\sigma_{ji}}{P_j} y_j(\tau,t) - \frac{\sigma_{ij}}{P_i} y_i(\tau,t) \right), \tag{5}$$

$$y_i(0,t) = \frac{\beta_i}{P_i} x_i(t) \int_0^{\bar\tau} y_i(\tau,t) g(\tau) d\tau, \tag{6}$$

where $x_i(t)$ and $y_i(\tau,t)$ are numbers of susceptibles and cases with duration τ at time t in city i, $\bar\tau$ is the maximal duration of a case, $g(\tau)$ is the distribution of duration of cases, β_i is the transmission coefficient in city i, P_i is the population of city i, σ_{ij} is the rate of migration from city i to city j. In early models σ_{ij} is assumed proportional to the populations P_i and P_j, and in later models it is estimated by real transportation data. After 1971, it is assumed that the transmission coefficient and the proportion of susceptible subpopulations are equal for every city during a particular year. They use the data of the early epidemics to predict timing, peak, and shape of the epidemic course in other cities. For details see Baroyan et al. (1977), Rvachev and Longini (1985), and Longini, Fine and Thacker (1986).

5.3 Models with Reed-Frost Transmission Rate

The model with bilinear transmission rate has a problem in its discrete time formulation: the number of new cases, given by the formula

$$C_{t+1} = bS_t C_t, \tag{7}$$

may be larger than the number of susceptibles at time t. This problem is more serious for small populations. The Reed-Frost transmission rate can be used to get around this problem. The basic equation is

$$E(C_{t+1}|C_t, S_t) = S_t(1 - (1-p)^{C_t}), \tag{8}$$

where the left hand side is the expected value of C_{t+1} for given C_t and S_t, and p is the probability that a susceptible has effective contact for transmission with a specified case in one time interval. Hence $1 - p$ is the probability that a susceptible fails to contact effectively a specified case in one time interval, and $(1 - p)^{C_t}$ is the probability that a susceptible fails to contact effectively any of the C_t cases in time interval t. Then $1 - (1 - p)^{C_t}$ is the probability that a susceptible effectively contacts at least one case in time interval t. Therefore the basic Reed-Frost equation makes sense.

Obviously C_{t+1} in this equation is never larger than S_t. Moreover, when p is very small, the Taylor expansion of the Reed-Frost basic equation is consistent with the equation for the bilinear transmission rate. However, the models with Reed-Frost rate are difficult to analyze, and most work is on numerical simulation.

Elveback, Fox, Ackerman and their colleagues have published a series of papers on this kind of model since 1952. Elveback et al. (1976) applied their model to the 1957 Asian (H2N2) and 1968 Hong Kong (H3N2) pandemic strains of influenza A in a small community of 1000 individuals. They assumed a particular age structure and a highly specific social structure in family or social groupings. Using a random number generator for the probabilities of infection and other probabilities, they kept track of the status of each individual on a daily basis. See also Ackerman, Elveback and Fox (1984).

5.4 Models of Two Virus Strains

The key reason why the influenza virus can cause epidemics and pandemics is its fast variation of surface antigens. The interaction of two viral strains is of basic importance for influenza modelling. The interference of viral agents was investigated by Elveback, Fox and Varma (1964) with a discrete model of the Reed-Frost rate, and by Dietz (1979) with a continuous model of bilinear rate. Castillo Chavez et al. (1988a, b) extend Dietz's model for a population with age structure and with partial cross-immunity among the various viral types. The model without age structure can be represented by the diagram:

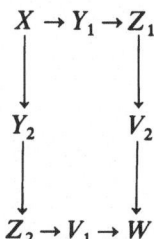

where X is the susceptible class, Y_i $(i = 1, 2)$ is the class infected by strain i but still susceptible to the other strain, Z_i is the class recovered from strain i with susceptibility to the other strain (j) reduced by the factor σ_j (the smaller the σ_j is, the stronger the cross immunity is), V_i is the class infected with strain i but recovered from the other strain and W is the class recovered from both strains. Let β_i be the transmission coefficient of strain i, γ_i the recovery rate from strain i, and μ be the constant mortality rate (and birth rate). The model can be written as:

$$X'(t) = -[\beta_1(Y_1 + V_1) + \beta_2(Y_2 + V_2)]X + \mu - \mu X \tag{9}$$

$$Y_i'(t) = \beta_i(Y_i + V_i)X - (\gamma_i + \mu)Y_i \tag{10}$$

$$Z_i'(t) = \gamma_i Y_i - (\sigma_j \beta_j(Y_j + V_j) + \mu)Z_i \tag{11}$$

$$V_i'(t) = \sigma_i \beta_i(Y_i + V_i)Z_j - (\gamma_i + \mu)V_i \tag{12}$$

$$W'(t) = \gamma_1 V_1 + \gamma_2 V_2 - \mu W \tag{13}$$

$$X(0) = X_0, \quad Y_i(0) = Y_{i0}, \quad Z_i(0) = Z_{i0}, \quad V_i(0) = V_{i0}, \quad W_i(0) = W_0 \tag{14}$$

where $j = 2$ if $i = 1$ and $j = 1$ if $i = 2$. The above system of equations is redundant since

$$X(t) + Y_1(t) + Y_2(t) + Z_1(t) + Z_2(t) + V_1(t) + V_2(t) + W(t) = 1 \tag{15}$$

and therefore can be reduced to a nonlinear system of seven first order equations. The contact number for strain i is

$$R_i = \frac{\beta_i}{\gamma_i + \mu}. \tag{16}$$

This model is similar to Dietz's model, except that his model has $\sigma_1 = \sigma_2 = 1$, so that recovery from one strain does not reduce an individual's susceptibility to the other strain; in other words, the two virus strains in Dietz's model have interference (an individual cannot be infected by both strains simultaneously), but have no cross-immunity.

The model has four equilibria G_i for $i = 1, 2, 3, 4$. The trivial equilibrium G_1 has $X = 1$ and all other variables equal to zero so that neither viral strain is present. If the contact numbers $R_1 \leq 1$ and $R_2 \leq 1$, then G_1 is the only equilibrium in the nonnegative orthant and all solutions in the nonnegative orthant approach G_1. The global asymptotic stability of G_1 in this case is shown by using the Lyapunov function $Y_1 + V_1 + Y_2 + V_2$ and the Lyapunov-LaSalle theorem (Hale, 1969).

If $R_1 > 1$, then there is a boundary equilibrium G_2 given by

$$G_2:(X, Y_1, Z_1, Y_2, Z_2, V_1, V_2) = \left(\frac{1}{R_1}, \frac{\mu}{\gamma_1 + \mu} \left(1 - \frac{1}{R_1} \right), \right.$$

$$\left. \frac{\gamma_1}{\gamma_1 + \mu} \left(1 - \frac{1}{R_1} \right), 0, 0, 0, 0 \right) \tag{17}$$

Analysis of the Jacobian of the system at G_2 reveals that G_2 is locally asymptotically stable if $R_1 > 1$ and

$$\frac{1}{R_2} > \frac{1 - \theta_1}{R_1} + \theta_1, \quad \text{where } \theta_1 = \frac{\sigma_2 \gamma_1}{\gamma_1 + \mu}; \tag{18}$$

the equilibrium G_2 is an unstable saddle if condition (18) is not satisfied. If $R_2 > 1$, then there is an analogous boundary equilibrium G_3.

If the two conditions

$$\frac{1}{R_2} < \frac{1 - \theta_1}{R_1} + \theta_1 \tag{19}$$

$$\frac{1}{R_1} < \frac{1 - \theta_2}{R_2} + \theta_2 \tag{20}$$

are both satisfied, then there is a nontrivial equilibrium G_4 at which both strains remain endemic. Note that θ_1 and θ_2 are both less than 1 because σ_1 and σ_2 are. Thus (19) and (20) guarantee $R_1 > 1$ and $R_2 > 1$ (see Fig. 2). The eigenvalues have been computed for different values of σ_1 and σ_2 that satisfy (19) and (20). In all cases, we (Castillo-Chavez et al., 1988a, b) found that the eigenvalues have negative real parts,

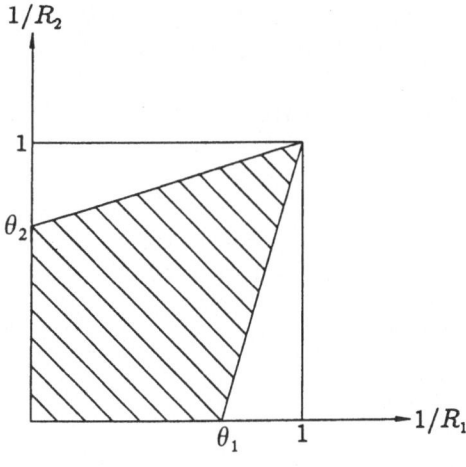

Fig. 2. $\dfrac{1}{R_2} < \dfrac{1-\theta_1}{R_1} + \theta_1$ and $\dfrac{1}{R_1} < \dfrac{1-\theta_2}{R_2} + \theta_2$

$(0 < \theta_1, \theta_2 < 1)$ imply that $R_1 > 1$ and $R_2 > 1$

although a proof that this must hold in all cases has eluded us. We note that the eigenvalue with largest real part usually has its real part around -10^{-4} and its imaginary part around 10^{-2} so the imaginary part is approximately 100 times the real part. Hence solutions are slowly damped as they approach equilibrium.

5.5 Model with Coupling of Two Host Populations

Since the first isolated swine influenza virus was proved antigenically similar to the virus causing the pandemic of Spanish flu in 1918–1919 (Shope, 1936), evidence for interaction among influenza viruses circulating in human and animal populations has accumulated. Moreover, if the entire population is assumed to be homogeneous and to mix at random, the critical population size for persistence of influenza viruses has been estimated to approach or exceed that of the whole human species (Yorke et al., 1979; and Thacker, 1986). Therefore it is reasonable to consider models for the interaction between two host populations, which can even be two host species. Two or more host models are discussed elsewhere by Hethcote (1978), Holt and Pickering (1986), Dobson and May (1986), Anderson and May (1986), Garnick (1986) and Liu (1987).

We consider two coupled SIRS models, each representing one host population whose total population size is assumed to be constant. Hence we can subdivide each host population into susceptible, infectious and recovered fractions. The system can be formulated as

$$\begin{cases} \dot{I}_1 = \lambda_1 I_1 S_1 - (\mu_1 + \gamma_1)I_1 + e_2(I_2)I_2 S_1, \\ \dot{R}_1 = \gamma_1 I_1 - (\mu_1 + \delta_1)R_1, \\ S_1 + I_1 + R_1 = 1, \\ \dot{I}_2 = \lambda_2 I_2 S_2 - (\mu_2 + \gamma_2)I_2 + e_1(I_1)I_1 S_2, \\ \dot{R}_2 = \gamma_2 I_2 - (\mu_2 + \delta_2)R_2, \\ S_2 + I_2 + R_2 = 1, \end{cases} \qquad (21)$$

where the term $e_2(I_2)I_2S_1$ stands for the transmission rate from population 2 to population 1, $e_1(I_1)I_1S_2$ stands for that from population 1 to population 2, and δ_i is the rate of immunity loss.

If there is no coupling, i.e., $e_i(I_i) = 0$ ($i = 1, 2$), it is well known that when the infectious contact number (or disease reproductive rate) $\sigma_i \equiv \lambda_i/(\mu_i + \gamma_i) > 1$, there is a globally stable nontrivial equilibrium $(\bar{I}_i, \bar{R}_i)$; otherwise the trivial equilibrium $(I_i, R_i) = (0, 0)$ is globally stable. The solution at most can show damped oscillations.

If there is coupling, but $e_1(I_1)$ and $e_2(I_2)$ are positive constants b_1 and b_2, i.e., the inter-population transmission rates are bilinear, then (21) is reduced to the form

$$
\begin{aligned}
\dot{I}_1 &= (\lambda_1 I_1 + b_2 I_2)(1 - I_1 - R_1) - (\mu_1 + \gamma_1)I_1, \\
\dot{R}_1 &= \gamma_1 I_1 - (\mu_1 + \delta_1)R_1, \\
\dot{I}_2 &= (\lambda_2 I_2 + b_1 I_1)(1 - I_2 - R_2) - (\mu_2 + \gamma_2)I_2, \\
\dot{R}_2 &= \gamma_2 I_2 - (\mu_2 + \delta_2)R_2.
\end{aligned}
\tag{22}
$$

A similar model, but without a recovered and immune class, is discussed by Holt and Pickering (1986). Hethcote (1978) considers a more general model than (22). Here we obtain and present our results about (22) in a more intuitive and direct way. For the system (22), there is at most one nontrivial equilibrium, which exists if and only if

(i) $\sigma_1 > 1$, or

(ii) $\sigma_2 > 1$, or

(iii) $\sigma_1 \leqq 1$, and $\sigma_2 \leqq 1$ and $(1 - \sigma_1)(1 - \sigma_2) < \dfrac{b_1 b_2}{(\gamma_1 + \mu_1)(\gamma_2 + \mu_2)}$ $\tag{23}$

(see Appendix)

When the nontrivial equilibrium exists, it is locally asymptotically stable and the trivial one is unstable. On the other hand, if

$$
\sigma_1 < 1, \quad \sigma_2 < 1, \quad \text{and } (1 - \sigma_1)(1 - \sigma_2) > \frac{b_1 b_2}{(\gamma_1 + \mu_1)(\gamma_2 + \mu_2)},
\tag{24}
$$

Table 1. Data of two SIRS systems for numerical calculations

	I	II
β	0.3766	0.4
δ	0.0001	0.0
γ	0.33333	0.33333
μ	0.00004	0.00028
$\bar{S}$	0.885210	0.834025
$\bar{I}$	0.000048	0.000139
$\bar{R}$	0.114742	0.165836
Re(χ)	−0.000079	−0.000168
IM(χ)	0.002459	0.004308
T	2555.3	1458.4
c_i	0.00008	0.0002
d_i	0.0001	0.00022

$(\bar{S}, \bar{I}, \bar{R})$ is the nontrivial equilibrium for each SIRS system, and χ is an eigenvalue of the Jacobian of the system at the nontrivial equilibrium. T is the period in days of the damped oscillation near the nontrivial equilibrium. The values in this line are respectively 7 and 4 years.

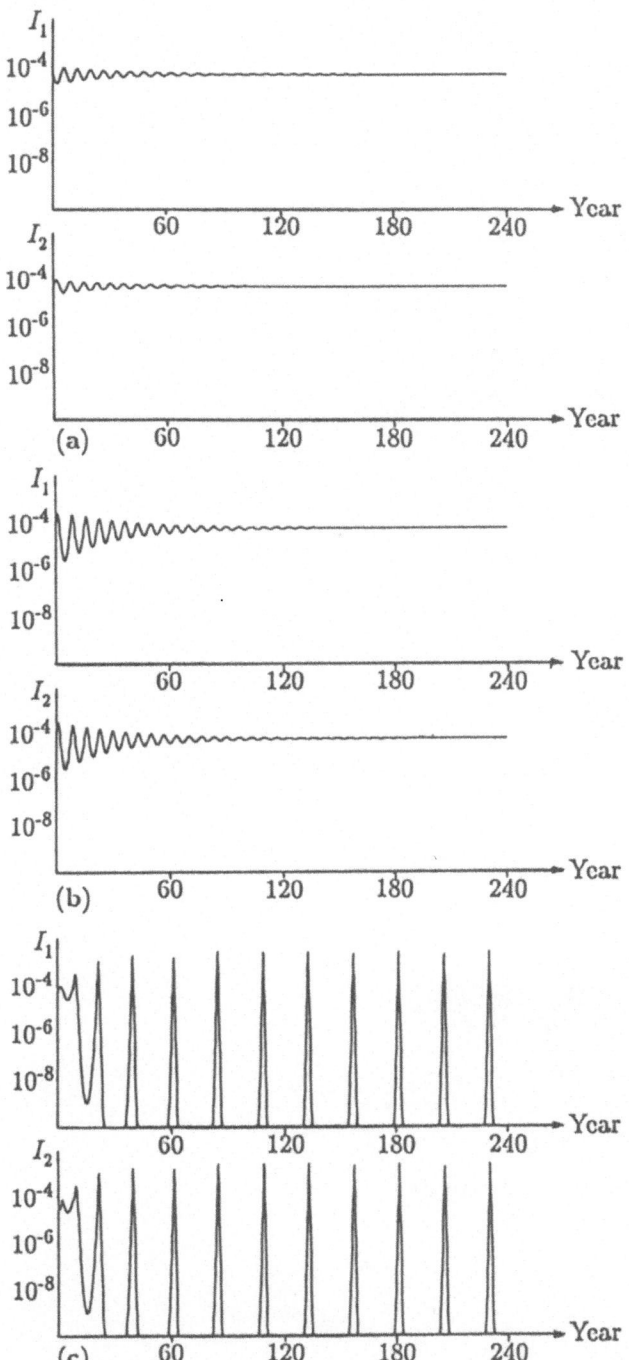

Fig. 3. Infectious fractions of two coupled SIRS subsystems with parameters in group I of Table 1, (a) damped oscillation in uncoupled system, (b) damped oscillation for bilinear coupling, (c) sustained oscillation for non-bilinear coupling

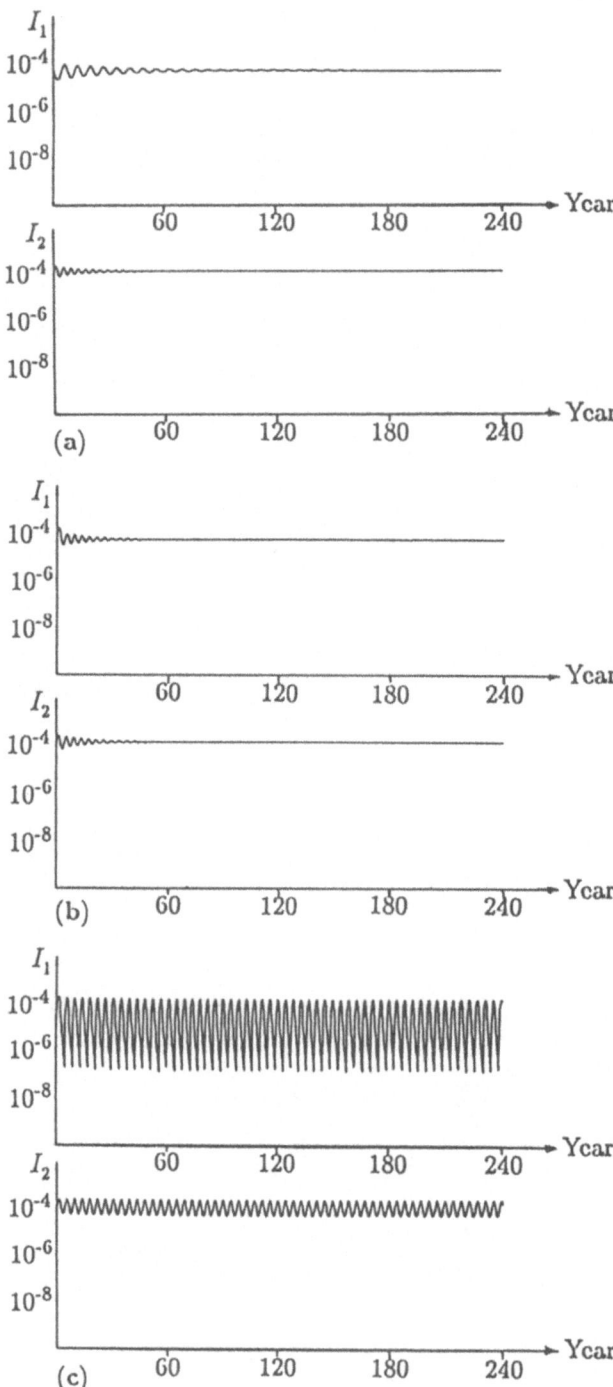

Fig. 4. Infectious fractions of two coupled SIRS subsystems with parameters in groups I and II of Table 1, (a) damped oscillation in uncoupled system, (b) damped oscillation for bilinear coupling, (c) sustained oscillation for non-bilinear coupling

there is no nontrivial equilibrium and the trivial equilibrium is locally asymptotically stable.

If coupling is one-way, i.e., $e_1(I_1) = 0$, but $e_2(I_2) \neq 0$, or vice versa, the nontrivial equilibrium is unique and locally asymptotically stable if it exists, and there is no possibility of Hopf bifurcation (see Appendix).

The effect of more complicated transmission rates can be investigated numerically. A reasonable assumption is that

$$e_i(I_i) = b_i j_i(I_i), \tag{25}$$

where b_i is a constant; and the function $j_i(I_i)$ equals 0 when I_i is less than a threshold value c_i and 1 when I_i is larger than d_i, and is a cubic function when I_i is between c_i and d_i. In our numerical calculations, two groups of parameters in Table 1 are used, group I represents a human population, and group II represents an animal population. Coupling I-I can be considered as the weak interaction between two relatively isolated human populations, and coupling I-II corresponds to the interaction between a human population and an animal population. The parameters c_1 and c_2 are chosen to be larger than the infectious fractions $\bar{I}_1$ and $\bar{I}_2$ at the nontrivial equilibria of the uncoupled subsystems, so that $(\bar{I}_1, \bar{R}_1, \bar{I}_2, \bar{R}_2)$ is a locally asymptotically stable nontrivial equilibrium of the coupled system. Initial conditions are chosen to be near this nontrivial equilibrium.

If there is no coupling, i.e., $b_1 = b_2 = 0$, the solution shows weak damped oscillations as we expect theoretically (Figs. 3a and 4a). If the coupling is bilinear, the solution also shows damped oscillation and the oscillations of two infectious fractions tend to be synchronous (Figs. 3b and 4b).

However, if the coupling is as given by (25), and the parameters b_1 and b_2, corresponding to the strength of transmission between two different populations, are larger than some critical values (but still much smaller than β_1 and β_2, the transmission coefficients within each population), then the attractive basin of the equilibrium $(\bar{I}_1, \bar{R}_1, \bar{I}_2, \bar{R}_2)$ is very small, and sustained oscillations are observed (Figs. 3c and 4c). Notice that in Fig. 3a the infectious fraction can reach a very low level, remain there for a long time and then increase again. This is reminiscent of the fact that some influenza viruses can cause pandemics, then seem to disappear for a long time, then reoccur and cause pandemics again. Therefore our model gives a possible mathematical explanation of the recurrent outbreaks of influenza.

6. Conclusion

The dynamics of influenza incidence is very complex. It has an annual flare-up of one virus subtype and outbreaks of new subtypes. Moreover, the old subtype which is replaced by other subtypes can reappear and cause new pandemics. So far, no single model can interpret all these aspects. However, based on studies of viral molecular structure and function and of the transmission rule of infection, mathematical models can now clarify some aspects, such as geographical spread of the disease, outbreaks in a structured population, and the recurrent outbreak of subtypes.

Acknowledgements. This work was done under the support of NSF Grant DMS-8406472 and McIntire Stennis Grant NYC-183568 to S.A. Levin. W. Liu's work was partly supported by the US Army Research Office through the Cornell University Mathematical Sciences Institute Fellowship. This research was conducted at the Cornell National Supercomputer Facility, Center for Theory and Simulation in Science and Engineering, which is funded, in part, by the National Science Foundation, New York State and IBM Corporation. Thanks are also due to Carlos Castillo-Chavez, Herbert W. Hethcote, Ira M. Longini, Jr. and Stephen B. Thacker for their helpful comments. This is publication ERC-143 of the Ecosystems Research Center, Cornell University.

References

Ackerman, E., Elveback, L.R., Fox, J.P. (1984) Simulation of Infectious Disease Epidemics. Thomas, Springfield, Ill.

Anderson, R.M., May, R.M. (1986) The invasion, persistence and spread of infectious diseases within animal and plant communities. Phil. Trans. Roy. Soc. Lond. *B314*, 533–570

Bailey, N.T.J. (1975) The Mathematical Theory of Infectious Diseases and Its Applications, 2nd edn. Griffin, London

Baroyan, O.V., Rvachev, L.A., Basilevsky, U.V., Ermakov, V.V., Frank, K.D., Rvachev, M.A. and Shashkov, V.A. (1971) Computer modelling of influenza epidemics for the whole country (USSR). Adv. Appl. Prob. *3*, 224–226

Baroyan, O.V., Rvachev, L.A. (1977) Mathematics and epidemiology. Moscow: Znanie (in Russian)

Beveridge, W.I.B. (1977) Influenza: the last great plague. Heinemann, London

Castillo-Chavez, C., Hethcote, H.W., Andreasen, V., Levin, S.A., Liu W. (1988a) Cross immunity in the dynamics of homogeneous and heterogeneous populations. Proceedings of the second autumn course on mathematical ecology

Castillo-Chavez, C., Hethcote, H.W., Andreasen, V., Levin, S.A., Liu W. (1988b) Epidemiological models with age structure and proportionate mixing

Choi, K., Thacker, S.B. (1981a) An evaluation of influenza mortality surveillance, 1962–1979 (I) Time series forecasts of expected pneumonia and influenza deaths. Amer. J. Epid. *113*, 215–226

Choi, K. Thacker, S.B. (1981b) An evaluation of influenza mortality surveillance, 1962–1979 (II) Percentage of pneumonia and influenza deaths as an indicator of influenza activity. Amer. J. Epid. *113*, 227–235

Couch, R.B., Kasel, J.A. (1983) Immunity to influenza in man. Ann. Rev. Microbiol. *37*, 529–49

Dietz, K. (1979) Epidemiologic interference of virus populations. J. Math. Biol. *8*, 291–300

Dobson, A.P., May, R.M. (1986) Patterns of invasions by pathogens and parasites. In: Mooney, H.A., Drake, J.A. (ed.) Ecology of biological invasions of North America and Hawaii. Springer, Berlin Heidelberg New York

Elveback, L.R., Fox, J.P., Varma, A. (1964) An extension of the Reed-Frost epidemic model for the study of competition between viral agents in the presence of interference. Amer. J. Epid. *80*, 356–364

Elveback, L.R., Fox, J.P., Ackerman, E., Langworthy, A., Boyd, M., Gatewood, L. (1976) An influenza simulation model for immunization studies. Amer. J. Epid. *103*, 152–165

Fine, P. (1982) Background paper: applications of mathematical models to the epidemiology of influenza: a critique. In: P. Selby (ed.) Influenza models: prospects for development and use. Sandoz Institute for Health and Socio-economic Studies, pp. 15–85

Friedman, R.M. (1981) Interferon: a primer. Academic Press, New York London Toronto Sydney San Franscisco

Garnick, E. (1986) A theoretical consideration of resource specialism vs. generalism in parasites and some related questions. Ph.D. Thesis, Cornell University

Hale, J.K. (1969) Ordinary differential equations. Wiley Interscience, New York

Hethcote, W.H. (1978) An immunization model for a heterogeneous population. Theor. Pop. Biol. *14*, 338–349

Holt, R.D., Pickering, J. (1986) Infectious disease and species coexistence: a model of Lotka Volterra form. Amer. Nat.

Kermack, W.O., McKendrick, A.G. (1927) Contributions to the mathematical theory of epidemics, pt. I. Proc. Roy. Soc. *A115*, 700–721

Kilbourne, E.D., (ed.) (1975) The influenza viruses and influenza. Academic Press, New York London Toronto Sydney San Franscisco

Liu, W. (1987) Dynamics of epidemiological models—recurrent outbreaks in autonomous systems. Ph.D. Thesis, Cornell University

London, W.E., Yorke, J.A. (1973) Recurrent outbreaks of measles, chicken pox and mumps, I. Seasonal variation in contact rates. Amer. J. Epid. *98*, 453–468

Longini, I.M., Fine, P.E.M., Thacker, S.B. (1983) Predicting the global spread of new infectious agents. Amer. J. Epid. *123*, 383–391

Palese, P., Young, J.F. (1982) Variation of influenza A, B, and C viruses. Science *215*, 1468–1474

Rvachev, L.A., Longini, I.M., Jr. (1985) A mathematical model for the global spread of influenza. Math. Biosci. *75*, 3–22

Selby, P. (ed.) (1982) Influenza models: prospects for development and use. Sandoz Institute for Health and Socio-economic Studies

Serfling, R.E. (1963) Methods for current statistical analysis of excess pneumonia-influenza deaths. Publ. Hlth. Rep. *78*, 494–506

Shope, R.E. (1936) The incidence of neutralizing antibodies for swine influenza virus in the sera of human beings of different ages. J. Exp. Med. *63*, 669–684

Spicer, C.C. (1979) The mathematical modelling of influenza epidemics. Brit. Med. Bull. *35*, 23–28

Stuart-Harris, C.H. Schild, G.C. (1976) Influenza, the viruses and the disease. Publishing Sciences Group, Littleton, Mass.

Thacker, S.B. (1986) The persistence of influenza A in human population. Epid. Rev. *8*, 129–142.

Webster, R.G., Laver, W.G., Air, G.M., Schild, G.C. (1982) Molecular mechanisms of variation in influenza viruses. Nature *296*, 115–121

Appendix

Here we prove the claims in Sect. 5.5 about system (22) with bilinear coupling, and those about system (21) with one-way coupling, i.e., $e_1(I_1) = 0$, but $e_2(I_2) \neq 0$, or vice versa.

A.1 Existence and Uniqueness of Nontrivial Equilibrium

First of all, we prove that condition (23) is a necessary and sufficient condition for the existence of the nontrivial equilibrium, which is unique if it exists. If there is a nontrivial equilibrium, it must satisfy

$$R_1 = h_1 I_1 \neq 0, \quad h_1 \equiv \frac{\gamma_1}{\mu_1 + \delta_1},$$

$$R_2 = H_2 I_2 \neq 0, \quad h_2 \equiv \frac{\gamma_2}{\mu_2 + \delta_2},$$

$$I_2 = \frac{I_1(\mu_1 + \gamma_1 - \lambda_1 S_1)}{b_2 S_1} = \frac{I_1(\mu_1 + \gamma_1 - \lambda_1(1 - I_1/H_1))}{b_2(1 - I_1/H_1)}, \quad I_1 < H_1 = \frac{1}{1 + h_1},$$

$$I_1 = \frac{I_2(\mu_2 + \gamma_2 - \lambda_2 S_2)}{b_1 S_2} = \frac{I_2(\mu_2 + \gamma_2 - \lambda_2(1 - I_2/H_2))}{b_1(1 - I_2/H_2)}, \quad I_2 < H_2 = \frac{1}{1 + h_2}.$$

The last two equations can be written as

$$I_2 = f_1(I_1) \equiv \frac{(\mu_1 + \gamma_1)I_1(\sigma_1 I_1/H_1 - (\sigma_1 - 1))}{b_2(1 - I_1/H_1)}, \tag{A1}$$

and

$$I_1 = f_2(I_2) \equiv \frac{(\mu_2 + \gamma_2)I_2(\sigma_2 I_2/H_2 - (\sigma_2 - 1))}{b_1(1 - I_2/H_2)}. \tag{A2}$$

Set

$$x_i = I_i/H_i$$

and

$$r_i = (\mu_i + \gamma_i)H_i/b_j H_j,$$

where $j = 3 - i$. Then (A1) and (A2) become

$$x_2 = r_1 x_1 \left(\frac{1}{1 - x_1} - \sigma_1 \right), \tag{A3}$$

and

$$x_1 = r_2 x_2 \left(\frac{1}{1 - x_2} - \sigma_2 \right), \tag{A4}$$

It is clear that both curves so defined pass through the origin. Furthermore, (A3) defines x_2 as a convex function of x_1, with initial slope $s_1 = r_1(1 - \sigma_1)$; and (A4) defines x_1 as a convex function of x_2, with initial slope $s_2 = r_2(1 - \sigma_2)$. Since $x_2 \to \infty$ as $x_1 \to 1$ in (A3), and $x_1 \to \infty$ as $x_2 \to 1$ in (A4), it is clear that the two curves cross for $0 < x_1, x_2 < 1$ unless s_1 and s_2 are both positive and $s_1 s_2 \leq 1$. That is, there is a nontrivial equilibrium provided $\sigma_1 > 1$, or $\sigma_2 > 1$, or $\sigma_1, \sigma_2 \leq 1$ and

$$r_1 r_2 (1 - \sigma_1)(1 - \sigma_2) < 1,$$

or equivalently,

$$(\mu_1 + \gamma_1)(\mu_2 + \gamma_2)(1 - \sigma_1)(1 - \sigma_2) < b_1 b_2, \tag{A5}$$

see (23).

A.2 Local Stability of Nontrivial Equilibrium

The nontrivial equilibrium (I_1, R_1, I_2, R_2) or (22) is locally asymptotically stable when it exists. The Jacobian matrix is

$$J = \begin{bmatrix} -x_1 k - y_1 & -y_1 & x_1 & 0 \\ \gamma_1 & -m_1 & 0 & 0 \\ x_2 & 0 & -y_2 - x_2/k & -y_2 \\ 0 & 0 & \gamma_2 & -m_2 \end{bmatrix} \tag{A6}$$

where

$$y_1 = \lambda_1 I_1 + b_2 I_2, \quad y_2 = \lambda_2 I_2 + b_1 I_1, \quad m_1 = \mu_1 + \delta_1, \quad m_2 = \mu_2 + \delta_2. \tag{A7}$$

and

$$x_1 = b_2 S_1, \quad x_2 = b_1 S_2, \quad k = \frac{I_2}{I_1}. \tag{A8}$$

Although the calculations are lengthy and tedious, the stability of the equilibrium follows in a straight forward manner from application of the Routh-Hurwitz criteria.

A.3 Local Stability of Trivial Equilibrium

The trivial equilibrium is locally asymptotically stable if (24) holds and is unstable if (23) holds.

The Jacobian matrix at the trivial equilibrium is

$$J_0 = \begin{bmatrix} -g_1 \xi_1 & 0 & b_2 & 0 \\ \gamma_1 & -m_1 & 0 & 0 \\ b_1 & 0 & -g_2 \xi_2 & 0 \\ 0 & 0 & \gamma_2 & -m_2 \end{bmatrix},$$

where

$$\xi_1 = 1 - \sigma_1, \quad \xi_2 = 1 - \sigma_2, \quad g_1 = \gamma_1 + \mu_1, \quad g_2 = \gamma_2 + \mu_2.$$

Thus, its eigenvalues are $-m_1$, $-m_2$, and the eigenvalues of the 2×2 matrix

$$\begin{pmatrix} -g_1 \xi_1 & b_2 \\ b_1 & -g_2 \xi_2 \end{pmatrix}.$$

Thus, by the Routh-Hurwitz criterion, the matrix is stable provided

$$g_1 \xi_1 + g_2 \xi_2 > 0$$

and

$$g_1 g_2 \xi_1 \xi_2 > b_1 b_2;$$

that is, provided

$$(1 - \sigma_1)(\gamma_1 + \mu_1) + (1 - \sigma_2)(\gamma_2 + \mu_2) > 0 \tag{A9}$$

and

$$(\mu_1 + \gamma_1)(\mu_2 + \gamma_2)(1 - \sigma_1)(1 - \sigma_2) > b_1 b_2. \tag{A10}$$

Considering the cases (i) $\sigma_1 \geq 1$, $\sigma_2 \geq 1$; (ii) $\sigma_1 \geq 1$, $\sigma_2 < 1$; (iii) $\sigma_1 < 1$, $\sigma_2 \geq 1$; and (iv) $\sigma_1 < 1$, $\sigma_2 < 1$ separately, we can show that in cases (i), (ii) and (iii) at least of the conditions (A9) and (A10) does not hold. Only in case (iv) it is possible for (A9) and (A10) to hold simultaneously, but (A9) is then implied by $\sigma_1 < 1$, and $\sigma_2 < 1$. One can also see that a necessary condition for the trivial equilibrium to be asymptotically stable is that

$$\frac{b_1 b_2}{(\gamma_1 + \mu_1)(\gamma_2 + \mu_2)} < 1.$$

A.4 One-Way Coupling

When $e_1(I_1) = 0$, and $e_2(I_2) \neq 0$, system (21) becomes

$$
\begin{cases}
\dot{I}_1 = \lambda_1 I_1 S_1 - (\mu_1 + \gamma_1) I_1 + e_2(I_2) I_2 S_1, \\
\dot{R}_1 = \gamma_1 I_1 - (\mu_1 + \delta_1) R_1, \\
\dot{I}_2 = \lambda_2 I_2 S_2 - (\mu_2 + \gamma_2) I_2, \\
\dot{R}_2 = \gamma_2 I_2 - (\mu_2 + \delta_2) R_2,
\end{cases}
\tag{A11}
$$

where

$$S_1 = 1 - I_1 - R_1, \quad S_2 = 1 - I_2 - R_2.$$

The last two equations of (A11) form an independent SIRS model. If and only if $\sigma_2 > 1$, it has a nontrivial equilibrium $(\bar{I}_2, \bar{R}_2)$. If $e_2(\bar{I}_2) = 0$, then the first two equations have a nontrivial equilibrium $(\bar{I}_1, \bar{R}_1)$ if and only if $\sigma_1 > 1$. If $e_2(\bar{I}_2) \neq 0$, we can substitute $R_1 = h_1 I_1$ and $S_1 = 1 - I_1/H_1$ into the first equation to obtain a quadratic equation for I_1 at the nontrivial equilibrium:

$$\sigma_1 I_1 (1 - I_1/H_1) - I_1 + e(\bar{I}_2)\bar{I}_2(1 - I_1/H_1)/(\mu_1 + \gamma_1) = 0. \tag{A12}$$

Since the coefficient of the quadratic term and the constant term have different signs, there are a positive root and a negative root of (A12). If this positive root is smaller than H_1, then it leads to a nontrivial equilibrium of system (A11); otherwise there does not exist a nontrivial equilibrium of (A11).

$$
J = \begin{bmatrix}
\lambda_1 S_1 - \mu_1 - \gamma_1 - \lambda_1 I_1 - e_2(I_2)I_2 & -\lambda_1 I_1 - e_2(I_2)I_2 & (e_2(I_2)I_2)'S_1 & 0 \\
\gamma_1 & -\mu_1 - \delta_1 & 0 & 0 \\
0 & 0 & \lambda_2 S_2 - \mu_2 - \gamma_2 - \lambda_2 I_2 & -\lambda_2 I_2 \\
0 & 0 & \gamma_2 & -\mu_2 - \delta_2
\end{bmatrix}
\tag{A13}
$$

The Jacobian matrix of (A11) at an arbitrary point is

$$
J = \begin{bmatrix}
-e_2(\bar{I}_2)\bar{I}_2\bar{S}_1/\bar{I}_1 - \lambda_1\bar{I}_1 - e_2(\bar{I}_2)\bar{I}_2 & -\lambda_1\bar{I}_1 - e_2(\bar{I}_2)\bar{I}_2 & (e_2'(\bar{I}_2)\bar{I}_2 + e(\bar{I}_2))\bar{S}_1 & 0 \\
\gamma_1 & -\mu_1 - \delta_1 & 0 & 0 \\
0 & 0 & -\lambda_2\bar{I}_2 & -\lambda_2\bar{I}_2 \\
0 & 0 & \gamma_2 & -\mu_2 - \delta_2
\end{bmatrix}
\tag{A14}
$$

The eigenvalues of this matrix are the eigenvalues of the pair of 2×2 diagonal block submatrices. Since each has negative trace and positive determinant, it follows that if the nontrivial equilibrium exists, all eigenvalues of J have negative real parts. Therefore the nontrivial equilibrium is locally asymptotically stable, and Hopf bifurcation is impossible.

Review of Recent Models of HIV/AIDS Transmission

Carlos Castillo-Chavez

Contents

HIV, the human immunodeficiency acquired immunodeficiency syndrome virus, is the etiological agent for AIDS (acquired immuno deficiency syndrome). In 1982 Gallo suggested that the cause of AIDS was likely to be a new human retrovirus and, in 1983, researchers at the Pasteur Institute under the direction of Montagnier were able to isolate a new retrovirus from a New York AIDS victim (see Barre-Sinoussi et al., 1983). In 1984, Gallo and his colleagues isolated the same type of retrovirus and proved it to be the etiological agent of AIDS (for more details see Gallo, 1986, 1987; Wong-Staal and Gallo, 1985). This virus has been estimated to kill at least 30% of those infected. By April 1988, about 58,000 individuals have died of AIDS in the United States, and the Coolfont Report (1986) predicts that by 1991 the lower bound for the cumulative number of AIDS cases will be 290,000 individuals in the United States alone. One of the biggest problems associated with HIV is that most infected individuals appear to be asymptomatic and infectious for long periods of time, with an average infectious period of at least 8 years. Furthermore, there is growing evidence that the infectiousness of individuals varies with time since infection; the amount of free virus is relatively high just after infection (Francis et al., 1984; Salahuddin et al., 1984), remains low for several years, and climbs again within a year or so of the onset of AIDS (Lange et al., 1986).

Some of the important factors in the dynamics of HIV relate to the heterogeneity of the host population. These considerations include sexual preference (homosexual, bisexual and heterosexual), degree and type of sexual activity (number of partners, length of partnerships, anal sex), age-structure (degree of sexual activity may be a function of age), intravenous drug use (sharing of contaminated needles), socio-economic factors (which affect the level of education and hence the degree of response to education programs, and, in the lower socio-economic classes, a higher incidence of prostitution, shared needles in intravenous drug use, etc.), and cultural factors (different degrees of sexual activity for males and females, acceptability and frequency of use of prostitutes, etc.). In addition, there are epidemiological factors that may be crucial to the dynamics of HIV, including latent periods, long periods of infectivity, variable infectivity, and a high percentage of asymptomatic carriers.

Mathematical models help investigate the relative importance of these factors. If this first step is accomplished with some degree of success, then we can proceed to determine the demographic and economic impact of this epidemic. This process may help us to develop reasonable, scientifically and socially sound intervention plans that are applicable under a variety of circumstances, and to determine the best ways to implement them rapidly. The series of models briefly reviewed in this chapter represent a first step in approaching these objectives, and also provide a complement to the more extensive and complete chapter by May and Anderson reprinted in this volume (or see May and Anderson, 1988a, b). While the following exposition suffers from inevitable incompleteness deriving from constraints of space, I hope this discussion of the modelling approaches to the spread of AIDS will attract more investigators into the field of mathematical epidemiology.

The Role of Long Periods of Infectivity

We (Castillo-Chavez et al., 1989a, Huang et al., 1989) have developed a series of methmatical models that begin by looking at AIDS as an exclusively sexually-transmitted disease. Our objective has been to evaluate the role of long periods of infectiousness in the dynamics of HIV. These models are natural extensions of those developed by Anderson and May (1987, 1988). Our approach has been to divide the population under discussion into social groups defined by criteria such as sex, sexual behavior, or age, and these groups can be extended to include socioeconomic categories of various types. As is customary in these models, differences among individuals of the same group are ignored but not those among groups; hence we assume a network of interactions of various strengths (see Fig. 1).

For our immediate purposes, only the simplest version of this model—that of a

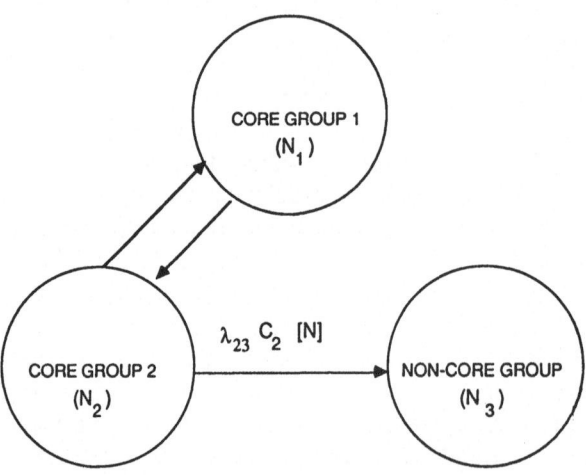

Fig. 1. Hypothetical 3 group network, λ_{ij} denotes the probability of transmission per unit time per partner between individuals of group i and j, $C_i[N]$ denotes the mean number of sexual partners that an average individual of group i has given that the population density is $N = N_1 + N_2 + N_3$

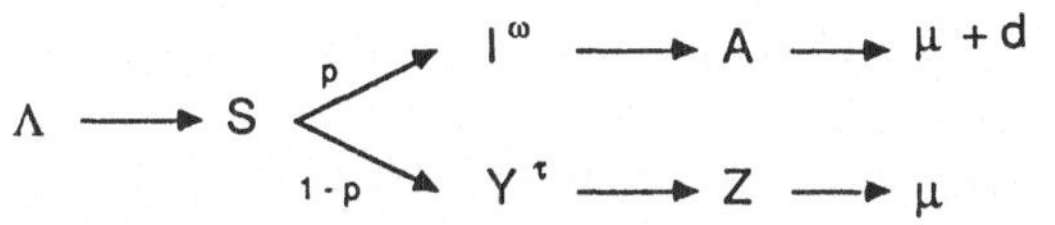

$$\Lambda \longrightarrow S \begin{array}{c} \overset{p}{\nearrow} \quad I^{\omega} \longrightarrow A \longrightarrow \mu + d \\ \\ \underset{1-p}{\searrow} \quad Y^{\tau} \longrightarrow Z \longrightarrow \mu \end{array}$$

Fig. 2. Single group model; for details see the text

single group—is included. We divide this population into five classes: S (susceptible), I (infectious that will go on to develop AIDS); Y (infectious that will not develop full-blown AIDS), Z (former Y that are no longer sexually active), and A (former I that have developed full-blown AIDS) (see Fig. 2). We assume that individuals become infectious immediately, and that individuals who develop full-blown AIDS have no sexual contacts and hence are not responsible for new infections. In order to explore the role played by the long period of asymptomatic infection we assume that the I- and Y-infectious individuals remain infectious for fixed periods of time: ω and τ units of time respectively.

We denote by $I_0(t)$ and $Y_0(t)$ those individuals that were in either class I or Y at time $t = 0$, and are still infectious; $Z_0(t)$, those individuals that were in class Z at time $t = 0$, and are still alive; and $A_0(t)$, those individuals who have already developed full-blown AIDS at time $t = 0$, and are still alive. Hence it is reasonable to assume that $Z_0(t)$ and $A_0(t)$ vanish for large t. In addition, since ω and τ denote the infectious periods, we assume that $I_0(t) = Y_0(t) = 0$ for $t > \max(\omega, \tau)$.

Λ denotes the "recruitment" rate into the susceptible class (defined as those individuals who become sexually active); μ, the natural mortality rate; d, the disease-induced mortality due to AIDS; p, that fraction of the susceptibles that become I-infectious [and therefore $(1 - p)$ the fraction of the susceptibles that become Y-infectious]. The function $H(x)$ denotes the Heaviside function, defined as being equal to 1 if $x > 0$ and zero otherwise. Following Castillo-Chavez et al. (1989a, b) and Anderson et al. (1988) and using Fig. 2, we arrive at the epidemiological model:

$$\frac{dS(t)}{dt} = \Lambda - \lambda C[T](t)\frac{S(t)W(t)}{T(t)} - \mu S(t),$$

$$I(t) = I_0(t) + \lambda p \int_{t-\omega}^{t} C(T(x))\frac{S(x)W(x)}{T(x)}H(x)e^{-\mu(t-x)}dx,$$

$$Y(t) = Y_0(t) + \lambda(1-p) \int_{t-\tau}^{t} C(T(x))\frac{S(x)W(x)}{T(x)}H(x)e^{-\mu(t-x)}dx,$$

$$A(t) = A_0(t) + \lambda p \int_{0}^{t-\omega} C(T(x))\frac{S(x)W(x)}{T(x)}H(x)e^{-\mu(t-x)-d(t-x-\omega)}dx,$$

$$Z(t) = Z_0(t) + \lambda(1-p) \int_{0}^{t-\tau} C(T(x))\frac{S(x)W(x)}{T(x)}H(x)e^{-\mu(t-x)}dx,$$

where $W(t) = I(t) + Y(t)$, and $T(t) = S(t) + W(t)$.

The function $C(T)$ denotes the mean number of sexual partners of an average individual per unit time, given that the population density is T, and λ (a constant) denotes the transmission probability per partner. More specifically (as in Hyman and Stanley 1988), $\lambda = i\phi$ where i denotes the probability of infection per sexual contact and ϕ denotes the average number of contacts per sexual partner. Hence $\lambda C(T)$ denotes the rate of transmission per unit time and $\lambda C(T)dt$ denotes the probability that a given sexual partner will transfer the disease to a particular individual in time dt. The factor W/T is the probability that a contact of a susceptible individual with randomly selected individual will be an infectious individual. We (1989a, b) have shown analytically that, under appropriate assumptions on $C(T)$, the local dynamics of this system is governed by the *reproductive number R*, the number of secondary infections produced by a single infected individual in a purely susceptible population. R is given by

$R = $ (prob. of transmission)

   $\times$ (mean number of sexual partners in a purely susceptible population)

        $\times$ (the mean infectious period)

or

$$ R = \lambda C\left(\frac{\Lambda}{\mu}\right)\left\{ p\frac{1-e^{-\mu\omega}}{\mu} + (1-p)\frac{1-e^{-\mu\tau}}{\mu} \right\}. $$

Briefly, the maintenance of the disease in the population at endemic levels can occur only if $R > 1$. Hence to eliminate the disease we should consider control strategies that could potentially reduce R below this critical value. Approaches of this type have already been used in the development of vaccination strategies for a variety of diseases (see this volume), or in educational programs for reducing transmission.

For AIDS, it is not sufficient to consider homogeneous populations, as the dynamics are affected critically by a variety of factors such as age, sexual behavior, geographical area, or other characteristics. Considering models that incorporate several interacting groups is therefore a priority. In this case, however, the ease of performing analytical computations disappears, and numerical simulations may be the only approach. Numerical computations for intergroup models also show the existence of threshold behaviors, allow estimation of those thresholds, and explore of parameter sensitivity.

Simulations are also useful in one-group models; for example, consider the role of education in the dynamics of HIV. Assuming that education becomes effective and parameters have been changed so that $R < 1$, then through numerical simulations we see that the time lag in system responsiveness assures a skewed concentration of infected individuals. This delay means that even if the disease eventually is eradicated, the number of infected individuals would continue to increase for a long time—perhaps several times the infectious period. Thus, the effects of education will not produce a decrease in the total number of infected individuals for many years. If the rate of decrease in the number of new cases is very slow, and the initial observation is a sharp increase in the number of cases of individuals who develop "full-blown" AIDS, then education could be perceived primarily as a cause of increased promiscuity, rather than as controlling the disease, thus having a serious negative impact on public policy.

Other models that concentrate on the effects of long infectious periods can be found in the works of Anderson and May (1987, 1988), Blythe and Anderson (1988), and Anderson (1988).

Effects of Partnership Dynamics

Study of the mathematical aspects of partnership models in a demographic context can be found in the works of Kendall (1949), Pollard (1973), and Fredrickson (1971). However, Dietz and Hadeler (1988) were the first to put these demographic models into an epidemiological framework. As Dietz and Hadeler [1988] point out, "The classical models for sexually transmitted infections assume homogeneous mixing either between all males and females or between certain groups of males and females with heterogeneous contact rates. This implies that everybody is all the time at risk of acquiring infection." Therefore, they ignore the periods of 'immunity' due to temporary monogamous partnership among uninfected individuals. These authors raise the question of the potential effects of these temporary periods of 'immunity' in the dynamics of sexually transmitted diseases. A simplified version of the Dietz/Hadeler model forms the background for the discussion that follows.

I will briefly describe this framework for a single homosexual population and expand it to include variable infectivity and variable AIDS-related mortality. This extension is more realistic, as it allows us to see AIDS as a progressive disease. Extensions are then possible to include a variety of sexual tendencies, age structures, multiple groups, and variable infectivity.

At time t we denote by $m(t)$ the number of uninfected single males, $M(t)$ the number of infected single males, Λ the constant "recruitment" rate into the m-category, and $P_{mm}(t)$, $P_{mM}(t)$, $P_{MM}(t)$, the number of **m-m**, **m-M** and **M-M** pairs

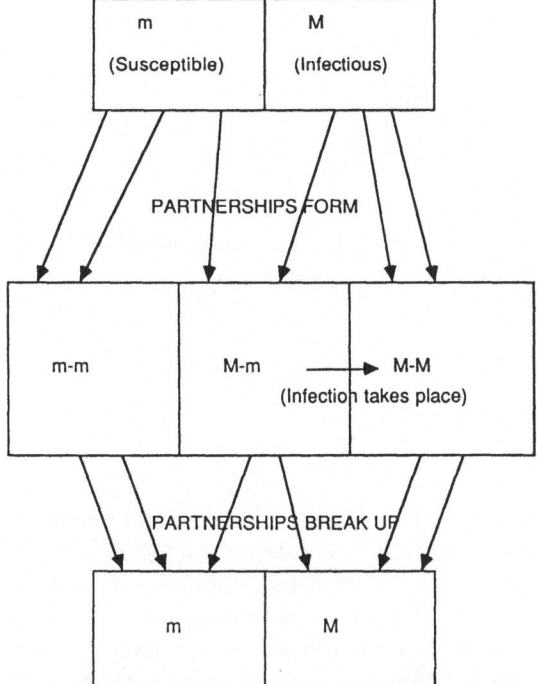

Fig. 3. Flow diagram for a single group partnership model

respectively. Furthermore, we denote by σ the constant break-up rate, μ the constant natural mortality rate for m-individuals, d the constant mortality rate for M-individuals, i the probability of infection per contact, and α the number of contacts per unit time while in a partnership. Since infection only takes place while in a partnership, we can make use of Fig. 3 to arrive at the following set of ordinary differential equations that describe the dynamics of HIV:

$$\frac{dm}{dt} = \Lambda - \mu m + 2(\sigma + \mu)P_{mm} + (d + \sigma)P_{mM} - 2\phi(m, m) - \phi(m, M),$$

$$\frac{dM}{dt} = -dM + (\mu + \sigma)P_{mM} + 2(d + \sigma)P_{MM} - \phi(m, M) - 2\phi(M, M),$$

$$\frac{dP_{mm}}{dt} = \phi(m, m) - (\sigma + 2\mu)P_{mm},$$

$$\frac{dP_{mM}}{dt} = \phi(m, M) - (\mu + \sigma + d)P_{mM} - \beta P_{mM},$$

$$\frac{dP_{MM}}{dt} = \phi(M, M) - (2d + \sigma)P_{MM} + \beta P_{mM},$$

$\beta = i\alpha$ and $\phi(x, y)$ is a non-negative nonlinear pair formation function that satisfies the following conditions as specified by Fredrickson (1971):

$$\phi(0, y) = 0 \qquad \text{for all } y \geq 0,$$
$$\phi(x, 0) = 0 \qquad \text{for all } x \geq 0,$$
$$\phi(\delta x, \delta y) = \delta\phi(x, y) \quad \text{for all } x, y, \delta \geq 0,$$
$$\frac{\partial\phi}{\partial x} \geq 0, \quad \frac{\partial\phi}{\partial y} \geq 0,$$

Following Dietz and Hadeler (1988), we assume symmetry; that is,

$$\phi(x, y) = \phi(y, x) \quad \text{for all } x \text{ and } y \geq 0.$$

For a discussion of several functional forms, see Keyfitz (1972), Kendall (1949), and Dietz and Hadeler (1988); however, the most common are

$$\phi(x, y) = \rho\frac{xy}{x + y},$$

and

$$\phi(x, y) = \rho\min(x, y):$$

the harmonic mean and the minimum function. A partial local stability analysis of a two-sex model that uses both pair-formation functions can be found in Dietz and Hadeler (1988). Their main conclusion is that endemic equilibria are possible only if the separation rate is large enough to ensure a sufficient supply of sexual partners. This result immediately suggests possible educational strategies to help reduce the incidence of sexually-transmitted diseases. They include the encouragement of longer monogamous partnerships or simply the encouragement of sequential (rather than overlapping) partnerships.

In the case of AIDS, other effects such as variable infectivity may have to be taken into consideration even in the case of a single population. We can formulate an extension of the Dietz and Hadeler's model that considers the now-serious possibility that AIDS is a progressive disease. The variables t and τ denote time and time since infection. The mortality rate for infected individuals $d = d(\tau)$ is a function of τ, $m(t)$ and $P_{mm}(t)$ have the same meaning as before. However, $M(t, \tau)$, $P_{mM}(t, \tau)$, $P_{MM}(t, \tau, \tau')$ now denote densities, so that for example:

$$\Delta t \int_{\tau}^{\tau + \Delta \tau} \int_{\tau'}^{\tau' + \Delta \tau'} P_{MM}(t, \tau, \tau') d\tau d\tau'$$

denotes the number of $M_\tau - M_{\tau'}$ pairs in time interval $[t, t + \Delta t]$. We still assume that μ and σ are constant and denote $\phi(x, y)$ by ϕ_{xy}. Again following Fig. 3, we arrive at the model:

$$\frac{dm}{dt} = \Lambda - \mu m - 2(\sigma + \mu)P_{mm} + \int_0^\infty (d(\tau) + \sigma)P_{mM}(t, \tau) d\tau - 2\phi_{mm} - \int_0^\infty \phi_{mM}(t, \tau) d\tau,$$

$$\frac{\partial M}{\partial t} + \frac{\partial M}{\partial \tau} = -d(\tau)M(t, \tau) + (\sigma + \mu)P_{mM}(t, \tau)$$

$$+ 2 \int_0^\infty (\sigma + d(\tau))P_{MM}(t, \tau, \tau') d\tau' - \phi_{mM}(t, \tau) - 2 \int_0^\infty \phi_{MM}(t, \tau, \tau') d\tau',$$

$$\frac{dP_{mm}}{dt} = \phi_{mm}(t) - (\sigma + 2\mu)P_{mm},$$

$$\frac{\partial P_{mM}}{\partial t} + \frac{\partial P_{mM}}{\partial \tau} = \phi_{mM}(t, \tau) - (\sigma + \mu + d(\tau))P_{mM} - \beta P_{mM},$$

$$\frac{\partial P_{MM}}{\partial t} + \frac{\partial P_{MM}}{\partial \tau} + \frac{\partial P_{MM}}{\partial \tau'} = \phi_{MM}(t, \tau, \tau') - (\sigma + d(\tau) + d(\tau'))P_{MM},$$

where $P_{MM}(t, \tau, 0) = \beta P_{MM}(t, \tau)$, $P_{MM}(t, \tau, \tau') = P_{MM}(t, \tau', \tau)$ and appropriate initial conditions are prescribed.

To use the harmonic mean to describe the process of pair formation, we set

$$\phi_{mm}(t) = \rho \frac{m(t)m(t)}{m(t) + M(t)},$$

$$\phi_{mM}(t, \tau) = 2\rho \frac{m(t)M(t, \tau)}{m(t) + M(t)},$$

$$\phi_{MM}(t, \tau, \tau') = \rho \frac{M(t, \tau)M(t, \tau')}{m(t) + M(t)},$$

where

$$M(t) = \int_0^\infty M(t, \tau) d\tau.$$

Whether or not the introduction of variable infectivity (and hence long periods of infectiousness) will affect Dietz and Hadeler's main conclusion is still an open question.

The Effects of Multiple Sexual Partners

Hyman and Stanley (1988) have developed some risk-based models based on the assumption that individuals with multiple sexual partners are usually infected first and tend to become the major source of spread into those groups with fewer sexual partners. In addition, they have also explored the role of variable infectivity in the context of their model. The simplest version of their model can serve as an example of this approach.

We let t and τ denote time and time since infection, μ denotes the natural mortality rate, $\delta(\tau)$ denotes the death rate τ-units after the onset of AIDS, $\beta(\tau)$ denotes the rate of developing AIDS τ-units after infection, $i(\tau)$ denotes the probability of infection from a contact with a person infected τ-units of time ago, and r denotes the average number of new sexual partners per year. In addition we denote by $S(t, r)$ the density of susceptibles according to their number of partners per year, $I(t, r, \tau)$ the density of infecteds according to the number of partners per year and the time since infection, $c(r, r')$ the number of contacts between people with r and r' partners per year, and $S_0(r)$ the density of people with r new partners per year before the introduction of the AIDS virus. Using this notation, Hyman and Stanley (1988) arrive at the following model:

$$\frac{\partial S(t, r)}{\partial t} = \mu(S_0(r) - S(t, r)) - \lambda(t, r)S(t, r),$$

$$I(t, 0, r) = \lambda(t, r)S(t, r),$$

$$\frac{\partial I(t, \tau, r)}{\partial t} + \frac{\partial I(t, \tau, r)}{\partial \tau} = (\beta(\tau) + \mu)I(t, \tau, r),$$

$$A(t, 0) = \int_0^\infty \int_0^\infty \beta(\tau)I(t, \tau, r)\, d\tau\, dr,$$

$$\frac{\partial A(t, \tau)}{\partial t} + \frac{\partial A(t, \tau)}{\partial \tau} = -\delta(\tau)A(t, \tau),$$

$$\frac{dA_T}{dt} = \int_0^\infty \int_0^\infty \beta(\tau)I(t, \tau, r)\, d\tau\, dr,$$

where

$$\lambda(t, r) = \frac{r\int_0^\infty C(r, r')r' \int_0^\infty i(\tau)I(t, \tau, r')\, d\tau\, dr'}{\int_0^\infty r''S(t, r'')\, dr'' + \int_0^\infty r''I(t, \tau, r'')\, d\tau\, dr''}$$

Stanley and Hyman observe that if $c(r, r')$ and $i(\tau)$ are constants, then partners are chosen at random from the population. Hence this model reduces to one of Anderson et al. (see May and Anderson's chapter in this volume). Preliminary numerical investigations show that the shape of the distribution of individuals according to risk can have a marked effect on the shape of the epidemic. In addition, the rate at which the susceptible population is infected seems to be very sensitive to small changes in the infectivity profile.

In the population genetics literature (see Crow and Kimura, 1970) the effects of different mating systems on gene flow are studied. An alternative approach to modelling the sexual transmission of HIV is through the mixing of the population genetics and epidemiological approaches, superimposing a 'mating' system, such as assortive mating, within the classical epidemiological framework. A starting point towards this approach may be that of Colgate, Hyman, and Stanley (1988). Their recent risk-based model, based on the observed cubic growth of AIDS cases, has built on the assumption that individuals with multiple sexual partners have sexual intercourse mostly with individuals with multiple sexual partners. That is, most contacts take place between individuals in the same risk group. An important question that Colgate et al. (1988) address is that of the speed of propagation of the disease from highly prosmiscuous individuals to those that are less prosmiscuous. Colgate et al. (1988) derive a diffusion equation when mixing is small. Their diffusion equation model has similarity solutions which are saturating waves moving from large risk groups to low risk groups with decreasing speed.

Conclusions

In this chapter we have introduce several models that begin to address several questions raised by the AIDS epidemic including the role of long periods of infectivity, the effects of temporary immunity due to the formation of sexual partnerships between non-infected individuals, and the overall effects of high risk (multiple sexual partners) behavior. May and Anderson's chapter in this volume addresses (among other issues) the possible demographic consequences of this epidemic.

Although progress has been made, there is still insufficient information to appropriately predict the dynamics of this epidemic. How will the dynamics be affected by education or intervention programs? How can we implement the use of drugs that may be developed to decrease the level of infectiousness and extend the life of HIV carriers? If a vaccine is developed, what is the optimal vaccination program?

Furthermore, we have to be aware that since many symplifying assumptions are made in the construction of these models, including the clumping or aggregation of many important variables and components, the results of these models cannot be used to circumvent the moral and ethical questions raised by this epidemic.

Acknowledgments. This work has been partially supported by NSF grant DMS-8406472 to Simon A. Levin, by The Center for Applied Mathematics and the Office of the Provost at Cornell University, as well as by a Ford Foundation Postdoctoral Fellowship for Minorities. I give my thanks to all of them.

References

Anderson, R.M. (1988) The epidemiology of HIV infection: variable incubation plus infectious period and heterogeneity in sexual activity (ms.)

Anderson, R.M., May, R.M. (1987) Transmission dynamics of HIV infection. Nature 326, 137–142

Anderson, R.M., et al. (1988) A preliminary study of the transmission dynamics of the human immunodeficiency virus (HIV), the causative agent of AIDS. IMA J. Math. Med. Biol. (in press)

Barre-Sinoussi, F., et al. (1983) Isolation of a T-lymphotropic retrovirus from a patient at risk for acquired immune deficiency syndrome (AIDS). Science 220, 868–70

Blythe, S.P., Anderson, R.M. Distributed incubation and infectious periods in models of the transmission dynamics of the Human Immunodeficiency Virus (HIV). IMA J. Math. Med. Biol. (in press)

Castillo-Chavez, C., Cooke, K., Huang, W., Levin, S.A. (1988a) The role of infectious periods in the dynamics of acquired immunodeficiency syndrome (AIDS). In: C. Castillo-Chavez, S.A. Levin, and C. Shoemaker (eds.) Proceedings International Symposium in Mathematical Approaches to Ecological and Environmental Problem Solving, Ithaca, 1987. Lecture Notes in Biomathematics. Springer, Berlin Heidelberg New York (to appear)

Castillo-Chavez, C., Cooke, K., Huang, W., Levin, S.A. (1988b) The role of long infectious periods in the dynamics of HIV/AIDS. Part 1 (in preparation)

Castillo-Chavez, C., Cooke, K., Huang, W., Levin, S.A. (1988c) The role of infectious periods in the dynamics of in the dynamics of HIV/AIDS. Part 2 (in preparation)

Colgate, S.A., Hyman, J.M., Stanley, E.A. (1988) A risk base model explaining the cubic growth in AIDS cases (unpublished manuscript)

Coolfont Report (1986) A PHS plan for prevention and control of AIDS and the AIDS virus. Public Health Rep. 101, 341–48

Crow, J.F., Kimura, M. (1970) An introduction to population genetics theory. Burgess Publishing Company. Minneapolis, Minnesota

Dietz, K., Hadeler, K.P. (1988) Epidemiological models for sexually transmitted diseases. J. Math. Biology (in press)

Francis, D.P. et al. (1984) Infection of chimpanzees with lymphadenopathy-associated virus. Lancet, 1, 1276–77

Fredrickson, A.G. A mathematical theory of age structure in sexual populations: random mating and monogamous marriage models. 1971. Math. Biosci. 10, 117–143

Gallo, R.C. (1986) The first human retrovirus. Scientific American, Dec. 88–89

Gallo, R.C. (1987) The AIDS virus. Scientific American, Jan.: 47–56

Hyman, J.M., Stanley, E.A., (1988) A risk base model for the spread of the AIDS virus. Math. Biosci. (in press)

Kendall, D.G. (1949) Stochastic processes and population growth. J. Roy. Statist. Soc. Ser. B11, 230–264

Keyfitz, N. (1972) The mathematics of sex and marriage. Proceedings of the Sixth Berkeley Symposium on Mathematical Statistics and Probability, vol: IV: Biology and Health, pp. 89–108

Lange, J.M.A., et al. (1986) Persistent HIV antigenaemia and decline of HIV core antibodies associated with transition to AIDS." Brit. Med. J. 293, 1459–62

May, R.M., Anderson, R.M., McLean, A.R. (1988a) Possible demographic consequences of HIV/AIDS: I, assuming HIV infection always leads to AIDS. Math. Biosci. (in press)

May, R.M., Anderson, R.M., McLean, A.R. (1988b) Possible demographic consequences of HIV/AIDS: II, assuming HIV infection does not necessarily lead to AIDS. In: Castillo-Chavez, C., Levin, S.A., Shoemaker, C. (eds.) Proceedings International Symposium in Mathematical Approaches to Ecological and Environmental Problem Solving, Berlin Heidelberg New York Ithaca, 1987. Lecture Notes in Biomathematics. Springer (to appear)

Pollard, J.H. (1973) Mathematical models for the growth of human populations, Chapter 7. The University Press, Cambridge

Suluhuddin, S.Z. et al. (1984) HTLV-III in symptom-free seronegative persons. Lancet, Dec. 22–29: 1418–20

Wong-Staal, F., Gallo, R.C. (1985) Human T-lymphotropic retroviruses. Nature 317, 395–403

The Transmission Dynamics of Human Immunodeficiency Virus (HIV)

Robert M. May and Roy M. Anderson

Contents

Abstract

The paper first reviews data on HIV infections and AIDS disease among homosexual men, heterosexuals, IV-drug abusers and children born to infected mothers, in both developed and devloping countries. We survey such information as is currently available about the distribution of incubation times that elapse between HIV infection and the appearance of AIDS, about the fraction of those infected with HIV who eventually go on to develop AIDS, about time-dependent patterns of infectiousness, and about distributions in rates of acquiring new sexual or needle-sharing partners.

Using this information, models for the transmission dynamics of HIV are developed, beginning with deliberately oversimplified models and progressing—on the basis of the understanding thus gained—to more complex ones. Where possible, estimates of the model's parameters are derived from the epidemiological data, and predictions are compared with observed trends. We also combine these epidemiological models with demographic

Reprinted from the Philosophical Transactions of The Royal Society (Discussion Meeting on the Epidemiology and Ecology of Infections Disease Agents, edited by Anderson and Thresh).

considerations, to assess the effects that heterosexually-transmitted HIV/AIDS may eventually have on rates of population growth, on age profiles, and on associated economic and social indicators, in African and other countries. The degree to which sexual or other habits must change to bring the "basic reproductive rate", R_0, of HIV infections below unity, is discussed. We conclude by outlining some research needs, both in the refinement and development of models, and in the collection of epidemiological data.

1. Introduction

As abundantly recorded in prose and picture throughout history, plague and pestilence have always excited more dread than the other three horsemen of the Apocalypse. Infectious diseases frighten us not simply because they can kill or cripple, but because the process of infection seems so insidious. Other chance events, such as car accidents (which in developed countries will many more people each year than does AIDS, although this could change in the USA in a few years), give the illusion of being under our control. But the inapparent nature of most infectious agents seemingly puts them outside our control, and gives rise to anxieties that are affected little by knowing, for example, that HIV is spread by a virus whose molecular sequence we know, rather than by some medieval miasma.

The Human Immunodeficiency Virus (HIV), the causative agent of Acquired Immunodeficiency Syndrome (AIDS), has several properties which make it especially liable to arouse such anxieties. For one thing, a high proportion of HIV infections (30% or more) lead to the disease AIDS and thus to death, with no cure and no vaccine in sight. For another thing, HIV infection as such is effectively asymptomatic in most cases, and the incubation interval to develop AIDS Related Complex (ARC) or AIDS is long and variable. Moreover, not enough is yet known about basic epidemiological parameters to make reliable long-term predictions about likely patterns of infection among particular groups (heterosexuals, homosexual males, IV-drug abusers, and so on) in particular countries or regions.

The epidemic spread or endemic maintenance of an infection depends on its basic reproductive rate, R_0, within the host population in question. For a "microparasitic" infection (sensu Anderson and May, 1979) such as HIV, R_0 is defined as the average number of secondary infections produced by one primary infection within a wholly susceptible population (Ross, 1911; Macdonald, 1956; Smith, 1970; Anderson and May, 1979, 1985). If R_0 exceeds unity, an epidemic will on average spread, expanding along chains of transmission in which each infection produces more than one "offspring"; if R_0 is less than unity, the infection dies out following each introduction. As discussed more rigorously below, for HIV/AIDS and other sexually transmitted diseases R_0 is found, from its definition, by multiplying together the probability of infecting any one partner, times the average number of new partners per unit time, times the average duration of infectiousness (which may, or may not, be the average time from requiring HIV to dying from AIDS). Each of these three basic components—incubation period and pattern of infectiousness, rates of acquiring new partners of a specified kind, and transmission probabilities—is currently ill-understood for HIV/AIDS.

Our paper deals with these and other aspects of the transmission dynamics of HIV/AIDS. We first survey and summarize what is currently known about the epidemiological processes outlined in the previous paragraph. We then present a range of simple mathematical models designed to capture the essentials of these processes, and we indicate what can be learned from the models. A separate section combines epidemiological with demographic considerations, to explore the possible effects of HIV/AIDS on age-profiles and other social and economic characteristics in countries, such as Africa, where the disease seems to be widely disseminated. We conclude by discussing the use of such models as guides to the kinds of data needed for long-term assessment, and as points of departure for the numerical exploration of more detailed and more realistic models (as approximate data become available). We also indicate how such models can be used to make inferences about individual-level parameters (such as transmission probabilities) from population-level data (such as the way seroprevalence changes over time).

2. Basic Epidemiological Factors

2.1 Groups at Risk for HIV/AIDS

In sub-Saharan Africa, HIV/AIDS appears to be mainly transmitted by sexual contacts among heterosexuals (Quinn et al., 1986). In developed countries, homosexual and bisexual males account for most (around 65–85% or more) cases of AIDS. IV-drug abusers also constitute a significant fraction of AIDS cases in developed countries, with significant variations among countries and among regions within a country; IV-drug abusers comprise about half the recently reported AIDS cases in northeastern regions of the USA.

HIV infections and AIDS cases resulting from blood transfusions in general, and among hemophiliacs in particular, amounted to a small fraction of the total in earlier years. In developed countries, there are now very few, if any, new infections in these categories, following the introduction of screening and testing of blood supplies. This transmission route is still open in some developing countries, however, and its role in the overall transmission pattern in such countries remains somewhat uncertain.

HIV/AIDS can be transmitted vertically to the offspring of infected mothers, and these "pediatric AIDS" cases are growing everywhere AIDS is reported. Although very uncertain, current estimates are that may be 30–50% or more of the children born to infected mothers will die of AIDS in their first few years. It is conceivable that HIV infection may also be transmitted vertically from males, in their sperm. We are not aware of any discussion of this contingency, but if there is such a possibility it could complicate the analysis of data, as well as adding to the significance of vertical transmission.

The age-specific incidence of AIDS cases in both developed and developing countries seems to rule out any significant transmission of HIV infection by insect vectors or by contaminated needles in public health programs. Were it otherwise, one would expect to see more cases in the 5–15 year age-range (although it is conceivable that multiple use of vaccination needles in developing countries is concentrated within age-cohorts, thus not propagating HIV among children so long as the prevalence remains very low among children; there is room for further empirical and theoretical studies of this question).

These facts lead us, in the first instance, to study models for the transmission of HIV/AIDS among homosexual males in large cities in the UK, USA and other developed countries, and among heterosexuals in Africa. Similar models pertain to transmission among IV-drug abusers. Public health planning will eventually require detailed studies involving more complex models, embracing all the transmission routes just discussed and assigning magnitudes to the parameters that characterize the linkages among different categories (Hethcote, 1988); little is yet known about these parameters.

The different patterns in different countries are reflected in ratios of AIDS cases among men to cases among women. The current overall incidence of AIDS cases in the USA is 13.0 times greater for males than for females (HHS, 1987), which accords with the ratios of 14–20 to 1 found in European countries (Anderson et al., 1986; May and Anderson, 1987). As summarized in Table 1, however, studies among particular groups in the USA give lower ratios, which seems reasonable given the nature of the groups. In Africa, by contrast, the sex ratio of Aids cases runs around 1:1, with perhaps a slight excess of female AIDS cases (Quinn et al., 1986).

Table 1. Ratio of AIDS cases or HIV prevalence among males to that among females, for some particular studies in the USA (from HHS, 1987)

Group	Statistic	Ratio of males to females
All AIDS cases	Cumulative number of reported AIDS cases, through 1987	13.0 to 1
Heterosexual adult and adolescent AIDS cases	Cumulative number assigned to this category, CDC data through 1987	2.9 to 1
Military recruit applicants	Seroprevalence of HIV; data from Department of Defense, 1985–1877	5.5 to 1
America Red Cross blood donors	Nationwide seroprevalence of HIV, 1986–87: first time donors repeat donors	4.6 to 1 4.6 to 1
Sentinel hospital patients in Midwest	Seroprevalence of HIV, unpublished CDC data	2.3 to 1
IV-drug abusers	New York City, 1985	0.9 to 1
IV-drug abusers	Four cities in Connecticut, 1986–1987	1.2 to 1

2.2 Prevalence of HIV Infection and AIDS Cases

As of November 1987, 126 countries had reported at least one case of AIDS. In the USA, around 55,000 cases of AIDS (resulting so far in 31,000 deaths) had been reported by March 1988. The corresponding numbers for the UK are around 1,200 cases and 700 deaths, as of January 1988. Figure 1 shows temporal trends in the numbers of AIDS cases for several countries.

Although complicated by lags in reporting and changes in the definition of what constitutes AIDS (Brookmeyer, 1988), the data for AIDS cases are much more abundant than that for HIV infections. HIV infections are usually asymptomatic, and many of those who are infected are unaware of their state. More than this, there is no registry for compiling data of HIV infections, and even if there were many of those infected would actively avoid such a registry. In many countries, plans for surveys of HIV seropositivity among unbiased samples of the population are being held up by worries about the feasibility of obtaining essentially 100% compliance (or at least unbiased noncompliance), by ethical dilemmas about how to counsel those found to be seropositive, and by technical problems about false positives and negatives. The kinds of studies we have all consequently involve groups that are biased in one way or another: blood transfusion recipients; attendees at clinics for sexually-transmitted diseases (STDs); pregnant women; military personnel or applicants for the armed services; female prostitutes; and others. Figure 2 shows data for the change in seroprevalence over time, for some of these groups.

In the USA as of November 1987, some 50 studies of homosexual and bisexual men, in different regions, show seroprevalence levels ranging from under 10% to as high as 70%, with most findings falling between 20% and 50%. Among IV-drug abusers, 88 surveys and studies find seroprevalence ranging from highs of 50–65% in the vicinity of New York City and in Puerto Rico, to rates which—although variable—are mostly below 5% in areas other than the East Coast. For individuals requiring treatment with clotting factor concentrates, seroprevalence runs around 70% for hemophilia A and around 35% for hemophilia B; the rates are uniform throughout the USA, reflecting the lack of any significant geographical factor in the distribution of hemophilia. Other special groups include prisoners and female prostitutes. Some 33 studies of HIV prevalence among prisoners showed levels ranging from 0 to 17%, which—as expected—was higher than among the general population but lower than among high-risk groups in the same general region. Some 19 studies of female prostitutes showed levels ranging from 0 to 45%, with the highest rates seen in large inner-city areas where IV-drug abuse is common; HIV prevalence was 3 to 4 times higher among prostitutes who acknowledged IV-drug abuse than among who did not (CDC, 1987; Johnson, 1988).

Among the USA population at large, such data as are available suggest HIV seroprevalence levels of the general order of 10^{-3}. For instance, Red Cross blood donors who have not previously been tested currently average 0.04% seropositive; applicants for military service (whom the US military believe to under-represent persons in high-risk groups, although we do not share their certainty) run around 0.15%; Job Corps entrants ("disadvantaged" 16 to 21 year-olds) average around

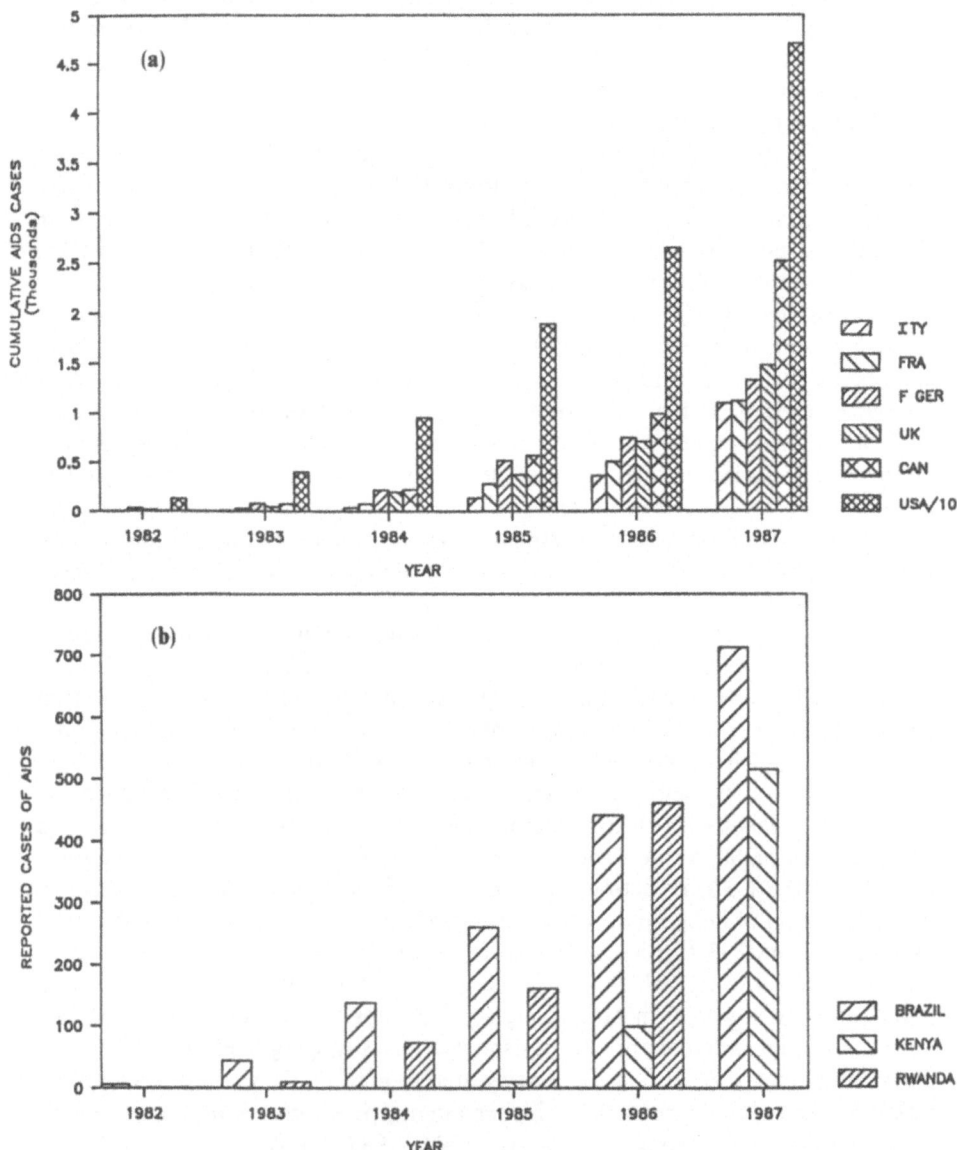

Fig. 1. (a) Cumulative number of cases of AIDS in European and North American countries (as reported to the World Health Organization). **(b)** Cumulative number of cases of AIDS in Brazil, Kenya and Rwanda (as reported to the World Health Organization)

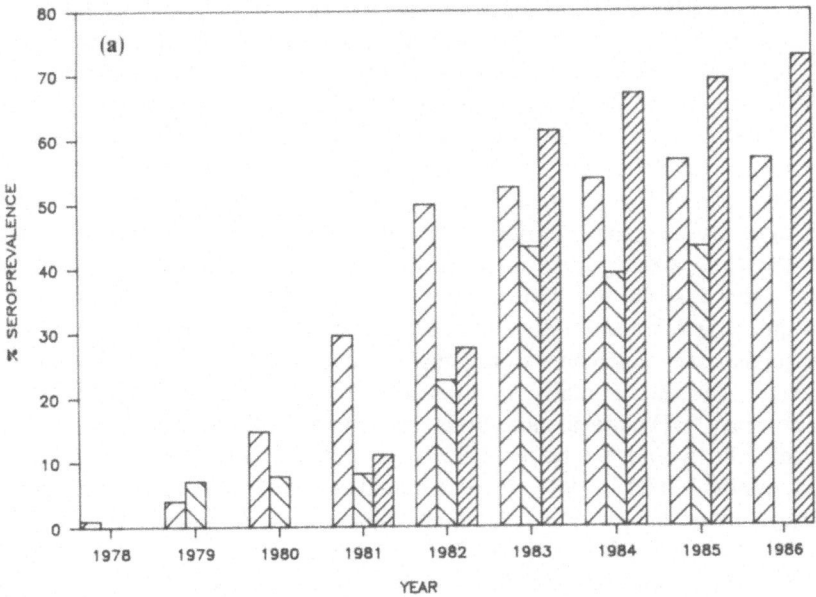

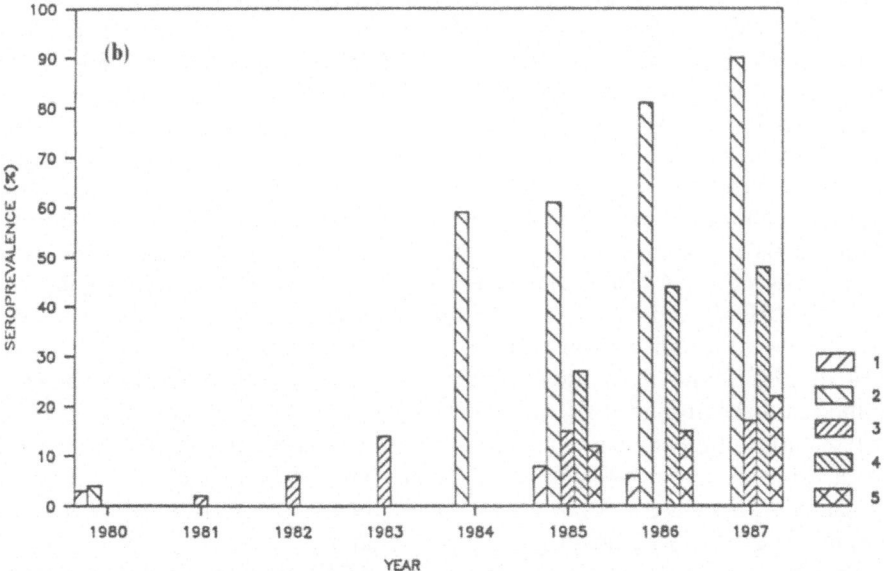

Fig. 2. (**a**) The rise in seropositivity to HIV antigens in cohorts of patients over the period 1978–1985. The studies in San Francisco, London and New York are of homosexual/bisexual males, and the study in Italy is of IV-drug abusers (for sources of these data, see May and Anderson, 1987). (**b**) Longitudinal changes in HIV seropositivity among people in various risk-groups in different countries in Africa: (*1*) pregnant women in Zaire; (*2*) female prostitutes in Nairobi, Kenya; (*3*) men with chancroid, Kenya; (*4*) men with STD infections, Rwanda; (*5*) pregnant women in Kinshasa, Zaire

0.33%; patients without AIDS-like conditions at 4 hospitals have seroprevalence levels around 0.32%; childbearing women in Massachusetts (tested anonymously through filter-paper blood specimens from their newborn infants) ran around 0.21%; 27 surveys in women's health clinics (excluding drug users) produced levels ranging from 0 to 2.6% (HHS, 1987). The overall figure accords with the independent estimate of around 1.5 million HIV infections, within the USA population of around 240 million. This rough figure is arrived at by adding up the estimated numbers in various categories of risk, along with rough estimates of seroprevalence levels in the various categories; the estimated number of homosexual males, for example, is based on the old Kinsey study.

Attempts have been made to refine the above USA data by removing those contributions to overall seroprevalence made by individuals in defined risk-groups (homosexual and bisexual males, IV-drug abusers, and so on). This leads to the conclusion that "HIV prevalence levels in persons without acknowledged or recognized risks would be below 0.021% in military applicants and 0.006% in blood donors" (HHS, 1987, p. 3). Limited studies of heterosexuals being treated for other STDs, but excluding those with identified risk factors by rigorous protocols of interviews, find seroprevalence levels ranging from 0 to 1.2% (in comparison with over 50% for homosexual males attending the same clinics).

In Africa, where transmission is thought to be mainly by heterosexual contacts, a different picture is emerging. In Nairobi, tests for HIV antibodies in blood sera stored from earlier studies (often for different purposes) show seropositivity levels among female prostitutes rising from 4% in 1980–81 to around 51% in 1983–84 and 59% in 1985–86, and from 1% to 14% and 18% at the corresponding times for males attending a clinic for sexually transmitted diseases (Quinn et al., 1986). Recent seroprevalence levels among high-risk groups, such as female prostitutes, have been reported to range from 27% to 88%, depending on socioeconomic status and geographic location. In a general survey of pregnant women in Nairobi, seropositivity levels rose from 0% in 1980–81 to 2% in 1985–86; a similar survey of pregnant women in Kinshasa shows seropositivity levels rising from 0.2% in 1970 to 3% in 1980–81 and 8% in 1985–86 (Quinn et al., 1986). A study of some 600 seronegative men and women working in a general hospital in Kinshasa from 1984 to 1985 found the annual incidence (or rate of seroconversion) of HIV antibodies to be around 0.8%, a figure which Quinn et al. (1986) think may be representative of the annual incidence of new HIV infections in Central and East Africa. Current studies of HIV seropositivity among blood donors in Central African countries suggest levels as high as 9% in Zaire, 11% in Uganda, 15% in Rwanda, and 18% in Zambia (Quinn, priv. comm.).

In short, we have much more data about AIDS cases in developed regions than about HIV infections, and much more data for HIV infections among high-risk groups in developed countries that for non-drug using heterosexuals in developed countries or for developing countries more generally.

Against this background, we now turn to examine the specific factors that influence the transmission dynamics of HIV.

2.3 Incubation Intervals and Infectiousness

One of the many things that makes HIV/AIDS different from most other infectious diseases is the long and variable incubation interval between infection with HIV and the collapse of the immune system that results in AIDS. Empirical information about the incubation interval, much less any fundamental understanding, is hard to get because it is not usually known when a given individual first acquired HIV infection. Most of the available data about incubation intervals come from transfusion-associated AIDS cases, where the date of the infection can usually be presumed to be the date of transfusion. Analysis of such data suggests an average incubation interval of around 8–9 years (Medley et al., 1987; 1988; earlier analyses gave answers ranging from around 4 to around 15 years). These authors also found average incubation intervals to be significantly shorter for infants (around 2 years for those aged 1–4 years) and somewhat shorter for older people (around 5–6 years among those over 60 years), although this latter effect may be associated with the confounding influence of other sources of mortality. Medley et al. emphasize that the average incubation interval they have inferred is much the same as the longest incubation interval in the data, and thus should not be seen as graven in stone. They also emphasize that transfusion-associated infections may not be character-istic of HIV/AIDS acquired by other routes; there are, indeed, some tentative indications that average incubation intervals are somewhat shorter for homosexual males and IV-drug abusers.

The cohort studies of Medley et al. (1987, 1988) are consistent with analyses of stored serum samples, taken from homosexual men in San Francisco as part of a study of HBV that reaches back to 1978. In this serendipitous longitudinal study, it is possible to see when individuals seroconverted. The current figures are that, of those who have seroconverted, none had AIDS after 3 years, 20% had AIDS after 6 years, and currently 36% have AIDS after 7–8 years. Taken together, the current data suggest that around 30–40% of HIV infectees go on to develop AIDS. But there is evidence of immunological deterioration in 80% of those who have been infected for 8–9 years, and it could well be that essentially all those infected with HIV will eventually go on to develop AIDS (unless they die from other causes) after characteristic incubation times that may be significantly greater than current estimates of the average period.

In general, the incidence of AIDS cases, the incidence of HIV infections, and the distribution of incubation intervals are connected by

$$dC(t)/dt = \int_0^t I(s)d(t-s)ds. \tag{2.1}$$

Here $C(t)$ is the cumulative number of AIDS cases up to time t since the infection first appeared (corrected for reporting lags and so on), so that dC/dt is the rate at which new cases appear; $I(t)$ is the corresponding rate at which new HIV infections appear; and $d(t)$ is the distribution of incubation times (that is, $d(t)$ is the probability that someone who acquired HIV infection at $t = 0$ will develop AIDS at time t). Eq. (2.1) is a standard Volterra integral equation of the first kind, and thus if we have complete information about any two of the quantities $C(t)$, $I(t)$, $d(t)$ then the third can be deduced. Specifically, we can use the convolution (or "faltung") theorem to express I in terms of C and d as follows (Morse and Feshbach, 1953, ch. 8):

$$I(t) = \frac{1}{2\pi i} \oint_C \frac{p\tilde{C}(p)e^{pt}\,dp}{\tilde{d}(p)}. \tag{2.2}$$

Here $\tilde{C}(p)$ and $\tilde{d}(p)$ are the Laplace transforms of $C(t)$ and $d(t)$, respectively, and the contour integral is over the standard Bromwich contour. The corresponding expression for $d(t)$ when $C(t)$ and $I(t)$ are known is obvious.

Unfortunately, we usually have insufficient information about both the incidence of HIV infection, $I(t)$, and the distribution in incubation intervals, $d(t)$. In essentials, what Medley et al. (1987, 1988) do is use the available data about $C(t)$ and $I(t)$, along with assumptions about the functional shape of $d(t)$, to estimate the characteristic parameters of the distribution in incubation times. Conversely, if an explicit assumption is made about the incubation distribution, we can assign some functional shape to $I(t)$ and then estimate the characteristic parameters; in this way, the HHS (1987) report arrives at the independent estimate that the total number of HIV infections in the USA at the end of 1987 is in the rough

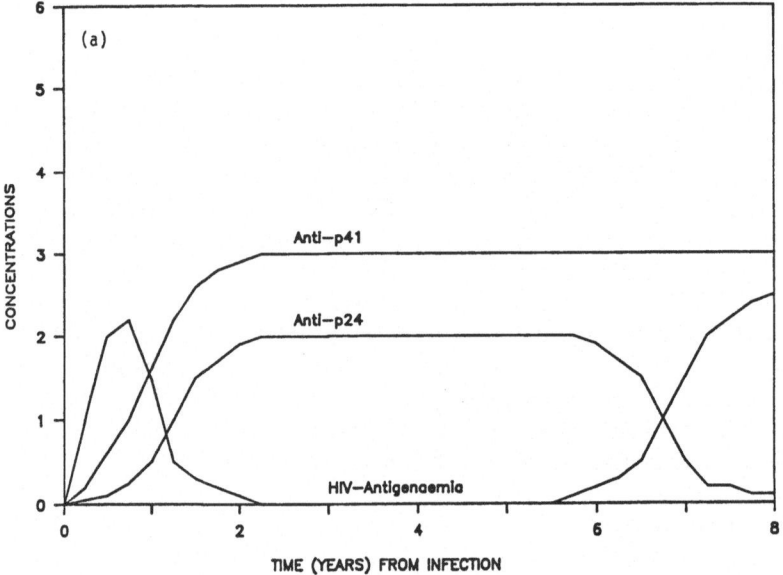

Fig. 3. (a) A schematic representation of changes in antigen and antibody concentrations, over the interval from HIV infection to the development of AIDS. The incubation period of AIDS was set at 8 years in this illustration (after Pedersen et al., 1987). (b) Average concentrations for antibodies to HIV antigens in serum samples taken from homosexual males infected with HIV, at various times (over the first year) after seroconversion. Sample sizes were: $n = 11$ at time 0, $n = 8$ at 3 months, $n = 4$ at 6 months, and $n = 4$ and 7–9 months (compiled from data surveyed in May et al., 1988c). (c) Antigen concentrations (in ng/ml) at specific times, in months, before conversion to AIDS in serum samples drawn from a male homosexual (see May et al., 1988c)

range 1 to 3 million (which is not inconsistent with the rough 1.5 million mentioned earlier).

One simple (and epidemiologically conventional) assumption is that infected individuals move on to develop AIDS at some constant rate, v. The distribution of incubation intervals is then

$$d(t) = v \exp(-vt), \tag{2.3}$$

and the average incubation interval is $1/v$ (Anderson et al., 1986). More generally,

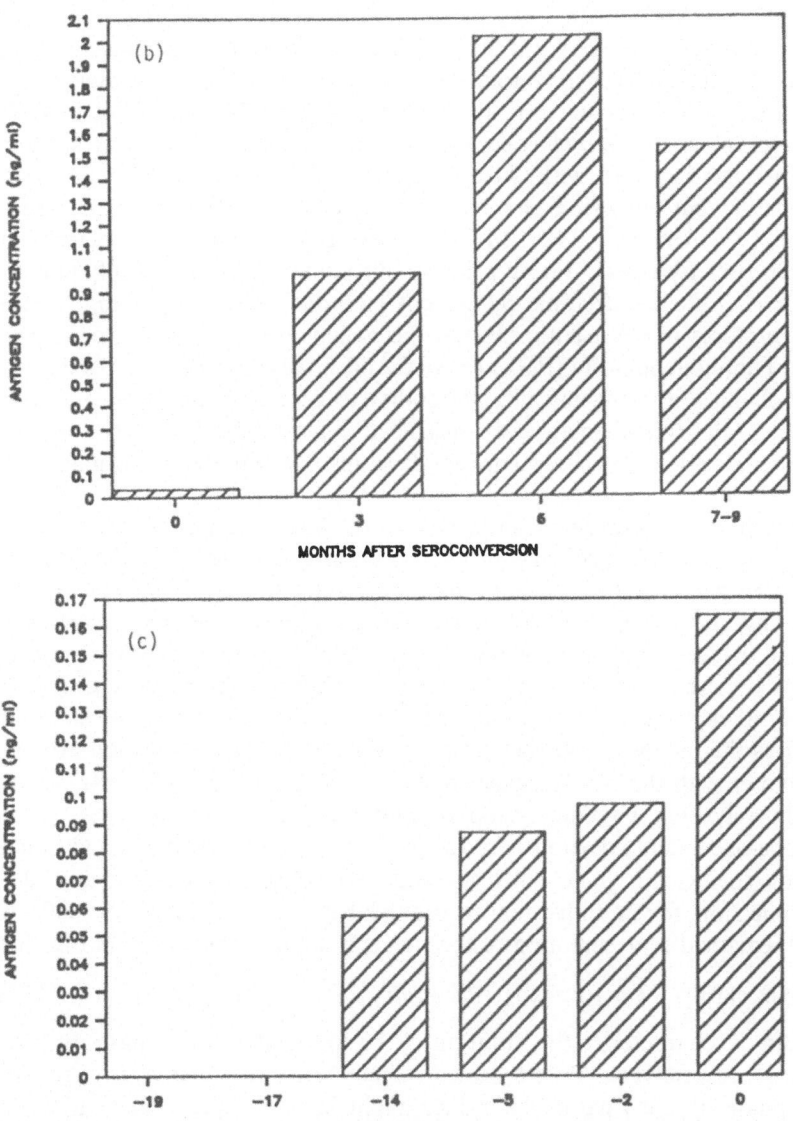

Fig. 3b, d.

suppose the probability, $v(t)$, of developing AIDS increases as the time, t, since acquiring HIV infection lengthens, according to $v(t) = \alpha t^v$. The result is a Weibull distribution in incubation intervals,

$$d(t) = \alpha t^v \exp\left[-\alpha t^{v+1}/(v+1)\right]. \tag{2.4}$$

The available data are fit reasonably well by such a Weibull distribution with $v = 1$, and this distribution is used in many of our numerical studies. The Weibull distribution gives an average incubation time of $(v + 1/\alpha)^{1/(v+1)}\Gamma(v + 2/v + 1)$, and coefficient of variation of the distribution is $CV = [\Gamma(v + 3/v + 1)/\Gamma^2(v + 2/v + 1)] - 1$ (for $v = 0$, the mean is $1/\alpha$ and the CV is 1; for $v = 1$, the mean is $(\pi/2\alpha)^{1/2}$ and the CV is 0.27...).

The simplest assumption is that infected individuals have some constant level of infectiousness throughout the duration of the incubation interval. Longitudinal studies of the fluctuations in viral abundance or in the concentration of HIV antigens in serum, cells or semen collected from infected patients, however, suggest a more complicated pattern, with two episodes of peak infectivity, one in the early stages of the incubation period and one in the late stages (Pedersen et al., 1987; Burger et al., 1986; Anderson, 1988). In these studies, the concentration of detectable HIV antigen rises to a peak during primary infection, and then typically falls to very low levels (often undectable by current methods) before beginning to rise again as symptoms of disease appear and the patient progresses from persistent generalized lymphadenopathy (PGL) to ARC and finally to AIDS. Concomitant with these changes in antigenaemia, antibodies to core antigens (Anti-p24) rise slowly, reach peak titres when antigen concentration is undectable or very low, and then fall to very low levels as clinical disease develops. These patterns are illustrated in Fig. 3.

There are practical problems associated with the quantification of antigen titres in serum samples, further problems in interpretation associated with the relation between antien concentration in serum and virus abundance in blood, excretions and secretions, and yet further problems in the assumption that infectiousness is proportional to levels of virus in the blood. The patterns in Fig. 3, however, suggest the tentative hypothesis that there are two phaser of infectiousness, one during and immediately after primary HIV infection (of duration 6–12 months or so), and the second as the patient progresses through ARC to AIDS (of duration around 1 year or more), with the two being separated by a period of low infectivity (which may be relatively long if average incubation intervals are around 8 years).

Such time-dependent infectiousness can be represented by defining a conditional transmission coefficient, $\gamma(t, \tau)$, which measures the infectiousness at time t after acquiring infection, for an individual whose incubation interval is $\tau (\tau > t > 0)$. A rough mathematical representation of the trends described above and in Fig. 3 is

$$\gamma(t, \tau) = \beta_0 \exp(-t/T_0) + \beta_1 \exp\left[-(\tau - t)/T_1\right]. \tag{2.5}$$

Here T_0 and T_1 characterize the duration of the first and second phases of high infectiousness, respectively, while β_0 and β_1 characterize the relative strengths of these two phases (typical parameter values might be $T_0 \simeq T_1 \simeq 1$ year and $\tau \simeq 8$ years; May et al., 1988).

The above discussion makes the implicit assumption that all HIV infections go on to produce AIDS. On current evidence, this may be the case, or it may not.

Suppose a fraction f do eventually develop AIDS, while the remaining fraction, $1 - f$, do not. In simple models with constant infectiousness, the non-AIDS-developing fraction could remain asymptomatic carriers for life (as is arguably the case for asymptomatic carries of hepatitis B virus, HBV). Alternatively we could assume this fraction, $1 - f$, revert to an uninfectious state at some characteristic rate v' (which my or may not be the same as the rate v of Eq. (2.3)). More generally, we could take the relatively realistic expression (2.5) and observe that the non-AIDS developing fraction, $1 - f$, manifest only the first phase of infectiousness (of characteristic duration T_0); for this fraction, effectively $\tau \to \infty$.

The possibility of some fraction never developing AIDS also affects the normalization of the distribution of incubation times. Eqs (2.3) and (2.4) assume the distribution integrates to unity; that is, the probability that an individual will incubate AIDS after some time interval is unity. More generally, the integral over all incubation times will come to f, while the remaining $1 - f$ represents individuals who never develop AIDS.

2.4 Transmission Probabilities

When considering the probability that an infected individual will infect a susceptible sexual partner, we may use one of two assumptions which are opposite extremes. One assumption is that there is, on average, some constant transmission probability, δ, per sexual contact, and that these probabilities compound independently randomly. The overall probability for an infected individual to infect a sexual partner then increases with the number of contacts, and this Poisson process gives, on average, a transmission probability per partnership of approximately $1 - \exp(-\delta n)$ after n contacts. The opposite extreme, which at first sight may seen less likely, is that the transmission probability is, on average, a constant for all partnerships of a given kind.

Figure 4 summarizes data from Peterman et al. (1988) which address this point. From a larger pool of individuals with transfusion-associated AIDS or HIV infection. Peterman et al. selected 106 index cases. Of these, 26 had had no sexual contact with their spouse since the transfusion, leaving a pool of 80 infected individuals—55 men and 25 women—and their spouses. Standard interviews were conducted with the index cases and their spouses, reviewing medical and social histories; the study excluded all individuals who had risk factors other than the infected spouse. In particular, wives/husbands were asked about their sexual practices, about the average number of sexual contacts with their infected husband/wife per month since the transfusion, and about changes in frequency of sexual contacts after the partner had been diagnosed as having HIV seropositivity or AIDS. In all, Peterman et al. found that 2 of the 25 husbands of infected wives, and 10 of the 55 wives of infected husbands, were HIV seropositive. Figure 4 shows the relation between transmission of HIV infection and the number of sexual contacts with the infected spouse. The figure points to the surprising conclusion

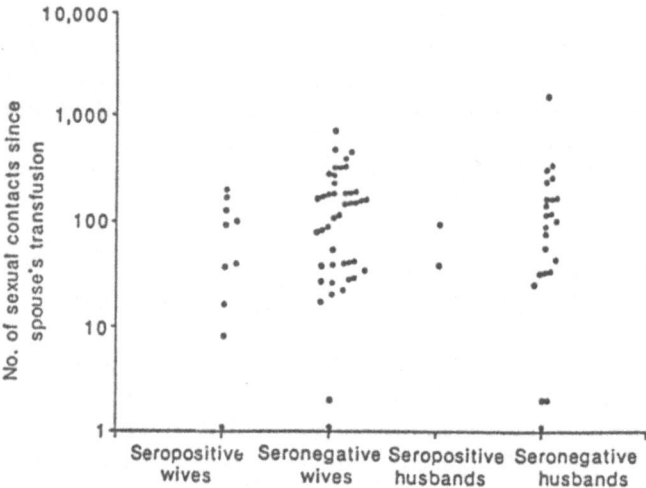

Fig. 4. For each of 80 individuals, this figure shows the number of sexual contacts with a spouse, after the spouse was (unknowingly) infected with HIV by blood transfusion. The 25 husbands and 55 wives of infected individuals are divided according to current serological status. The figure shows that, in this study group, transmission probability is uncorrelated with number of sexual contacts. (After Peterman et al., 1988.)

that transmission probability is unrelated to the actual number of sexual contacts, for partnerships ranging from one to several thousand contacts. Insofar as there is any statistical correlation between transmission probability and number of contacts, it is negative (although not to a statistically significant degree): one infected woman had had only a single sexual contact, and another had only 8; 11 of 55 wives and 5 of 25 husbands remained uninfected after more than 200 sexual contacts with their infected spouses. For further discussion, including caveats and speculations about the reasons for the apparent lack of correlation between transmission probability and number of sexual contacts, see Peterman et al. (1988) and May (1988).

Given the small sample sizes, Peterman et al's estimate of an average transmission probability of around 0.1 (2/25) female-to-male and around 0.2 (10/55) male-to-female is consistent with other estimates of transmission probabilities in heterosexual partnerships, summarized in Table 2. For many of the studies in Table 2, it is by no means clear that infection was acquired by sexual contact with the heterosexual index case. None of these other studies reported data about the correlation between transmission probability and number of sexual contacts.

It is more difficult to use partner tracing to estimate transmission probabilities in homosexual partnerships, because of the relatively high number of multiple partners among homosexual men (Grant et al., 1987). These authors have used a mathematical model, in conjunction with seroprevalence studies and information about the sex history of individual homosexual men, to estimate that the average transmission probability per partnership (which they call the infectivity), associated with the practice of unprotected receptive anal intercourse, is around 0.09, with a

Table 2. HIV seroprevalence in heterosexual partners of individuals infected with HIV, from studies summarized by HHS (1987, where references are given). Unlike the study by Peterman et al. (1988), which is not included here, none of these studies provide information abot the correlation between transmission probability and number of sexual contacts

Country	Number of partners tested	Sex of partner at risk	Fraction seropositive (%)	Source of HIV seropositive partners
USA	21	F	10	hemophiliacs
USA	19	F	21	hemophiliacs
France	148	F	7	hemophiliacs
England	36	F	8	hemophiliacs
USA	24	F	17	hemophiliacs
England	14	F	0	hemophiliacs with ARC or AIDS
USA	21	F	10	hemophiliacs with ARC or AIDS
USA	4	F	25	transfusion recepients
USA	55	F	24	bisexual men
USA	7	F	43	bisexual men
USA	12	F	42	IV-drug users
USA	69	F	46	IV-drug users
USA	5	M	60	IV-drug users
USA	11	M	55	individuals with AIDS
USA	45	M and F	58	individuals with AIDS
USA	42	M and F	48	individuals with ARC or AIDS
USA	8	F	0	various
USA	35	F	35	various
USA	22	M and F	36	various

confidence interval from 0.04 to 0.15. Within this group, there is a pronounced correlation between number of partners and seropositivity levels, as suggested by the mathematical model and as discussed in more detail below. The rough equivalence between this transmission probability and those just discussed for heterosexual partnerships is surprising. The evidence of Fig. 4 to the contrary, we think it may be that a higher transmission probability per sexual act among homosexual men could be counterbalanced by the longer average duration of the heterosexual partnerships that were studied.

A fully accurate mathematical model for the transmission dynamics of HIV by sexual contacts of a given kind would need to keep track of the formation and break-up all partnerships, taking account of the infection status of each partner and of the possible dependence of transmission probabilities upon the duration of partnerships. This would be a formidable undertaking. Existing models are based on various kinds of approximations.

One approach, pioneered by Dietz (1988), is to take explicit account of the formation and dissolution of partnerships (either heterosexual or homosexual), but to allow an individual to have at most one partnership at any one time. While this may be a reasonable basis for approximating heterosexual transmission of HIV in a fairly monogamous society, even in this circumstance we think it likely to underestimate the rate at which infection spreads (because individuals who have several concurrent sexual relationships are likely to play a disproportionate role in the dynamics of transmission). Dietz also assumes some average transmission probability per contact, or per unit time, compounding in Poisson fashion. The basic approach can, however, easily be modified to treat the transmission probability as being roughly a constant per partnership, independent of its duration (which Fig. 4 suggests may be closer to reality), or as having some intermediate dependence on partnership duration (along lines suggested by Hyman and Stanley, 1988).

A more phenomenological approach, which has been used successfully by Hethcote and Yorke (1984) for modelling gonorrhea, is to assume the probability (per unit time) for a susceptible individual to acquire infection is equal to the number of sexual partners, i, times the probability of being infected by any one partner, λ. In turn, λ is given by the probability that a partner is infected times the probability (per unit time) that infection will be transmitted from such a partner. It is clear that this rough approximation compounds the instantaneous transmission probabilities over successive time intervals, paying no attention to possible correlations among the sexual partners in successive time intervals. As such, the approach may be more accurate when there are relatively large numbers of short-lived partnerships (as for highly-active homosexual males or, *mutatis mutandis*, needle-sharing IV-drug abusers).

The same mathematical expression for the rate at which susceptibles acquire infection, $i\lambda$, can be given an alternative biological interpretation. Instead of taking i to be the average number of partners and λ the infection probability per partner per unit time, we may interpret i as the rate at which new sexual partners are acquired (that is, the number of new partners per unit time) and λ as the probability that any one such newly-chosen partner will transmit infection (over the duration of the partnership). This interpretation has the virtue of conforming to the data in Fig. 4, in that transmission probabilities do not compound in Poisson style (tending to saturate to unity in long-lasting partnerships). The concomitant fault in the approach is the mathematical inconsistency of having an instantaneous probability of infection (or probability per unit time) that involves overall transmission probabilities per partnership. To make this approach more accurate, we would need to take account of the formation and break-up of partnerships, thus moving toward the models of Dietz but allowing for several concurrent partnerships. If, however, infection is acquired within partnerships on time scales less than, or of the order of, other dynamically relevant time scales, then the above approach may be a reliable approximation.

For transmission among IV-drug abusers, high rates of encounter with new individuals seem likely. Models that deal with rates of sharing needless times the probability that a given needle will cause infection (compounding Poisson-style) are thus likely to be accurate.

These different mathgmatical approaches can be put in some perspective by considering the basic reproductive rate, R_0, for HIV infections within a specified risk-group. From the definition given above, we have that R_0 is equal to the average number of new partners acquired per unit time by an infected individual multiplied by the average duration of infectiousness (to give the average number of new sexual partners acquired while infectious), all multiplied by the average transmission probability per partnership. This is commensensical, and we could thus obtain a rough esimate of R_0 for any particular risk-group from the kinds of data shown in Fig. 4 (for heterosexual transmission) or in the study by Grant et al. (1987, for homosexual males), along with estimates of average rates of acquiring partners and duration of infectiousness. The difficulties enter, however, in giving more rigorous specification to what is meant by "average" in the above statements. This will emerge in our discussion of the mathematical models below.

2.5 Rates of Acquiring New Partners

There have, until recently, been surprisingly few studies of average rates of acquiring new heterosexual or homosexual partners by men and women. Even less information is available about the distribution in such rates. Reviews of such data as are available for patterns in developed countries are given by May and Anderson (1987), Anderson (1988) and Johnson (1988); what follows is a sketchy summary.

In San Francisco, a study of an unbiased sample of 814 homosexual/bisexual men gave an average of 10.8 partners in 1984 (with 27% having more than 10 partners per year; Winkelstein et al., 1987). A similar study in London gave an average of 10.5 partners per year in 1986, falling to 4.8 in 1987, following education compaigns on television and elsewhere (Andeson and Johnson, 1988). As illustrated in Fig. 5, these and other data show high variability in rates of acquiring

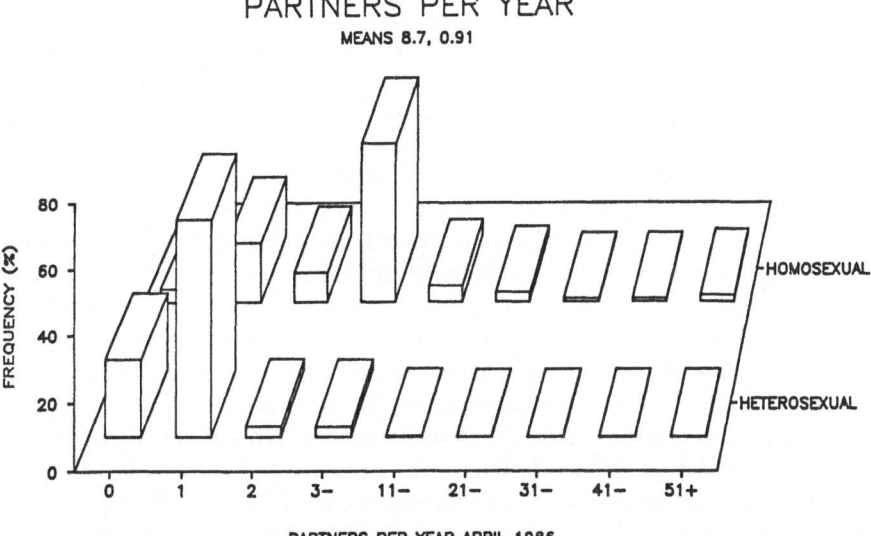

Fig. 5. Distributions in numbers of sexual partners per year (as of April 1986) among a survey group of homosexual males and among a survey group of heterosexual males and females, in Britain (data from BMRB, 1987; see also Anderson, 1988)

new homosexual partners, with the distributions typically having coefficients of variation (ratios of standard deviations to means) in excess of unity. Both surveys about sexual habits and the recorded incidence of other STDs, such as gonorrhea, suggest significant decreases in the average rates at which new homosexual partners are acquired in the USA and the UK. These average rates, however, appear still to be significantly higher than the corresponding rates among heterosexuals.

It is probable that bisexual men currently represent an important source of transmission of HIV infection to women. There are, unfortunately, very few studies of the pertinent rates of acquiring female partners by bisexual men. In one recent study of homosexual men, 30% reported themselves to be bisexual, and of these roughly one-third reported more than two female sexual partners over the past year; this seems to us to be a disturbing figure from the point of view of HIV epidemiology (Anderson, 1988).

Studies of heterosexual men and women in developed countries consistently show the average rate of acquiring new partners, or the average number of partners per year or per lifetime, to be significantly smaller than for homosexual males. For example, the above-mentioned study by Winkelstein et al. (1987) also included 212 heterosexual men: they reported an average of 2.8 female partners in 1984 (with only 2.8% reporting more than 10 partners). In the UK, several independent surveys of heterosexuals suggest an average of around 1 partner per year for both men and women, and average lifetime totals of around 3–4 for women and around 4–12 for men (Wadsworth et al., 1988); see Fig. 6. Studies of men and women attending STD clinics tend to give higher average values. Again, as illustrated in Figs. 5 and 6, all these studies tend to exhibit considerable variability in rates of acquiring new partners within any one study group.

Studies that combine information about seropositivity with information about the sexual habits of homosexual males consistently show significant correlations between rates of acquiring new homosexual partners and seropositivity levels (Winkelstein et al., 1987; Johnson, 1988; May and Anderson, 1987). Although less abundantly documented, similar correlations have been found among heterosexuals.

There is very little data about rates of partner change in Central African countries. It has been suggested, however, that the patterns of sexual activity of males and females may differ more than is typically the case in developed countries, with the majority of females having relatively monogamous marriages or "union libres" (persistent cohabitation without formal marriage), but where many of the male partners in such relationships are less monogamous, with the books kept in balance by a cardre of young female prostitutes (Quinn et al., 1986).

For IV-drug users, the corresponding information would seem to be the typical size of needle-sharing groups, and the typical rate at which individuals enter and leave such groups. A study in Edinburgh, for example, suggests needles may be shared among 10–20 drug abusers (Robertson et al., 1986). The ethnography of needle-sharing, however, differs greatly from place to place. Among some groups there is apparently a practice of drawng blood into the syringe, to mix it with the drug before injecting; the ritual can be to leave some blood in the syringe (a token

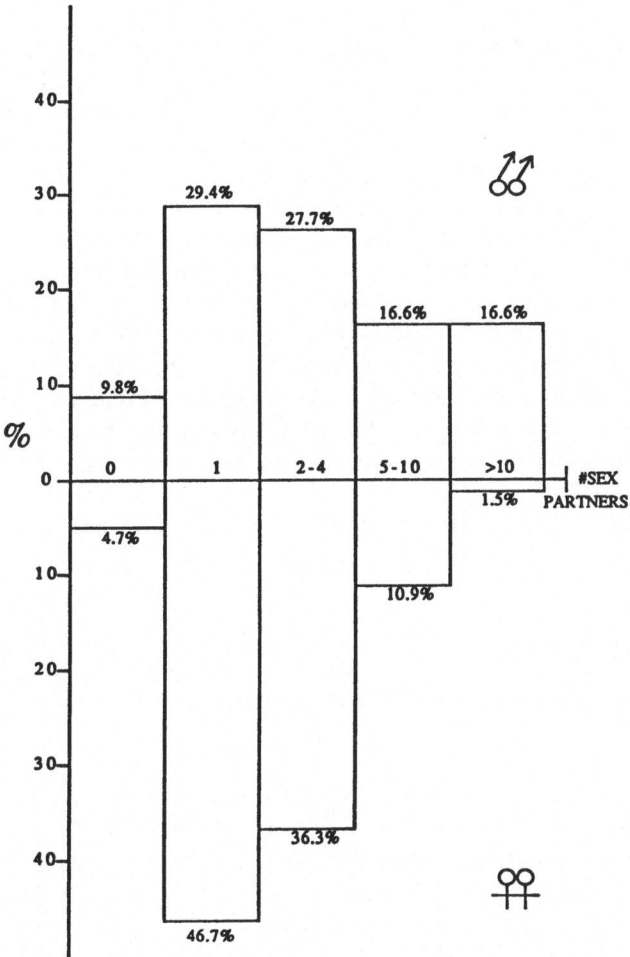

Fig. 6. Distributions in reported lifetime numbers of sexual partners for a random sample of 296 heterosexual males and 405 heterosexual females, between the ages of 16 and 64, carried out in Britain in early Autumn 1987 (Wadsworth et al., 1988). Notice that males report a significantly higher average number of lifetime partners; this imbalance may derive partly from age-structure effects within the samples or from inaccurate reporting, but it probably arises mainly because very promiscuous females (such as prostitutes) are not represented in this relatively small sample

of sharing, as it were). In this event, the total number sharing the needle can be less important than the HIV status of the individual immediately preceding you in using the needle. These complexities matter. One of the most important public health measures could, for instance, be to persuade the managers of the illegal drug trade that it is in their long-term interest to ensure sterile needles in "shooting galleries".

3. A Basic Model

We begin by considering a very simple model, to which various kinds of refinement will be added.

Let the total population under consideration be $N(t)$ at time t. This population may be homosexual males in a large city, or the total population of heterosexuals (with male-to-female and female-to-male overall transmission rates assumed equal, so that the model effectively has the structure of a one sex model), or any other defined group. Let $N(t)$ be subdivided into $X(t)$ susceptibles and $Y(t)$ infecteds, who are assumed also to be infectious. It is further assumed that all infecteds move at a constant rate v (that is, after an average incubation time $1/v$) to develop full-blown AIDS, at which point they are regarded for the purposes of the model as being effectively removed from the group under study (that is, $N(t) = X(t) + Y(t)$); in fact, the average life expectancy of diagnosed AIDS patients is 12.6 months in the UK (Reeves and Cox, 1988) and 11.4 months in the USA (Rothenberg et al., 1987). Deaths from all other causes occur at a constant per capita rate μ, and new susceptibles appear at the overall rate $B(t)$.

This simple system is described by the pair of first-order differential equations (May et al., 1988a; Anderson et al., 1988);

$$dX/dt = B - (c\lambda + \mu)X, \tag{3.1}$$

$$dY/dt = c\lambda X - (v + \mu)Y. \tag{3.2}$$

The dynamics of the total population, $N = X + Y$, thus obeys

$$dN/dt = B - \mu N - vY. \tag{3.3}$$

Here $c\lambda$ is the "force of infection", representing the probability per unit time that a given susceptible will become infected. As discussed in Sect. 2, we may take c to be the average rate at which ne partners are acquired, and λ to be the probability of acquiring infection from any one partner (Hethcote and Yorke, 1984). Further, we can write

$$\lambda = \beta Y/N. \tag{3.4}$$

Here β is the probability of acquiring infection from any one infected partner, and Y/N is the probability that a randomly-chosen partner will be infected. Finally, the input of new susceptibles is assumed to be proportional to the size of the population:

$$B(t) = vN(t). \tag{3.5}$$

This is a natural first approximation if we are dealing with (symmetrical) heterosexual transmission of HIV in the total population, in which case v is the overall average per capita birth rate. More generally, v is a rate characterizing the entry of new members into the group in question.

The above system of equations now reduces to a pair of equations for $N(t)$ and $Y(t)$, which can be written

$$dY/dt = [\Lambda - \beta c(Y/N)], \tag{3.6}$$

$$dN/dt = [r - v(Y/N)].$$ (3.7)

Here we have introduced the definitions Λ and r:

$$\Lambda = \beta c - \mu - v,$$ (3.8)

$$r = v - \mu.$$ (3.9)

As we shall see, Λ is the initial exponential growth rate of the infection (from very low values) within the population, and r is the rate at which the population grows in the absence of AIDS.

Eqs (3.6) and (3.7) can be solved analytically. The fraction infected at time t is (May et al., 1988a)

$$\frac{Y(t)}{N(t)} = \frac{\Delta \exp(at)}{1 + (b/a)\Delta[\exp(at) - 1]}.$$ (3.10)

and the total population size is

$$N(t) = N(0)e^{rt}[1 + (b/a)\Delta(e^{at} - 1)]^{-v/b}.$$ (3.11)

Here Δ is the initial fraction infected, at $t = 0$. The quantities a and b have been defined for notational convenience: $a \equiv \Lambda - r, b \equiv a + v$.

Two things are immediately apparent.

First, let us neglect demographic parameters in comparison with epidemio-

Table 3. Doubling times, t_d, based on AIDS case reports to the World Health Organization

Country	Time period	Doubling time, t_d, in months
Australia	1986–87	12.9
Austria	1983–85	15.6
Bahamas	1986–87	12.7
Brazil	1983–85	8.5
Canada	1983–85	12.7
Dominican Republic	1986–87	6.4
France	1983–85	11.2
Greece	1986–87	7.8
Jamaica	1986–87	5.8
Kenya	1986–87	4.9
Mexico	1986–87	7.5
Netherlands	1983–86	11.2
Portugal	1986–87	11.8
South Africa	1986–87	11.8
Spain	1983–86	6.3
Sweden	1983–86	9.9
Switzerland	1983–86	10.0
Tanzania	1986–87	8.3
United Kingdom	1983–86	14.0
United States	1981–86	13.0
Venezuela	1986–87	11.2
Zambia	1986–87	13.9

logical ones by assuming the per capita birth and death rates (v and μ) are significantly less than the transmission and incubation rates (βc and v), so that $a \simeq \Lambda \simeq \beta c - v$. Eq. (3.10) then shows that the fraction of the population infected with HIV initially grows exponentially at the rate Λ. These results have an intuitive explanation: infected individuals on average transmit HIV, with probability β, to c new partners each year, while a fraction v of infected individuals go on to develop AIDS; hence the exponential growth rate of infection among the population is $\beta c - v$. That is, we expect the incidence of HIV infection (and thus, very roughly, the subsequent incience of AIDS) initially to grow exponentially with a doubling time $t_d \simeq (\ln 2)/\Lambda$. Table 3 summarizes such doubling times for various risk groups in different countries. In particular, early doubling times for the incidence of infection among male homosexuals in developed countries are consistently around 0.5 to 1 years. That is, $\Lambda \sim 1 \text{ yr}^{-1}$ for such groups. Finally, if the average incubation time is indeed around 8 years or so, this very simple model, in combination with the data summarized in Table 3, suggests that for homosexual males in developed countries the product βc is around 1 yr^{-1}. This wholly independent estimate, based on population-level data, is consistent with the estimates $\beta \sim 0.1$ and $c \sim 10 \text{ yr}^{-1}$ for homosexual males, as discussed in Sects. 2.4 and 2.5, respectively.

Second, if the demographic parameters are retained, then Eq. (3.10) shows that we require $\Lambda > r (a > 0)$ for HIV/AIDS to establish itself; for $r > \Lambda > 0$, HIV infects an increasing number, but a decreasing proportion, of the exponentially growing population. Given that the infection can establish itself ($a > 0$), Eq. (3.11) shows it will have a demographic impact, reducing the asymptotic rate of population growth from the disease free rate, r, to some lower rate, ρ, given by

$$\rho = r - v(\Lambda - r)/(\Lambda + \mu). \tag{3.12}$$

We see that HIV/AIDS can even lead to population decline (negative rates of population growth, $\rho < 0$), provided Λ is large enough ($\Lambda > r(v + \mu)/(v - r)$). A more realistic discussion of the demographic implications of HIV/AIDS is given in Sect. 6.

4. HIV Epidemics in Heterogeneous Populations

We now extend the basic model to take account of various kinds of heterogeneity within the risk-group under consideration. In much of this analysis we consider the transmission dynamics of HIV in closed groups, with no recruitment of new susceptibles (that is, $B = 0$).

4.1 Heterogeneity in Rates of Acquiring Sexual Partners

Consider a closed population of homosexual males (or, *mutatis mutandis*, IV-drug abusers, or—under certain symmetry assumptions—heterosexuals), whose magnitude is $N(t)$ at time t. This population is divided into sub-groups, N_i, whose numbers on average acquire i new sexual partners per unit time. Initially we have $N_i(0) = N(0)p(i)$, where $p(i)$ is the initial probability distribution in rates of acquiring partners. The number of susceptible, infected (and infectious), and no-longer-infectious individuals in the ith class are defined to be X_i, Y_i, and Z_i, respectively, so that $X_i + Y_i + Z_i = N_i$ (it may, of course, be that $Z_i = 0$). If we ignore deaths from causes other than AIDS in this closed population (that is, $\mu = 0$), then Eqs. (3.1)–(3.3) are replaced by

$$dX_i/dt = -i\lambda X_i, \tag{4.1}$$

$$dY_i/dt = -i\lambda X_i - vY_i, \tag{4.2}$$

$$dN_i/dt = -fvY_i. \tag{4.3}$$

Here we have assumed that a fraction, f, of those infected with HIV go on to develop AIDS—whereupon they are effectively removed from the population under consideration—at a constant rate, v (corresponding to an average incubation time of $D = 1/v$). The remaining fraction, $1 - f$, are assumed to become uninfectious at the same rate. As discussed more fully in Sec. 2.3, the realities are more complicated. In particular, it may be that those who do not develop AIDS remain infectious indefinitely, or it may be that they are infectious only for the initial phase shown in Fig. 3; in the absence of detailed knowledge, the preliminary assumption that such individuals lose infectiousness at the same rate as those who develop AIDS seems a sensible point of departure.

The infection probability per partner, λ, now is given by generalizing Eq. (3.4)

$$\lambda = \beta \sum iY_i / \sum iN_i. \tag{4.4}$$

In Eq. (3.4) the probability that any one partner is infectious is simply Y/N; in Eq. (4.4) partners are weighted according to their degree of sexual activity, i. By assuming that partners are chosen randomly (apart from the activity levels characterized by the weighting factor i), we may be overestimating the contacts of less active individuals with those in more active categories, and thus overestimating the spread of infection among such less active sub-groups. Conversely, the transmission probability β may be higher for longer-lasting partnerships (the data in Fig. 4 to the contrary), so that use of a constant β may tend to underestimate the spread of infection among less active people. The net effect of these countervailing refinements is hard to guess.

Studies of the dynamical behavior the system of Eqs. (4.1)–(4.4) is facilitated by defining the quantity

$$\phi(t) = \int_0^t \lambda(s)ds. \tag{4.5}$$

If we take the initial seed of infection to involve a negligible number of individuals, the initial value of X_i is $X_i(0) = N_i(0) = Np(i)$, and Eq. (4.1) for $X_i(t)$ can be integrated to give

$$X_i(t) = Np(i)e^{i\phi}. \tag{4.6}$$

The factor i in the exponent means that susceptibility is depleted faster in the more highly active groups, as intuition would suggest. By substituting from Eq. (4.3) for Y_i in Eq. (4.4) for $\lambda(t)$, and integrating over time, we can obtain the useful result

$$\sum i N_i(t) = N \langle i \rangle \exp(-f v \phi / \beta). \tag{4.7}$$

Here we have defined $\langle i \rangle$ to represent the expectation value of i (the initial mean rate of acquiring partners) over the distribution $p(i)$; more generally, we define

$$\langle F(i) \rangle = \sum p(i) F(i). \tag{4.8}$$

Finally, we can add Eqs. (4.1) and (4.2), integrate over time, substitute the resulting expression for Y_i in Eq. (4.4) for λ, and then use Eqs. (4.6) and (4.7) to arrive at a first-order differential equation for the dynamical variable $\phi(t)$:

$$(d\phi/dt)\exp(-f v \phi / \beta) = \beta \langle i [1 - \exp(-i\phi)] \rangle / \langle i \rangle$$
$$- (\beta / f)[1 - \exp(-f v \phi / \beta)] + \lambda(0). \tag{4.9}$$

Here $\lambda(0)$ is the initial value of the infection probability, calculated from Eq. (4.4) with the very small numbers of infectious "seeds", $Y_i(0)$. From the definition (4.5), ϕ has the initial value $\phi(0) = 0$. A more detailed derivation of this and other results is given by May and Jose (1988).

Once the epidemiological parameters β, v, f, $\lambda(0)$ and the initial distribution $\{p(i)\}$ are specified, $\phi(t)$ can be calculated, and thence any other epidemiological variable can be evaluated. In particular, the overall fraction of the original population to have experienced infection by time t, $I(t) = 1 - X(t)/N$, is given from Eq. (4.6) as

$$I(t) = \langle 1 - e^{-i\phi} \rangle. \tag{4.10}$$

The cumulative number of AIDS cases up to time t, $C(t)$, is by definition given by $C(t) = f v \sum \int_0^t Y_i(s) ds$. By using Eqs. (4.2), (4.6) and (4.10), it can be seen that $C(t)$ may be derived from $I(t)$ via the differential equation

$$dC(t)/dt + v C(t) = f v N I(t), \tag{4.11}$$

with the initial condition of course being $C(0) = 0$.

Before presenting some numerical examples and discussing their general properties, we consider some limiting cases and the biological insights they provide.

4.1.1 Early Stages of the Epidemic

In the early stages of the epidemic, $\lambda(t)$ and consequently $\phi(t)$ will be small. Differentiating Eq. (4.9) in this limit (remembering $\lambda = d\phi/dt$, from Eq. (4.5)), we have

$$d\lambda/dt = \Lambda \lambda + \mathcal{O}(\lambda^2). \tag{4.12}$$

Here the early rate of exponential growth, Λ, is defined as

$$\Lambda = \beta \langle i^2 \rangle / \langle i \rangle - v. \tag{4.13}$$

This result can also be obtained directly from Eqs. (4.2) and (4.4) by putting $X_i \simeq N_i$ for the early phases of the epidemic.

Note that Eq. (4.13) conforms with the earlier definition of Λ, Eq. (3.8), provided we define c as

$$c \equiv \langle i^2 \rangle / \langle i \rangle. \tag{4.14}$$

That is, for epidemiological purposes the effective value of the average number of new partners per unit time is not the mean of the distribution, but rather is the ratio of the mean-square to the mean. This result simply reflects the disportionate role played by individuals in the more active groups, who are both more likely to acquire infection and more likely to spread it. Eq. (4.14) can alternatively be written

$$c = m + \sigma^2 / m, \tag{4.15}$$

where m is the mean and σ^2 the variance of the distribution $\{p(i)\}$. Thus the effective value of the average number of sexual partners for epidemiological purposes can be significantly larger than the simple mean, if the variance is high. To put it another way, efforts directed toward reducing the rate of partner change among the highly active groups are likely to be disportionately effectively in reducing transmission (Anderson et al., 1986; May and Anderson, 1987).

4.1.2 Asymptotic Fraction Ever Infected

For this closed population, the value of ϕ in the limit $t \to \infty$ can be found by putting $d\phi/dt = 0$ in Eq. (4.9). Ignoring $\lambda(0)$ (the very small initial level of infection that gets the epidemic started), we can obtain the asymptotic value of $\phi, \alpha \equiv \phi(\infty)$, from the transcendental equation

$$\alpha = -(\beta/fv)\ln\{1 - f\langle i(1 - e^{-i\alpha})\rangle / \langle i \rangle\}. \tag{4.16}$$

With α thus determined, it is a simple matter to calculate the fraction ever infected, $I(\infty)$, from Eq. (4.10), and other such quantities.

Eq. (4.16) has a non-trivial solution only if the quantity R_0 exceeds unity, where R_0 is defined as

$$R_0 \equiv \beta c / v. \tag{4.17}$$

Here c given by Eq. (4.14), and R_0 is the basic reproductive rate for HIV, as discussed in more intuitive terms in Sect. 2.

In the limit of a non-lethal infection ($f \to 0$) in a homogeneous population (where on average all individuals have the same number of partners, c), Eqs. (4.10) and (4.16) reduce to the Kermack-McKendrick (1927) result, $I = 1 - \exp(-R_0 I)$. More generally, however, for any specified values of f and R_0, the asymptotic fraction ever infected decreases as the heterogeneity in degrees of sexual activity within the population increases (that is, as the coefficient of variation, CV, of the partner-change distribution increases). This is essentially because, other things being equal, the epidemic tends to burn itself out among those in the highly active classes, thus driving the effective value of the reproductive rate of HIV below unity, before a large fraction of those in the low activity classes have been infected. The greater the

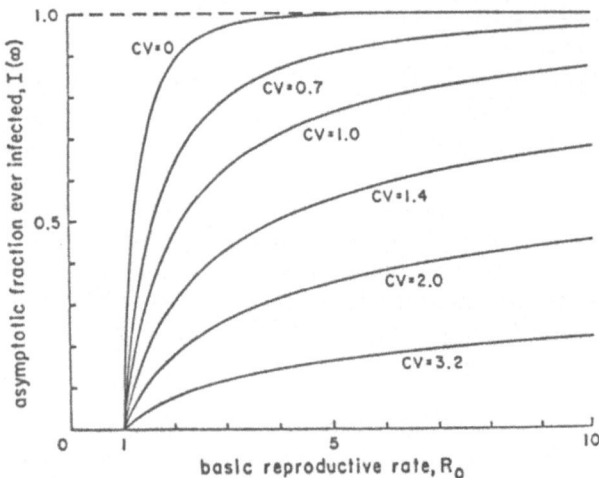

Fig. 7. Within a closed population of homosexual males, the fraction ever infected during an HIV epidemic, $I(\infty)$, is shown as a function of the basic reproductive rate, R_0, of the infection. The distribution in rates of acquiring new sexual partners within the population is taken to be a gamma distribution, with the coefficient of variation, $CV = \sigma/m$, having the values 0 (classic Kermack-McKendrick epidemic), 0.7, 1, 1.4, 2, 3.2, as shown. It is assumed that 50% of those infected with HIV eventually die from AIDS (for further details, see the text and May, 1987)

variability in degrees of sexual activity, the more pronounced this effect will be.

Figure 7 shows the asymptotic fraction of the population ever to experience infection, $I(\infty)$, as a function of R_0, assuming 50% of those infected go on to develop AIDS ($f = 0.5$). The initial distribution in rates of acquiring new partners is assumed to be a gamma distribution, and the different curves show $I(\infty)$ versus R_0 for specified values of the distribution's CV (for details, see Appendix A).

Figure 7 bears out the comments made above about the relation between heterogeneity within the population and the asymptotic level of infection for specified R_0. This kind of figure can, moreover, be used to make an indirect inference about lower bounds to the value of R_0. As reported in Sec. 2, around 50% of homosexual men in some large cities in the USA are already HIV seropositive. Survey data shows significant heterogeneity in rates of partner acquisition in the early 1980s, corresponding to coefficients of variation of at least unity, and probably more. These two observations are hard to reconcile with Fig. 7 unless R_0 for transmission among such populations is around 5 or more. Insofar as sexual habits have changed, making for lower average rates of partner acquisition in these populations, we would tend to infer higher values for R_0 in the early stages of this epidemic.

4.1.3 Dynamics of the Epidemic

Using Eq. (4.9) with a gamma distribution for the initial distribution $\{p(i)\}$, we can obtain numerical results for the fraction seropositive, the incidence of new AIDS cases, cumulative AIDS cases, and so on, as functions of time.

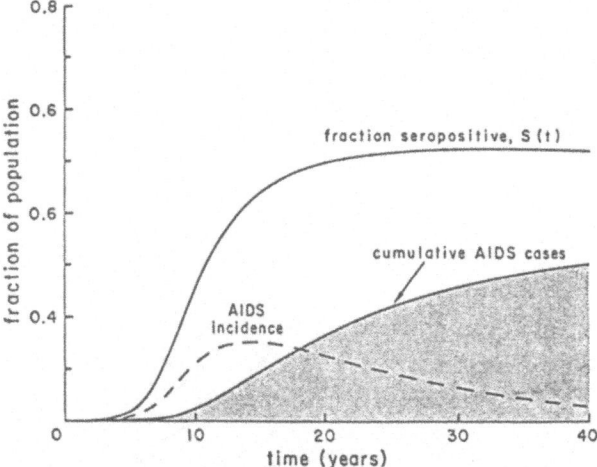

Fig. 8. This figure shows the fraction seropositive $(S(t) = [NI(t) - C(t)]/[N - C(t)])$, the cumulative number of AIDS cases (as a fraction of the original population, $C(t)/N$, and the rate at which new AIDS cases appear (dC/dt divided by N, and multiplied by 10), as functions of time. These illustrative curves are for a closed population of homosexual males, as described by the mathematical model defined in Sect. 4.1 and with a gamma distribution for $\{p(i)\}$; the parameters are $R_0 = 10$, $v = 0.1 \, \text{yr}^{-1}$ (so that $\beta c = 1 \, \text{yr}^{-1}$), $f = 0.5$ and $CV = 1.4$ ($v = 0.5$) for the sexual-activity distribution. The initial "seed" of infection is $c\lambda(0) = 0.001 \, \text{yr}^{-1}$

One such result is illustrated in Fig. 8. The fraction seropositive at first increases exponentially, but this very soon gives way to a more gradual pattern of increase as the highly active classes begin to saturate and new infections come more from the slower dissemination of infection to less active classes. This pattern is significantly different from the classic epidemic of, for instance, measles (where the exponential phase lasts longer), but it accords with the kinds of data discussed in Sect. 2 (May and Anderson, 1987).

These features are made more explicit in Fig. 9, which shows the progress of the epidemic among separate groups, differentiated according to rates of acquiring new partners. It can be seen that prevalence levels peak relatively soon for the more sexually active categories, and significantly later among the less active categories (May, 1986; Anderson et al., 1987).

4.2 Other Kinds of Heterogeneity

On the road to further realism, suppose that—in addition to the heterogeneity in sexual habits just discussed—individuals may also be subdivided into categories, labelled by an index k ($k = 1, 2, \ldots, n$), which differ in their reaction to exposure to HIV infection. These categories may be associated with "cofactors" such other STDs, intrinsic geneic differences, or other things. If we assume these categories are not correlated with sexual activity levels (which, as discussed further in Sect. 4.5, would not be true with the categories derived from other STDs), we may write $N_{i,k}$

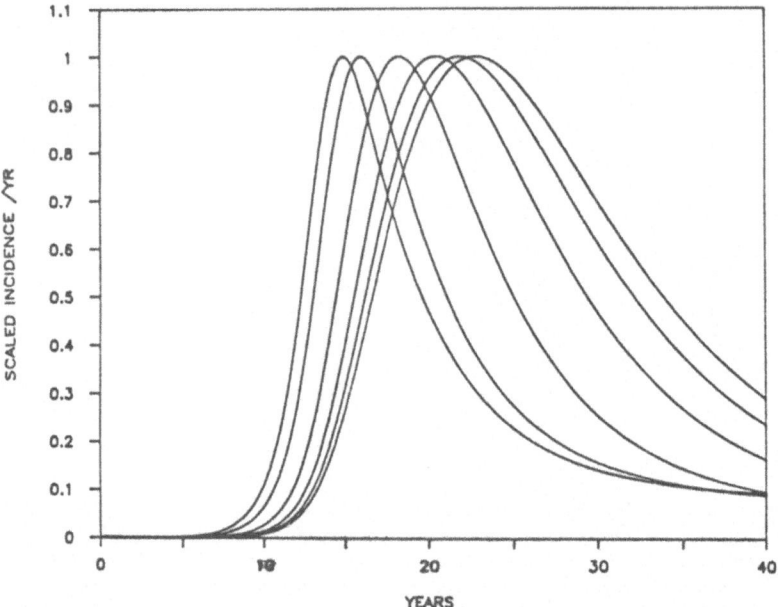

Fig. 9. This figure shows HIV incidence, or rate at which new HIV infections appear, as a function of time, for each of 6 categories of sexual activity within a population of homosexual males. From right to left, the 6 curves are for the subgroups with 0–1, 2–5, 6–10, 11–50, 50–100, > 100 new sexual partners per year, respectively. The underlying mathematical model is as defined in Sec. 4.1, with a gamma distribution for $\{p(i)\}$; the parameters here are $R_0 = 5$, $v = 0.2\,\mathrm{yr}^{-1}$ (so that $\beta c = 1\,\mathrm{yr}^{-1}$), $f = 0.3$, and $CV = 1.4\,(v = 0.5)$. For each of the 6 groups, the HIV incidence curve has been scaled to have magnitude 1 at its peak, in order to facilitate comparison among the subgroups (on an absolute scale, the curves for the high-activity groups would be scarcely discernible, because these groups include relatively few individuals). It can be seen that the incidence of HIV infection peaks much sooner among the groups with high rates of acquiring new sexual partners, with peak incidence in the lowest-activity group occurring a decade later than the peak for the highest-activity group. For further discussion, see Anderson et al. (1986, 1987)

to denote the number of individuals in the ith sexual-activity class and kth epidemiological category; initially the proportion of the population in the kth category is $q(k)$, so that $N_{i,k}(0) = Np(i)q(k)$. $N_{i,k}$ may be further partitioned into susceptible, infected, and possibly no-longer-infectious groups, $X_{i,k}$, $Y_{i,k}$, and $Z_{i,k}$, as before.

The dynamical behavior of $X_{i,k}$, $Y_{i,k}$, and $N_{i,k}$ will again be given by Eqs. (4.1)–(4.3), with the difference that the rate of moving out of the infectious class is v_k (average incubation time $1/v_k$), the fraction going on to develop AIDS is f_k, and the infection probability is λ_k, each of which may be specific to members of the kth category. Eq. (4.4) for λ becomes

$$\lambda_k = \sum_l \beta_{kl} \sum_i iY_{i,l} / \sum_i iN_i. \tag{4.18}$$

Here β_{kl} is the transmission probability for a partnership in which the infected partner is in the lth category, and the susceptible in the k-th.

For any given set of assumptions about the epidemiological parameters v_k, f_k, β_{kl}, and the initial sexual-activity distribution $\{p(i)\}$, this set of equations can be solved numerically.

Some general insight can, however, be obtained by considering the early phase of the epidemic where, as discussed in Sect. 4.1.1, we may put $X_{i,k} \simeq N_{i,k}$. The dynamical variables in the ensuing set of linear differential equations may now have their time-dependence factored out as $\exp(\Lambda t)$, with Λ given by

$$\sum_l \left\{ \frac{c\beta_{kl}q(l)}{\Lambda + v_l} - \delta_{kl} \right\} \lambda_l = 0. \tag{4.19}$$

Here c is as defined by Eq. (4.14); for details of the derivation, see May and Jose (1988). The terms in the curly brackets represent the elements of an $n \times n$ matrix, whose determinant must vanish if Eq. (4.19) is to be satisfied. A more explicit result for Λ can be obtained if we make the reasonable assumption that the transmission probability can be written as a product, $\beta_{kl} = g_k h_l$, where h_l characterizes the infectiousness of individuals in the l-th category, and g_k the susceptibility in the k-th. In this event, the requirement that the determinant of the matrix defined by Eq. (4.19) should vanish gives

$$\sum c\beta_{kk}q(k)/(\Lambda + v_k) = 1. \tag{4.20}$$

Eq. (4.20) will give growth rates, Λ, with positive real parts only if $R_0 > 1$, where the definition (4.17) for R_0 is generalized to

$$R_0 \equiv \sum c\beta_{kk}q(k)/v_k. \tag{4.21}$$

For a derivation and further discussion of this result, see May and Jose (1988) and May (1987).

Eq. (4.21) may be used to illustrate the connection between f, the fraction of HIV infectees who go on to develop AIDS and die from it, and the total number dying in a HIV epidemic in a closed population. This connection can be somewhat paradoxical, in that increasing f does not necessarily mean a larger total of deaths. To see this, let us assume a fraction f of those infected develop AIDS after an average incubation time $1/v_1$, while the remaining fraction, $1 - f$, remain infectious for a characteristic time $1/v_2$. Both groups are assumed to be equally infectious, with transmission probability β; this assumption can easily be generalized. We assume, however, that $1/v_1$ is significantly smaller than $1/v_2$ (with these times possibly being something like 8 years and 30 years, respectively), so that v_1/v_2 significantly exceeds unity. Then R_0 has the form

$$R_0 = [c\beta/v_1][f + (1 - f)(v_1/v_2)]. \tag{4.22}$$

We see that R_0 can be significantly larger for smaller values of f than for larger ones, by virtue of the factor v_1/v_2. The overall fraction of the original population experiencing infection, $I(\infty)$, is larger for larger values of R_0. The total number of AIDS deaths is proportional to f times $I(\infty)$; the first of these two factors obviously increases with increasing f, but the second decreases with increasing f. Thus the total number of AIDS deaths does not bear a simple relationship to f in this case. Numerical studies suggest the total number of AIDS deaths initially increases roughly linearly with increasing f, but remains roughly independent of f for f-values above 50% or so; if v_1/v_2 is big enough, it can even be that total deaths decline somewhat for f-values approaching 100% (May and Anderson, 1987; May and Jose, 1988). All this makes sense: if a substantial proportion of infected individuals remain asymtomatic carriers effectively for life, more infections will be produced. This phenomenon, of course, depends on the existence of long-lived asymptomatic carriers. Current uncertainties about these matters are a major obsticle to determining whether R_0 is likely to exceed unity for heterosexual transmission in developed countries.

4.3 Distributed Incubation Times and Variables Infectiousness

Returning to Sect. 2.3, we recall that the rate of progressing from HIV infection to AIDS, $v(\tau)$, is not a constant (as our models have assumed up to this point) but rather depends on the time, τ, since infection was acquired. In addition, transmission probabilities may depend on τ, with the conditional transmission probability for an individual whose incubation time is s being $\beta(\tau; s)$ at time τ since infection.

This makes for substantial complications in the analysis, some of which are now sketched. For a more full discussion of the biological implications see May et al. (1988c) and Anderson (1988), and for mathematical details see May and Jose (1988) and Castillo-Chavez et al. (1988).

To begin, let the index k of Sect. 4.2 label individuals according to the duration of their incubation times, with individuals in the k-th category having incubation times of length k; this corresponds to putting $v_k(\tau) = 0$ for $\tau < k$ and $v_k(\tau) = \infty$ for $\tau > k$, for individuals in the k-th category. We now define $Y_{i,k}(t, \tau)$ to be the number of individuals who have been infected for a time-interval τ (at time t), and who are in the ith sexual-activity class and have incubation times of duration k (Anderson et al., 1986). This quantity obeys the partial differential equation obtained by generalizing Eq. (4.2),

$$\partial Y_{i,k}/\partial t + \partial Y_{i,k}/\partial \tau = -v_k(\tau)Y_{i,k}(t, \tau). \tag{4.23}$$

One boundary condition is given by the rate which new susceptibles appear,

$$Y_{i,k}(t, 0) = i\lambda X_{i,k}(t). \tag{4.24}$$

The other boundary condition specifies $Y_{i,k}$ for all τ at some initial time $t = 0$. The infection probability λ obeys the appropriate generalization of Eq. (4.18):

$$\lambda = \sum_k \sum_i i \int_0^k \beta(\tau; k) Y_{i,k}(t, \tau) d\tau / \sum iN_i. \tag{4.25}$$

We have assumed all individuals are equally susceptible.

Once the functional form of $\beta(\tau; k)$ and the distribution of incubation times are specified, the dynamical system defined above can be studied numerically for any chosen set of epidemiological parameters.

Again, however, some general insights can be gained by considering the early phase of the epidemic in which $X_{i,k} \simeq N_{i,k} = Np(i)q(k)$. As before, time dependences can be factored out as $\exp(\Lambda t)$, and Eq. (4.23) can then be integrated to get

$$Y_{i,k}(\tau) = i\lambda N p(i) q(k) e^{-\Lambda \tau}, \tag{4.26}$$

for $\tau < k$; $Y_{i,k}$ is zero for $\tau > k$. A more rigorous solution of Eq. (4.23), using Laplace transform techniques, is presented elsewhere (May and Jose, 1988). Eq. (4.26) may now be substituted into Eq. (4.25), to give a "dispersion relation" for Λ:

$$1 = c \sum_k q(k) \int_0^k \beta(\tau; k) e^{-\Lambda \tau} d\tau. \tag{4.27}$$

Finally, we recall from Sect. 2.3 that the probability of developing AIDS after an incubation interval of duration k, $q(k)$, may be expressed in terms of the time-dependent rate process, $v(k)$, as $q(k) = v(k) \exp[-\int_0^k v(s) ds]$. Substituting this into Eq. (4.27), we have the early exponential growth rate, Λ, given by

$$1 = c \int_0^\infty v(k) \exp[-\int_0^k v(s) ds] \int_0^k \beta(\tau; k) e^{\Lambda \tau} d\tau dk. \tag{4.28}$$

If $\beta(\tau)$ has no conditional dependence on the incubation interval (which is unlikely), Eq. (4.28) can be brought into simpler form

$$1 = \int_0^\infty c\beta(\tau) \exp[-\Lambda \tau - \int_0^\tau v(s) ds] d\tau. \tag{4.29}$$

This is the so-called Euler equation of mathematical demography. This result has a sensible interpretation, with $c\beta(\tau)$ being the "fecundity" of HIV infections of "age" τ, $\exp[-\int v ds]$ the age-specific survivorship function, and Λ the rate of growth of the HIV "population". .More generally, however, we have Eq. (4.28), in which the fecundity of a given individual is at all times contingent upon the life expectancy of that individual.

In Sect. 2.3 we discussed the possibility that there may typically be two phases of peak infectivity, separated by a relatively long episode of lower infectiousness. This can have significant implications for the interpretation of temporal trends in the incidence of AIDS and for the estimation of epidemiological parameters. Suppose, for example, we took $\beta(\tau; k)$ to be given by the phenomenological Eq. (2.5) discussed earlier; if the two episodes of infectiousness are short in relation to the total incubation period, and if the douling time of the epidemic is around 1 year or so, then to a good approximation the early growth rate is given by $\Lambda \simeq c\beta_0 - 1/T_0$. But the basic reproductive rate involves both episodes of peak infectiousness, $R_0 \simeq c(\beta_0 T_0 + \beta_1 T_1)$. This is in marked contrast with the simpler models with constant infectiousness, where comparison of Eqs. (4.13) and (4.17) show Λ and R_0 to have a very direct relationship to each other, $\Lambda = v(R_0 - 1)$. The above observation that Λ may often depend mainly on the parameters characterizing the first phase of infectiousness, while R_0 involves both phases, is intuitively reasonable: in the terminology of conventional demography, early "births" count more than later ones toward population growth rates, but all 'births" are relevant to the total number of offspring (here meaning infected people).

More thorough consideration of these complications (May et al., 1988c) suggests that, if indeed there are two peak phases of infectiousness which are both significantly shorter than the average incubation period for AIDS, then current estimates of the transmission coefficients (based on knowledge of the doubling time of the epidemic and the assumption that individuals are infectious over the entire incubation period of 8 years or more) are likely to be significant underestimates of the true likelihood of transmission during any given infectious episode. Such fluctuations in infectiouness, moreover, can induce complex temporal patterns in the epidemic curves, which make it harder to use the models to analyze data and to make predictions.

4.4 Changes in Sexual Habits Over Time

All the above models assume that, on average, individuals do not change their levels of sexual activity over time. At least for homosexual males this is transparently not the case, as discussed in Sect. 2.5.

The basic model of Sect. 3, which dealt simply with an average rate of acquiring new partners, c, can easily be modified to take account of changing sexual patterns, by letting $c(t)$ be time-dependent. Such changes may be deduced from survey data or inferred from data about other STDs. Numerical calculations can then be carried out for any specific assumption about how $c(t)$ changes over time.

For the basic model defined by Eqs. (3.6) and (3.7), which combines epidemiology with demography, it happens that the analytic expression (3.10) for the fraction seropositive at time t can be generalized for time-dependent $c(t)$ to give

$$\frac{Y(t)}{N(t)} = \frac{\Delta \exp[A(t)]}{1 + \Delta\{\exp[A(t)] - 1 + v \int_0^t \exp[A(s)]\,ds\}}. \tag{4.30}$$

Here Δ is again the initial fraction infected (who start the epidemic), and $A(t)$ is defined as $A(t) = \int_0^t [\beta c(s) - v - v]\,ds$; $A(t)$ reduces to the at of Eq. (3.10) if c is a constant.

Figure 10 compares the pattern of seropositivity over time given by the basic model with constant c, with that given by Eq. (4.30) for a population where the average rate of acquiring new partners falls as $c(t) = c(t_0)/[1 + (t - t_0)/T]$ for times $t > t_0$; here t_0 is the time when seropositivity first exceeds 5%. Figure 10 shows results for several values of the parameter T, which measures the average time taken for $c(t)$ to halve its original value. Remember, $f = 1$ in these models, so that Y/N measures the fraction seropositive. It is apparent that such systematic reductions in levels of sexual activity can produce epidemic curves that depart from the classic, measles-like, exponential growth pattern, to show early deceleration.

In Figs. 8 and 9 we noted that observed trends in HIV seropositivity appear to move from early phases of roughly exponential growth into slower (and more nearly linear) growth phases, and we indicated how this is to be expected if there is significant heterogeneity in levels of sexual activity. In these heterogeneous models, the more rapid spread of HIV, and thence AIDS, among more sexually active groups

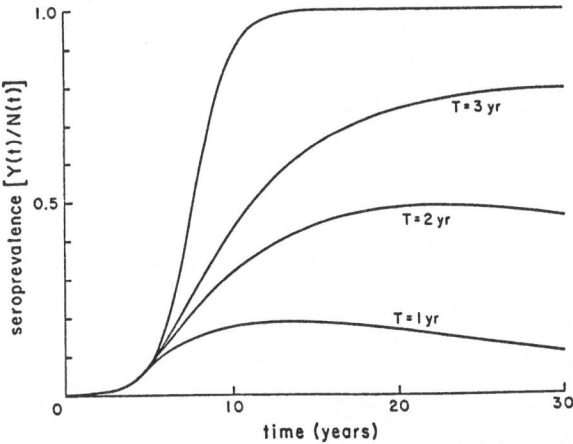

Fig. 10. The pattern of seropositivity, $S(t)$, over time in a closed population of homosexual males for the basic model of Sect. 3 is contrasted with the patterns that arise when average levels of sexual activity change over time (as discussed in Sect. 4.4); in these models, $S(t) = Y(t)/N(t)$. Specifically, this figure compares the basic model with an unchanging rate of partner change, c (the top curve), with models in which numbers of partners decrease as $c(t) = c(t_0)/[1 + (t - t_0)/T]$ for $t > t_0$, where t_0 is the time at which seropositivity first exceeds 5%. As shown, the curves are for $T = 1, 2, 3$ yr; the other epidemiological parameters in Eqs. (3.10) and (4.30) are $\beta c = 1 \text{ yr}^{-1}$, $v = 0.1 \text{ yr}^{-1}$, $v = \mu = 0$, and the "seed" of infection is $\Delta = 0.001$

results in average rates of partner-change decreasing over time (as more active individuals are removed), even though individuals as such as not assumed to change their activity levels. Indeed, we can obtain an explicit expression for the change in the mean number of new partners per unit time, $m(t)$, for the models of heterogeneous transmission in a closed population that were discussed in Sect. 4.1:

$$m(t) \equiv \sum iN_i(t)/\sum N_i(t). \tag{4.31}$$

The numerator on the RHS is given in terms of the dynamical variable $\phi(t)$ by Eq. (4.7), and the denominator can be expressed as $N - C(t)$, where N is the original population size and $C(t)$ is the cumulative number of AIDS cases, given by Eq. (4.11). Assuming a gamma distribution for the initial $\{p(i)\}$, we can calculate the overall change in average numbers of partners, over time, using the same epidemiological parameters are used in Figs. 7–9. The results are shown in Fig. 11, and it can be seen that the dynamics of the epidemic can—by differentially removing more active individuals—produce a marked decrease in overall levels of sexual activity, even though individuals do not change.

In other words, the epidemic curves in Fig. 10 are based on a homogeneous model, in which the average rates of partner-change of individuals decrease over time, in response to social changes. By way of contrast, Figs. 8 and 9 are based on heterogeneous models, in which individuals do not change their sexual habits; here, overall average rates of partner-change fall as a result of the dynamics of the epidemic. In reality, both processes are operating. While we believe that changes in patterns of sexual activity among populations of homosexual men come predomi-

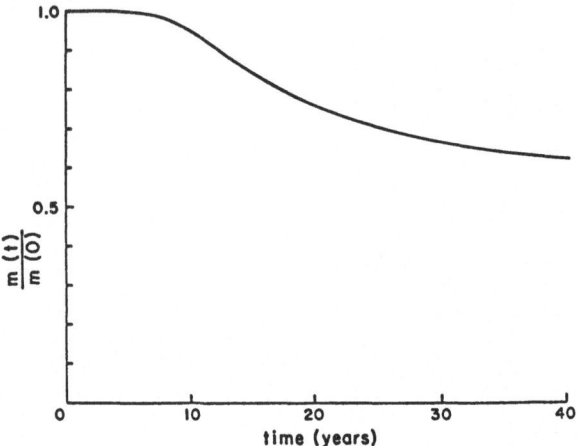

Fig. 11. This figure shows the change in the average rate at which new sexual partners are acquired (expressed as a ratio to the initial rate, $m(t)/m(0)$), as a function of time, t. Even though no individuals are assumed to change their sexual habits, this change in the overall average comes about because individuals who are more sexually active are more likely to acquire HIV infection, and thus to be removed from the population by dying from AIDS. This illustrative example is based on the mathematical model discussed in Sect. 4.1, with a gamma distribution for $\{p(i)\}$; the parameters are $R_0 = 10$, $v = 0.1 \text{ yr}^{-1}$, $f = 0.5$ and $CV = 3.2$ ($v = 0.1$)

nately from changes in individual behavior, the additional changes produced by the epidemiological dynamics may not be negligible, and deserve to be recognized.

4.5 HIV Infection in Conjunction with Other STDs

It is increasingly believed that heterosexual transmission of HIV in Africa and elsewhere may be facilitated by the presence of other STDs, which can cause sores or ulcers, or otherwise help infectious agents to penetrate the skin. Insofar as this is the case, we would expect such other STDs to be differentially present among those with high rates of partner-change, further enhancing such individuals' disproportionate role in acquiring and transmitting HIV infection. There is the further implication that intervention programs aimed at reducing the incidence of these other STDs could retard the spread of HIV. A clearer understanding of the interplay between the transmission dynamics of HIV and other STDs can help us evaluate the likely efficacy of such a program. (In passing, we note that a high prevalence of HIV infection among those with other STDs does not necessarily mean that such STDs indeed facilitate HIV transmission; we could easily be seeing correlation without causation.)

We now sketch the essential conclusions of such an analysis. The details will be presented elsewhere. Suppose that, as in Sect. 4.1, the population at risk is divided into sub-groups, N_i, in which an average of i new sexual partners are acquired per unit time. The endemic prevalence of a given STD (or some aggregation of STDs) can now be estimated; this prevalence will, of course, be higher among the more sexually active groups. Now we introduce HIV. As a simplest first approximation, we assume the transmission probability for HIV, β, in enhanced by the factor a ($a > 1$) if either the infectious or the susceptible partner has another STD, and by a^2 if both do. This dynamics of this system can now be studied, for any given set of epidemiological parameters.

As ever, it is easier to obtain analytic results for the early stages of the epidemic, when the equations can be linearized and time-dependences factored out as $\exp(\Lambda t)$. When the model of Sect. 4.1 is generalized to include the concomitant effects of some other STD which is endemic, we find—after the dust has settled—that Λ is given by

$$1 = \frac{\beta}{\langle i \rangle} \left\langle \frac{i^2[(\Lambda + \mu + v + \gamma + ia\zeta)(\mu + \gamma) + (\Lambda + \mu + v + i\zeta + \gamma/a)i\zeta a^2]}{(\mu + \gamma + i\zeta)(\Lambda + \mu + v)(\Lambda + \mu + v + \gamma + i\zeta)} \right\rangle. \tag{4.32}$$

Here β, a, Λ and v are as defined immediately above or earlier, μ is the death rate from causes other than AIDS (note that in Sect. 4.1 we put $\mu = 0$), γ is the rate at which those infected with the other STD revert to an uninfectious but susceptible state (it may be that $\gamma = 0$), and ζ is the probability that a randomly-chosen partner will infect a susceptible individual with this STD (analogous to the λ of Sect. 4.1 for HIV; ζ, in turn, may be calculated from the transmission parameters for the other STD, in the usual way). As in Eq. (4.8), the sharp brackets denote averages over the initial distribution in rates of partner-change, $\{p(i)\}$.

If the other STD as no effect on HIV transmission, we have $a = 1$ and Eq. (4.32) reduces to $\Lambda = \beta c - v - \mu$, with c defined by Eq. (4.14). This is just Eq. (4.13) for Λ (except that $\mu = 0$ in Sect. 4.1). More generally, however, if the enhancing effects of other STDs are strong, $a \gg 1$, and if we assume $\gamma = 0$ (lifelong infection), Eq. (4.32) reduces to the rough approximation.

$$\Lambda \simeq \frac{\beta a^2}{\langle i \rangle} \left\langle \frac{i^3 \zeta}{\mu + i\zeta} \right\rangle - v. \tag{4.33}$$

The factor a^2 here means that concomitant STD infection has a determining influence on HIV transmission. There are at least two strategies that can be employed here to reduce the rate, Λ, at which HIV infection spreads. Programs aimed at reducing rates of partner-change, particularly among the most active groups, will as before reduce HIV transmission rates and will eventually reduce the prevalence of endemic STDs also, but the latter effect will be long- rather than short-term. Programs aimed at treating other STDs can, if successful, increase the effective value of γ (the rate of converting to the uninfected state), and possibly also reduce the value of a (by reducing somewhat the incidence of symptomatic ulcers and chancres). Both these methods of reducing Λ by decreasing the incidence of other STDs will be more effective when a is large; neither will have any effect if $a = 1$.

5. Two-Sex Models for Heterosexual Transmission of HIV

Up to this point, all our models have been for HIV transmission in a single-sex population. For heterosexual transmission, we must in general deal with two distinct populations, N_1 of males and N_2 of females. The male-to-female transmission probability, β_1, is thought to be higher than that for female-to-male, β_2, possibly by a factor 2 or more (see Fig. 4 and Table 2); it does appear that $\beta_1 > \beta_2$ for STDs such as gonorrhea, although this fact is of doubtful relevance to HIV (for a review of the data, see May, 1988). The initial distribution in rates of partner-change, $\{p_1(i)\}$ and $\{p_2(i)\}$ for males and females, respectively, may also be different, subject to the obvious constraint that the mean rates must be equal, $m_1 = m_2$ (assuming the sex-ratio is initially 50:50 among the relevant age classes). As we shall see below, the average rate for male-to-female transmission is characterized by the parameter combination $\beta_1 c_1$, where c_1 is the mean-square to mean value for the males' partner-acquisition distribution (see Eq. (4.14)), and the corresponding female-to-male transmission rate is characterized by $\beta_2 c_2$. If these characteristic rates $\beta_1 c_1$ and $\beta_2 c_2$ are roughly equal, the system is symmetrical and we may collapse two-sex models back to effectively single-sex systems (note that males and females may have different distributions, so that c_1 and c_2 may be significantly different, even though the means must be equal). Thus there can be some justification for studying single-sex approximations to the heterosexual transmission of HIV.

In general, the overall transmission rate from males to females will not be identical with those from females to males. A two-sex version of the model of Sect. 4.1 is then

$$dX_{\alpha,i}/dt = B_{\alpha,i} - (\mu + i\lambda_\alpha)X_{\alpha,i}, \tag{5.1}$$

$$dY_{\alpha,i}/dt = i\lambda_\alpha X_{\alpha,i} - (\mu + v)Y_{\alpha,i}. \tag{5.2}$$

Here the index α ($\alpha = 1$ for males, 2 for females) labels the two sexes, and we have re-introduced the demography along the lines of Sect. 3. $B_{\alpha,i}$ represents the rate at which new males or females enter the susceptible class with sexual activity level i. The probability that a male will acquire infection from any one female partner is

$$\lambda_1 = \beta_2 \sum i Y_{2,i} / \sum i N_{2,i}, \tag{5.3}$$

and vice versa for λ_2.

There are some new complications in the study of such two-sex models. As previously remarked, by ignoring any effects of age-structure and assuming the initial sex ratio is 50:50, we will have male and female contacts in balance before the advent of HIV/AIDS so long as $m_1 = m_2$. But AIDS will not necessarily remove equal numbers of males and females, especially if transmission parameters differ for the two sexes. If this happens, the patterns in the distributions of acquiring new sexual partners must change over time, in such a way as to keep the total number of male and female contacts equal. Hyman and Stanley (1988) have discussed this problem in more detail, and have suggested a phenomenological modification of equations similar to Eqs. (5.1) and (5.2) that preserves the male-female balance. A more complicated alternative is to work with the full distributions for males and for females acquiring new partners, and to allow these distributions to change over time in appropriate ways (that preserve the balance in contacts). Once it has been decided how to deal with this problem, the epidemiological dynamics can be explored numerically for any chosen set of parameters.

Yet again, interesting results can be gained by considering the early stages of the epidemic, when the equations may be approximately linearized and time-dependences characterized by $\exp(\Lambda t)$. Neglecting μ in comparison with v, we obtain from Eq. (5.2) the expression $Y_{\alpha,i} \simeq i\lambda_\alpha N_{\alpha,i}$. Substituting this into Eq. (5.3) gives an equation for λ_1 in terms of λ_2:

$$\lambda_1 = \beta_2 c_2 \lambda_2 / (\Lambda + v). \tag{5.4}$$

Here c_2 is given by Eq. (4.14), with the averages taken over the distribution of rates of partner-change for females. A similar equation for λ_2 can be obtained by reversing the indices 1 and 2 in Eq. (5.4). The requirement that this pair of equations be consistent then gives the early growth rate of the epidemic as

$$\Lambda = (\beta_1 c_1 \beta_2 c_2)^{1/2} - v. \tag{5.5}$$

The corresponding ratio of incidence of HIV infection, and thus approximately of AIDS cases, among men to that among women can be seen to be $\lambda_1 m_1 N_1 / \lambda_2 m_2 N_2$ in the early stages of the epidemic. But we have noted that $m_1 N_1$ and $m_2 N_2$ must be equal (each sexual contact involves one man and one woman), so the early case ratio is λ_1 / λ_2. Using Eqs. (5.4) and (5.5) we thus arrive at the approximate result

$$\frac{\text{HIV/AIDS among men}}{\text{HIV/AIDS among women}} \simeq \left(\frac{\beta_2 c_2}{\beta_1 c_1}\right)^{1/2} \tag{5.6}$$

It is often asserted that the roughly 1:1 ratio of AIDS cases among men and women in Central Africa constitutes some kind of proof of heterosexual transmission. In fact, Eq. (5.6) shows there is no reason to assume such a 1:1 ratio if transmission efficiencies differ between men and women. Insofar as the male-to-female transmission probability, β_1, may be significantly larger than that for female-to-male, β_2, we might expect more cases among females in the early stages. But we explained in Sect. 2.5 that it is thought that the variance in the partner-change distribution in Africa may be significantly higher for females than for males, by virtue of the cadre of female prostitutes who preserve the overall balance in a population where males tend to be more promiscuous than the typical female. If this is so, then (see Eq. (4.15)) we could have c_2 significantly larger than c_1. From an epidemiological standpoint, it is as if females on average had more sexual partners than males, because of the disproportionate role played by female prostitues. Such effects could counter-balance any tendency for β_1 to exceed β_2, resulting in case ratios being roughly 1:1 for the two sexes. But any such 1:1 ratio is essentially a problem to be explained, not an automatic consequence of heterosexual transmission.

It is also possible to obtain expressions for the asymptotic fraction of each sex ever infected as the epidemic spreads in a closed population, along the lines of Sect. 4.1.2. As could be deduced from Eq. (5.5), this analysis shows the basic reproductive rate for heterosexual transmission to be

$$R_0 = (\beta_1 c_1 \beta_2 c_2)^{1/2} / v. \tag{5.7}$$

That is, the parameter combination βc of the single-sex models is replaced by the geometric mean of $\beta_1 c_1$ and $\beta_2 c_2$ for the two separate populations. Further discussion of two-sex models is given by Hyman and Stanley (1988), Dietz (1988) and May et al. (1988a).

We do not know enough to make any reliable predictions about the average value of R_0 for heterosexual transmission in developed countries. But some very tentative remarks can be made. Suppose we take Peterman et al's (1988) estimates to indicate that the transmission probabilities per partnership are very roughly $\beta_1 \sim 0.2$ and $\beta_2 \sim 0.1$; see Fig. 4 and Table 2. Suppose also that distributions in rates of acquiring new sexual partners are roughly the same for men and women in developed countries, so that $c_1 \simeq c_2 \simeq m + \sigma^2/m$ (where m is the mean, and σ^2 the variance, of the distribution in the numbers of new sexual partners acquired each year). Finally, suppose the duration of infectiousness, $1/v$, is roughly given by the average incubation time of 8–9 years. Thus *very* approximately we have $R_0 \sim (m + \sigma^2/m)$. That is, R_0 may exceed unity among heterosexual groups where new partners are typically acquired more often than annually, or for lower mean rates if the variance is large enough. This rough estimate is subject to the very important caveats given in Sect. 2.3 and 4.3: if infectiousness varies significantly over the duration of the long and variable incubation interval, simple estimates based on expressions such as Eq. (5.7) can be misleading. If indeed infectiousness is effectively confined to two phases, each about 1 year long, at the onset of HIV infection and again at the onset of AIDS, then R_0 for heterosexual transmission is likely to require c-values in excess of one per year (Eq. (5.7) can, however, no longer be used).

6. Demographic Consequences of HIV AIDS

The basic model in Sect. 3 combined epidemiology with demography, to show that HIV/AIDS can reduce overall rates of population growth, and can even lead to population decline under some circumstances; see Eq. (3.12). The basic model, however, was very simple. It ignored all age-structure (dealing with average birth and death rates per capita), took no account of the possible effects of vertical transmission of HIV, and thence AIDS, to the offspring of infected mothers, and assumed all HIV infections eventually produced AIDS. We now give a somewhat more detailed account of the possible demographic effects of HIV/AIDS in developing countries such as Africa, using models that incorporate age-structure and vertical transmission of HIV infection, and where a fraction f ($1 \geq f \geq 0$) of those infected go on to develop AIDS. The models, however, do retain the symmetry assumption that overall transmission rates male-to-female and female-to-male are roughly equal ($\beta_1 c_1 \simeq \beta_2 c_2$), so that we may deal with an effectively single-sex population (Anderson et al., 1988; May et al., 1988a, b).

6.1 A Basic Model with Age-Structure

We define $N(a, t)$ to be the total number of individuals of age a, at time t. These total numbers may, as before, be subdivided into susceptible, infected-and-infectious, and no-longer-infectious categories, $X(a, t)$, $Y(a, t)$, and $Z(a, t)$, respectively. The probability, per unit time, that a given susceptible will acquire infection is $c\lambda(a, t)$, which now depends explicitly on age; as in Sect. 3, we have neglected heterogeneity in degrees of sexual activity, and let c represent some appropriate average rate of acquiring new partners. We assume that death from causes other than AIDS occurs at the age-dependent per capita rate $\mu(a)$, and births at the per capita rate $m(a)$. All other rate processes are as in Sect. 3. The dynamical behavior of this system is described by the following set of partial differential equations:

$$\partial X/\partial t + \partial X/\partial a = -[c\lambda(a, t) + \mu(a)]X(a, t), \tag{6.1}$$

$$\partial Y/\partial t + \partial Y/\partial a = c\lambda X - [v + \mu(a)]Y(a, t), \tag{6.2}$$

$$\partial N/\partial t + \partial N/\partial a = -\mu(a)N(a, t) - fvY(a, t). \tag{6.3}$$

These equations have as one boundary condition the requirement $X(0, t) = N(0, t) = B(t)$ and $Y(0, t) = 0$, where the birth rate, $B(t)$, is

$$B(t) = \int m(a)[N(a, t) - (1 - \varepsilon)Y(a, t)]\,da. \tag{6.4}$$

Here ε represents the probability that a child born to an infected mother will survive, while the remaining fraction $(1 - \varepsilon)$ die of AIDS in the first few years of life. As discussed in Sect. 2.1, ε is currently thought to be around 0.3–0.5 or more, although this number is not certain. The other boundary condition is given by specifying $X(a, 0)$, $Y(a, 0)$, and $N(a, 0)$; that is, by specifying the age-specific numbers in each category at some initial time, $t = 0$.

To complete the description of this system, we generalize Eq. (3.5) to define $\lambda(a, t)$ as

$$\lambda(a, t) = \beta \int p(a, a')\gamma(a', t)\,da' / \int p(a, a')N(a', t)\,da'. \tag{6.5}$$

The ratio of integrals in Eq. (6.5) gives the probability that any one partner will be infected; $p(a, a')$ is the probability that a susceptible of age a will choose a partner of age a'. May et al. (1988a) have explored the asymptotic properties of this system of equations under the two extreme assumptions that all ages mix homogeneously ($p(a, a') = $ constant independent of a and a', so long as both ages lie in the sexually-active range) and that partners are restricted to be from the same age cohort ($p(a, a') = \delta(a - a')$). These two assumptions tend to represent opposite extremes, bracketing reality. May et al. found qualitatively similar results under these two extreme assumptions. In detail, the age-specific prevalence of HIV infection understandably tends to rise more slowly in the age-restricted model than in the homogeneously mixed one, resulting in HIV/AIDS having somewhat less demographic impact— other things being equal—if sexual pairings are age-restricted rather than homogeneously mixed. In what follows, we restrict attention to the homogeneously mixed case. For a more general analysis, see May et al. (1988a, b).

This system of age-structured, partial differential equations can now be solved numerically (Anderson et al., 1988). Starting with some specified set of age profiles for $N(a, 0)$, $X(a, 0)$, and $Y(a, 0)$, and some specified set of demographic and epidemiological parameters, we compute the initial birth rate, $B(0)$, and force of infection, $\lambda(a, 0)$. Eqs. (6.1)–(6.3) then give the age profiles one time step later, and so on.

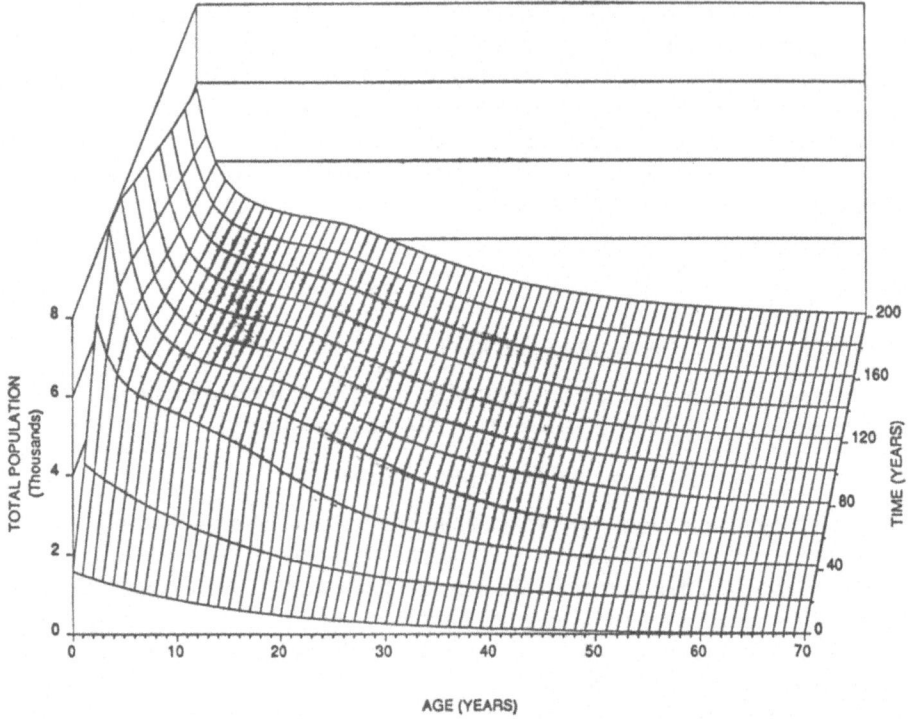

AGE (YEARS)

Fig. 12. The changes that HIV/AIDS can produce in a population's age structure—numbers of individuals in the different age classes, from 0 to 70 years—are shown as a function of time (for the first 200 years after the onset of the epidemic). This illustrative example uses Eqs. (6.1)–(6.5), with the assumption that partners are chosen at random within the sexually-active age-range (taken to be 15–50 years). The other demographic and epidemiological parameters in the model are: $\mu = 1/52\,\mathrm{yr}^{-1}$; reproduction occurs at a constant rate for females between 15 and 50 years of age (the rate being such that $r = 0.04\,\mathrm{yr}^{-1}$ before AIDS); $v + \mu = 1/15\,\mathrm{yr}^{-1}$; $\Lambda = \beta c - (v + \mu) = 0.233\,\mathrm{yr}^{-1}$ (see Table 3); ε (survival probability for babies born to infected mothers) $= 0.3$; $f = 1$ (all HIV infections eventually lead to AIDS). The features of this figure are as discussed in the text

Figure 12 shows the results of one such computation, for a representative set of parameter values. In this example, the population continues to grow for some 50 years after AIDS first appears, but then begins slowly to decrease. As this happens, the age profiles change markedly, from the simple "pyramidal" pre-AIDS profile to more complicated profiles at later times.

Figure 13 summarizes a collection of other examples, showing the dynamical behavior of the population as a whole under various assumptions about f, the fraction of those infected who go on to die from AIDS. Again we see that the long-term effects of HIV/AIDS can take several decades to show up (for a more detailed discussion, see May et al., 1988a, b). Of course, on such long time scales many other things may happen. It could even be, as Lee (1987) and others have suggested, that "homeostatic" mechanisms could come into play to resist such a decline.

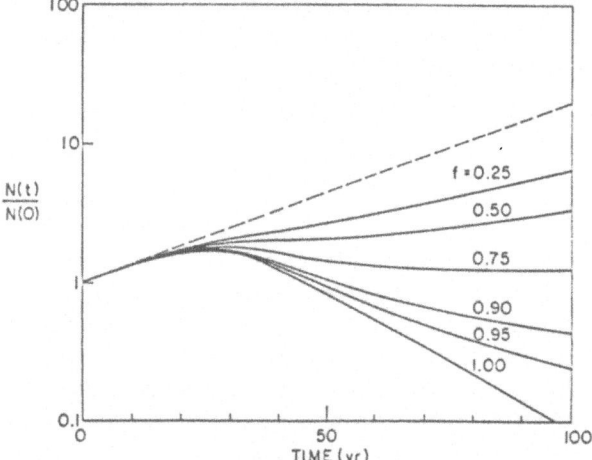

Fig. 13. The magnitude of the total population is plotted (on a logarithmic scale, as a ratio to the initial magnitude) as a function of time, t, for a variety of f-values, as shown. The dashed curve illustrated the AIDS-free rate of purely exponential growth; the other curves have the features discussed in the text. These curves are derived from Eqs. (6.1)–(6.5) with the epidemiological and demographic parameters having the values $\Lambda = 0.2$, $v = 0.06$, $\mu = 0.02$, $r = 0.03$ (all in yr^{-1}), $\varepsilon = 0.5$; at $t = 0$, $Y(0)/N(0) = 0.01$

6.2 Asymptotic Dynamics and Age Profiles

As $t \to \infty$, the time dependence in $N(a, t)$ and other such quantities can be factored out, as $\exp(\rho t)$. As explained in May et al. (1988a), this is essentially because $\lambda(a, t)$ depends on the ratio of infecteds to total numbers, and thus becomes independent of time as $t \to \infty$. It follows that, after a sufficient length of time has elapsed, $N(a, t)$, $X(a, t)$, and $Y(a, t)$ will in general tend to exhibit stable age profiles, whose shapes do not change even though the total numbers increase or decrease.

We now analyze this asymptotic behavior, and present numerical results for the asymptotically stable age profiles, and for social and economic indicators that can be derived from them.

The asymptotic age profiles can be obtained by first defining:

$$N(a, t) = n(a)G(a)e^{\rho t}, \tag{6.6}$$

$$X(a, t) = x(a)G(a)e^{\rho t}, \tag{6.7}$$

$$Y(a, t) = y(a)G(a)e^{\rho t}. \tag{6.8}$$

Here the function $G(a)$ is the stable age profile before the advent of AIDS, except that ρ replaces the pre-AIDS rate of population growth, r:

$$G(a) = \exp\left[-\rho a - \int_0^a \mu(s)\,ds\right]. \tag{6.9}$$

The functions $n(a)$, $x(a)$, and $y(a)$ describe the demographic effects of HIV/AIDS upon the shapes of age profiles, as distinct from its effects upon the overall growth rate ρ (and thus upon $G(a)$). Substituting Eqs. (6.6)–(6.9) into (6.1)–(6.3), we see that these functions obey the set of ordinary differential equations

$$dn/da = -fvy, \tag{6.10}$$

$$dx/da = -c\lambda x, \tag{6.11}$$

$$dy/da = c\lambda x - vy. \tag{6.12}$$

Eqs. (6.10)–(6.12) can now be integrated, to obtain analytic expressions for $n(a; \lambda, \rho)$, $x(a; \lambda, \rho)$ and $y(a; \lambda, \rho)$ as functions of age, a, and the parameters λ and ρ. We then return to Eq. (6.4) for $B(t)$ and Eq. (6.5) for $\lambda(t)$, and substitute the explicit asymptotic expressions for $N(a, t)$ and $Y(a, t)$ from Eqs. (6.6) and (6.7), to end up with two relationships between the quantities λ and ρ. Eliminating λ, we can in principle thus obtain an explicit expression for the asymptotic rate of population growth, ρ, in terms of basic epidemiological and demographic parameters.

May et al. (1988b) have obtained such an explicit formula for ρ, under the simplifying assumptions that the per capita death rate, μ, is constant and that the per capita birth rate, $m(a)$, is the same for all ages above τ, where τ characterizes the onset of adult sexual activity:

$$\rho = -(v + \mu) + v \left(\frac{\beta c}{\beta c - \theta} \right) \left(\varepsilon + \frac{[1 - f]v}{\rho + \mu} \right). \tag{6.13}$$

Here v is the overall average birth rate per capita, which can be defined in terms of the pre-AIDS rate of population growth, r, as $v = (r + \mu) \exp[(r - \rho)\tau]$; May et al. (1988b) give an intuitive explanation of this result along with its derivation. The quantity $\theta \equiv fv + (1 - \varepsilon)v$ essentially represents the rate of adult deaths from horizontally-transmitted AIDS, and infant deaths from vertically-transmitted AIDS. All the other parameters are as defined previously. Eq. (6.13) embodies the further approximation that the upper age limit for sexual activity, ξ, is effectively infinite (some subtleties arise when the limits $f \to \infty$ and $\xi \to \infty$ are both taken; these non-uniform limiting processes are discussed in May et al. (1988b)).

Eq. (6.13) makes it plain that asymptotic rates of population growth can be driven negative by HIV/AIDS, depending on the relative magnitudes of the various epidemiological and demographic parameters. For a population whose per capita growth rate is r before the advent to AIDS, the curves in Fig. 14 show the value of f—the fraction of those infected who go on to develop AIDS—that will eventually produce zero population growth ($\rho = 0$), for three different assumptions about the survival probability for offspring born to infected mothers ($\varepsilon = 0, 0.5, 1$). It appears that populations with reasonably large initial rates of population growth may eventually stop growing, if f is large enough.

Once ρ and λ have been found in terms of basic epidemiological and demographic parameters, the asymptotic age profiles for total numbers and numbers infected, $N(a) = n(a)G(a)$ and $Y(a) = y(a)G(a)$, respectively, can be calculated. Figure 15 shows such an asymptotic age profile in total numbers (and in numbers infected), and contrasts it with the pre-AIDS profile.

It is immediately apparent that deaths from horizontally- and vertically-

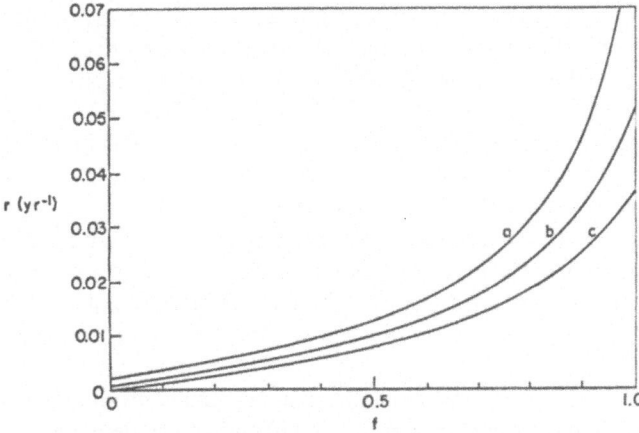

Fig. 14. This figure shows the growth rate in the AIDS-free population, r, which can asymptotically be brought exactly to zero (stationary population) for a specified value of f. The other epidemiological and demographic parameter values are $\Lambda = 0.1, v = 0.1, \mu = 0.02$ (all in yr^{-1}), and, as in Figs. 12 and 13, within the age-range 15–50 years sexual partners are chosen randomly and reproduction takes place at a constant rate. The curves labeled a, b, c are for $\varepsilon = 0, 0.5, 1.0$, respectively ($\varepsilon$ is the survival probability for offspring born to infected mothers). These results are obtained from the critical version of Eq. (6.13) in which $\rho = 0$

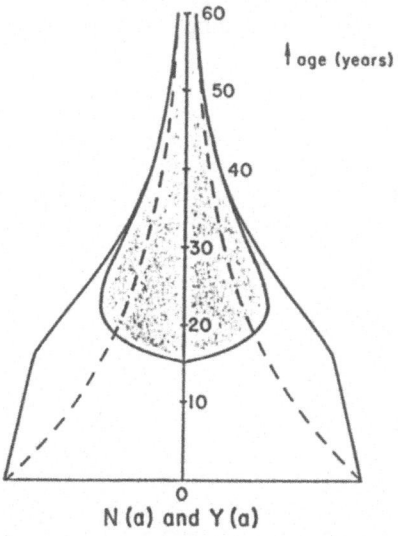

Fig. 15. The asymptotic age-profile of the total population, $N(a)$, and of the infected fraction, $Y(a)$ (the solid outer curves and the hatched inner region, respectively), are shown as functions of age, a (the vertical axis), under the assumption that sexual contacts are "homogeneously mixed" among all age groups above 15 years. The demographic and epidemiological parameters here have the values $\Lambda = 0.2, v = 0.06, \mu = 0.02, r = 0.03$ (all in yr^{-1}), $\varepsilon = 0.5$ and $f = 1.0$. The dashed age-profile is for the original, AIDS-free population, growing steadily at the rate $r = 0.03$ per annum

transmitted HIV infections have two countervailing effects upon the asymptotic form of the age profile for total numbers. On the one hand, increasing death rates and effectively decreasing birth rates cause the overall rate of population growth to decrease, which means—other things being equal—that age profiles tend to be less steep than is characteristically the case at present in developing countries. This effect, by itself, tends to decrease the ratio of numbers of children to numbers of adults. On the other hand, adult deaths from AIDS tend to steepen the age profile at older ages, and thus to increase the ratio of numbers of children to numbers of adults. It is not intuitively obvious which of these opposing tendencies—adult deaths tending to steepen the age profile or slowed population growth rates tending to make it less steep—will predominate for any specific set of demographic and epidemiological parameters.

One rough measure of, as it were, the ratio of tax consumers to tax producers in a developing country is the "child dependency ratio", CDR, which we define as the fraction of the total population who are below the age of 15 years. Figure 16 shows the value of this ratio, as a function of f, for several different values of the per capita rate of population growth before AIDS (and a representative set of other demographic and epidemiological parameters). In this instance, the opposing tendencies described in the previous paragraph roughly cancel, leaving the CDR essentially unchanged. That is, despite the pronounced changes in the asymptotic age profiles (see Figs. 12 and 15), the overall fraction below the age of 15 years remains relatively unchanged.

In short, the demographic effects of HIV/AIDS in developing countries are not simple. The analyses and examples presented above are based on grossly oversimplified models, but we believe they provide a basic understanding and a point of departure for more realistic numerical computations.

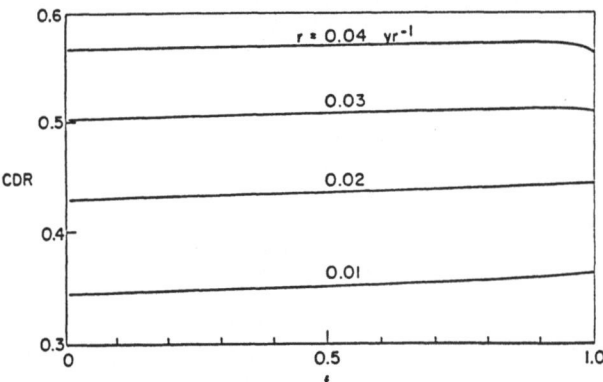

Fig. 16. This figure shows asymptotic values of the "child dependency ratio", CDR, defined here as the fraction of the total population who are below the age of 15 years. CDR is shown as a function of f (the fraction of HIV infectees who go on to develop AIDS), for several different values of the growth rate in the AIDS-free population, r, as shown. This figure is for $\Lambda = 0.4$, $v = 0.1$, $\mu = 0.02$ (all in yr^{-1}), and $\varepsilon = 0.5$. As discussed in the text, CDR is roughly unaffected by HIV/AIDS for most values of the epidemiological parameters

7. Conclusion

Although short-term predictions can sensibly be made by extrapolating current trends, long-term predictions about the prevalence of HIV and AIDS within any particular group requires an understanding of the nonlinear dynamics of transmission. The epidemiology of HIV has many unusual features. Amongst other things, we need to know more about the distribution of incubation times for those who do develop AIDS, the fraction of those infected who will eventually go on to develop AIDS, the patterns of infectiousness over time, the overall transmission probability within a given kind of sexual or other relationship, and the distribution in rates of acquiring new sexual or needlesharing partners. The kinds of information that are currently available are surveyed in Sect. 2, partly for their own sake and partly as a basis for the epidemiological models developed in the rest of the paper.

Given the current uncertainties about so many biological and sociological aspects of HIV transmission, we believe it is sensible to begin by exploring relatively simple models that caricature the transmission dynamics, with a view to understanding qualitative features of the epidemic. Beginning with a very simple model, we have incorporated heterogeneities in levels of sexual activity, distributed incubation times and time-dependent patterns of infectiousness that are conditional on the incubation interval, changes in patterns of sexual behavior over time, the possible effects of other STDs upon HIV transmission, and the interplay between HIV/AIDS and broad demographic processes.

The conclusions defy any brief summary. We have seen how the epidemiologically appropriate average number of new partners per unit time, c, can significantly exceed the simple mean of the distruibution (Sect. 4.1), and how this possibly relates to the ratio of AIDS cases among men and women in Africa (Sect. 5). Rough estimates of the product of the transmission probability, β, times the average rate of partner-change, c, may be obtained from the doubling time of the epidemic in its early stages; these estimates are consistent with more directly-derived estimates of β and c separately, although we emphasize that many such estimates may be compromised by significant time-dependence in patterns of infectiousness (Sect. 4.3).

The shape of observed epidemic curves for HIV seropositivity and AIDS incidence depart from those for more classical, measles-like, epidemics, in that early patterns of rapid growth soon give way to slower increases. We believe the observed patterns can be explained in terms of the substantial variabilities in rates of acquiring new partners within given risk-groups (Sect. 4.1). For populations having substantial heterogeneity in levels of sexual activity, the transmission dynamics of HIV will cause average rates of partner-change to fall over time, even if no individuals change their habits; this effect could interfere with some indirect methods for assessing changes in the behavior of individuals with respect to rates of partner-change (Sect. 4.4).

Once fully established, HIV/AIDS is likely to have two countervailing effects on age profiles in developing countries. On the one hand, deaths from horizontally- and vertically-transmitted infections have the indirect effect of reducing rates of population growth, resulting in less steep age profiles (and a smaller fraction under the age of (say, 15 years). On the other hand, the direct effect of adult deaths from AIDS makes for steeper age profiles among adults, and a higher fraction under the age of 15 years). Simple models suggest these opposing tendencies may roughly cancel out, for representative values of the epidemiological and demographic parameters that may pertain in Africa; this conclusion is, however, very tentative (Sect. 6).

Acknowledgements. We have benefited greatly from conversations with J.L. Aron, S. Blythe, C. Castillo-Chavez, J.M. Hyman, M. John, S.A. Levin, A.R. Mclean, G.F. Medley, E.A. Stanley, and many others. This work was supported in part by the Sloan Foundation and the NSF under grant DMS87-03503 (RMM), and by the Overseas Development Agency, UK, PANOS and the Medical Research Council (RMA).

References

Anderson, R.M. (1988) The epidemiology of HIV infection: variable incubation plus infectious periods and heterogeneity in sexual activity. J. Roy. Stat. Soc. Ser. A *151*, 66–98

Anderson, R.M., Johnson, A.M. (1988) Rates of sexual partner change in homosexual and heterosexual populations in the United Kingdom. In: Boller, B. (ed.) Kinsey Institute Conference on Sexual Behavior in Relation to AIDS. Oxford University Press, New York. (In press)

Anderson, R.M., May, R.M. (1979) Population biology of infectious diseases. Nature *280*, 361–367 and 455–461

Anderson, R.M., May, R.M. (1985) Vaccination and herd immunity to infectious diseases. Nature *318*, 323–329

Anderson, R.M., May, R.M., McLean, A.R. (1988) Possible demographic consequences of AIDS in developing countries. Nature 000

Anderson, R.M., Medley, G.F., Blythe, S.P., Johnson, A.M. (1987) Is it possible to predict the minimum size of the acquired immunodeficiency syndrome (AIDS) epidemic in the United Kingdom? Lancet (i) 1987, 1073–1075

Anderson, R.M., Medley, G.F., May, R.M., Johnson, A.M. (1986) A preliminary study of the transmission dynamics of the human immunodeficiency virus (HIV), the causative agent of AIDS. IMA J. Math. Appl. Med. Biol. *3*, 229–263

BMRB (1987) AIDS Advertising Campaign: Report for Surveys During the First Year of Advertising, 1986–87. British Market Research Board, London

Brookmeyer, R. and Damiano, A. (1989) Statistical methods for short-term projections of AIDS incidence. Stat. in Med., *8*, 1–20

Burger, M. et al.. (1986) Transmission of LAV/HTLVIII in sexual partners: seropositivity does not predict infectivity in all cases. Am. J. Med., *81*, 5–10

Castillo-Chavez, C., Cooke, K., Levin, S.A. (1988) The role of long incubation periods in the dynamics of acquired immunodeficiency syndrome (AIDS). In press

CDC. (1987) Antibody to human immunodeficiency virus in female prostitutes. MMWR, *36*, 157–161

Dietz, K. (1988) On the transmission dynamics of HIV. Math. Biosci. (In press)

Grant, R.M., Wiley, J.A., Winkelstein, W. (1987) Infectivity of the Human Immunodeficiency Virus: estimates from a prospective study of homosexual men. J. Infect. Dis. *156*, 189–193

Hethcote, H. (1987) AIDS modeling in the USA. In: Future Trends in AIDS. HMSO, London, pp. 35–41

Hethcote, H.W., Yorke, J.A. (1984) Gonorrhea Transmission Dynamics and Control Lecture Notes in Biomathmatics, Vol. 56. Springer, Heidelberg Berlin New York

HHS (1987) HIV infections in the United States: a review of current knowledge and plans for expansion of HIV surveillance activities (a report of the Domestic Policy Council). Department of Health and Human Services, Washington, D.C.

Hyman, J.M., Stanley, E.A. (1988) A risk-based model for the spread of the AIDS virus. Math. Biosci. (In press)

Johnson, A.M. (1988) Social and behavioral aspects of the HIV epidemic: a review

Kermack, W.O., McKendrick, A.G. (1927) A contribution to the mathematical theory of epidemics. Proc. Roy. Soc. A *115*, 700–721

Lee, R.D. (1987) Population dynamics of humans and other animals. Demography *24*, 443–465

Macdonald, G. (1956) Theory of the eradication of malaria. Bull. Wld. Hlth. Org. *15*, 369–387

May, R.M. (1987) Nonlinearities and complex behavior in simple ecological and epidemiological models. Ann. N.J. Acad. Sci. *504*, 1–15

May, R.M. (1988) HIV infection in heterosexuals. Nature *331*, 655–656

May, R.M., Anderson, R.M. (1987) Transmission dynamics of HIV infection. Nature *326*, 137–142

May, R.M., Anderson, R.M., Johnson, A.M. (1988c) The influence of temporal variation in the infectiousness of infected individuals on the transmission dynamics of HIV. (under review)

May, R.M., Anderson, R.M., McLean, A.R. (1988a) Possible demographic consequences of HIV/AIDS: I, assuming HIV infection always leads to AIDS. Math. Biosci. *90*, 475–505

May, R.M., Anderson, R.M., McLean, A.R. (1988b) Possible demographic consequences of HIV/AIDS epidemics: II, assuming HIV infection does not necessarily lead to AIDS. In: Castillo-Chavez, C., Levin, S.A., Shoemaker, C. (eds.) Proceedings International Symposium in Mathematical Approaches to Ecological and Environmental Problem Solving. Lecture Notes in Biomathematics, vol. 82. Springer New York Heidelberg Berlin

May, R.M., Jose, M. (1988) Epidemic propagation of sexually transmitted infections in behaviorally heterogeneous populations. J. Theor. Biol. (In press)

Medley, G.F., Anderson, R.M., Cox, D.R., Billard, L. (1987) Incubation period of AIDS in patients infected via blood transfusions. Nature *328*, 719–721

Medley, G.F., Anderson, R.M., Cox, D.R., Billard, L. (1988) The distribution of the incubation period of AIDS. Proc. Roy. Soc. B *233*, 367–377

Morse, P.H., Feshbach, H. (1953) Methods of Theoretical Physics. McGraw-Hill, New York

Pedersen, C., Nielsen, C.M., Vestergaard, B.F., Gerstoft, J., Krogsgaard, K., Nielsen, J.O. (1987) Temporal relation of antigenaemia and loss of antibodies to core antiens to development of clinical disease in HIV infection. Brit. Med. J. *295*, 567–569

Peterman, T.A., Stoneburner, R.L., Allen, J.R., Jaffe, H.W., Curran, J.W. (1988) Risk of HIV transmission from heterosexual adults with transfusion-associated infections. JAMA *259*, 55–63

Quinn, T.C., Mann, J.M., Curran, J.W., Piot, P. (1986) AIDS in Africa: an epidemiologic paradigm. Science *234*, 955–963

Reeves, G., Cox, D.R. (1988) Average survival times for AIDS patients in the UK. (Submitted for review)

Robertson, J.R., Bucknall, A.B.V., Welsby, P.D. (1986) Epidemic of AIDS related virus (HTLVII/LAV) infection among intravenous drug abusers. Brit. Med. J. *292*, 527–529

Ross, R. (1911) The Prevention of Malaria, 2nd edn. Murray, London

Rothenberg, R., et al. (1987) Survival with acquired immunodeficiency syndrome. New Eng. J. Med. *317*, 1297–1302

Smith, C.E.G. (1970) Prospects for the control of infectious disease. Proc. Roy. Soc. Med. *63*, 1181–1190

Johnson, A.M., Wadsworth, J., Elliott, P., Blower, S., Wallace, P., Miller, D., Adler, M.W., Anderson, R.M. (1989) A pilot study of sexual lifestyles in a random sample of the Great Britain. AIDS, *3*, 135–142.

Winkelstein, W., Lyman, D.M., Padian, N., Grant, R.M., Samuel, M., Wiley, J.A., Anderson, R.E., Lang, W., Riggs, J., Levy, J.A. (1987) Sexual practices and risk of infection by HIV. JAMA *257*, 321–325

Appendix A

In this Appendix, we sketch the assumptions about the initial distribution, $\{p(i)\}$, and the consequent analysis, that underlie the illustrative Figs. 7, 8, 9 and 11.

As discussed more fully by Anderson et al. (1986), we assume the initial distribution in rates of acquiring new sexual partners obeys a continuous gamma distribution, such that the proportion of the risk-group having between i and $i + di$ new partners per unit time is given by

$$p(i)di = \alpha^v i^{v-1} e^{-i\alpha} di/\Gamma(v). \tag{A.1}$$

It follows that $\langle i \rangle = v/\alpha$, $\langle i^2 \rangle = v(v+1)/\alpha^2$, and thence the variance of the distribution is $\sigma^2 = v/\alpha^2$. Rather than characterize this initial distribution by the parameters v and α of Eq. (A.1), we prefer to use the parameters c and CV: here c is the effective average number of partners defined by Eq. (4.14) or (4.15), and CV represents the coefficient of variation of the distribution, $CV = \sigma/m$. These parameters c and CV are related to the α and v of Eq. (A.1) by the relations

$$c = (v+1)/\alpha, \tag{A.2}$$

$$CV = v^{-1/2}. \tag{A.3}$$

The expectation values that enter into Eqs. (4.9) and (4.10) can now be evaluated explicitly, to get

$$\langle 1 - e^{-i\phi} \rangle = 1 - (1 + \phi/\alpha)^{-v}, \tag{A.4}$$

$$\langle i(1 - e^{-i\phi}) \rangle/\langle i \rangle = 1 - (1 + \phi/\alpha)^{-v-1}. \tag{A.5}$$

The dynamics of the epidemic are now described by Eq. (4.9), with the explicit form (A.5) for the relevant expectation value. The dependence of the dynamics upon the basic epidemiological parameters can be more explicitly seen if we replace $\phi(t)$ by $\psi(t)$, with the definition

$$\psi(t) \equiv (v+1)\phi/\alpha = c\phi(t). \tag{A.6}$$

The basic dynamical Eq. (4.9) then becomes

$$(d\psi/dt)\exp(-f\psi/R_0) = vR_0\{1 - [1 + \psi/(v+1)]^{-v-1}$$
$$- [1 - \exp(-f\psi/R_0)]/f\} + c\lambda(0). \tag{A.7}$$

The computational procedure is now straightforward. First, we specify the epidemiological parameters R_0 (the basic reproductive rate), f (the fraction of HIV infectees who eventually proceed to develop AIDS), v (related to the coefficient of variation of the initial sexual-activity distribution by $CV = v^{-1/2}$), and $c\lambda(0)$ (the magnitude of the initial infection that "seeds" the epidemic); the parameter $v(v = 1/D$, where D is the average incubation period) only enters in setting the time scale. Once these fundamental epidemiological parameters are specified, we can integrate Eq. (A.7) to find $\psi(t)$ as a function of time, t.

All the other epidemiological quantities displayed in Figs. 7, 8, 9 and 11 can be calculated in terms of $\psi(t)$ and the parameters just specified. In particular, the fraction of the initial population to have experienced infection by time t, $I(t)$, is now given by

$$I(t) = 1 - [1 + \psi/(v+1)]^{-v}. \tag{A.8}$$

The cumulative number of AIDS cases, $C(t)$, is then obtained by integrating the first-order differential equation

$$dC/dt = v[fNI(t) - C(t)]. \tag{A.9}$$

Here N is the initial size of the risk-group in question, and dC/dt is the rate at which new AIDS cases appear, at time t. Seropositivity at time t, $S(t)$, is given by

$$S(t) = [NI(t) - C(t)]/[N - C(t)]. \tag{A.10}$$

Finally, the average number of new partners (per unit time) at time t, as given by Eq. (4.31), is

$$m(t) = \frac{m(0)\exp(-f\psi/R_0)}{1 - C(t)/N}. \tag{A.11}$$

Here $m(0)$ is the initial rate, $m = \langle i \rangle = v/\alpha$.

Figures 8, 9, and 11 are obtained by performing the above analysis. The asymptotic results in Fig. 7 are obtained more simply in the limit when $d/dt \to 0$, whereupon the relation between the asymptotic fraction ever to experience infection, $I(\infty)$, and the basic reproductive rate, R_0, can be seen to be given implicitly (for any specified value of f and $CV = v^{-1/2}$) by

$$I(\infty) = 1 - [1 + \psi_\infty/(v+1)]^{-v}, \tag{A.12}$$
$$R_0 = -f\psi_\infty/\ln(1 - f\{1 - [1 + \psi_\infty/(v+1)]^{-v-1}\}). \tag{A.13}$$

Part IV. Ecotoxicology

Models in Ecotoxicology: Methodological Aspects[1]

Simon A. Levin

Contents

The science of ecotoxicology is based to a large extent on extrapolation—extrapolation from one system and one stress to another, extrapolation from laboratory tests and microcosm studies to field situations, extrapolation across scales. Such extrapolation must be based on some underlying model or models; thus, models are an essential and ineluctable component of ecological risk assessment.

Models come in a variety of forms, and serve a variety of purposes. They may be the means for explicit prediction, may serve as screening tools, may provide understanding of the mechanisms underlying observed patterns, or may be part of an integrated adaptive management scheme. Because models are called on to serve so many purposes, it would be foolhardy to expect a simple model to meet all objectives; different criteria must be applied depending on the uses to which one intends to put the model. Some models are deliberately oversimplified in order to isolate parts of a complex system, and to provide a means for investigating the implications of hypothetical relationships. Often, to understand why some ecological relationship is what it is, it is necessary to embed that relationship within a broader framework of possibilities, and then to ask why one form of a relationship is observed more often than others. In other situations, such as models used for prediction of fish stock dynamics, parameter estimation is a fundamental consideration, and the ease and reliability of estimating parameters must influence the form of the model. Other population models, however, including some fisheries models, are intended primarily for pedagogical purposes, and are designed without regard for the parameter estimation problem. It is when such models have been misapplied, taken outside of the contexts in which they were developed, that abuses

[1] This is adapted, with only minor changes, from Levin, S.A., et al. (eds.) (1989) Ecotoxicology: Problems and Approaches. Springer Advanced Texts in Life Sciences. Springer, New York Berlin Heidelberg London Paris Tokyo

have occurred. The result has been misunderstanding about the power of models (Limburg et al., 1986). The most important conclusion is that one must recognize the multiplicity of purposes models serve; understand the assumptions underlying particular models, and their capabilities; and select models appropriate to the purposes at hand rather than trying to fit a square peg into a round hole.

Ecosystem models are perhaps the best choices for illustrating a familiar problem for modelers, that of appropriate detail; but it is a problem that pertains to any class of models. Many of the models developed for ecosystems are highly detailed, and include hundreds of components. Such models serve as tableaux, upon which can be laid out a panoply of interactions and flows; but these models are useless as tools for prediction. Apart from the fact that the level of detail in such models must necessarily be arbitrary (any grouping could be subdivided further and further); they contain too many parameters to be estimated robustly, and to provide reliable prediction. The problem of appropriate detail is a fundamental one in modeling, and one to which I shall return later. The large ecosystem models do serve a purpose in displaying relationships; but more simplistic relationships such as the Vollenweider loading curves (Vollenweider, 1968) have been far more useful and influential as predictive devices (Peters, 1986). A challenge to modelers, however, is to replace these regressions with mechanistic models that do not err on the side of mindless detail, but provide sufficient biological and physical detail to explain why the observed relationships hold and to predict when they won't (Lehman, 1986).

In general, in ecotoxicology, it is essential to recognize that no single model can meet all objectives, and that one must interface models that contain different levels of detail. This is evident in the models of fate, transport, and effects of chemicals discussed by Thomann and Hallam in the following two chapters of this section. Generic models of fate and transport incorporate basic mechanisms, such as diffusion, advection, reaction, and depuration. These components must always be present, and generic models provide the basic structure for incorporating them. However, application of these models to specific cases involves fine-tuning them to particular environments, and fixing the parameters either from basic considerations or by calibration. Many general questions are best answered by consideration of the more abstract and generic models; others require detailed consideration of local geometry and topography, and can be investigated only by computer simulations. Both kinds of models—and hybrid versions—are essential in ecotoxicology; and much has been learned from the full spectrum of investigations, ranging from abstract treatment of partial differential equations to very detailed site-specific three-dimensional simulations.

Physical and Biological Scales

The earliest applied models in mathematical ecology, those introduced by Volterra (1926) in considering the fisheries of the Adriatic, shared a number of features:

1) They considered only the interactions among populations, ignoring details of the physical environment.
2) They were deterministic and autonomous; that is, they expressed relationships that did not depend on time.
3) They treated time as continuous.

Each of these features subsequently has been relaxed by various investigators, but it is to the first restriction that I direct attention here. The dichotomy between models that emphasize biological versus physical factors has divided population ecologists from ecosystem ecologists, and unnecessarily so (Kingsland, 1985). As is usually the case with scientific dichotomies, the truth lies somewhere between the extremes, and is inseparable from the question of scales. In the oceanographic literature, it is often taken as a starting point that temporal scales are biologically determined, while spatial scales are physically determined (Denman and Powell, 1984). But the situation is more complicated than that. Oceanic species are patchily distributed on almost every scale of investigation. On broad scales, consideration of Fourier spectra indicates coherence between the distribution of physical markets such as temperature and the distribution of biological species. This suggests that on the broadest scales the spatial distribution of the biota indeed is determined by physical factors. On finer spatial scales, however, on the order of up to 50 km, such coherence often breaks down, especially at higher trophic levels (Weber et al., 1986) indicating that biological processes and interactions must be important in determining fine-scale patchiness. Such correlations, often viewed as being synonymous with prediction (Peters, 1986), cannot be the stopping point; Lehman (1986) elegantly has reminded us of the need to move from correlation to causation, for otherwise we cannot know the limits to predictability. However, correlations certainly are suggestive, and can serve an important role in helping us to structure a system of models, especially with regard to choosing appropriate scales.

In the development of predictive models, two factors have been paramount in preventing the integration of physics and biology. One is the desire for simplicity and reliability: a model that has only two or three parameters has little to hide, whereas a complex computer code is virtually impenetrable. Furthermore, as already discussed, detailed models are harder to parameterize and may be less reliable for prediction than simpler models. Secondly, highly detailed models can be investigated only via computer simulations, and exhaustive simulations can be expensive and cumbersome to carry out. Recent advances in computing, however, and in hierarchical approaches to modeling provide us the opportunity to overcome these obstacles, and to develop the requisite combined models. Component biological models, such as those describing the movements and searching behavior of organisms on small scales, can be integrated to provide the behavior of aggregates at higher scales; these can be coupled with models of physical processes. Parallel processing facilitates incorporation of environmental heterogeneity and systematic exploration of parameter regimes. It is to be expected that the next few years will see great advances in our ability to develop integrated physical and biological models, and improvements in methods for examining the interrelationships among phenomena on different scales.

Aggregation, Simplification, and the Problem of Dimensionality

There is no correct level of description for an ecosystem. Just as Mandelbrot (1983) has educated us to the endless spectrum of detail that can be observed at finer and finer levels of description, so can we find an endless spectrum of detail in the taxonomic or functional organization of an ecosystem (O'Neill et al., 1986; Levin, 1987). The analysis of the structure of food webs, one of the hottest topics in ecological theory, is frustrated by the arbitrariness with which descriptions of food webs must be amassed. Any investigator will, according to personal taste, lump certain groups of species while breaking others into age classes. The decision is based inevitably on the particular investigator's perspective, and on what is of special interest in a particular study. There can be no unique way to describe an ecosystem. Each biological grouping is an attempt at simplification, focusing on the average properties of an ensemble rather than on the variance within the ensemble. Partitioning ensembles into subunits accounts for some of the heterogeneity within the ensemble, but at the cost of a reduced capability to make deterministic statements about the subunits. This paradox is not unique to biology; it confronts investigators studying any system, and underlies such basic concepts as the measurement of temperature in physical systems.

Given that there is no correct level of detail, it is important to learn how variability changes across scales, and what the consequences are of lumping components previously treated as independent.

For the dynamical systems common in ecological modeling, in which the dynamics of system components are assumed to be governed by autonomous equations of the form

$$dX_i/dt = f_i(X_1,\ldots,X_n) \quad i=1,\ldots,n, \tag{1}$$

the formal aggregation problem is that of finding a reduced set of variables.

$$Y_j = g_j(X_1,\ldots,X_n) \quad j=1,\ldots,m, m < n \tag{2}$$

that satisfy their own autonomous dynamics

$$dY_j/dt = F_j(Y_1,\ldots,Y_m) \quad j=1,\ldots,m \tag{3}$$

In general, such reduction is not possible; the dynamics of any reduced set of variables Y_j is likely to depend not only on their own values, but also on other (hidden) variables. However, in some cases such reduction can be made (Iwasa et al., 1987). If one defines the Jacobian matrix $B = (B_{jk})$, where $B_{jk} = \partial g_j/\partial X_k$, and the second matrix $A = (A_{jk})$, where $A_{jk} = \partial(\sum_l B_{jl} f_l)/\partial X_k$, then (Iwasa et al., 1987) the formal condition that (2) leads to an autonomous (perfect aggregation) scheme is that

$$AB^+B = A \quad \text{for all } X, \tag{4}$$

where B^+ indicates the generalized inverse of the matrix B (Penrose, 1955).

In general, it is unreasonable to expect such perfect aggregation to be possible, and one turns to approximate schemes that minimize the error associated with dimensional reduction. This is closely related to the investigations of Schaffer and his collaborators (Schaffer et al., 1986), who apply the techniques of Takens (1981) to determine the fractal dimensionality of data sets. To the best of my knowledge, no attempt has been made to relate these two lines of inquiry, which are complementary to one another.

The need for simplification is widely acknowledged, but techniques for achieving it are scarce. Levin (1987) argues that "the choice of a useful model must be governed by the dynamic behavior of candidate models, by their parametric sensitivity, and by the tradeoffs between uncertainty in knowledge of model structure and uncertainty in acquiring data. .....The general approach is to begin with some particular system description, to study the behavior of the dynamics associated with that description, and to compare the behavior with that of reduced (e.g. aggregated) descriptions, with reference to some set of indicator variables." For ecological models, a few such studies have been carried out (e.g. Ludwig and Walters, 1985; Gardner et al., 1982); but we are just beginning to learn how to proceed. With the need in ecotoxicology to focus on ecosystem processes, and on the ecosystem context for population processes, in assessing the responses of ecosystems to stress, we need to develop simplified descriptions of the critical processes and components. We need to know which species are crucial to maintaining system structure and function, and which are exchangeable parts of functional groupings. Paine's perceptive Tansley Lecture to the British Ecological Society (Paine, 1980), elaborating on notions of tight linkage and the modular organization of ecosystems, introduces a valuable way to think about this problem. Ultimately, such questions cannot be resolved without performance of experimental manipulations of natural systems; but computer and analytical experiments with mathematical models can suggest and guide such experiments.

Equilibrium and Variability

A final point must be made about the interrelationship between variability and scale, and about the need to escape from the straitjacket imposed by the assumption that ecological systems are homogeneous and equilibrial. The idea that a system, once disturbed, will tend asymptotically to an equilibrium state has its roots both in classical vegetation theory (Clements, 1916), and in the mathematical theory of dynamical systems. Indeed, it is congruent with our intuition that the effects of initial conditions should become less and less important as time passes; mathematically, this is equivalent to the observation that the asymptotic dynamics of the system (1) should have lower fractal dimension than does the full space of initial states, and this principle underlies the methods of Takens (1981) and Schaffer et al. (1986) referred to earlier. Yet the classical mathematical ecological literature has carried this observation to the extreme by focusing virtually entirely on the lowest dimensional attractors — equilibrium points and limit cycles — and by ignoring more complicated autonomous behavior, non-autonomous influences, stochasticity, and delays.

Implicit in the classical mathematical approach, based on systems of the form (1), also is the notion of homogeneity, since spatial distributions are ignored entirely. A substantial recent literature exists on distributed systems (see Levin, 1976), with a considerable focus on regional coexistence through local variability. I have emphasized earlier that in oceanographic systems, patchiness can be detected on virtually every scale of examination. Similar observations hold for terrestrial, freshwater aquatic, and intertidal systems, in many of which localized disturbances continually disrupt competitive interactions and underlie the maintenance of ever-changing spatio-temporal mosaics (Watt, 1947; Levin and Paine, 1974; Whittaker and Levin, 1977). Systems may seem highly variable on small scales, and less so on larger scales, although the interplay between temporal and spatial variability complicates this expectation.

The principal conclusion from this discussion is that variability is a property of scale, and thus that observed variability is affected by the level of description the observer imposes on the system. We must proceed beyond the homogeneous descriptions implicit in (1) (or discrete-time analogues), by considering their spatially distributed relatives. These may still take the form of systems of ordinary differential equations, formally indistinguishable from (1) if the geometry is fixed and patchy; or may be partial differential equations of the form discussed in the chapter by Thomann; or may take some other form if the patch itself is taken as the unit of description. Whatever description is chosen, we must develop methods for measuing and examining spatial pattern on multiple scales, both to understand how our conclusions depend on the choice of scale and to relate to techniques ranging from remote sensing to detailed small-scale investigations that provide information on widely differing scales (Cushman 1986; Levin 1989; Milne 1988). Furthermore, only by understanding how processes change across scales will we develop the requisite ability to scale up from laboratory and microcosm studies, and provide the basis for a predictive ecotoxicology.

Acknowledgments. This publication is ERC-163a of the Ecosystems Research Center, Cornell University, and was supported by the U.S. Environmental Protection Agency Cooperative Agreement Number CR812685-01. Additional funding was provided by Cornell University.

The work and conclusions published herein represent the views of the author and do not necessarily represent the opinions, policies or recommendations of the Environmental Protection Agency.

References

Clements, F.E. (1916) Plant succession: an analysis of the development of vegetation. Carnegie Inst Wash Publ. 242. 512 pp
Cushman, J.H. (1986) On measurement, scale, and scaling. Water Resources Research *22*, 129–134
Denman, K.L., Powell T.M. (1984) Effects of physical processes on planktonic ecosystems in the coastal ocean. Oceanogr. Mar. Biol. Ann. Rev. *22*, 125–168
Gardner, R.H., Cale, W.G., O'Neill, R.V. (1982) Robust analysis of aggregation error. Ecology *63*, 1771–1779
Iwasa, Y., Andreasen, V., Levin, S.A. (1987) Aggregation in model ecosystems. I. Perfect aggregation. Ecological Modelling *37*, 287–302
Kingsland, S.E. (1985) Modeling Nature. The University of Chicago Press, Chicago, 267 pp.

Lehman, J.T. (1986) The goal of understanding in limnology. Limnol. Oceanogr. *31*, 1160–1166

Levin, S.A. (1976) Population dynamic models in heterogeneous environments. Ann. Rev. Ecol. System. 7, 287–331

Levin, S.A. (1987) Scale and predictability in ecological modeling. In: Vincent, T.L., Cohen, Y., Grantham, W.J., Kirkwood, G.P., Skowronski, J.M. (eds.) Modeling and Management of Resources Under Uncertainty. Lecture Notes in Biomathematics, vol. 72. Springer, Berlin Heidelberg New York London Paris Tokyo, pp. 2–8

Levin, S.A. (1989) Challenges in the development of a theory of community and ecosystem structure and function. In: Roughgarden, J., May, R.M., Levin, S.A. (eds.) Perspectives in Ecological Theory. Princeton University Press, Princeton NJ 394 + vii pp.

Levin, S.A., Paine, R.T. (1974) Disturbance, patch formation, and community structure. Proc. Nat. Acad. Sci. USA. *71*, 2744–2747

Limburg, K.E., Levin, S.A., Harwell, C.C. (1986) Ecology and estuarine impact assessment: lessons learned from the Hudson River (USA) and other estuarine experiences. J. Environ. Manage. *22*, 255–280

Ludwig, D., Walters, C.J. (1985) Are age-structured models appropriate for catch-effort data? Can. J. Fish. Aquat. Sci. *42*, 1066–1072

Mandelbrot, B.B. (1983) The Fractal Geometry of Nature. Freeman, W.H. and Co., San Francisco, 468 pp.

Milne, B.T. (1988) Measuring the fractal geometry of landscapes. Appl. Math. Computation. *27*, 67–79.

O'Neill, R.V., DeAngelis, D.L., Waide, J.B., Allen, T.F.H. (1986) A Hierarchical Concept of Ecosystems. Monographs in Population Biology, vol. 23. Princeton University Press, Princeton NJ

Paine, R.T. (1980) Food webs: linkage, interaction strength and community infrastructure. The Third Tansley Lecture. J. Animal. Ecol. *49*, 667–685

Penrose, R. (1955) A generalized inverse for matrices. Proc. Cambridge Philos. Soc. *51*, 496–513

Peters, R.H. (1986) The role of prediction in limnology. Limnol. Oceanogr. *31*, 1143–1159

Schaffer, W.M., Ellner, S., Kot, M. (1986) Effects of noise on some dynamical models in ecology. J. Math. Biol. *24*, 479–523

Takens, F. (1981) Detecting strange attractors in turbulence. In: Rand, D.A., Young, L.S. (eds.) Dynamical Systems and Turbulence. Warwick 1980. Lecture Notes in Mathematics, vol. 898. Springer, Berlin Heidelberg New York, pp. 366–381

Vollenweider, R.A. (1968) Scientific Fundamentals of the Eutrophication of Lakes and Flowing Waters, with Particular Reference to Nitrogen and Phosphorus as Factors in Eutrophication. Rep. Organisation for Economic Cooperation and Development, DAS/CSI/68.27, Paris. 274 pp.

Volterra, V. (1926) Variazionie fluttuazioni del numero d'individui in specie animale conviventi. Mem. R., Accad. Nazionale del Lincei (Ser. 6) *2*, 31–113

Watt, A.S. (1947) Pattern and process in the plant community. J. Ecology *35*, 1–22

Weber, L.H., El-Sayed, S.Z., Hampton, I. (1986) The variance spectra of phytoplankton, krill and water temperature in the Antarctic Ocean south of Africa. Deep-Sea Research *33*, 1327–1343

Whittaker, R.H., Levin, S.A. (1977) The role of mosaic phenomena in natural communities. Theor. Pop. Biol. *12*, 117–139

Deterministic and Statistical Models of Chemical Fate in Aquatic Systems[1]

Robert V. Thomann

Contents

This paper has several purposes: (a) to summarize the basic models of the steady state transport and fate of chemicals in aquatic systems including uptake and distribution in the aquatic food chain, (b) to illustrate the deterministic time variable behavior of chemical fate models with several applications to the Great Lakes and (c) to develop some statistical models of chemical variability in aquatic organisms, specifically, the fish.

The ability to analyze and predict the transport of potentially toxic chemicals is one of the central requirements of risk assessment and subsequent risk management. Steady state models can be of specific value in the early stages of chemical screening for generic problem contexts and to elucidate basic principles of chemical fate and uptake into the food chain. Time variable models are particularly useful for predicting recovery times of aquatic systems following some abatement program of chemical control. These steady state and time variable models essentially estimate the average or deterministically varying chemical exposure concentration to aquatic organisms. Risk assessment also requires some evaluation of the stochastic behavior of chemicals both in the water and in fish. The paper is therefore divided into four parts: 1) the basic theory and associated equations; 2) steady state simplifications; 3) deterministic time variable models and 4) analytical and numerical models of statistical behavior of chemicals in fish.

---

[1]This article also appears in Levin, S.A., et al. (eds.) (1988) Ecotoxicology: Problems and Approaches. Springer Advanced Texts in Life Sciences. Springer, New York Heidelberg Berlin London Paris Tokyo, chap. 10, pp. 245–277.

I. Theory

A. Physical-Chemical Fate and Transport Model

The principal components of the physical-chemical fate and transport model framework are reviewed in Thomann and Mueller (1987), Delos et al. (1984), Thomann and Di Toro (1983), and Di Toro et al. (1981), among others.

The development can begin by considering a simple one-dimensional river as shown in Fig. 1. The chemical in the water column is transported by the flow Q. Losses of chemical may occur as a result of microbial degradation, volatilization or other pathways. The sediment however in all of the models discussed in this paper is not considered to be moving. There is a transfer of chemical from the sediment to the water column and vice versa via settling and resuspension of particulate chemical forms and sediment diffusion of dissolved chemical.

The one-dimensional mass balance equation for any form of the chemical (dissolved or particulate) is for the water column

$$\frac{\partial c_1}{\partial t} = -\frac{1}{A}\frac{\partial}{\partial x}(Qc_1) + \frac{1}{A}\left[\frac{\partial}{\partial x}EA\frac{\partial c_1}{\partial x}\right] + \text{sources} - \text{sinks} \tag{1}$$

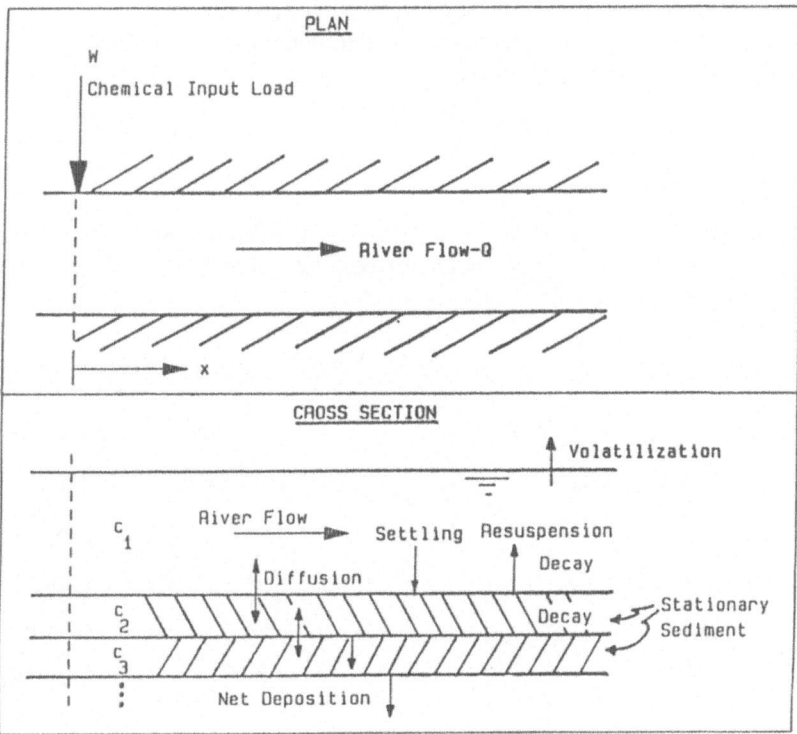

Fig. 1. Notation for physio-chemical fate model in streams

and for the surface sediment

$$\frac{\partial c_2}{\partial t} = \text{Sources} - \text{Sinks} \tag{2}$$

where c_1 and c_2 are the chemical concentrations in the water column and sediment $[M_T/L^3; M_T = \text{mass of toxicant}, L^3 = \text{bulk volume of solids plus water}]$, Q is the river flow $[L^3/T]$, A is the cross-sectional area $[L^2]$, E is the longitudinal dispersion coefficient $[L^2/T]$, x is distance down-stream and t is time.

The chemical in the models discussed herein is assumed to be composed to two forms: 1) the dissolved form, c_d' $[M_T/L_w^3; L_w^3 = \text{volume of water}]$, and 2) the particulate form, c_p $[M_T/L^3]$, i.e. the toxicant sorbed onto particulate matter in the water column or sediment. The total chemical concentration is then

$$c_T = c_p + \phi c_d' \tag{3}$$

where ϕ is the porosity $[L_w^3/L^3]$.

Eq. (3) is
$$c_T = c_p + c_d \tag{4}$$

where

$$c_d = \phi c_d' \tag{5}$$

for c_d $[M_T/L^3]$ as the porosity corrected dissolved concentration.

With the general framework described, the detailed equations for the various forms of the chemical can be presented.

Dissolved Chemical

An explicit finite differencing of Eq. 1 together with sources and sinks of the dissolved chemical in a temporarily constant control volume (V_1) of the water column is given by

$$V_1 \frac{dc_{d1}}{dt} = [(Qc_{d1}^+) - Q_1 c_{d1} + E'((c_{d1})^+ - c_{d1}) + E'((c_{d1})^- - c_{d1})] \qquad \text{(Transport)}$$

$$+ k_1 V_1 c_{p1} - k_{u1} m_1 V_1 c_{d1} \qquad \text{(Sorption-desorption)}$$

$$+ K_{f12} A(c_{d2}' - c_{d1}') \qquad \text{(Sediment diffusive exchange)}$$

$$- K_{d1} V_1 c_{d1} \qquad \text{(Decay and losses)}$$

$$- k_{l1} A(c_g/H_e - c_{d1}') \qquad \text{(Volatilization)}$$

$$+ W_{d1}. \qquad \text{(Input)}$$

$$\tag{6}$$

The group of terms in brackets represents the transport and dispersion of the dissolved toxicant. Superscript $+$ indicates the upstream direction and superscript

– indicates the downstream direction. The net transport flows, Q, are written in an equivalent backward difference approximation to the underlying partial differential equation (E1.1). The dispersion or mixing between segments of length Δx is given by the bulk dispersion coefficient which in turn is related to the dispersion coefficient by

$$E' = \frac{EA}{\Delta x}. \tag{7}$$

The second line on the left of Eq. (6) is the balance between the desorption of the chemical in the particulate phase $(k_{d1} V_1 c_{p1})$ which increases the dissolved form (the desorption rate is k_{d1} $[1/T]$, and the adsorption from the dissolved phase onto the particulates given by $k_{u1} m_1 c_{d1}$. (The sorption rate is k_{u1} $[L^3/M_s - d]$ and the solids concentration is m_1 $[M_s/L^3]$. Note that this latter term depends on the mass of solids available for sorption from the dissolved phase.

The third line of Eq. (6) represents the diffusive exchange between the sediment dissolved chemical concentration c_{d2}' in the interstitial water and the dissolved chemical concentration in the water column, c_{d1}'. The sediment-water diffusive transfer coefficient, K_{f12} $[L/T]$ can be considered as an overall interfacial transfer coefficient relating to the diffusion of the toxicant across the sediment-water interface.

Decay and loss mechanisms such as biodegradation, photolysis etc. of the dissolved form are included in the fourth line of the equation. Therefore, K_{d1} $[1/T]$ represents the sum of individual rates, some of which in turn may represent rather complex mechanisms. Note that for this model all the loss rates are assumed to be first order.

Volatilization of the dissolved toxicant is given by the fifth line of Eq. (6) where c_g represents the gas phase of the chemical $[M_T/L_g^3; L_g^3 =$ volume of gas] which may or may not be zero, and H_e is the Henry's constant for the chemical $[M_T/L_g^3 \div M_T/L_w^3]$.

The last line represents all external sources or inputs of dissolved chemical, W_{d1} $[M_T/T]$ from point direct discharge sources as well as non-point and tributary inputs.

An equation similar to Eq. (6) can be written for the dissolved chemical in the sediment layer underneath the typical water column segment 1. This layer is designated with the subscript 2. Thus

$$V_2 \frac{dc_{d2}}{dt} = k_{d2} V_2 c_{p2} - k_{u2} m_2 V_2 c_{d2}$$

$$+ K_{f12} A (c_{d1}' - c_{d2}')$$

$$- K_{d2} V_2 c_{d2}$$

$$- v_{d2} A c_{d2}$$

$$+ K_{f23} A (c_{d3}' - c_{d2}'). \tag{8}$$

The first three lines of the right side of Eq. (8) have already been discussed relative to the water column. The fourth line of Eq. (8) expresses the "burial" or transfer down into the sediment of the dissolved toxicant due to net sedimentation or build-up of the sediment layer at a net sedimentation rate of v_{d2} $[LT]$. The last line of Eq. (8) is the diffusive exchange of dissolved toxicant between the first and second sediment layers under the water column. Similar equations can be written for each successive sediment layer. Note that there are no dissolved transport terms for the sediment thereby indicating that the sediment is assumed to be stationary in the horizontal direction. Also, mechanical mixing of sediment layers (due, for example, to bioturbation) is not included, but is readily added with an additional mixing term.

Particulate Chemical

The mass balance equation for the chemical sorbed onto the particulates in the water column segment 1 is given by

$$V_1 \frac{dc_{p1}}{dt} = [(Qc_{p1})^+ - Q_1 c_{p1} + E'((c_{p1})^+ - c_{p1}) + E'((c_{p1})^- - c_{p1}) \text{ (Transport)}$$

$$- k_{d1} V_1 c_{p1} + k_{u1} m_1 V_1 c_{d1} \qquad\qquad \text{(Desorption-sorption)}$$

$$- v_s A c_{p1} \qquad\qquad\qquad\qquad \text{(Particulate settling)}$$

$$+ v_u A c_{p2} \qquad\qquad\qquad\qquad \text{(Particulate resuspension)}$$

$$- K_{p1} V_1 c_{p1} \qquad\qquad\qquad\qquad\qquad \text{(Decay)}$$

$$+ W_{p1}. \qquad\qquad\qquad\qquad\qquad\qquad \text{(Input)} \quad (9)$$

The first line of this equation is the transport of the particulate chemical due to net advection (Q) and dispersion (E'). The particulate chemical is assumed to be transported in the same manner as the dissolved form. The second line is the sorption-desorption mechanism discussed above and as can be noted for the particulate form, sorption is a source and desorption is a sink of toxicant. The third and fourth lines are respectively the particulate settling of the chemical from the water column and the resuspension of particulate chemical from the sediment into the water column. The settling velocity, $v_s [L/T]$ and the resuspension velocity $v_u [L/T]$ are functions of particle type (sand, silt, organics) and the hydrodynamics of the water-sediment interface. The fifth line represents any decay mechanisms (e.g. bacterial degradation) of the chemical on/in the particulates at a rate $K_{p1} [1/T]$ and the last line is the external mass input of particulate toxicant, $w_{p1} [M_T/T]$.

The particulate chemical in the sediment is given by an equation similar to Eq. (9) except that, as noted, the sediment is assumed to be stationary in the horizontal direction. That is, bed load transport or sediment movement horizontally throughout the water body is not considered.

The particulate chemical equation for the sediment segment underlying the water column segment 1 is then given by

$$V_2 \frac{dc_{p2}}{dt} = - k_{d2} V_2 c_{p2} + k_{u2} m_2 V_2 c_{d2}$$

$$+ v_s A c_{p1} - v_u A c_{p2}$$

$$- K_{p2} V_2 c_{p2}$$

$$- v_d A c_{p2}. \qquad\qquad\qquad\qquad\qquad\qquad (10)$$

The first three lines of this equation parallel the equivalent mechanisms in the water column (sorption-desorption, settling-resuspension and decay, at rate K_{p2}). The fourth line represents the net down-ward flux of sediment particulate toxicant due to the net sedimentation velocity v_d. Again, mixing of the sediment due to factors such as bioturbation or deep sediment mixing is not included, but can be added as an additional mixing term.

B. Local Equilibrium Equations

Eqs. (6) and (8) for the dissolved component and Eqs. (9) and (10) for the particulate component in the water column segments and sediment segments respectively represent a set of interactive, differential equations, one for each control volume of the finite difference grid. Note that the coupling of the dissolved and particulate components is through the reaction kinetics of sorption and desorption. For some chemicals, these reaction kinetics tend to be "fast" (i.e. completion times on the order of hours) compared to the kinetics inherent in other mechanisms of the problem. These latter mechanisms include bacterial decay, net loss rates to the sediment and sedimentation rates that have reaction times on the order of days to years.

The "fast" kinetics of sorption-desorption indicate that for time scales of days to years, there will be a virtually continuous equilibration of the dissolved and particulate forms depending on the local solids concentration. This partitioning between the two components permits the specification of the fraction of dissolved and particulate chemical to the total. The dissolved and particulate chemical are therefore assumed to be always in a "local equilibrium" with each other. Assuming that the kinetics are reversible and that the sorption/desorption kinetics are linear, then a partition coefficient $\P[M_T/M_s \div M_T/L_w]$ can be defined as follows:

$$\P = r/c_d' \tag{11}$$

or since $c_d' = c_d/\phi$

$$\P' = \P/\phi = r/c_d \tag{12}$$

for $\P'$ as $[M_T/M_s \div M_T/L^3]$ and r as the chemical concentration on a solids basis $[M_T/M_s]$.

The particulate toxicant concentration relative to the bulk volume is given by

$$c_p = rm. \tag{13}$$

The fraction of the total that is dissolved, f_d, is given by

$$f_d = (1 + \P'm)^{-1} \tag{14}$$

and the particulate chemical as a fraction of total chemical (f_p) is given by

$$f_p = \frac{\P'm}{1 + \P'm}. \tag{15}$$

The local equilibrium assumption therefore permits specification at all times and places of the fraction of the total toxicant in the dissolved and particulate form. It should be stressed again here that this local equilibrium assumption assumes complete reversibility between the solid and liquid phases. There is evidence (e.g. Di Toro, et al. 1982a, and Di Toro, 1985) that this is not the case for certain chemicals.

Also in these relationships it is assumed that the partition coefficient does not depend on the concentration of the sorbing solids. There is considerable evidence, however, as given by O'Connor and Connolly (1980) and Di Toro (1985) who indicate that the partition coefficient does apparently depend on the concentration of solids. The development continues here on the assumption of a constant partition coefficient.

With this assumption, attention can then be focused solely on the mass balance equation for the total chemical. The total chemical in the water column or sediment is given by Eq. (3). Adding the water column equations for dissolved chemical (Eq. 6) and particulate chemical (Eq. 9) and using Eqs. 14 and 15 gives

$$V_1 \frac{dc_{T1}}{dt} = [Qc_{T1}^+ - Q_1 c_{T1} + E'(c_{T1}^+ - c_{T1}) + E'(c_{T1}^- - c_{T1})$$

$$+ K_f A(f_{d2} c_{T2}/\phi_2 - f_{d1} c_{T1}) - (K) V_1 c_{T1}$$

$$+ k_{l1} A[(c_g/H_e) - f_{d1} c_{T1}] - v_s A f_{p1} c_{T1} + v_u A f_{p2} c_{T2} \qquad (16)$$

where $K_1 = K_{d1} + K_{p1}$.

Note that the kinetics of sorption-desorption do not appear in this equation because it represents a mass balance of the total. The net loss rates and exchanges that are dependent on the form of the toxicant do however, remain.

A total chemical equation for the sediment segement (subscript 2) can be obtained in a similar manner. Thus adding Eqs. (8) and (10) gives:

$$V_2 \frac{dc_{T2}}{dt} = - K_f A(f_{d2} c_{T2}/\phi_2 - f_{d1} c_{T1}) - (K_2) V_2 c_{T2}$$

$$+ v_s A f_{p1} c_{T1} - v_u A f_{p2} c_{T2} - v_d A f_{p2} c_{T2}$$

$$+ K_f A(f_{d3} c_{T3}/\phi_3 - f_{d2} c_{T2}/\phi_2) \qquad (17)$$

where $K_2 = K_{d2} + K_{p2}$.

Eqs. (16) and (17) are the fundamental equations used in the succeeding analyses. These equations are coupled parametically to the suspended solids and sediment solids concentrations (see Eqs. 14 and 15). These concentrations can be specified externally as an input or the mechanisms of solids settling, resuspension and deposition can be explicitly modeled. In addition, an independent tracer can be used to calibrate these parameters (see Thomann and Di Toro, 1983, for the use of plutonium-239, 240 as a tracer).

C. Food Chain Model

The transfer of a chemical in the aquatic food chain occurs through two principal routes:

1) direct uptake from the water
2) accumulation due to consumption of contaminated prey.

The uptake of a chemical directly from water through transfer across the gills as in fish or through surface sorption and subsequent cellular incorporation as in phytoplankton is an important route for transfer of chemicals. This uptake is often measured by laboratory experiments where test organisms are placed in aquaria with known (and fixed) water concentrations of the chemical. The accumulation of the chemical over time is then measured and the resulting equilibrium concentration in the organism divided by the water concentration is termed the bioconcentration factor (BCF). A simple representation of this mechanism is given by a mass balance equation around a given organism. Thus,

$$\frac{dv'}{dt} = k_u w c'_d - K v' \tag{18}$$

where v' is the whole body burden of the chemical (M_T), k_u is the uptake sorption and/or transfer rate $(L^3/T \cdot M(w); M(w) =$ mass of organism, wet weight)), w is the weight of the organism $(M(w))$, c'_d is the dissolved water concentration (M_T/L^3), K is the desorption and excretion rate $(1/T)$ and t is time. This equation indicates that the mass input $(\mu g/d)$ of toxicant given by $k_u wc$ is offset by the depuration mass loss rate $(\mu g/d)$ given by Kv'. The whole body burden v' is given by

$$v' = vw \tag{19}$$

where v is the concentration of the chemical $(M_T/M(w))$. Substitution of (19) into (18) gives, after simplification

$$\frac{dv}{dt} = k_u c - K'v \tag{20}$$

where

$$K' = K + G \tag{20a}$$

for $G(1/T)$ as the net growth rate of the weight of the organism. At equilibrium or steady state,

$$v = \frac{k_u c}{K'} \tag{21}$$

and the BCF is given by

$$N_w = \frac{v}{c} = \frac{k_u}{K + G} . \tag{22}$$

The ratio N_w, the bioconcentrations factor, is in units $M_T/M(w) \div M_T/L^3$, e.g., $\mu g/kg \div \mu g/1 (= 1/kg)$.

For organic chemicals, the BCF is conveniently defined on a lipid normalized basis, i.e. $M_T/M(lip) \div M_T/L^3$, e.g., $\mu g/kg(lipid) \div \mu g/1$. The lipid normalization assumes that the lipid compartment of the organism is the principal receptor of the hydrophobic organic chemical.

The octanol-water partition coefficient (K_{ow}) of a chemical is a useful ordering parameter to express the tendency of the chemical to partition into the lipid pool.

At equilibrium then for organic chemical BCF, to first approximation,

$$N_w = K_{ow} \tag{23}$$

for the laboratory case of no organic growth and N_w as the lipid normalized BCF. Thomann (1987) suggests the following expression for the field BCF as a function of K_{ow}

$$N_w = K_{ow}\left[1 + \frac{10^{-6}K_{ow}}{E(K_{ow})}\right]^{-1} \tag{24}$$

where $E(K_{ow})$ is an efficiency of chemical transfer across the gills as a functin of K_{ow} and can be approximately expressed as

$$\log E = -1.5 + 0.4 \log K_{ow} \quad \text{for } \log K_{ow} = 2\text{–}3$$
$$E = 0.5 \qquad\qquad\qquad \text{for } \log K_{ow} = 3\text{–}6 \tag{25}$$
$$\log E = 1.2 - 0.25 \log K_{ow} \quad \text{for } \log K_{ow} = 6\text{–}10$$

D. Age-Dependent Model

The general age-dependent model utilizes a mass balance of chemical around a defined compartment of the aquatic ecosystem. In the most general case, a compartment is defined as a specified age class of a specified organism or in steady state simplification, a compartment is considered as an "average" age class or range of ages for a given organism. Fig. 2 schematically shows the compartments. As indicated in Figure 2a, each age class of a given trophic level is considered as a compartment and a mass balance equation can be written around each such age

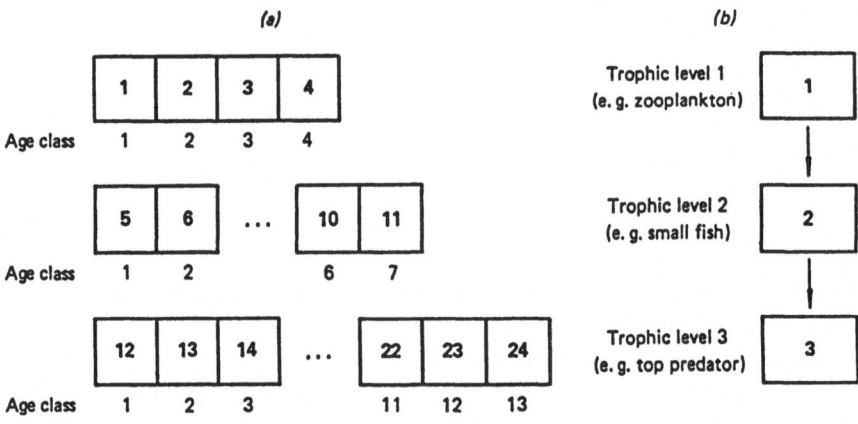

00 = Compartment number

Fig. 2a, b. Schematic of compartment definition for (a) age dependent model, (b) simplified steady state model

class. The zero trophic level is considered to be the phytoplankton-detritus component representing one of the principal sorption mechanisms for incorporating toxicants into the food chain.

Consider then the phytoplankton, detrital organic material, and other organisms, all of size approximately $< 100\,\mu m$ as the base of the food chain. An equation for this compartment is given by a simple reversible sorption-desorption linear equation as:

$$\frac{dv_0}{dt} = k_{u0}c - K_0 v_0 \tag{26}$$

where all terms have been defined, the subscript zero refers to the base of the food chain, and t is real time.

For a compartment above the phytoplankton/detritus level, the mass input of the toxicant due to ingestion of contaminated food must be included. This mass input will depend on a) toxicant concentration in the food, b) rate of consumption of food and c) the degree to which the ingested toxicant in the food is actually assimilated into the tissues of the organisms.

The general mass balance equation for the whole body burden for a given compartment, i, is then similar to Eq. (18) for water uptake but with the additional mass input due to feeding. Therefore,

$$\frac{dv_i'}{dt} = \frac{d(vw)_i}{dt} = \frac{w_i v_i}{dt} + \frac{v_i dw_i}{dt} = k_{ui}w_i c - K_i v_i'$$

$$+ \sum_j p_{ij}\alpha_{ij}C_i v_j w_i \quad i = 1\ldots m \tag{27}$$

where α_{ij} is the chemical assimilation efficiency (M_T absorbed/M_T ingested), C_i is the weight-specific consumption of organism i ($M(w)$ predator/$M(w)$ prey-d), p_{ij} is the food preference of i on j, and t is real time (days). Consider now a simple case of a sequential food chain where predation is only on the next lowest trophic level.

An equation for the individual organism weight is

$$\frac{dw_i}{dt} = (a_{i,i-1}C_i - r_i)w_i, \quad i = 1\ldots m \tag{28}$$

where $a_{i,i-1}$ is the biomass assimilation efficiency ($M(w)$ predator/$M(w)$ prey) and r_i is the respiratory weight loss ($1/T$) due to routine metabolism, swimming, and other activities. The weight change is therefore

$$G_i = \frac{dw_i}{dt}\bigg/ w_i = (a_{i,i-1}C_i - r_i) \tag{29}$$

Equation (27), for a food chain in contrast to a food web, can then be written as

$$\frac{dv_i}{dt} = k_{ui}c + \alpha_{i,i-1}C_i v_{i-1} - K_i'v_i; \quad i - \ldots m \tag{30}$$

where i is the predator and $i-1$ is the prey.

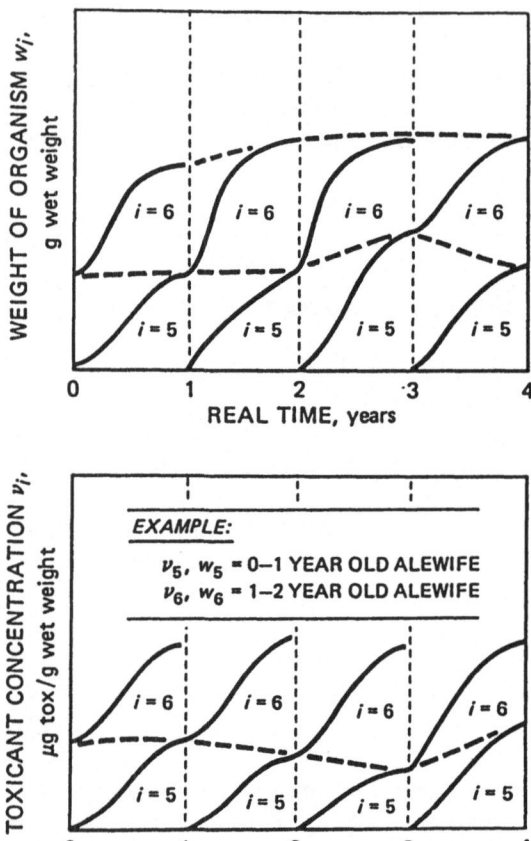

Fig. 3. Illustration of meaning of $w_i(t)$ and $v_i(t)$ showing as an example a 0–1 year old and 1–2 year old alewife

The interpretation of w_i and v_i in Eqs. (28) and (30) is further explained in Fig. 3. The variation of the weight (and chemical organism concentration) of a given compartment (i.e. a given age class of a given organism) is shown with real time. If as an example, w_5 is a 0–1 year old alewife then it is seen that the weight of this age class may vary from year to year. Similarly, the distribution of the chemical may change from year to year for a given compartment depending on, for example, the variation in the water column toxicant concentration. The specification of the boundaries of a given compartment depends on the life cycle of the organism.

It should be recognized therefore that several of the biological and chemical parameters of the weight change equation (29) and the food chain equation (30) are functions of organism weight within the time interval defined by the compartment. See Thomann and Connolly, 1984, for an application of the age dependent model to PCB accumulation in the lake trout of Lake Michigan.

II. Steady State Simplification for Rivers

In spite of the complexity of Eqs. (16) and (17), it can be shown (see (e.g. Thomann and Mueller, 1987) that under steady state and constant spatial parameters that for a single source to first approximation, the maximum concentration c_0 is at the outfall. Thus

$$c_0 = (Q_u c_u + Q_e c_e)/Q$$

where Q_u and c_u are the upstream flow and concentration, respectively, Q_e and c_e are the effluent flow and concentration respectively, and Q is the total river flow.

Since the maximum generally occurs at the outfall, there are generally two situations where one would be interested in estimating the downstream fate of a discharged chemical:

a) There is a critical water use point downstream of a single discharge and the concentration at the point of use (e.g., a water supply withdrawal) needs to be estimated,
b) there are several inputs of the same chemical along the length of the river and the total concentration must be estimated.

The downstream fate of the chemical or mixture depends on:

a) the properties of the river such as the depth, velocity and dilution downstream due to groundwater infiltration or tributary inflow,
b) the chemical properties, such as volatilization, biodegradation, or partitioning onto the solids.

Thomann and Mueller (1987) discuss these factors in some detail. A simplified summary (from Thomann and Salas, 1986) is given here. An important point for the computation of the downstream fate of a chemical or chemical mixture is that the calculation is very similar, indeed for preliminary analyses, the computation is identical to classical stream water quality calculations. The basic equation under steady state conditions is given by

$$c = c_0 \exp[-[K_T + q]t = c_0 \exp\left[-\left(\frac{v_T}{HU} + q\right)x\right] \tag{31}$$

$$= c_0 \exp[-(K + g')t^*]$$

where c is the concentration as a function of distance downstream, c_0 is the chemical concentration or toxicity in the river after mixing of the outfall, v_T is the net loss of chemical expressed as a velocity $[L/T]$, t^* is the time of travel ($= x/u$), q is the slope of the natural logarithm of river flow with distance, H is river depth, U is river velocity and q' is the river flow slope on a travel time basis.

The calculation of the downstream fate of a toxic substance or toxicity of an effluent or effluents depends then on the estimation of the dilution of the river and the loss rate of the chemical. The discharge of a conservative substance with no dilution would result in a constant concentration in the downstream direction, i.e., from Eq. (31) with K_T and $q' = 0$, $c = c_0$.

The determination of whether there is an downstream infiltration of ground-water or overland drainage can be made by examining the downstream distribution of a known conservative substance such as chlorides or total dissolved solids or other tracer.

If dilution exists then a conservative substance may exhibit apparent non-conservative behavior. The for $K_T = 0$, but $q' \neq 0$, Eq. (31) is

$$c = c_0 \exp(-qt^*) \tag{32}$$

which indicates an exponential decline in the conservative substance due to distributed downstream dilution.

Finally, if the toxic substance is non-conservative, i.e., the chemical undergoes biodegradation or volatilization or other losses and dilution is also occurring then the net loss rate of the chemical, K_T must be estimated.

Table 1 provides some guidelines for a preliminary assessment of downstream fate. Toxicity in this Table is whole-effluent toxicity of the chemical mixture to standard organisms and is expressed on a toxic unit basis. A toxic unit is defined

$$Tu = 100/LC50 \text{ or NOEL} \tag{33}$$

where LC50 is the lethal concentration to 50% of the organisms and NOEL is the no observed effect level. As indicated, both the heavy metals and the toxicity measure are assumed to be a conservative variable for first approximations. Thus, only dilution need be considered for these variables. For the organic chemicals a somewhat arbitrary division has been made based on the water solubility of the chemical.

The rationale is that at solubilities less than about 1µg/l, the chemical will partition onto the solids because of a relatively high partition coefficient (about $10^4 - 10^6$ l/kg). Also, the general tendency will be for such chemicals to biodegrade and volatilize to a lesser degree than the more soluble chemicals. For first approximations, however, such low solubility chemicals may be assumed conservative.

Table 1. Guideline for estimating downstream loss rate of chemicals and toxicity

Group	Guideline[1]
Heavy metals	Conservative ($K_T = 0$) and additive
Toxicity	Conservative ($K_T = 0$) and additive
Organic chemicals Water solubility > 1 µg/l	Conservative ($K_T = 0$) and additive
Organic chemicals Water solubility > 1 µg/l	Estimate loss rate (Eq. 34)

[1] In all cases, dilution in the downstream direction must be included
Source: Thomann and Salas (1986)

Table 2. Approximate fraction of total chemical in dissolved form and on particulates

Organic chemicals

Chemical solubility	f_d – Fraction of chemical in dissolved form[1]		Ratio of particulate conc. to total conc. $t/c_T[\mu g/g \div \mu g/l]$[2]	
	Range	Approxi. mean	Range	Approx. mean
> 100	0.5–1.0	0.7	0–50	10
10–100	0.3–0.9	0.5	0.1–70	50
< 10	0.3–0.8	0.4	0.1–70	60
Heavy metals				
	0.6–1.0	0.8	0.4–16	2.5

[1] Approximated from solids dependent partition coefficient relationships of Di Toro (1985), solids range $10 \rightarrow 1000$ mg/l

[2] $r/c_T = (1 - f_d)/(.01 \rightarrow 1.0)$

Source: Thomann and Salas, 1986

For organic chemicals with solubilities greater than about 1 µg/l, the loss rate must be estimated. For first approximations, the net loss velocity $v_T(= K_T H)$ can be estimated from

$$v_T = K_T H = (K_d H + k_l) f_d + v_n f_p \tag{34}$$

where $K_d[T^{-1}]$ is the decay rate of the chemical due to processes such as biodegradation or photolysis, k_l is the loss due to volatilization $[L/T]$, v_n is the net loss velocity of the solids in the river $[L/T]$ and f_d and f_p are the dissolved and particulate fractions of the total (see Eqs. 14 and 15).

For rivers, the net loss of solids often, although not always, can be assumed equal to zero. Thus, $v_n = 0$ and the chemical loss rate depends only on the degradation rate, volatilization rate and fraction dissolved.

Table 2 provides some guidelines for estimating the fraction of the chemical that is in the dissolved form. These guidelines employ a more complex interaction of chemical partitioning and solids concentration as given in Di Toro (1985) and discussed in Thomann and Mueller (1987). The f_d can also be approximated with Eq. 14. Figure 4 shows the range of the volatilization loss rate as a function of the river depth and reaeration characteristics. This figure is for substances with Henry's constant $> 10^{-4}$ atm-m^3/mol for which the volatilization rate is estimated from the oxygen transfer rate K_L. Thus,

$$k_l \approx \left(\frac{32}{M}\right)^{1/4} K_L \tag{35}$$

where

$$K_L = \left(D_L \frac{U}{H}\right)^{1/2}, \text{ for } D = \text{oxygen diffusivity } (0.000181 \text{ m}^2/\text{d}). \tag{36}$$

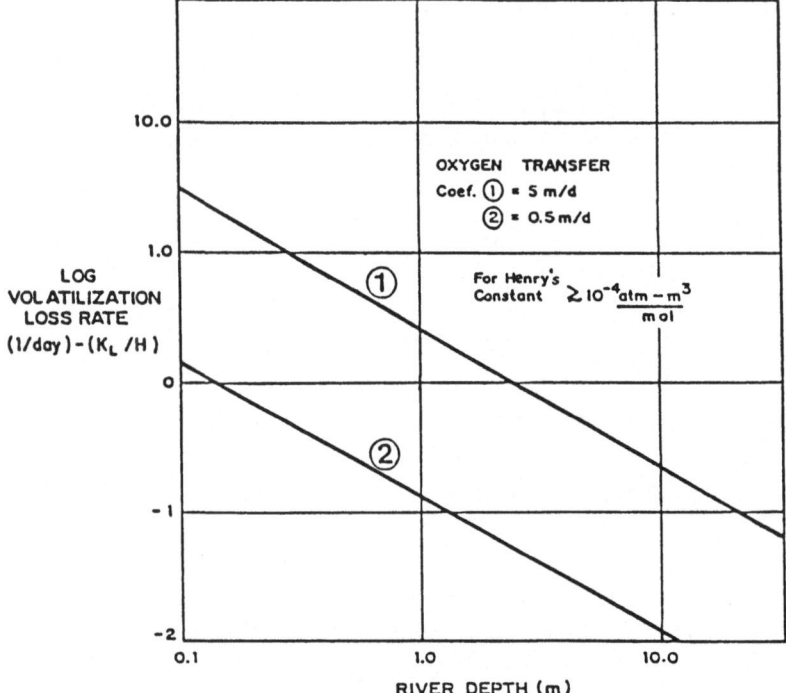

Fig. 4. Range of volatilization loss rate as a function of river depth for different oxygen transfer rates

The loss rate, K_d, is generally site-specific and chemical specific and no general simplification is available.

Table 2 also gives the approximate ratio of the chemical adsorbed to the suspended particulates to the total water concentration. Therefore,

$$\pi = r/c_T \tag{37}$$

for π in µg/g(d) ÷ µg/l, r in µg/g(d) and c_T in µg/l (g(d) = grams dry weight).

It can also be shown that for $\P_2 = \P_1$ and $K_{d2} = 0$, the sediment particulate concentration is equal to the water column particulate concentration, i.e.,

$$r_2 = r_1 \tag{38}$$

where r_2 is the sediment particulate concentration and r_1 is the water column particulate concentration both in µg/g(d).

In summary, it is seen that the calculation of the fate of the chemical or mixture is similar to the procedure for conventional water quality variables. For many chemicals including the toxicity measure, the assumption that the chemical is conservative is appropriate for first approximations. If such an assumption cannot be made, then an estimate of the downstream loss rate must be made using the preliminary guidelines discussed here.

III. Time Variable Models of Benzo(a)pyrene and Cadmium in the Great Lakes

In this section, the fully time-variable model (Eqs. (16) and (17)) is applied to two chemicals: a) benzo(a)pyrene, a polycyclic aromatic hydrocarbon (PAH), and b) cadmium, a representative metal. The model uses the Great Lakes segmentation of Thomann and Di Toro (1983) shown in Fig. 5 and details are in Thomann and Di Toro (1984).

Benz(a)pyrene

The distribution of this chemical, one of the PAH compounds resulting from incomplete combustion of organic materials has been widely studied (e.g., Neff, 1979) because of its potential carcinogenicity. The fate of benzo(a)pyrene (BaP) in the Great Lakes has recently been evaluated in a series of papers by B.J. Eadie (Eadie et al., 1982; Eadie, 1983; Eadie et al., 1983). In that work, data are presented for the range of concentration of BaP in the water column and surficial sediments as well as preliminary data on the BaP concentration in the pore water of the sediment. It is those data (together with estimates of loading) that can be used as an application of the physico-chemical model.

BaP is sparingly soluble in water $0.172 \, \mu g/L$ (Neff, 1979) and as such would be expected to have an affinity for solids. The partitioning onto particulates and in addition, the extent of volatilization of the BaP must be estimated. Although BaP is

Fig. 5. Great Lakes and Saginaw Bay and sediment segmentation used in model

Table 3. Comparison of calculated and observed BaP for great lakes under different solids partition assumptions

	Calculated range of BaP across all lakes[a]		Observed mean of BaP	Ref.
	¶ = 10,000 l/kg	¶ = 100,000 l/kg		
Total water (Conc. (ng/l))	5–6	1–2	12 ± 8[b]	Eadie et al. (1983)
Surficial sediment Conc. (ng/g(d))	38–60	46–133	Mich: 480 ± 246(7)[c] Erie: 255 ± 152(3) Sup. : 28(1) Hur. : 294(1) Ont. : 306(1)	Eadie (1983)
Sediment pore Water conc. (ng/l)	3–5	0.5–1.3	850 ± 1260	Eadie et al. (1983)
Particulate Conc. in Water Column (ng/g(d))	46–64	46–165	Mich: 200–400	Eadie (1983)

[a] After 20 years of loading
[b] Mean ± Std. Dve.
[c] () = No. of samples

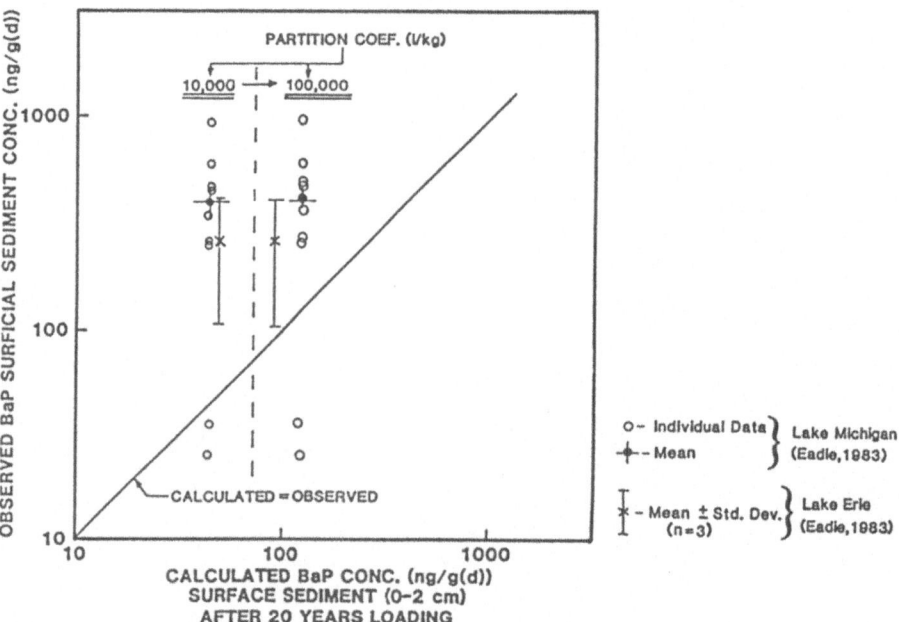

Fig. 6. Comparison between calculated surface sediment BaP concentration after 20 years and observed concentration

known to undergo photolysis (Neff, 1979) this pathway is not considered in this application.

The estimated atmospheric loading on an areal basis is about 95 g BaP/km²-yr across all of the lakes but as noted by Eadie (1983), all BaP load estimates are based on quite limited data and therefore may vary as additional information becomes available. From the data given in Eadie (1983) for Lake Michigan, and other data, the model for BaP was run for ¶ from 10,000 to 100,000 l/kg, thereby providing a range of one order of magnitude in the solids partition coefficient.

The time variable calculation using a steady loading for 20 years is shown in Table 3 and in Figs. 6 and 7. As shown in Table 3, the lake to lake variation in BaP concentrations either in the water column or the sediment differs by less than about a factor of two. The highest concentration of BaP in the surface sediments for ¶ = 10,000 is in Saginaw Bay. It can also be noted in Table 3 and Fig. 6 that at ¶ = 10,000 the calculated surface sediment concentration for Lakes Michigan and Erie is about 45 ng/g(d) or about one order of magnitude lower than the observed data. Figure 7 shows the calculated time history under the two partition coeffi-

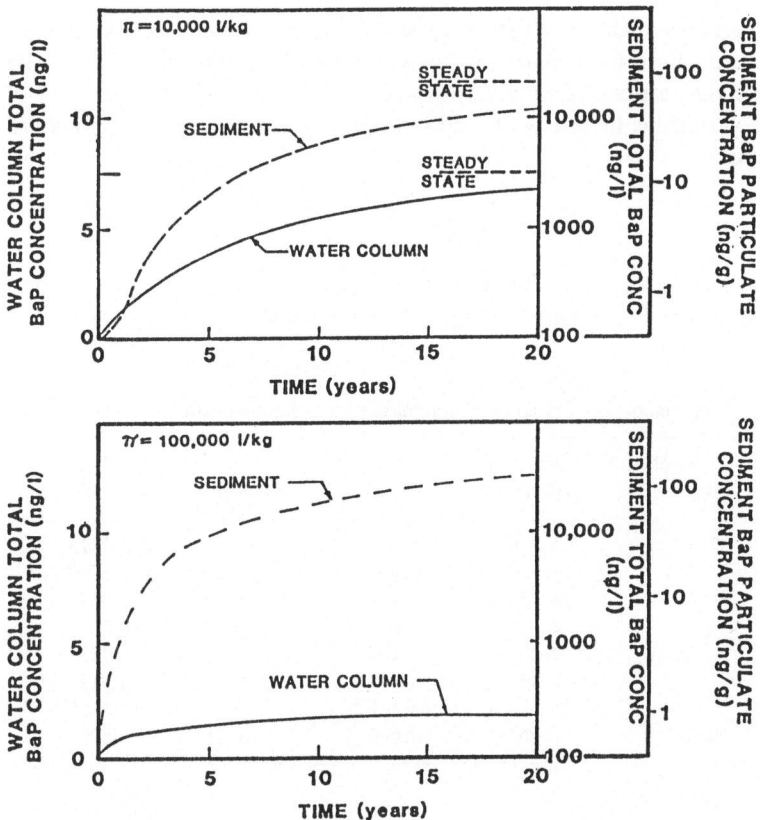

Fig. 7. Time variable BaP response in Lake Michigan under two partition coefficients with 20 year constant loading

cients. For $\P = 10,000\,l/kg$, the water column and sediment are at about 80 and 60% of steady state respectively while for $\pi = 100,000\,l/kg$, the water column and sediment are at about 20% of steady state.

Figure 6 and Table 3 indicate that a more favorable (but not totally desirable) comparison to observed data is obtained at the higher BaP partition coefficient of 100,000 l/kg. The results also indicate the need to determine the partition coefficient for BaP, as a representative PAH for Great Lakes solids concentrations. On the basis of this application of the physico-chemical model to BaP in the Great Lakes, it is concluded that:

1) the estimate of the BaP partition coefficient obtained from published empirical relationships is probably low by about an order of magnitude for the Great Lakes system.
2) with an increased BaP partition coefficient and assuming loss due to volatilization, the physico-chemical toxic substances model of the Great Lakes approximate observed BaP water column and sediment data only to order of magnitude,
3) the model confirms that on a lake-wide scale, the principal external source of BaP is the atmosphere,
4) for the larger lakes such as Michigan, the 50% response time of the lake to external loads is about 6–10 years for the water column-sediment system while for Lake Erie the response time is about two years.
5) lake to lake variations in BaP water column and sediment concentrations are less than a factor of two.

Cadmium

Time variable model calculations for cadmium were made using the low and high load estimate of Table 4 and two assumptions on the cadmium partition coeffi-

Table 4. Summary of contemporary external cadmium loads (not including upstream loads)

Lake/Region	Atmospheric[1] mt/yr	Tributary[3] mt/yr	Mun. + Ind.[4] (mt/yr)	Total mt/yr	Total g/km²-yr
Superior	41–108	13–126	0–0.3	54–234	651–2823
Michigan	12–120	9–92	0.4–4	21–216	363–3739
Huron	23–57	14–136	0.1–1	37–194	644–3378
Sag. Bay[2]	3	2	0	5	1184
Erie-West	1–18	2–20	1–8	4–46	1321–15202
-Central	3–94	2–16	0.2–2	5–112	318–7126
-East	1–38	1–12	0.1–0.5	2–51	320–8155
Ontario	10–44	7–66	0.6–6	18–116	924–5953

[1] From Allena and Halley (1980)
[2] From Dolan and Bierman (1982)
[3] Assuming tributary flow at total cadmium conc. of 0.2–1.0 µg/l
[4] At 0.5–5.0 µg/l for direct municipal point sources

Table 5. Estimate of approximate background cadmium concentration for great lakes

Late/Region	Range in total Background water Col. conc. (ng/l)[1]	Background conc. used in load estimate (ng/l)	Background total load (mt/yr)[2]	External background load (mt/yr)[3]	g/km²-yr	Ratio: Contemp. load Background load
Superior	2.8–5.6	4.0	6.1	6.1	37	18–76
Michigan	2.8–5.6	4.0	3.1	3.1	54	7–69
Huron	2.8–5.6	4.0	5.2	4.8	83	8–41
Sag. Bay	6.5–13.0	10.0	0.6	0.4	105	11
Erie-West	12.5–25.0	20.0	10.3	8.6	283	5–54
Central	5.0–10.0	8.0	22.0	10.0	635	0.5–11
East	5.0–10.0	8.0	12.0	4.7	754	0.4–11
Ontario	2.8–5.6	4.0	4.0	2.6	132	7–45

[1] At assumed background sediment conc. of $0.5-1.0\ \mu gCd/g(d)$; $r_2 = r_1$ and $\P = 200,000\ l/kg$

[2] Includes any upstream or exchange loads

[3] Not including any upstream or exchange loads

cient: a) variable partition with solids concentration as given in HydroQual (1982)
by

$$\P = (3.52 \cdot 10^6) m^{-0.92} \tag{39}$$

and b) a constant partition coefficient of $2 \cdot 10^5$ l/kg. Cadmium was assumed to be
conservative. For all calculations, zero initial conditions were assumed and the
loads were inputted as constant over time. It became apparent from initial runs
that the time to steady state especially for the upper lakes is long so that the
computation was carried out for a 100 year period. The computation therefore
represents the response of the Great Lakes system to a constant loading and such
loading can be viewed as a loading in addition to the background loading and
cadmium concentrations shown in Table 5. If increased loading of cadmium is
assumed to have begun in approximately the 1920's then the output from the model
calculation at $t = 50$ years would be representative of the 1970's, the period during
which some reliable data are available.

Figure 8 shows the comparison of calculated and observed surface sediment
cadmium data for $t = 50$ years. As shown, the calculation is reasonable to order
magnitude. Figure 9 (high load estimate) show the full 100 year calculation for
Lake Michigan and central Lake Erie as illustrations. The sensitivity of the
calculation to the assumptions in the partition coefficient is shown. As indicated,
the effect of the solids dependent partition coefficient is to greatly increase the time
to steady state as a result of the diffusive flux of cadmium from the sediment due to

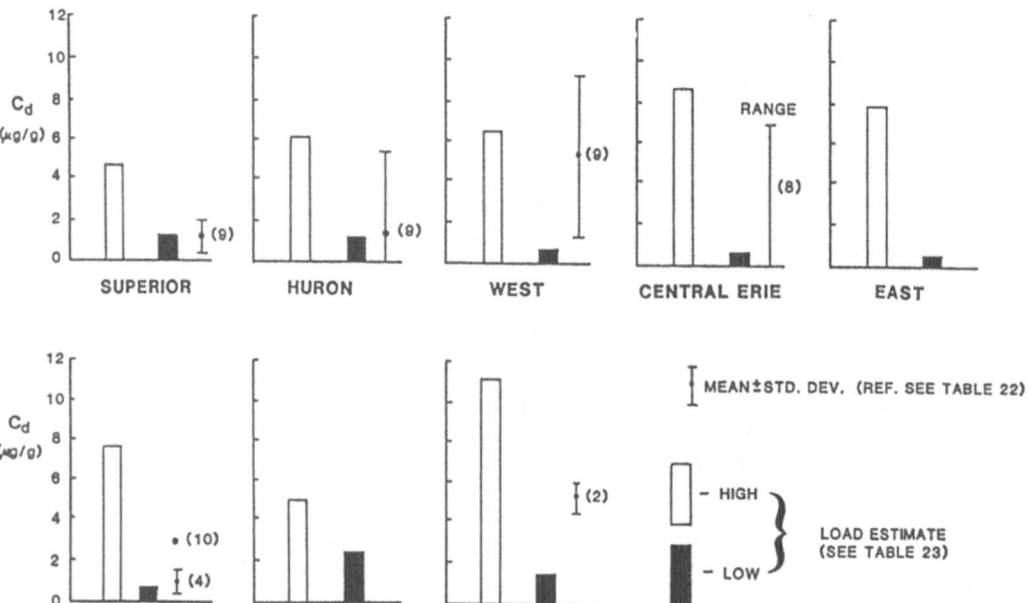

Fig. 8. Calculated surface sediment concentration (μg/g(d)) at $t = 50$ years with partition coefficient
as a function of solids concentration

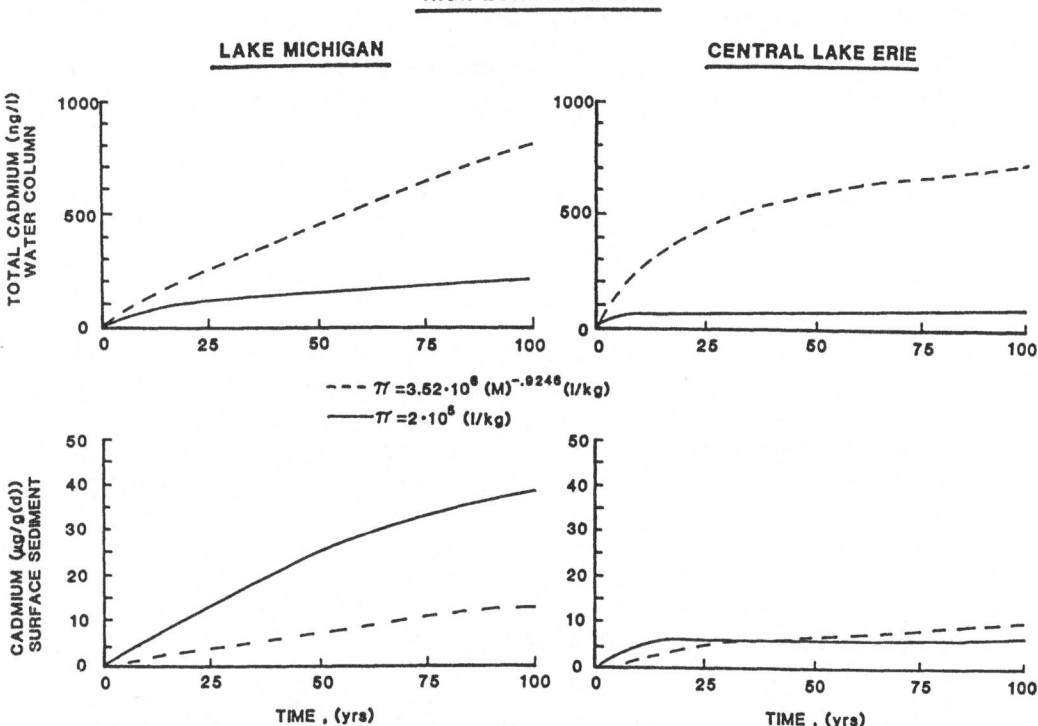

Fig. 9. Comparison of calculated cadmium concentration under two assumptions on partition coefficient. High load estimate

the lower partition coefficient. For Lake Michigan, the surface sediment concentration decreases with the variable partition coefficient but for Lake Erie the surface sediment concentration increases. The continual increase in concentration for central Lake Erie reflects the non-equilibrium condition of the upstream Lakes.

If a constant partition coefficient is assumed (as in Muhlbaier and Tisue, 1981), then it is seen that for the water column a steady state is reached in about 25 years for Lake Michigan and about 10 years for Lake Erie. A calculation then for Lake Michigan that attempts to calibrate to a mean concentration of 27 ng/l (Muhlbaier and Tisue (1981)) then is simply a matter of estimating the approximate average load and may not reflect a non-steady state condition as concluded by Muhlbaier and Tisue (1981). However under a solids dependent partition coefficient for cadmium, the Great Lakes are not in equilibrium with the external load and for all practical purposes never reach a steady state condition. Clearly then, under this model construct, it is important to determine the solids dependence of cadmium for the range of solids encountered in the Great Lakes water column and sediments (i.e. 0.5–240,000 mg/l). If however it is assumed that a solids dependent partition coefficient as given by Eq. (?) is applicable to the Great Lakes, then the system is not in equilibrium with respect to the external load.

It is concluded from this application of the time variable physico-chemical model to cadmium in the Great Lakes that

1) The degree of any dependence of the cadmium partition coefficient with solids has a market effect on time to steady state and interstitial cadmium concentration.
2) Under a solids dependent cadmium partition assumption, the Great Lakes, especially the upper Lakes, do not reach a steady state condition after 100 years of constant loading.
3) Under a constant partition coefficient for cadmium, the Lakes do reach an equilibrium condition varying from about 25 years for Lake Michigan to years for Lake Erie.
4) The concentration of cadmium in the Lakes would be expected to increase by about 60% over the next 50 years if the average cadmium loading for the preceding 50 years continues.
5) Based on assumed sediment cadmium concentrations for Lake Erie, it is estimated that the cadmium concentration in the water column is about an order of magnitude higher than the other Lakes.

IV. Statistical Variation in Fish

The statistical behavior of a chemical in fish is of interest for at least three reasons:

a) U.S. Food and Drug Administration action limits for fish have resulted in the closing of commercial fisheries because of excessive concentrations of chemicals in predators such as the striped bass in the Hudson estuary and surrounding waters. The ability to predict not only the mean value of chemical in a fish but also the variance of the concentration is of importance in control strategies needed to reopen a fishery.
b) The relationships between variable exposure concentrations in the water column and resulting variability in fish and other organisms is also related to the subsequent acute and chronic toxicity effects on the organism, and, as such, a framework for predicting organism chemical variance is of value in elucidating toxic effects on aquatic animals.
c) The integration of the physico-chemical modeling framework (which includes external inputs of the chemical to the water column) and the biological modeling framework in a time variable sense to predict statistical properties in the water and fish is of particular value in a generalized risk assessment determination.

The variability of chemicals in water over time is large. Figure 10 from the Mississippi River is an example. Recognizing this variability in the water chemical concentration of the food chain model of Sect. I-C, analytical models of expected response in the fish can be developed.

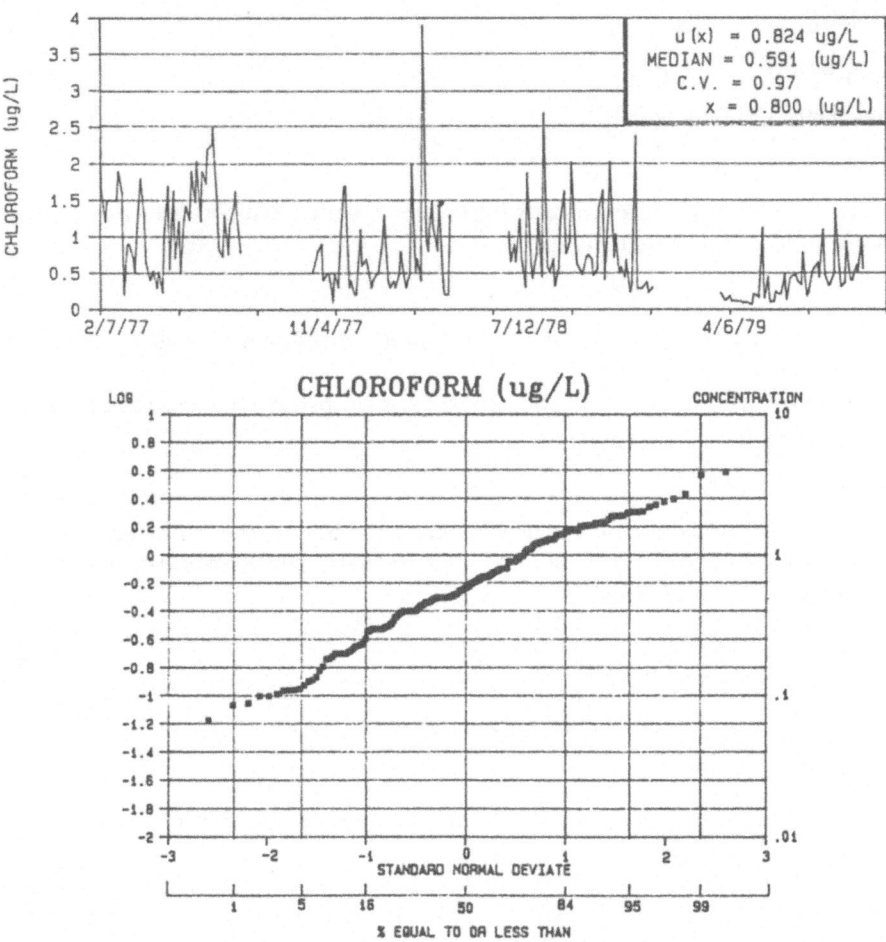

Fig. 10. Organic chemical data from Mississippi River, Jefferson Parish station, Chloroform (*top*) time series, (*bottom*) cumulative frequency distribution. (Compiled from data of USEPA, 1986)

Analytical Model of Statistical Variation of Organic Chemical in Fish—Water Uptake Only

Consider the case where the water concentration to which the fish is exposed is represented by a first-order autoregressive process. Thus, the autocorrelation function for the water concentration is assumed as

$$\rho_{ck} = \exp(-ak); k = 0, 1, 2 \ldots m \tag{40}$$

where ρ_{ck} is the normalized autocovariance, k is the lag number, m is the maximum number of lags and a is the exponential decay rate of the autocorrelation.

Now

$$\rho_{ck} = \rho_{c1}^k \tag{41}$$

and then from (39) and (40),

$$\rho_{c1} = \exp(-a). \tag{42}$$

For the synthetic generation of a time series x_t with a first-order autoregressive input, Bras and Iturbe (1985) give

$$(x_t - \mu) = \rho_1(x_{t-1} - \mu) + \sigma_x(1 - \rho_1^2)^{1/2} Z(t) \tag{43}$$

for mean μ, variance σ_x^2 and $Z(t)$ a standard normal deviate (mean $= 0$, variance $= 1$).

Thus if a $\gg 0$, i.e. the autocorrelation function drops repidly to zero at about lag one, $\rho_1 \to 0$ and Eq. (42) gives

$$x_t - \mu = \sigma_x Z(t)$$

or a normally distributed uncorrelated random variable, the "white noise" case.

For the autocorrelation given by Eq. (39), Bendat and Piersol (1971) show that the spectrum for frequency f is

$$G_c^N(f) = \frac{4a}{a^2 + 4\pi^2 f^2} \tag{44}$$

where $G_c^N(f)$ is a normalized spectrum. Therefore,

$$G_c(f) = \sigma^2 G_c^N(f). \tag{45}$$

The variance of the fish concentration can then be shown to be (Thomann, 1987)

$$\sigma_v^2 = \sigma_c^2 k_u^2 4a \int_0^\infty (K'^2 + 4\pi^2 f^2)^{-1/2}(a^2 + 4\pi^2 f^2)^{-1/2} df. \tag{46}$$

Completing the integration, the ratio of the coefficients of variation between the fish and the water is

$$r = \sqrt{\frac{K'}{a + K'}} \tag{47}$$

where

$$r = \frac{vv}{v_c}$$

for vv and v_c as the coefficients of variation of the chemical in the fish and in the water, respectively. This remarkably simple result indicates the significance of the depuration rate plus growth rate, K', as the principal controlling factors in generating relative variability in the fish concentration.

In Eq. (46), r depends primarily on K' because at low excretion rate, pulses in water concentration tend to be retained by the fish over a longer period of time than at high excretion rates. Similarly, as a fish increases in weight, concentration variability will tend to shift into the low frequency end of the variance spectrum.

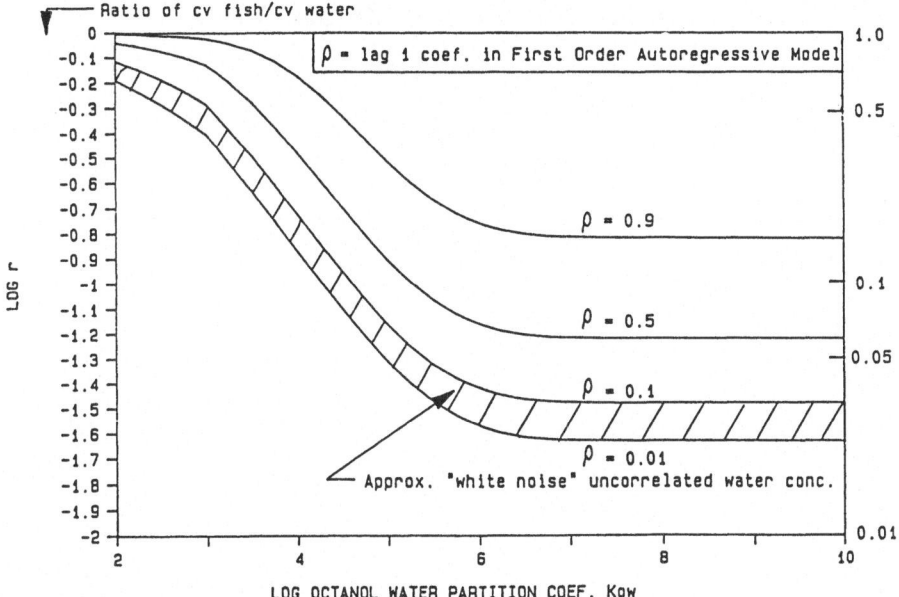

Fig. 11. The effect of the first order autocorrelation parameter of the water concentration on the relationship between r and K_{ow}. Growth rate variable, wt. = 1000 g

Figure 11 shows the behavior of r from Eq. (46) and indicates that correlated water concentrations increases the variability of the fish concentration relative to water. This is a consequence of correlated inputs introducing more "low frequency" variations in water concentration which are not dampened by the fish. Thus additional variance propagates through the fish and is reflected in the increase in the coefficient of variation.

Laboratory data for evaluating this analytical development come from two papers of Oliver and Niimi (1983, 1985). In this work, rainbow trout were exposed to chlorinated and brominated organic chemicals for up to about 100 days.

In Oliver and Niimi (1983), the data reported include the individual fish concentrations and associated water concentrations and the accompanying statistics of mean and standard deviation. For Oliver and Niimi (1985) only the BCF are reported. To use the data in this research, the fish concentrations were calculated from the individual BCF values and the mean water concentration. These data are therefore only an approximation to the actual fish concentration.

As noted frequently by the authors, several of the chemicals did not reach equilibrium during the test. The preceding statistical development assumes a dynamic equilibrium has been reached. The computation of the variance would be severely biased by the increase in concentration to a dynamic equilibrium. With a few exceptions, those chemicals excluded by the authors have also been excluded herein.

Figure 12 (top) shows the results of compiling these data and calculating the r ratio: v_v/v_c. As a function of $\log K_{ow}$, it is seen that there is a general downward

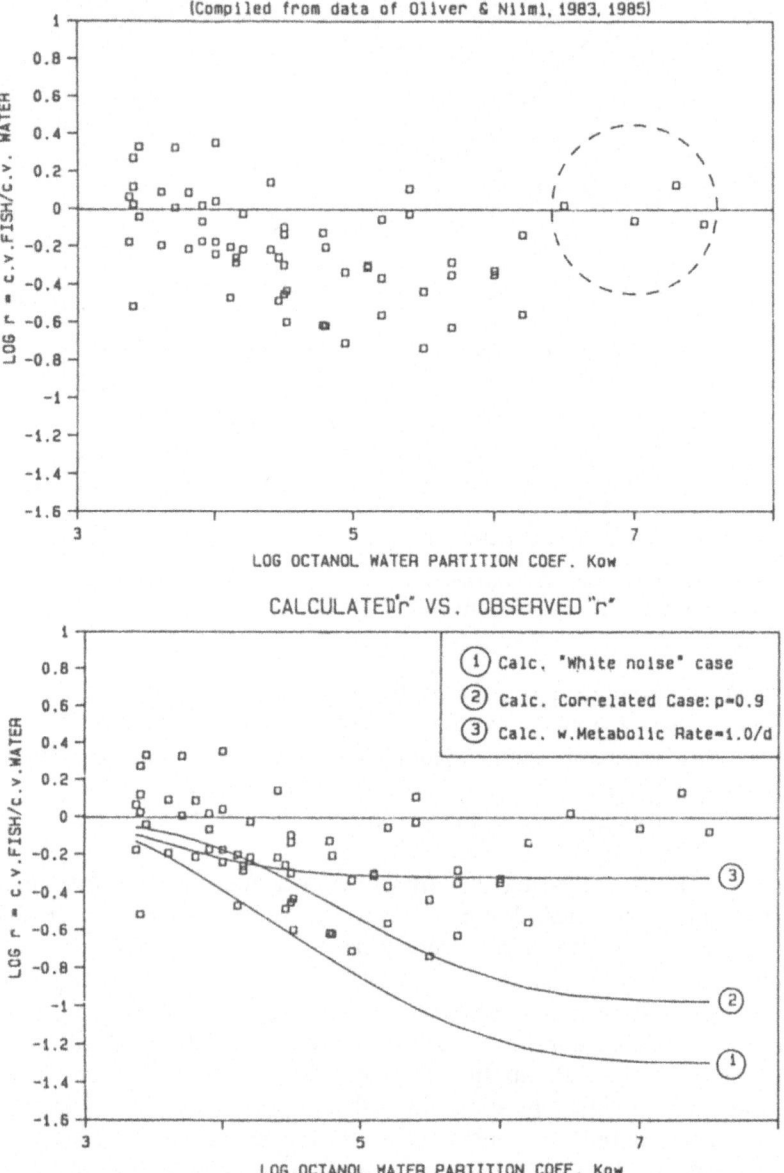

Fig. 12. (*Top*) Compiled laboratory *r* values from experiments of Oliver and Niimi (1983, 1984), (*bottom*) comparison of calculated and observed *r* values under assumptions on water correlation structure and metabolic rate

trend to the data from initial values of > 1 at low $\log K_{ow}$. The four circled points at the high $\log K_{ow}$ value of > 6 are exceptions. The chemicals are pentabromotoluene, pentabromoethylbenzene, hexabromobenzene and octachloronaphthalene. These points were also noted by Oliver and Niimi (1985) in their evaluation of the BCF data. No ready explanation is available for why these chemicals exhibit less relative variability than expected, although there are no data in this $\log K_{ow}$ range where r values are low. Indeed, the data indicate a trend towards decreasing r values to a minimum at about $\log K_{ow}$ of about 5.5 and then an increase to higher r values with increasing $\log K_{ow}$. This is not consistent with the previously developed theory as shown in Fig. 12 (bottom).

This figure shows the calculation of the theoretical r ratio under three different assumptions. For these calculations the growth rate of the rainbow trout was calculated from the reported data at an average level of about $0.01/d$. Also, the excretion rate is given Eqs. (22) and (23) and the uptake rate as a function of the efficiency of transfer E. Thus,

$$K = \frac{10^3 w^{-0.2} E/p}{K_{ow}} \tag{48}$$

where p is the fraction lipid and E is given by Eqs. (25).

The first calculation assumed the water concentration was statistically "white noise", i.e. uncorrelated in time. As noted, this calculation underestimates r by a significant amount.

The second calculation using Eq. (46) assumed a correlated first order process $(a = 0.9)$ for the water concentration. Now, the lower bound of the r ratios is captured although at $\log K_{ow}$'s of > 6 the model deviates from the data.

Case 3 is a calculation assuming some metabolism at a high rate of $1.0/d$. This is essentially an additive factor to the excretion rate and as seen, the calculation now approximates the data somewhat better. As noted however, this calculation is simply an hypothesized mechanism for increasing the loss rate of the chemical. In the theoretical development, the only parameter that influences the shape of the r function is the excretion ($+$ metabolism) rate. To capture the four circled points with the theory would require an increase in the effective excretion $+$ metabolism rate to levels of > 1.0 day.

In summary, the laboratory data of Oliver and Niimi show a general downward trend of the r ratio with $\log K_{ow}$, but with several notable exceptions at high $\log K_{ow}$. The theory approximately duplicates the observed data only with a high level of first order correlated water input at to about $\log K_{ow}$ of about 5.5–6.0. At higher $\log K_{ow}$'s, increased metabolism and/or excretion would be necessary to reproduce the observed data.

V. Conclusions

The physio-chemical and food chain model structures discussed herein provide a basis for understanding and predicting the fate and transport of chemicals in surface water systems. While the fully time variable equations appear formidable, steady state simplifications permit rapid assessment of chemical fate, both in the water column and in the food chain. Such initial screening models have the same structure as "classical" water quality models; notably a linear response to external inputs of chemicals.

The interaction of the sediment and water column plays a significant role and again under some reasonable assumptions, rapid estimates can be made of chemical concentrations in the sediment.

In a similar setting, the age dependent food chain models can be simplified at steady state and rapid screening assessments of chemical bioconcentration can be made. For organic chemicals, the octanol water partition coefficient is a useful ordering parameter. Lone term time variabe deterministic calculations for the Great Lakes indicate the significance of the sediment as a reservoir of the chemical. Response times to changes in external load are long (e.g. years) and are significantly increased when sediment partition coefficients are assumed to be different from water column partition coefficients.

Stochastic variability in chemical concentration is high. Analytical models of the resulting variability of chemical concentration in fish indicates the importance of the excretion rate of the chemical. However, additional research is necessary to more fully describe the observed variance in laboratory experiments especially for chemicals with log octanol water partition coefficients greater than about 6.5.

Acknowledgements. The research directed towards the long term calculations in the Great Lakes was partially supported through a Cooperative Agreement between the USEPA (Large Lakes Research Station) and Manhattan College. Grateful acknowledgement is made to Mr. William Richardson of the USEPA for his support and enthusiasm in the work. The research directed towards the stochastic variability analyses was supported by a grant from the National Science Foundation to Manhattan College. Grateful thanks are offered to Mr. Ed Bryant of NSF for his encouragement of that research.

Colleagues at the Manhattan College Environmental Engineering and Science Graduate Program deserve special mention for their constant willingness to share insights and thoughts on the research. Special and grateful thanks are therefore given to Donald J. O'Connor, Dominic M. Di Toro and John P. Connolly. Finally, the typing of the manuscript by Eileen Lutomski is gratefully appreciated.

References

Allen, H.E., Halley, M.A. (1980) Assessment of airborne inorganic contaminants in the Great Lakes. Appendix B of 1980 Annual Report Appendix-Background Reports. Grt. Lks. Sci. Adv. Bd., IJC, Windsor, Ont., 160 pp

Bendat, J.S., Piersol, A.G. (1971) Random data: analysis and measurement procedures. Wiley-Interscience, New York, 407 pp

Bras, R.L., Rodriguez-Iturbe, I. (1985) Random functions and hydrology. Addison-Wesley Pub. Co., Reading, MA, 559 pp

Delos, C.G., Richardson, W.L., DePinto, J.V., Ambrose, R.B., Rodgers, P.W., Rygwelski, K., St. John, J.P., Shaughnessy, W.J., Faha, T.A., Christie, W.N. (1984) Technical Guidance Manual for Performing Waste Load Allocations, Book II. Streams and Rivers, Chapter 3, Toxic Substances. Off. of Water Reg. and Stds., Monitoring and Data Support Div., Water Qual. Anal. Br., USEPA, Washington, D.C., 203 pp + Appendix. EPA-440/4-84-022

Di Toro, D.M., O'Connor, D.J., Thomann, R.V., St. John, J.P. (1981) Analysis of fate of chemicals in receiving waters, phase 1. Chemical Manufact. Assoc., Washington, D.C., Prepared by HydroQual, Inc., Mahwah, N.J., 8 Chapters + 4 Appendixes

Di Toro, D.M., Horzempa, L.M., Casey, M.M., Richardson, W. (1982a) Reversible and resistant components of PCB adsorption-desorption: Adsorbent concentration effects. J. Great Lakes Res. 8(2), 336–349

Di Toro, D.M., O'Connor, D.J., Thomann, R.V., St. John, J.P. (1982b) Simplified model of the fate of partitioning chemicals in lakes and streams. In: Dickson, D.S., Maki, A.W., Cairns, J., Jr. (eds.) Modeling the fate of chemicals in the aquatic environment. Ann Arbor Science, Ann Arbor, MI, pp. 165–190

Di Toro, D.M. (1985) A particle interaction model of reversible organic chemical sorption. Chemosphere 14(10), 1503–1538

Dolan, D.M., Bierman, V.J., Jr. (1982) Mass balance modeling of heavy metals in Saginaw Bay, Lake Huron. J. Grt. Lks. Res. 8(4), 676–694

Eadie, B.J., Faust, W., Gardner, W.S., Nalepa, T. (1982) Polycyclic aromatic hydrocarbons in sediments and associated benthos in Lake Erie. Chemosphere 11(2), 185–191

Eadie, B.J., Rice, C.P., Frez, W.A. (1983) The role of the benthic boundary in the cycling of PCBs in the Great Lakes. In: Mackay, D., Paterson, S., Eisenreich, S.J., Simmons, M.S. (eds.) Physical behavior of PCBs in the Great Lakes, Ann Arbor Science, Ann Arbor, Michigan, pp. 213–228.

Eadie, B.J., Faust, W.R., Landrum, P.F., Morehead, N.R., Gardner, W.S., Nalepa, T. (1983) Bioconcentrations of PAH by some benthic organisms of the Great Lakes. 7th PAH Vol., Battelle Mem. Inst

HydroQual, (1987) Report on metal partition coefficients. Prepared for USEPA, Large Lakes Research Sta., Grosse Ile, MI

Muhlbaier, J., Tisue, G.T. (1981) Cadmium in the southern basin of Lake Michigan. Water, Air and Soil Poll. 15, 45–59

Neff, J.M. (1979) Polycyclic aromatic hydrocarbons in the aquatic environment. Applied Sci. Pub. Ltd., London, Eng., 262 pp

O'Connor, D.J., Connolly, J.P. (1980) The effect of concentration of adsorbing solids on the partition coefficient. Water Research 14, 1517–1523.

Oliver, B.G., Niimi, A.J. (1983) Environ. Sci. Technol. 17, 287–291

Oliver, B.G., Niimi, A.J. (1985) Environ. Sci. Technol. 19, 842–849

Thomann, R.V., Di Toro, D.M. (1983) Physico Chemical Model of Toxic Substances in the Great Lakes. J. Great Lakes Res. 9(4), 474–496

Thomann, R.V., Di Toro, D.M. (1984) Physico chemical model of toxic substances in the Great Lakes. Project Report to USEPA, Large Lakes Res. Sta., Grosse Ile, MI, 163 pp

Thomann, R.V., Connolly, J.P. (1984) Age dependent food chain model of PCB in Lake Michigan lake trout. Env. Sci. & Tech., Feb. 1984

Thomann, R.V., Salas, H.J. (1986) Manual on Toxic Substances in Surface Waters. Pan American Health Organization, CEPIS, Lima, Peru

Thomann, R.V. (1987) Statistical model of environmental contaminants using variance spectrum analysis. Final Report to the National Science Foundation, 161 pp. + Append

Thomann, R.V., Mueller, J.A. (1987) Principles of surface water quality modeling and control. Harper and Row Pub. Inc., N.Y., N.Y., 644 pp

Thomann, R.V. (1987) Bioaccumulation model of organic chemical distribution in aquatic food chains. Draft manuscript

USEPA, (1986) Private correspondence. Municipal Environmental Research Lab., Cincinnati, OH

Effects of Toxicants on Aquatic Populations

Thomas G. Hallam, Ray R. Lassiter, and *S.A.L.M. Kooijman*

Outline

I. Introduction: A Strategy for Risk Assessment

In the United States, the Environmental Protection Agency regulates chemicals under several legislative acts, two of which are the Toxic Substances Control Act (TSCA) and the Federal Insecticides, Fungicides, and Rodenticides Act (FIFRA). The Agency regulates chemicals under these acts using, in part, the assessment of risks both to humans and to the environment. A scientifically based methodology of high utility to assist in evaluating environmental risk posed by the introduction of chemicals is currently under development. The purpose of this article is to provide indications of past developments, of current theoretical research, and of directions of environmental risk assessment of chemical stress on populations. Fate and effects at the community and ecosystem level are, at this stage, only speculative.

There are potential misinterpretations of the word "environment" in the context of risk assessment. The misconception of equating ecology with environmentalism by the public does not differentiate the science "ecology" from the social and political movement "environmentalism". The term ecological risk assessment is preferable to environmental risk assessment because we will explicitly include abiotic and biotic components and their interactions as objects of concern in risk assessment.

Optimally, regulation of chemicals occurs prior to their environmental release. That is, based upon information available prior to registration (for pesticides) or during the period after receipt of a premanufacture notification (for industrial chemicals), a regulatory decision is made that protects the environment and does not unduly restrict commerce. For the regulatory decisions to precede environmental release, the information about a chemical and the environment that might be exposed to it must be used to predict outcomes of alternative actions. The ability to extrapolate and predict successfully requires that mathematical models be the primary tools used. In general, predictive needs are a prognosis of ecological effects from a chemical released into the environment under various regulatory options. The three key ideas—prognosis, ecological effects, and the environment—of this statement are now interpreted to present our perspective.

Preventive regulation—the regulation of chemicals to prevent damage to the environment and to avoid costly operations (such as spill or depository, cleanups, areas closed to fishing or recreation)—requires knowledge of the environmental behavior and response of ecological systems for each chemical and combinations of chemicals that are considered for regulation. Without such information, regulatory actions would always be problematic. Knowledge of this sort, however, is not factual; rather, it is available only inferentially. Likely behaviors and responses are inferred on the basis of other, usually more certain, knowledge. The basic knowledge, itself, is often mixed as to the certainty with which it is held. Some of it is as well established as the principles of chemistry and some of it is as poorly documented as the specific structure and function of particular ecosystems that will be exposed to the chemical.

Wide variation in the degree of certainty of knowledge of elements on which prognoses are made has led to the often expressed view that prognoses need to be accompanied by statements of uncertainty expressed probabilistically. This view, although not without merit where such statements are possible (i.e., where uncertainties can be quantified), improperly likens the assessment of risks to be expected from new toxic chemicals to something akin to a weather forecast. For new toxicants, however, in contrast to weather forecasting or similar predictions, there is often no general, previously existing base of experience on which to calculate probabilities.

At least two classes of uncertainty are associated with useful prognostic statements for ecological risk assessments. One class is the state of knowledge on which to make predictions of fate and effects of chemicals. This includes the question of how to incorporate known uncertainties in the values of model parameters into interval estimates of model output, as well as questions of model interval validity (Bartell et al., 1983; Gardner et al., 1982; O'Neill et al., 1982). The other class is the uncertainty of the degree to which the calculations and considerations utilized in a risk assessment are sufficient. In other words, what are the most useful kinds of information that can be obtained for risk assessments and at what point is an assessment adequate? A major obstacle for risk assessment is that knowledge of the nominal operating behavior of a system is often unavailable. Recognition of deviations from the nominal behavior is impossible in these circumstances.

For most new chemicals, whether pesticides or industrial compounds, it must be anticipated that they will be used widely. Analyses should relate to the range of ecological and environmental conditions potentially to be affected. "The environment" includes all potential environments of concern. The magnitude of work involved in such an analysis is quite large and, therefore, an organized approach is needed. This problem has been discussed recently by Lassiter (1986) who proposes a staged analysis in which simpler calculations are used first to permit categorizing chemicals into those for which significant risk can be discounted, those that pose a clear risk, and those that merit more detailed analysis. At any stage in the analysis, the ecosystem is represented with the least degree of detail necessary to carry out the analysis. For the analysis of chemical fate, the simplest environmental descriptors have been employed; these simple descriptions are termed "canonical environments". Use of simplified descriptions of systems is necessary to prevent the analysis from becoming impossibly complex, and follows a sort of Ockam's razor principle for ecological risk assessment.

In this approach, detailed prognoses for any specific system are deferred in order first to obtain a broader picture of the fate and effects of a chemical over a range of conditions. Use of models in this manner is a learning exercise in which a basis of expectation of chemical fate and effects is developed.

In this article, we deal primarily with the area of "response to the chemical". What is needed for practical risk assessments are identification and quantitative estimates of effects that would be likely to occur in natural systems under given exposure conditions. To our knowledge, no concepts have been advanced whereby all types of potential effects can be identified. Effects that are of ecological significance are those that alter behavior of populations, including microbial as well as macro-organism communities. Such effects that may span the spectrum include alternations in direct mortality, reproductive rates, response to environmental stimuli, mobility, and biosynthetic rate or rates of energy transduction. These effects are typically considered to be direct effects. Their occurrence in natural systems, however, leads to additional effects through indirect pathways, and these indirect effects can be expressed in ways that are difficult to anticipate. For example, reproduction rate can be indirectly reduced by a wide variety of factors including some that are directly observable.

We refer to all of the above effects, both direct and indirect, as ecological effects. Any response to chemical exposure is an effect and is often computed via mathematical models as relative differences between a quantity obtained under exposed and unexposed conditions. With this view of effect, all models, from the simplest to the most complex, can be treated very similarly by considering differences in output for situations with and without toxicant.

Risk assessment needs in a staged analysis could be structured by the following sequence.

1) Models to calculate the maximal body burden attainable for the chemical of concern via food and environmental pathways.
2) Models for the response of populations to toxicant exposures.
3) Models for the response of interacting populations and, ultimately, for ecosystem responses.

For the models at each level to be operational, the strategy necessarily involves developing representations of chemical transport, distribution, metabolism of toxicants, and response of individuals. This assessment program strategy is viewed as both holistic and reductionistic. It is goal oriented and, therefore, structured in a holistic manner. Most current bioassay results are premised upon the tenet that toxicants affect individuals often directly in terms of physiological processes. If bioassays are to be utilized in assessing risk, this resolution level must be maintained in the model to study toxicant stress at higher organizational levels. Thus, this direction of research is regarded as reductionistic because low level, high resolution processes are considered.

It is our intent to survey models of effects as they fall into the above categories and strategy. Our goals are to indicate some current directions for research on the effects and the theoretical basis on which to build predictive models for risk assessments. First, we will indicate some background approaches that affect current risk assessment methodology at the population level.

II. Some Generic Population-Toxicant Models

Our current activities have been influenced and, to some extent, anticipated by previous efforts. We discuss briefly a few of these approaches.

Wallis (1975) employed a difference equation model to study the impact of waste on a fish population. His formulation included both deterministic and stochastic components in the model formulations for birth, death, and dispersal. Variation of the deterministic growth processes studied include the classical exponential, the Beverton-Holt (1957), the Ricker (1954), and the logistic formulations. With simulation studies, Wallis investigated the exponential model, did a comparative study on the logistic model, and studied sensitivity of population change to changes in environmental disturbances.

The focus of Wallis' article is on the impact of waste releases. Because chemical impact is observed not only on the numbers of organisms in the population but also on different sized organisms, Wallis also presents a difference equation model that includes a length structure and a stochastic recruitment formulation. Effects of exposure to waste is reflected in the model via prediction that an increase in mortality results from an increased magnitude of discharge.

The general ideas espoused by Wallis are consonant with many current developments in risk assessment. The sensitivity structures that are fundamental for risk assessment include formulation of structural sensitivity in both deterministic and stochastic components as accomplished by comparison of exponent and logistic deterministic parts and employing several distributions for the environment fluctuations. Classical sensitivity of parameters and sensitivity to size structure are important model perspectives. Effects of the anthropogenic stress are measured with a sensitivity type measure—increasing the waste discharge increases mortality. The foundational ideas in Wallis' approach are sound, but the simplicity of the model formulations lead to predictions that might be suspect for many reasons.

Another rudimentary model, on a continuous domain rather than the discrete problem considered by Wallis, develops the effects of a toxicant on a Daphnia population (Hallam and deLuna, 1984). The population model, a modification of one developed by Smith (1963), is

$$\frac{dx}{dt} = x\left[\frac{r(C_T)B(c + r_0 - a) - acx}{B(C_T + r_0 - a) + ax}\right]$$

where $r(C_T)$ denotes the growth rate of the population as a function of total body burden C_T; B is the carrying capacity of the population (B need not be an equilibrium); a is a measure of the population's response to stress effects; $r_0 = r(0)$ is the growth rate of the population in the absence of the toxicant; and c is the rate of replacement of mass in the population at saturation level.

An important difference between the work of Hallam and de Luna and that of Wallis is that in the former the population model is coupled with an uptake model so the interaction between the chemical and the population can be studied.

The uptake model has the form:

$$\frac{dC_T}{dt} = a_1 C_W + \frac{d_1 C_F \beta}{a_1} - l_1 C_T - l_2 C_T$$

where C_W is the concentration of toxicant in the environment, a_1 is the rate of toxicant uptake per unit mass of the population; C_F is the constant toxicant concentration in the resource; β is the average rate of food intake per unit organismal mass; d_1/a_1 reflects the nonadditivity between environmental and food chain pathways; and l_1 and l_2 are egestion and depuration rates respectively. The population and uptake equations are related by a dose-response formulation $r = r(C_T) = r_0 - H(C_T)$ for the population growth rate. A conservation of mass argument shows that the dynamics of environmental concentration of toxicant are governed by

$$\frac{dC_W}{dt} = g(t) + k_1 l_1 C_T x - k_1 a_1 C_W x - k_2 C_W$$

Here g represents exogenous input of toxicant into the environment; the second term represents total toxicant egested; the third term represents total toxicant uptake from the environment; and the last term models losses due to detoxifying processes such as photolysis, hydrolysis, volatilization, etc..

Several difficulties that arise with this approach are addressed, in part, in the following sections. This formulation requires knowledge of a population level dose-response function. Such information is usually not available from toxicological bioassays. Another deficiency is the treatment of uptake with the traditional bioconcentration equation with constant coefficients. The more mechanistic work of Barber et al. (1988) on this subject will be explored later. The population model regards all individuals in the population as the same average prototype individuals. Susceptibility clearly differs for individuals and some factor must account for this feature of population-toxicant interactions.

Another rudimentary approach is based upon energy budget considerations for individuals and leads to a growth equation of the von Bertalanffy type (Kooijman and Metz, 1984). An advantage of this procedure over some other rudimentary approaches is that most of the hypotheses can be related to biologically sound information and data. The derivation of the model follows from the assumption that energy gain per food particle is constant so that the total energy intake can be written as $[\dot{A}_m]W^{2/3}[x/(K + x)]$, where x is the food density, W is the weight of the animal, K is the saturation constant, and $[\dot{A}_m]$ is the surface area specific assimilation rate.

Respiration energy is assumed to be a fraction κ of the total assimilation energy intake. This respiration energy can be divided into maintenance, which is assumed proportional to weight, and energy spent on tissue growth, which is assumed proportional to weight change per unit time. The mathematical model is

$$\frac{\kappa[\dot{A}_m]xW^{2/3}}{K + x} = \zeta W + \eta\frac{dW}{da},$$

where ζ is the weight-specific maintenance energy, η is the conversion factor for energy to weight, and a is age. This equation can be transformed to an equation, linear in length, by employing the isometric relationship $W = \alpha L^3$. This resulting von Bertalanffy equation for organismal growth reflects Daphnia growth data extremely well (Kooijman, 1986a).

Kooijman and Metz considered population growth in the situation where food density is constant and the internal concentration of toxicant is in equilibrium with a constant environmental concentration. Under these conditions, it is possible to show that a stable age distribution exists so that the population (ultimately) grows exponentially in numbers:

$$N(t) = N(0)\exp\{rt\};$$

(see Metz and Diekmann, 1986). Here t is time and r is the intrinsic growth rate of the population. The population parameter, r, is related to survival, F, and reproduction, R, through the characteristic equation (often called Lotka's equation):

$$\int_0^\infty \exp\{-ra\}F(a)R(a)da = 1.$$

Kooijman and Metz investigated the effects of chemicals on a population by sensitivity studies on parameters related to chemical impacts. By employing reproduction as determined by the individual model, Lotka's equation provides a means of relating individual characteristics with a population level parameter, the growth rate. It is important in risk assessment to be able to extrapolate from individual effects to population effects.

The prototype models, taken as a unit, exhibit many properties that a population risk assessment protocol should contain. Individual dynamics are needed (Kooijman and Metz, 1984). An exposure model is needed (Hallam and de Luna, 1984) and should be coupled to the model of the individual. A method to determine dose and associated individual or population response is important (Hallam and de Luna, 1984). A method of inferring effects at the population level from individual effects is needed (Kooijman and Metz, (1984). Sensitivities of the model to structural and parametric perturbations, which cover a biologically reasonable range of forms whether they are deterministic or stochastic, are important features of risk assessment (Wallis, 1975). From an ecotoxicological perspective, it is necessary to deal with individuals as input-output systems. If only rudimentary models, where characteristics of individuals are averaged, are used to model populations, the behavior of populations in the more complex settings encountered in natural environments remains obscure.

We propose to assess the effects of a chemical on a population by a procedure based upon physiological characteristics of the individual and chemical properties of the toxicant. An overview of our approach is presented in order that the details of the subsequent sections do not mask our objectives.

A basic premise of our approach is to focus upon the appropriate structural level for the problem taking into account information available as inputs, the characteristics of the biological and chemical interactions, and the ultimate goals of the investigation. Chemicals stress individuals and through this interaction affect higher organizational levels such as populations. Bioassay results focus primarily upon individual organisms. An implication of these statements is that individual dynamics are fundamental to assessing risk of exposure to a toxic chemical. In sect. V, we address the individual component of our assessment procedure by discussing two physiologically based dynamic models of individuals.

To determine the effects of a toxic exposure to an individual, the dose to that individual must be ascertained and the response to that dosage quantified. The ideas behind the exposure model of Barber and coworkers are discussed in Sect. III. The model output is the concentration of chemical in the various phases (aqueous, lipid, structural) of the individual. The dynamics of the population are needed to determine the effects of the chemical exposure on the population. We briefly discuss the manner in which the population evolves based upon individual dynamics in Sect. VII.

III. Transport, Distribution, and Metabolism

Models that accurately represent transport of a chemical between an organism and its environment, the internal distribution of the chemical, and its metabolism are necessary for the successful evaluation of the effects of a toxicant on a system. Most of our discussion will be restricted to organic nonelectrolytes. This is a large and important group of compounds that includes most industrial chemicals. Concepts for evaluation of effects of toxicants in general are well illustrated by this category, and many complexities are avoided, such as those associated with speciation, identification of chemical species that are exchanged, and complex modes of action.

The classical representation of uptake and depuration is

$$\frac{dC_T}{dt} = k_1 C_W - k_2 C_T$$

where C_T is the concentration per unit mass of a whole organism, C_W is the concentration of the chemical in the environment, k_1 is the uptake parameter, and k_2 is the depuration parameter. Often k_1 and k_2 have been assumed to be constant; however, only for certain chemicals and exposure time scales is this a valid hypothesis. We follow a derivation of Barber et al. (1988) to indicate the functional representations of k_1 and k_2 (see also Lassiter and Hallam, 1988).

Hypotheses imposed in the model development include the following. An individual organism is assumed to be comprised of three chemical phases— aqueous, structural, and lipid. The structural phase is generally viewed as being composed of (physiologically active) protein.

A second hypothesis is that the chemical is assumed to cause its effect via a reversible mode of action. The narcotics are a general class of chemicals that act by a reversible action mode. Chemicals in this category, in varying degree, may be hydrophilic or lipophilic and represent a range exceeding eight orders of magnitude of hydrophobicity (as measured by the octanol to water partition coefficient). From a modelling perspective, it is necessary to decompose the organism conceptionally into compartments corresponding in part to the chemical phases (water and lipid phases to accommodate the representation of degree of hydrophobicity) and the remainder (primarily protein and other structural components to represent the biotic aspects of the organism).

Direct toxicant exchange between organism and environment is assumed to be primarily across an external, chemically permeable membrane. (Toxicant uptake from food will be discussed later.) The time scales for the exchange across the boundary membrane is regarded as being much slower than the distribution of chemical within the organism.

For the present, we assume that the chemical is not metabolized by the organism. Metabolism of chemicals, however, is of potential importance in interactions between the chemical and the individual. Metabolic characteristics of chemicals can be contraindicative. Vaughan et al. (1982) state that 80% of the phenol and 40% of the aniline taken up by plants was metabolized to high molecular weight fractions. Anthracene, on the other hand, was catabolized to lower molecular weight compounds (Edwards et al., 1982). Metabolic decomposition can maintain low internal levels of chemical despite high exposure levels. Unfortunately, a widely applicable, theoretical basis for developing metabolic rate constants is not presently available.

An exposure model that contains many influential factors, with the exception of metabolism, is the GETS model recently developed by Barber et al. (1988). The model, based upon transport driven by thermodynamic potential, represents the chemical exchange between fish and the aqueous environment that occurs across gill membranes. Although GETS was developed for fish, the general form of the uptake model is similar to the classical bioaccumulation model but includes dilution of toxicant due to organism growth:

$$\frac{dC_T}{dt} = k_1 C_W - k_2 C_T - \frac{1}{W_T} \frac{dW_T}{dt} C_T. \tag{1}$$

The coefficients k_1 and k_2 are specified by

$$k_1 = SkW_T^{-1}, \quad k_2 = SkW_T^{-1}(P_A + P_L K_L + P_S K_S)^{-1}$$

where k is the conductance per unit area of the exposure membrane; P_A, P_L and P_S are fractions of the organism that are aqueous, lipid, and protein, respectively; K_L

and K_S are partition coefficients of the chemical in organismal lipid: water and protein: water, respectively; S is the chemically permeable surface area of the exposure membrane, and W_T is the total weight of the organism. The unit conductance k may be calculated from the molecular weight of the chemical and K_{OW}. Barber et al. (1988) show that $k \sim N_{Sh} D_W h_W^{-1}$ where N_{Sh} is the Sherwood number, D_W is the toxicant's diffusion coefficient (a function of the molecular weight of the chemical), and h_W is the characteristic dimension of interlamellar channels. The weight of the organism is contained in the expressions for each term in the exposure model (1) and it is used in the computation of k.

In addition to external exposure to toxic chemicals, aquatic organisms exchange chemicals across their gut walls. To complete the model for overall exchange, therefore, an expression for exchange with the food is needed. One can assume, as was assumed for gill exchange, that exchange occurs between the chemical dissolved in the aqueous environment and that in the aqueous blood. Here, however, the aqueous environment is the aqueous phase of the gut contents, and the membrane across which exchange occurs is the gut wall. This assumption can be incorporated simply by including a term for gut exchange analogous to that for gill exchange:

$$\frac{dC_T}{dt} = k_W(C_W - C_A) + k_F(C_{AG} - C_A). \tag{2}$$

where C_W, C_A, and C_{AG} are the concentration of toxicant in the environment (water), in the aqueous phase of the organism, and in the food in the aqueous phase of the gut, respectively. k_W and k_F are the unit conductances of the chemical from the environment to the organism's aqueous phase and the chemical from the aqueous phase of the gut to the organism's aqueous phase, respectively. The formulation in (2) is derived from the simple assumption that exchange occurs passively.

Application of this model would be thwarted, however, by complexities of the feeding and digestive physiology and anatomy. Organisms typically feed intermittently. To incorporate this behavior would require at least knowledge of the fraction of time during which food is available in the intestine for exchange. During the passage of food through the intestine, the gut distends, so that the exchange area varies, as does the thickness of the gut wall. These and all the other transient events of the digestive process make the exchange gradient model (2) very difficult to apply. The simplest assumption that avoids most of these difficulties seems to be that gut content equilibrates chemically with the body of the organism prior to defecation. Application of a model based on this assumption would require knowledge of the average quantity of food eaten per unit of time, the concentration of the chemical of concern in the food, the fecal production, and the partition coefficient of chemical between water and fecal material. As with the model for gill exchange, of course, knowledge of the mass of the organism's body and its composition are required.

The basic assumption of this model, that of equilibration of chemical between the organism's body and the gut contents, is not necessarily true, of course. It will be shown later, however, that this assumption is a worst case assumption during increasing body concentration when exposed to contaminated food (i.e., no more chemical could be taken up under any thermodynamically consistent assumption that would be taken up when food and body equilibrate). During depuration, however, this assumption leads to predicted minimum depuration times; that is, any other thermodynamically consistent assumption would lead to longer predicted depuration times. For toxicity evaluations, this would usually not be considered the worst case. The blood-to-gut equilibrium assumption can be incorporated by first expressing total exchange using a mass balance term for gut exchange as in the following equation:

$$\frac{dC_T}{dt} = k_W(C_W - C_A) + \frac{1}{W}(FC_F - EC_E) \tag{3}$$

in which W is weight of the organism; F is weight of food eaten per unit time; E is weight of material defecated per unit time; C_F is concentration of chemical in the food; and C_E is concentration of chemical in the feces.

The equilibrium assumption is

$$C_E = K_E C_A$$

in which K_E is the partition coefficient of chemical to excrement. When incorporated into Eq. (3) this assumption leads to:

$$\frac{dC_T}{dt} = k_W C_W + \frac{F}{W}C_F - \left(k_W + \frac{EK_E}{W}\right)C_A. \tag{4}$$

Implications of the incorporation of gut exchange can be evaluated using appropriate assumptions about the chemical content of the food, etc. The simplest assumption about the quantity of toxicant in the food is that it is at chemical equilibrium with the water, which in turn requires knowledge of the composition of the food. We assume that the food, like the consumer, consists of three components, water, lipid, and structural materials. When the concentrations in the environment and aqueous food phases are equal

$$C_F = P_{AF}C_W/D_{AF}$$

where P_{AF} and D_{AF} are the fraction of the food that is in the aqueous phase and the fraction of chemical that is contained in the aqueous phase, respectively.

Using this assumption for chemical concentration in the food, and assuming equilibrium of the chemical within the organism the chemical concentration in the organism at the steady state can be obtained:

$$C_T = \left(\frac{k_W + \dfrac{F}{W}\dfrac{P_{AF}}{D_{AF}}}{k_W + \dfrac{EK_E}{W}}\right)\left(\frac{P_A}{D_A}\right)C_W,$$

where D_A represents the fraction of the chemical contained in the aqueous phase of the organism. The biomagnification factor, BMF, as defined here, indicates accumulation beyond the level that would be reached via simple equilibrium of the whole organism concentration with that of the aqueous environment. If simple chemical equilibrium between organism and the environment were achieved, then the aqueous concentration in the organism would equal exactly the aqueous environmental concentration, i.e. $C_A = C_W$. We define BMF as C_A/C_W, and thus only when $BMF > 1$ is there biomagnification. The expression is

$$BMF = \frac{k_W + \dfrac{F}{W}\dfrac{P_{AF}}{D_{AF}}}{k_W + \dfrac{EK_E}{W}}.$$

Here, in contrast to exchange via gills alone, it is apparent from differences in the numerator and denominator that BMF can differ from 1, and therefore, that inclusion of this representation for the feeding and digestion processes does, indeed, influence model predictions. To evaluate the probable limits to BMF would require considerable analysis of probable values of the variables defining BMF and their covariance. Here we take values of the variables that we believe to be reasonably near the limits of their range and assume that they vary independently to obtain BMF limits. That is, we choose values from the set that give the lowest and the highest calculable BMF and consider these values over the range of chemical hydrophobicity. We also use the approximation $P_{AF}/D_{AF} \approx P_{LF}K_{ow}$ where P_{LF} is the fraction of food in the lipid phase. Ranges of the variables and their biological meanings are given in Table 1. Table 2 gives the calculated lower and upper limits on BMF.

Although not definitive, the values given in Table 2 present two interesting phenomena. BMF (upper) values indicate the widely expected food chain biomagnification. (This is a two link food chain, water to food organism, and food organism to consumer.) It is perhaps somewhat surprising that potential biodilution can occur. It is indicated by values less than one in the BMF (lower) column.

Food chain biomagnification here occurs with the assumption that chemical transport is entirely passive. The only work occurring is that of the digestive process plus that implicit in the assumption that assimilation of food occurs in the presence

Table 1. Ranges of values of variables determining upper and lower limits on BMF

Range	Explanation
$0.01 < F/W < 0.05$	Daily food ingestion rates range between 1% and 5% of body weight (Swenson and Smith, 1973)
$0.1K_{ow} < K_E < 0.2K_{ow}$	Partition coefficient of chemical between water and fecal material is in the indicated range, which includes the organic carbon-to-water coefficient (here referenced to organic matter) as the upper limit (Karickhoff, 1981)
$0.03 < P_L < 0.25$	Fraction lipid in the food organism is in the indicated range
$0.1F < E < 0.5F$	Fecal production ranges between 0.1 and 0.5 of the food intake rate, probably corresponding to piscivores at the lower value and to herbivores feeding at a high rate at the upper value
$200 < k < 800$	The first order coefficient for gill exchange normalized to gill surface and body weight is in the indicated range (possibly somewhat low)

Table 2. Calculated values of upper and lower limits to BMF for chemicals with octanol-water partition coefficients up to 10^7

$\log_{10}(K_{ow})$	BMF(lower)	BMF(upper)
< 4	1	1.1
4	1	1.6
5	.8	6.1
6	.6	19.4
7	.5	25.7

of zero growth rate. The latter quantity of work, i.e., that done to offset the intake of food to maintain a stable body size, is assumed for convenience. In reality, most aquatic organisms with food intake and assimilation rates that were assumed in obtaining the upper limits would be growing, and therefore, diluting the body burden of chemical. Although not discussed in detail here, kinetic numerical studies, in which the diluting effect of growth is considered, indicate that the magnitude of this effect is slight. It is probably much more important in contributing to high estimates of maximum *BMF* that extreme parameter values were selected and assumed to be independent. It was our purpose, in these calculations, however, to obtain bounding values.

It would be improbable that biodilution would be observed in natural situations. When sampled from natural environments, organism exposure history is unknown, and a very low body concentration probably would be interpreted as resulting from low exposure. In addition, biodilution would pose no danger to organisms and would not generally be of interest to anyone conducting a field study. It would be of scientific interest, however, because if organisms could be identified that meet the criteria of *BMF* (lower), experiments with them would provide a useful and critical test of the theoretical concepts involved in the model used for these predictions.

One point in the steady state analysis remains to be discussed. It was noted above that the steady state assumption of equilibration of gut content with body concentration of chemical is a worst case assumption with respect to uptake of chemical. Equilibration is one extreme case. The other extreme is approached with extremely low gut residence times, perhaps during times of heavy feeding. In reality this case is not subject to analysis apart from experimentation, but a bound can be set, and will suffice for the present purpose. The bounding condition would occur if feeding were so intense that only an insignificant portion of the food was digested and, as a result, that fecal composition was undetectable from the food. The organism would be exposed to food and its associated toxicant concentration throughout the entire intestinal tract. In this case in the describing equation, the term for exchange with food simply disappears, because food intake, composition, and toxicant concentration are identical to fecal production, composition, and toxicant composition. This case, therefore, reduces to the model for exchange via gill alone, where $BMF = 1$.

The steady-state assumption is useful for determining the tendencies of *BMF* for various feeding regimes. It is not always a valid assumption, however, because transient conditions may dominate the period of interest. These transient conditions can be examined by using the model to obtain time-dependent results. Results of this sort have indicated agreement with the steady state analyses. In addition, greater biomagnification (or biodilution) is obtained with longer food chains, as has been commonly reported. And as is also commonly reported, periods of intense activity, particularly when associated with low feeding rates, can cause transient and significant biomagnification above that already experienced under less extreme physiological conditions.

IV. Effects of Chemicals on Individuals

The effects of chemicals on individuals are determined from bioassays. To model the effects, we need to couple the uptake model with models for the mode of action and models for concentration-response relations. The uptake model, used to compute concentration of chemical in the individual, can be employed to find the individuals susceptible to a particular environmental concentration. The toxicological bioassay results present baseline information to determine the effect concentration. We now indicate a procedure that employs these ideas to measure the effect of a toxicant on an individual.

It is convenient, but not necessary, to consider an individual that does not grow during the exposure period since the growth term in the uptake equation (1) vanishes; hence,

$$\frac{dC_T}{dt} = SkW_T^{-1}[C_W - C_T(P_A + P_L K_L + P_S K_S)^{-1}],$$

$$= SkW_T^{-1}[C_W - C_T(d_1 + d_2 P_L)^{-1}], \tag{5}$$

where the last equality defines d_1 and d_2. For hydrophobic chemicals, the parameter P_L is an important focal point since it represents the compartment for which the chemical has high affinity. We have written the second part of Eq. (5) to recognize the role of P_L. The first order linear equation can be solved explicitly; when C_W is constant one obtains

$$C_T(t) = \exp\left[-\frac{Sk}{W_T}(d_1 + d_2 P_L)^{-1}t\right][C_T(0) - C_W(d_1 + d_2 P_L)] + C_W[d_1 + d_2 P_L].$$

The total concentration of toxicant can be converted to aqueous portion through multiplication by the fraction of chemical in the phase:

$$C_A(t) = C_T(d_1 + d_2 P_L)^{-1}$$

$$= C_W\{1 - \exp[-SkW_T^{-1}(d_1 + d_2 P_L)^{-1}t]\}.$$

This formula assumes that there is no toxicant in the organism initially.

For a single individual, an effect occurs when concentration of chemical in the aqueous component reaches a critical level. (This effect, although measured independently of the concentration of toxicant in the other phases of the organism, is, of course, related via the assumption of internal chemical equilibrium to the concentrations in the other phases.) Individuals differ, however, and this method of assessing effects allows the variation in levels of high affinity partitioning components (lipid for hydrophobic chemicals) to vary the consequences of exposure among individuals by varying their degree of buffering against the chemical's effects. For a given exposure concentration and duration, larger lipid fractions provide greater protection from transient exposures to hydrophobic chemicals by lengthening the time to effect.

Equation (5) can be solved to obtain the lipid fraction, P_L^e, that will be associated with an effect occuring at an effect concentration, C_{We}, with duration of exposures, τ:

$$P_L^e = \frac{-\left[d_1 \ln \left(1 - \frac{C_{We}}{C_W} \right) + SkW_T^{-1}\tau \right]}{d_2 \ln \left(1 - \frac{C_{We}}{C_W} \right)}. \tag{6}$$

Note that P_L^e is a linear function of q. When exposure duration τ satisfies $\tau \leqq -d_1 \ln(1 - C_{We}/C_W)W_T S^{-1}k^{-1}$ then no effect is observed for any individual. If $\tau > -d_1 \ln(1 - C_{We}/C_W)W_T S^{-1}k^{-1}$ then an organismal lipid level less than or equal to P_L^e (given by (6)) will result in the individual exhibiting the effect by time τ.

In general, susceptibility in a population is determined by a variable—in the above instance, individual lipid mass—that can change considerably even on daily time scales. It is important to have a model of an individual that will reflect the desired component dynamics.

V. A Model of an Individual: An Energy Budget Approach

There are many reasons to base ecotoxicological models upon individuals viewed as input-output systems. Effects of some compounds on individuals depend upon their chemical composition and dosage, which, in part, is a function of the recent history of the resource uptake of the exposed organism. Another reason, and probably the main one, is that population dynamics depend upon reproductive behavior, which also depends on resource uptake. Two biological tenets that should be reflected in individual models are that resource levels are typically dynamic variables and that energy storage is a necessary component of energy budget formulations.

Kooijman (1986a) has followed the approach of developing a model including energy resources and keeping the number of parameters as sparse as possible. The general perspective is that energy resources, consisting of carbohydrates, proteins, lipoproteins, and lipids, function as resources of building materials to be used in the continuous reconstitution of enzymes and membranes and, when relevant, in size increase. The individual is assumed to have hormonal control over the chemical composition of its energy reserves, primarily adipose tissue. There is good physiological evidence for this assumption; for example, carbohydrates can be converted into lipids and, in particular, glucose can be converted into triglyceride (Newsholme and Start, 1973). Carbohydrate and fatty acid metabolisms are intimately related to provide a homeostatic control mechanism. See Randle et al. (1963) for a discussion of the glucose/fatty acid cycle.

The way in which hormonal control is achieved has not yielded to modelling efforts. Current research by Zonneveld, Doucet, and Kooijman on the pond snail *hymnaea stagnalis* is directed toward this goal. The strategy is to initially regard lipid content of an individual as a simple function of total energy resources. For example, as a first approximation, the organismal lipid can be assumed to be a fraction of total resources. Stored energy dynamics has been deduced from feeding, growth, and reproduction behavior in various feeding regimes and is a relatively simple nonlinear function of feeding rate and size of the animal.

Hence, to discuss the effects of toxicants on populations, for many chemicals it is necessary to model the storage of an exposed individual. Some observations about representations in Kooijman's (1986a) model that will be employed in a subsequent analysis of effects are now given.

He found that, at constant food density x, the storage density, $[S]$, asymptotically satisfies $[S] = f[S_m]$ where f is the scaled functional response, $f = x/(K + x)$, where K is the saturation constant, and $[S_m]$ the maximum storage energy density. The reproduction rate (in number of eggs per unit time) developed by Kooijman can be written in the form

$$R = \frac{1-\kappa}{E} \frac{f}{f + \frac{\eta}{\kappa[S_m]}} \frac{\kappa^2[\dot{A}_m]^3}{|\dot{\zeta}|^2} \left\{ \frac{\eta}{\kappa[S_m]} \frac{W^{2/3}\dot{\zeta}^2}{\kappa^2[\dot{A}_m]^2} + \frac{W\dot{\zeta}^3}{\kappa^3[\dot{A}_m]^3} \right\}. \tag{7}$$

Thus, R is proportional to $fl^2(\alpha + l)/(\alpha + f)$ where $\alpha = \eta/\kappa[S_m]$ and l is a scaled length $l = W^{1/3}\dot{\zeta}/\kappa[\dot{A}_m] = W^{1/3}/W_m^{1/3}$.

In Eq. (7), $[\dot{A}_m]$ denotes the maximum size specific assimilation energy rate; η is the energy requirement for a unit increase in weight; κ is the proportion of utilized energy spent on growth and routine metabolism; $\dot{\zeta}$ is the maintenance energy consumption rate per unit of weight; E is the energy investment per egg; W is the weight of the individual; and W_m is the maximum weight of the organism.

Energy flows into development as long as the weight of the animal satisfies $W < W_j$ (or $l < l_j$, l_j is the scaled length at which reproduction starts) where W_j is the weight at onset of reproduction. Whenever the weight W exceeds W_j, the maintenance of a degree of development of magnitude $\dot{\zeta}((1 - \kappa)/\kappa)W_j$ is required but the remainder is devoted to reproduction. The quantity

$$\frac{1-\kappa}{E} \frac{\dot{\zeta}}{\kappa} W_j = \frac{1-\kappa}{E} \frac{\kappa^2[\dot{A}_m]^3}{\dot{\zeta}} l_j^3 \tag{8}$$

is thus subtracted from the reproduction rate leading to the result that R is proportional to $fl^2(\alpha + l)/(\alpha + f) - l_j^3$. Thieme (1986) has proposed that this energy drain is connected with maintaining the state of maturity. There is recent experimental support for this proposition (Zonneveld and Kooijman, 1987) in the pond snail *Hymnaea stagnalis*. According to Kooijman (1986a), size at constant food density is given by

$$W^{1/3}(a) = W_\infty^{1/3} - (W_\infty^{1/3} - W_b^{1/3})e^{-\gamma a}$$

where $\gamma = \dot{\zeta}/\kappa/3(\eta/\kappa + f[S_m])$; W_b is the weight at birth; and $W_\infty^{1/3} = f\kappa[\dot{A}_m]/\dot{\zeta}$.

This leads to a representation of length as

$$l(a) = f - (f - l_b)e^{-a\beta/(3(\alpha + f))} \tag{9}$$

where $\beta = \dot{\zeta}/\kappa[S_m]$. These representations of an individual will be employed when we discuss effects.

Hallam et al. (1989) follow a different strategy in their approach to modelling individuals. In formulating a theory, "Survival of the Fattest" (a name attributed by Lassiter to S. Karickhoff), for the effects of a lipophilic chemical on an organism, it is necessary to have a dynamic representation of the lipid fraction of the organism. Here, because of the rather weak excuse of a paucity of information about carbohydrate dynamics in *Daphnia* but motivated by a lack of utilization of carbohydrates in fish, they represent the uptake of carbohydrates, protein, and lipids from resource as occuring independently, and refrain from modelling any conversions inside the individual. The ideas of Hallam and coworkers about modelling individuals from a physiological perspective are not currently widely available. Because we wish to employ them subsequently to discuss effects of toxicants on populations, they are reviewed here.

The dynamic structural and compositional changes of an individual during various time scales ranging from daily to lifetime relate to the effects of a chemical on that organism. We now describe a model, based upon energy considerations, that meets initial requirements for assessing the effects of hydrophobic chemicals.

This work by Hallam et al. (1988) modelled an individual employing energy budget techniques much like those suggested by Kooijman and Metz (1984) and Kooijman (1986a). Although the model is formulated for *Daphnia*, it seems to be somewhat generic in applicability.

The flow diagram, in terms of energy, of an individual is given in Fig. 1. The flow diagram of Kooijman (1986a) is closely related to Fig. 1. A brief discussion focusing upon compartments and flows is now presented.

INDIVIDUAL MODEL

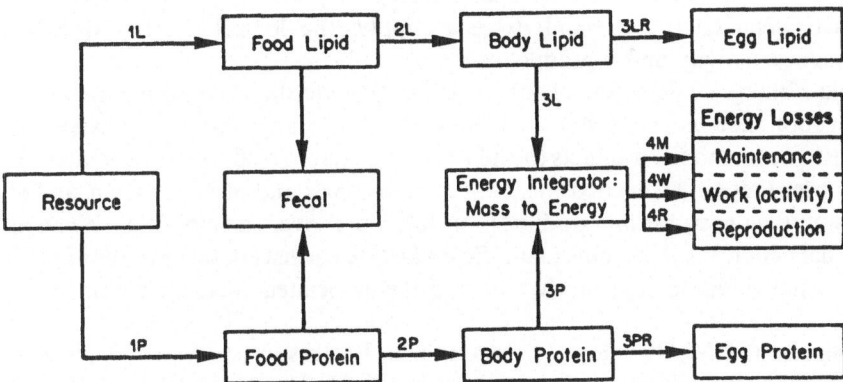

Fig. 1. Compartment and flow diagram for the individual model

Model Compartments for Daphnia

Gut. The gut is represented only implicitly in the model. The residence time for food in the gut is generally brief in comparison to individual and population time scales. For example, the gut residence time for *Daphnia magna* is 28 to 54 minutes at 15°C with the type, the amount of food (in this instance, algae), and the size of the water flea determining the residence time. Hence, we assume that ingested food is instantaneously partitioned to the model components, protein or lipid. For the latter component, this assumption is not unreasonable since algal lipid is readily assimilated. After daphnids withdraw lipids from an algal cell, they withdraw proteins and carbohydrates. Carbohydrate dynamics are ignored here not because of a lack of importance (although this may be the situation) but because of a general lack of knowledge (Hallam et al., 1988).

Protein. The protein compartment is assumed to be composed of labile and nonlabile portions. The nonlabile portion consists of somatic tissue and is viewed as protected structure that may not be utilized as an energy source under any circumstance. This model variable, m_{pp} (mass of protected protein), is nondecreasing as a function of age. The labile portion of the protein pool is represented by $m_p - m_{pp}$, where m_p designates the mass of the protein.

Lipid. The lipid compartment, viewed as storage, is decomposed into two subcompartments, labile and nonlabile lipids. The nonlabile lipid, represented in the model as εm_{pp}, is a proportion of the lipid compartment associated with protein in cell membranes and other fine subcellular structure and is not available to the organism even under conditions of starvation. The labile storage is $m_L - \varepsilon m_{pp}$, where m_L designates the mass of the individual's lipid pool.

Energy Integrator. Labile lipid and labile protein are utilized as energy sources, in an indiscriminant manner, to support the energy demands of an individual. The energy integrator processes the mass of labile protein and lipid and converts it to energy, and supplies the energetic sinks of the individual—work, maintenance, and reproduction.

The energy integrator acts to meet energy demand unless demand is above the capacity of the supplier, the labile subcompartments, in which case the energy integrator mobilizes energy at the maximal rate. A priority allocation scheme is operative when energy demand exceeds supply; the logical priority order is maintenance, activity and reproduction.

Sinks. Energetic demands of the organism include the sinks of maintenance, reproduction, and work. *Maintenance* consists of the energy costs of transporting nutrients and metabolic products to and from cells, that is, respiration, net materials used in tissue repair, associated energy of cellular repair and reconstruction, and of maintaining concentration gradients of nutrients over membranes. *Work* is individual activity such as movement. *Reproduction* consists of the transfer of mass from the individual for egg production and the associated energetic costs.

Assimilation and Mobilization: Flows 1 and 2. The resource for the *Daphnia* is assumed to consist of fraction of protein, *PP*, and lipid, *PL*, so that *PPx* and *PLx* are the input densities of protein and lipid in the resource. The assimilation rate, *AR*, is assumed to be proportional to ingestion rate, IR; hence, $AR = \alpha_1 IR$ for some constant α_1. The ingestion rate is assumed to be of the form

$$IR = \frac{x}{F_m^{-1} + xI_m^{-1}}$$

where F_m and I_m are the maximum filtering rate and maximum ingestion rate, respectively. This hyperbolic representation of uptake has F_m controlling assimilation for x small and I_m determining assimilation for x large. These hyperbolic uptake representations are abundant in the literature (Rashevsky, 1959; Lassiter, 1986).

The inputs determining growth of lipid and protein are now found by assuming that all lipid in the resource is assimilated and that a constant proportion of the resource protein is assimilated (Palohiemo et al., 1982). Furthermore, F_m and I_m should depend upon the size of the organism. We require that F_m be proportional to the surface area of the organism and that I_m be proportional to the volume of the organism. The protein pool determines each of these quantities; thus, the form of F_m and I_m are $F_m = \alpha_2 m_p^{2/3}$ and $I_m = \alpha_3 m_p$, for some constants α_2 and α_3.

The input of lipid represented by Flow 2L is of the form

$$GL = \frac{A_0 PLxm_p}{A_1 m_p^{1/3} + A_2 x}$$

while the input of protein as represented in Flow 2P is

$$GP = \frac{A_3 PPxm_p}{A_1 m_p^{1/3} + A_2 x}.$$

The remainder of the mobilization flow consists of indigestibles and nonassimilated proteins. For example, blue-green algae, a poor quality resource for *Daphnia*, and gelatinous sheathed-algae often pass through the gut undigested (Goulden et al., 1982).

Energy Integrator Flows: Flows 3 and 4. The energy integrator compartment supply flow, 3, is set by energetic demands of the individual. The maximum available energy (MAE) supply is a function of labile lipid and labile protein: $MAE = 37.68 A_4 (m_L - \varepsilon m_{pp}) + 16.75 A_5 (m_p - m_{pp})$. The factors 37.68 and 16.75 (Joules per milligram) are conversions for lipid and protein mass, respectively, to energetic equivalents.

If the total energy demand, TED, as represented by the sum of the demand flows 4 to the sinks of maintenance, reproduction, and activity exceeds the available energy, AE, then the supply flow will consist of all the energy the organism can mobilize. The lipid supply flow 3L is then $A_4 (m_L - \varepsilon m_{pp})$. Similarly, the protein supply flow will be $A_5 (m_p - m_{pp})$. If, however, AE exceeds TED then the supply is the fraction from each component necessary to meet TED. The flows 3L and 3P become $A_4 (m_L - \varepsilon m_{pp}) TED/AE$ and $A_5 (m_p - m_{pp}) TED/AE$, respectively.

Total energy demand is determined by the allocation to the sinks of work, maintenance, and reproduction. The flow 4W is assumed to be specified by

Gerritsen (1984) who describes the activity of free-swimming aquatic animals. For
Daphnia, both viscous and inertial forces are important. The formulation in energy
units is of the form $W = A_6(m_L + m_p)^{1/3} + A_7(m_L + m_p)^{2/3}$, for constants A_6 and A_7.
For maintenance, we assume this is of the form $A_8 m_L$ and $A_9 m_p$, where these terms
represent energetic equivalents of masses required to maintain each living cell.
Kooijman (1986b) has indicated in his work on egg development that energy
resources do not contribute to maintenance costs. It should be noted that there are
other forms used to represent maintenance in the literature. Most of these are of the
form aW^b for constants a and b (Gabriel, 1982; Paloheimo et al., 1982; Buikema,
1972; Richman, 1958). When these forms are employed, activity is not often
modelled independently. Kooijman's (1986a) analysis includes a rationale for b to lie
in the range $2/3 \le b \le 1$.

Whereas maintenance and activity are continuous losses in this model,
reproduction and carapace formation are regarded as discrete time losses at
reproductive periods. Protein allocated to carapace formation is assumed to
be $PLCF = (0.017)(m_L + m_p)^{1.5}$. This quantity is extrapolated from Lynch et al.
(1986).

Reproduction in this model concentrates upon the number of eggs produced and
the energy required to make those eggs. The reproductive capability of an individual
is assumed constrained by both labile lipid storage and labile protein.

Variability in egg size is set by egg lipid content. Because protein per egg is
assumed to be constant, the interplay between protein and lipid in eggs is
accomplished by utilizing the variable, $MEGP$, the maximum number of eggs per
available protein. $MEGP$ is constrained by organism size and labile protein in the
following manner:

$$MEGP = \min \left\{ \max \left[0, S(x) \left(\frac{m_p^{1/3}}{A_9} - 2.0 \right) \right], (m_p - m_{pp})/0.0082 \right\}.$$

The first term is a relationship, linear in length, for the number of eggs that a given
size individual can produce (Taylor, 1986). The second expression is the number of
eggs that may be produced by utilizing available labile protein; the constant 0.0082 is
the prescribed amount (mg) of protein in each egg (Goulden and Henry, 1982). The
number of eggs, REG, is

$$REG = \min \left\{ MEGP, \frac{(m_L - \varepsilon m_{pp})[A_{10}MEGP + (m_L - \varepsilon m_{pp})]}{e_m + e_L A_{10}MEGP} \right\}$$

where e_m and e_L are the maximum and minimum lipid content, respectively, of eggs.
The second expression in REG is a measure of lipid available for each egg although
the computation for the final lipid per egg, LEG, is

$$LEG = \frac{(e_m - e_L)(m_L - \varepsilon m_{pp})}{A_{10}REG + (m_L - \varepsilon m_{pp})} + e_L$$

$$= \frac{e_M + e_L A_{10}REG}{A_{10}REG + (m_L - \varepsilon m_{pp})}.$$

The Individual Model

On a continuous time scale, the model of an individual daphnid consists of the two coupled differential equations, one equation governing lipid mass and one governing protein mass. The dynamics of the lipid mass is

$$\frac{dm_L}{dt} = \frac{A_0 PLxm_p}{A_1 m_p^{1/3} + A_2 x} - \begin{cases} A_4(m_L - \varepsilon m_{pp}) & TED > AE, \\ A_4(m_L - \varepsilon m_{pp})\dfrac{TED}{AE} & TED \le AE, \end{cases}$$

where available energy

$$AE = 37.68 A_4(m_L - \varepsilon m_{pp}) + 16.75 A_5(m_p - m_{pp})$$

and total energy demand

$$TED = A_6(m_L + m_p)^{1/3} + A_7(m_L + m_p)^{2/3} + 37.68 A_8 m_L + 16.75 A_9 m_p.$$

The dynamics of the protein mass is

$$\frac{dm_p}{dt} = \frac{A_3 PPxm_p}{A_1 m_p^{1/3} + A_2 x} - \begin{cases} A_5(m_p - m_{pp}) & TED > AE, \\ A_5(m_p - m_{pp})\dfrac{TED}{AE} & TED \le AE. \end{cases}$$

The hormones governing reproduction are assumed to be regulated in a periodic manner once the individual has reached adult stage. In *Daphnia*, reproductive state is attained by size not age (Kooijman, 1986a). The reproductive losses include both lipid and protein mass losses to eggs and the energy required for that deposition, the protein mass loss for carapace formation, and the necessary energy required to form the exoskeleton. The numerical implementation of this model imposes a priority allocation scheme at reproductive periods. If sufficient energy and mass are available, then first, a carapace is formed and, second, eggs are produced. Starvation is computed from size of the organism and is related to the nonlabile protein.

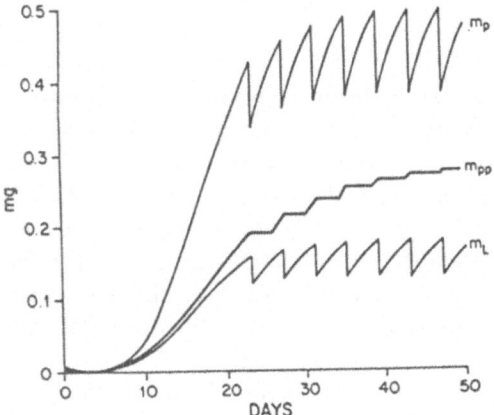

Fig. 2. Lipid (m_L) and protein (m_p) cycles in the individual model. The dynamics illustrate the decrease in size while in the brood pouch, exponential growth as a juvenile, and the molt cycle. The nonlabile protein, m_{pp}, is a nondecreasing function of age

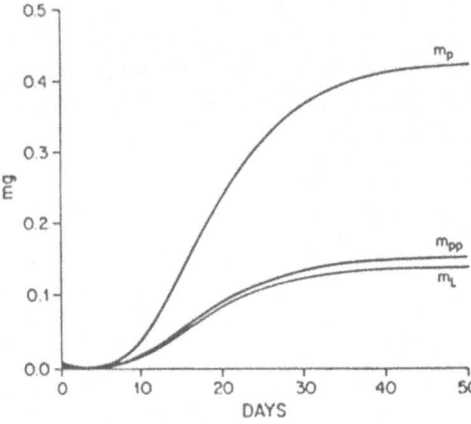

Fig. 3. Individual growth where reproductive size is not attained over an individual's life span

Examples of the dynamic growth of an individual are given in Figs. 2 and 3. The numerical solutions indicate that, at constant resource level, individual mass grows exponentially in the early juvenile stages, a fact consistent with the experimental work of Tessier and Goulden (1987). The bounded behavior of the individual model may be demonstrated analytically (Hallam et al., 1989).

This model of an individual daphnid has components chosen with risk assessment purposes in mind. For example, the effects of lipophilic chemicals on an individual cannot be adequately assessed unless a representation of its lipid pool is available.

We now describe how individual dynamics can be incorporated into a population model. Then, an indication of the ways to transfer the effects of toxicants on individuals to an aggregate of those individuals, in order determine the effects of a chemical on a population, is presented.

VI. Some Population Models and Their Utility for Risk Assessment

Classical population dynamics that appear useful in assessing risk of exposure models are of two types—the Leslie-Lefkovitch model, a discrete time formulation, and the McKendrick-von Foerster model, a continuous time formulation (see Frauenthal, 1986; Nisbet and Gurney, 1986; Metz and Diekmann, 1986). The Leslie-Lefkovitch model has been used the most widely in analyzing populations because the theoretical underpinnings are reasonably well understood, at least for the density independent cases. It is, however, not obvious that the individual physiological characteristics necessary for risk assessment can be incorporated into such a setting.

There is a continuous time analogue of the Leslie model (Hastings, 1983; Li, 1987), but the associated theoretical developments are not sufficiently complete to be of broad utility.

The most important population model from a risk assessment perspective seems to be an extension of the McKendrick (1926)–von Foerster (1959) formulation. This model, a hyperbolic partial differential equation, includes representations for the dynamics of individuals. The extended, physiologically structured model is of the general form

$$\frac{\partial p(t, x)}{\partial t} + \sum_i \frac{\partial}{\partial x_i} [g_i(t, x)p(t, x)] = -\mu(t, x, p)p(t, x) \tag{10}$$

where p is the population density function; t is time; x is a vector of individual physiological properties such as age, length, lipid mass, protein mass; $g_i = dx_i/dt$ is the growth rate for the ith physiological property; and μ is the mortality function. This partial differential equation must be coupled with the renewal equation $p(t, x/\text{age} = 0)$, which is a description of the birth process (Frauenthal, 1986). This model has existed for some time (Sinko and Streifer, 1967), but only recently has the potential of the model been widely recognized (see Metz and Diekmann, 1986, for a historical perspective). Even though recognition of the model's utility has occurred, applications are not yet flourishing primarily because the underlying theoretical developments are not strong, because insufficient numerical techniques exist, and because computational power has not been available.

The individual model constructed in the previous section can easily be incorporated into the coefficients of the McKendrick–von Foerster model as growth rates of physiological variables. The relationship of this problem to effects of toxicants on populations is described briefly in the next section.

VII. Theoretical Studies at the Population Level

Analytical Methods

The problem of assessing the effects of a toxicant on a population is complex and one should expect neither the models nor the model analysis to be simple. Metz and Diekmann (1986) present a state of the art overview of physiologically structured models. The degree of complexity of the population model can be moderated by hypotheses about the characteristics of the resources. For example, in the case of constant food density, a characteristic equation approach can be used to relate the rate at which the population is growing exponentially to models for individuals (Kooijman and Metz, 1984; and the next section). In other instances, we might think of a dynamic situation as a sequence of different constant food densities to which the characteristic equation method applies. Generally, however, we will need a more elaborate approach that, at least initially, has to rely on computer simulation studies.

If the exposure duration is short, the population is approximately constant. For this situation, a theory—survival of the fattest—has been proposed to determine the effects of a chemical on a population (Lassiter and Hallam, 1989).

The effects of a chemical on an individual are influenced by many chemical and biological factors. For example, for lipophilic chemicals, the principal component of effects determination is the individual compartment of high chemical affinity, the lipid compartment. The distribution of lipid within various population classifications forms the basis of an approach to study population effects. We have recently initiated investigations into a hierarchical set of population structures. For short, acute exposures, the population distribution might be viewed as static. For longer exposures, the population should be considered to be dynamic. Work on the static population approach has been done by Lassiter and Hallam (1989) and work is currently in progress on some dynamic population techniques by each of the authors. These approaches will be briefly described although the work must be regarded as preliminary.

Effects of Acute Exposures on Populations

To assess the effects of a lipophilic chemical on a population, Lassiter and Hallam (1989) propose the following procedure, which utilizes the uptake model (5). Baseline data include information about the chemical, the individual organism, and the population. The molecular weight, the environmental concentration, and the octanol-water partition coefficient, K_{OW}, must be known for the chemical. The active uptake surface area of the organism's environmentally exposed membrane, the total weight of the organism, and the proportion of the structure that is equal to the aqueous phase also are needed. In addition to information about the individual,

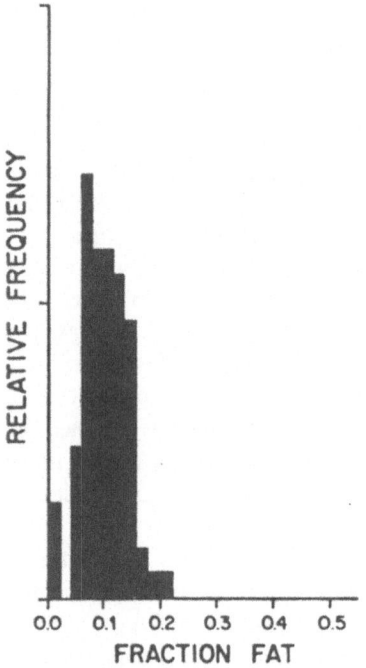

Fig. 4. Observed distribution of fraction lipid in a laboratory population of cichlids

knowledge is required about the distribution of lipid in the population. Complete information of this type is seldom available even for laboratory populations, although Brockway (1972) presented data to permit construction of a lipid distribution for a laboratory population of fish, all of about the same size (Fig. 4). We proceed by assuming that the static distribution of lipid in the population is known, whether it is determined by data collection or established through theoretical constructs is not important at this stage.

Because the exposure period is, by definition, short (the population distribution is regarded as being static), the effects region of the population can be determined. The effect concentration is determined from bioassay relationships that relate K_{OW} to effect concentrations. (Figure 5 is a graph that includes data from Konemann (1981) and Veith et al. (1983).)

The uptake model (5) is used to compute the concentration of chemical in the aqueous phase of the organism and the lipid fractions that yield the effect at the exposure concentration. All organisms for which the lipid fraction is less than or equal to P_L^e, the maximal lipid fraction that yields the effect, exhibit the effect (see Fig. 6). The truncation of the distribution represents the fraction of the population that has been affected at the effects level. If the capacity for a toxicant of an organism's lipid fraction exceeds P_L^e, its capacitance for a toxicant is assumed to prevent its aqueous concentration from reaching toxicant effects level by the end of the exposure period. A detailed example illustrating the above discussion may be found in Lassiter and Hallam (1989).

Simulation Studies

When the chemical exposure is chronic, knowledge of the dynamics of a population is fundamental to evaluation of survival. Indeed, most risk assessments should account for the dynamics of the population; unfortunately, few presently do. For

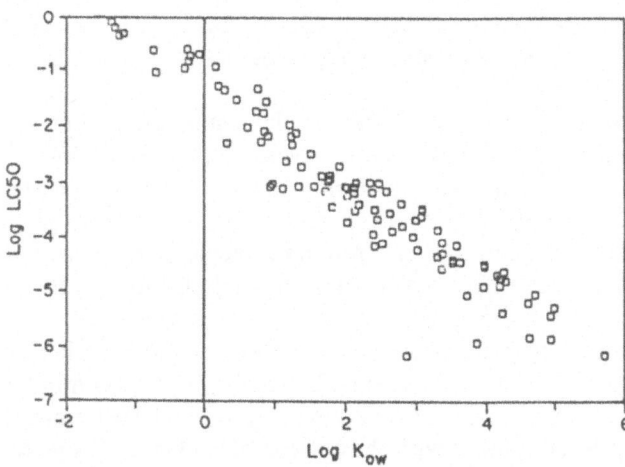

Fig. 5. Relationships between $\log LC_{50}$ and $\log K_{ow}$. Data from Veith et al. (1983) and Konemann (1981)

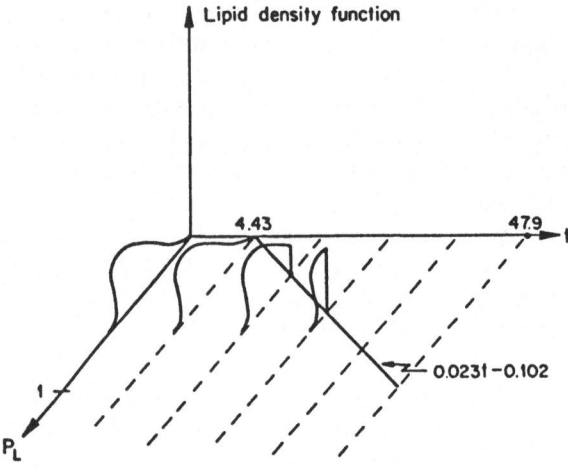

Fig. 6. Theoretical lipid distribution in an exposed population with truncated portion representing the individuals exhibiting the effect of the chemical

lipophilic chemicals, we have indicated that lipid dynamics in the population is probably an important consideration in determining risk.

As far as we know, there are two basic approaches to simulate population dynamics. The most efficient procedure, the frequency method, is to numerically integrate the partial differential equations representing the population. (de Roos (1987) provides an efficient algorithm.) The method seeks to evaluate the population size as well as the frequency distribution of individuals over the individual state space. For this purpose, one has to distinguish discrete classes for state variables. The computational effort depends heavily upon the number of classes that are selected.

Another approach, the family tree method, keeps track of each individual and the computational effect is a function of the number of individuals in the population. The optimum is a compromise between computing time and memory use at high numbers contrasted with stochastic phenomena at low numbers. Although this method is significantly less efficient than the frequency method, it is much more flexible. It can easily be adapted for differences between individuals such as when parameter values differ slightly between different individuals. This can be crucial for dynamics generated by a broad class of energy budget models (Kooijman et al., 1987).

In Kooijman's work, the number of parameters in the simulation studies (of scaled variables) is very small. From a physiological viewpoint, there is one for the constant food supply rate, one for the ingestion process, three for energy transduction, two for size stages, four for survival characteristics, one for biological variability; hence, a total of twelve parameters. From an ecological perspective, it would make computations of integrated systems more feasible if the number of parameters were condensed even more.

To derive a risk assessment tool that would be of utility to managers, a family tree structure has been implemented by Lassiter and Hallam for personal computer application. This development requires that population composition be determined by two parameters: resource level (coupled with the fraction of lipid in the resource) and the filtering rate. The general approach requires baseline prototype individuals whose life history is completely known. The individual model for lipid and protein compartments described in Section V is utilized here. All newborn are referenced to one of these prototype individuals through feeding level of the mother and through filtering rate. The family tree approach is basically a numerical scheme that keeps track of numbers of individuals, their ages, their lipid masses, and their protein masses at a given time by assuming an initial lipid distribution and following the offspring, the offspring of the offspring, etc., through a family tree. Output of this population model includes the distribution of lipid in the population at a given time.

VIII. Effects at Population and Ecosystem Levels

A crucial observation about effects of chemicals on populations is that no population effects occur if no individual effects occur. When experimental toxicological work with populations began a decade ago, it was thought that essential effects on individuals might easily be missed. Therefore, one should study populations and experimental ecosystems because, at these levels of resolution, effects on individuals are "integrated" in a manner that is relevant to ecological assessment. This idea applied not only to the exhibited variety of effects but it also reflected the possibility that small effects at the individual level might result in large effects at the population level (e.g., Halbach et al., 1983). Although these thoughts had and still have reasonable support, experiments with populations and ecosystems generally do not demonstrate the expected sensitivity (Sloff, 1985). Various conjectures exist for the discrepancy between expectations and experimental findings.

One explanation of the discrepancy concerns biological availability of the compounds which, because of complex fate processes, is less in the environment than in single species tests in the laboratory. Another aspect of this problem relates to the processes of adaptation and selection for resistant individuals (van Cappelleveen, 1987). Again, this might occur for natural populations but might not be included in single species tests unless specific precautions are taken to maintain natural population vigor. The consequence of both of these comments is that compounds appear less toxic in the environment than in laboratory situations. In principle, additional information can be obtained to explicate these perceived difficulties.

An antipodal perspective is that ecosystems might appear less sensitive than individuals when, in fact, they are not. Often the only variables observed in ecosystems are aggregated ones like total biomass as a measure of size or total rate of CO_2 incorporation as a measure of the general rate of photosynthesis. Effects might escape notice because it may not be possible to measure the subtle changes (such as species composition).

Another direction concerns the error of second kind in statistical testing. Although there are effects, we fail to recognise them because of the occurrence of stochastic fluctuations in the development of (experimental) ecosystems. This is combined with a poor insight into the development of uncontaminated ecosystems. Populations are notorious for their erratic dynamics. A possible explanation of this

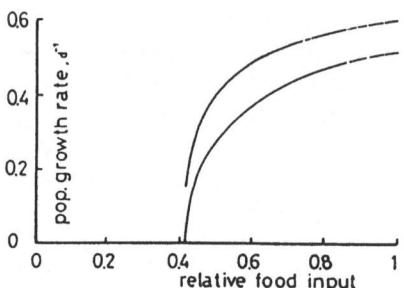

Fig. 7. The population growth rate as a function of relative food input. The difference in the two curves is the total energy invested in development

erratic behavior can be obtained from the representation of the population growth rate as a function of (constant) food density (Fig. 7). Figure 7 implies that at the low food densities, which frequently occur in natural populations, a small change in food density can result in a large change in the population growth rate. The effects of inclusion and exclusion of the term (8) is depicted by Fig. 7. The difference between laboratory and field toxicity might be accounted for by a good accounting of the chemical fate, particularly its sorption to particles. We do not expect to find a consistent relationship to total chemical in the environment apart from the environmental composition in a manner analogous to internal chemical fate inside an organism to lipid, structure, and aqueous phases. Another, more complicated, reason that natural populations seem less sensitive than expected relates to the circumstance that the sensitivity of populations depends on the food density. This can be illustrated by an analysis of the characteristic equation (valid at constant food densities for large time).

The subsequent discussion concerning chemically stressed populations relies heavily on Kooijman and Metz (1984), Kooijman (1986a), and Kooijman et al. (1987). Details about model formulations and motivation may be found in Kooijman (1986a, b, c). Although the modelling is primarily based upon the parthenogenetically reproducing water flea *Daphnia magna*, the principles appear to be quite general in application (Kooijman, 1987; Zonneveld and Kooijman, 1987). The subsequent results on the characteristic equation are similar to those in Kooijman and Metz (1984), but the present analysis accounts for energy reserves and uses a Weibull survival probability (instead of a prescribed survival up to a fixed age).

Two observations are useful in obtaining a characteristic equation representation of the components of a population. First, if age since hatching, a, is less than, a_j, the age at which $l(a_j) = l_j$, then $R(a) = 0$. Second, the incubation time can be considered as a time lag imposed on reproduction so in the characteristic equation, the incubation time a_b is added to the age since hatching to yield

$$\int_{a_j}^{\infty} \exp\{(a + a_b)r\} F(a) R(a) \, da = 1.$$

It is assumed that $F(a) = \exp\{-(a/a_m)^g\}$ is the Weibull survivorship function for aging with time parameter a_m and shape parameter g. The reproduction rate $R(a)$ is proportional to $f l^2(a)[\alpha + l(a)]/(\alpha + f)$ with length $l(a) = f - (f - l_b)\exp\{-\alpha\beta/(3\alpha + 3f)\}$ as was indicated in the individual models in Sect. V (see Eqs. (7) and (9)).

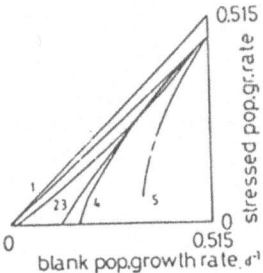

Fig. 8. The population growth rate of individuals (which are changed so that the maximum growth rate is 90 percent of the blank) as a function of the blank population growth rate. Type of effects: *1* reproduction, *2* growth, *3* assimilation, *4* survival, *5* maintenance

We now indicate how the stressed population growth rate relates to the uncontaminated population growth rate when chemicals affect the energy transfer rates. Suppose there is abundant food and the chemical exposure modifies the rate of conversion of food into feces and young in such a way that a small decrease of the population growth rate occurs. If the unstressed population growth rate is known, the characteristic equation can be solved for food input. This value is substituted into the characteristic equation for the chemically stressed population (which differs from the unstressed population only by changes in one of the energy budget parameters), and the resulting stressed population growth rate is determined. Figure 8 relates uncontaminated versus stressed population features such as growth rate, reproduction rate, assimilation rate, survival rate, and maintenance rate as affected by chemical insult. If we focus on maintenance rate and view growth rates as determined by food densities, we find that hard-to-detect effects are exhibited at higher food densities whereas relatively larger effects occur at lower food densities. For effects on reproduction rate we have the opposite situation. Significant effects may occur at high food densities but effects may not be noticed at low food densities. Hence, if we do not observe toxic effects on a population, we can not be certain that, at the same level of toxic stress, there will be no effects if feeding conditions change.

The effects of food limitation and predation can be compared in this setting. If a population is controlled by predation, it is likely to live at high food densities. In the absence of predators, the population probably is governed by food limitation. In this situation, there might be a low mean food density. In populations that are at equilibrium (i.e. the growth rate is zero) if the toxicant directly affects reproduction (see Fig. 8), no effects are observed at the population level. This corresponds to the situation where, in a food limited environment, food dictates whether young are born. A limited change in reproductive capacity does not manifest itself at the population level. If predation pressure removes the population from food limitation, effects on reproduction again become apparent. This has ramifications for the interpretation of data from field or experimental ecosystems. Observations must account for the spectrum from food limitation to predator control. Data from Kooijman (1986a) on the population growth rates of the rotifer *Brachionus rubens* presented for sodium metavanadate, which primarily affects the maintenance rate, and for 3, 4 dichloro-aniline, which affects the reproduction rate, confirm the above analysis.

Extensive simulation studies using the family tree approach has been done by Kooijman et al. (1987) for a constant food supply rate to the population. The results indicate that the population oscillates apparently due to a synchronization of life cycles. The cause of this synchronization is still under study but it seems to depend upon the fact that the ultimate size of an individual increases with food density. Because of the population oscillations, food density is also oscillatory so the analysis above that employs the characteristic equation is not applicable here. By averaging over a cycle, the population is constant. Here again it can be demonstrated that a reduction of up to 78% of potential reproduction rate hardly affects the population.

Lassiter, Hallam, Li and others at the University of Tennessee and the USEPA's Environmental Research Laboratory, Athens, Georgia, have done simulation studies utilizing the family tree approach. The work is based upon the extended McKendrick–von Foerster population model (10) with physiological growth rates determined by the individual model of Section V. Results of the research are preliminary and will be announced in a subsequent publication.

IX. Summary

The premise that ecological risk assessment must focus on the interface between biological and chemical components implies that biological models of individuals are mandated for ecotoxicological problems. We have discussed some recent efforts to model individuals from an energetics perspective.

Assessment of effects of particular chemicals on individuals requires that the components of the individual model should be selected with modelling objectives and properties of the chemical in mind. For example, to adequately assess the effects of a lipophilic chemical, a compartment that reflects lipid dynamics in an individual must be included in the model.

Determination of the exposure of an individual to a toxic chemical is another important portion of the risk assessment process. Many models currently used for risk assessment in the United States do not employ adequate evaluation of exposure of individuals nor do they adequately model the response of a population to a given exposure. For example, most population or community models used in risk assessment require a population level dose-response formulation to indicate the proportion of the population that is removed by the stress; unfortunately, such representations are not available from traditional bioassay results. We have described an exposure model GETS (and FGETS, the gill exchange model plus food uptake) which has proved useful in exposure analyses (Barber et al., 1988).

The exposure model and the individual model can be coupled and effects determined provided a quantitative effects assessment method is utilized. One proposal discussed above uses bioassay structure activity relationships for the effects assessment purpose when lipophilic chemicals with octanol-water partition coefficients ranging from approximately 10 to 10^7 are studied.

We concluded with a brief introduction to our approaches to assess the effects of chemicals on a population. The modelling efforts on this problem incorporated individual dynamics into a population setting in a manner such that the structure of the population is determined by the totality of the individuals.

References

Barber, M.C., Suarez, L.A., Lassiter, R.R. (1988) Modeling bioconcentration of nonpolar organic pollutants by fish. Envir. Tox. Chem. 7:545–558

Bartell, S.M., Gardner, R.H., O'Neill, R.V., Giddings, J.M. (1983) Error analysis of predicted fate of anthracene in a simulated pond. Envir. Tox. Chem. 2, 19–28

Beverton, R.J.H., Holt, S.J., (1957) On the dynamics of exploited fish populations. U.K. Min. Agric. and Fish. Fish. Invest. Ser 2, 19

Brockway, D.L. (1972) The uptake, storage and release of dieldrin and some effects of its release in the fish cichlasoma bimaculatum (Linnaeus). Dissertation, University of Michigan, Ann Arbor, MI

Buikema, A.L., Jr. (1972) Oxygen consumption of cladoceran daphnia pulex, as a function of body size, light, and acclimation. Comp. Biochem. Physiol. 42A, 877–888

De Roos, A. (1987) Numerical methods for structured population models. The escalator boxcar train. Num. Methods Part. Diff. Equations (to appear)

Edwards, N.T., Ross-Todd, B.M., Garver, E.G. (1982) Uptake and metabolism of ^{14}C-anthracene by soybean (Glycine S.). Environ. Exptl. Botany 22, 349–357

Frauenthal, J.C. (1986) Analysis of age-structure models. In: Hallam, T.G., Levin, S.A. (eds.) Mathematical Ecology: An Introduction Biomathematics, vol. 17. Springer Berlin Heildelberg New York London Paris Tokyo pp. 117–148

Gabriel, W. (1982) Modelling reproductive strategies of Daphnia. Arch. Hydrobiol. 95, 69–80

Gardner, R.H., Cale, W.G., O'Neill, R.V. (1982) Robust analysis of aggregation error. Ecology 63, 1771–1779

Gerritsen, J. (1984) Size efficiency reconsidered: A general foraging model for free swimming aquatic animals. Am. Nat. 123, 450–467

Goulden, C.E., Comotto, R.M., Hendrickson, Jr., J.A., Hornig, L.L., Johnson, K.L. (1982) Procedures and recommendations for the culture and use of Daphnia in bioassay studies. In: Pearson, J.G., Foster, R.B., Bishop, W.E. (eds.) Aquatic Toxicology and Hazard Assessment: Fifth Conference. ASTM STP 766. American Society for Testing and Materials, Washington, D.C., pp. 139–160

Goulden, C.E., Henry, L.L. (1982) Lipid energy reserves and their role in Cladocera. Am. Assoc. Adv. Sci. Symp.

Halbach, U., Siebert, M., Westermayer, M., Wissel, C. (1983) Population ecology of rotifers as a bioassay tool for ecological tests in aquatic environments. Ecotox. and Envir. Safety. 7, 484–513

Hallam, T.G., de Luna, J.L. (1984) Effects of toxicants on population: A qualitative approach III. Environmental and food chain pathways. J. theor. Biol. 109, 411–429

Hallam, T.G., Lassiter, R.R., Li, J., Suarez, L.A. (1989) Modelling individuals employing an integrated energy response: Application to Daphnia Ecology (to appear)

Hastings, A. (1983) Age dependent predation is not a simple process. I. Continuous time models. Theor. Popul. Biol. 23, 347–362

Karickhoff, S.W. (1981) Semi-empirical estimation of sorption of hydrophobic pollutants on natural sediments and soils. Chemosphere 10(8), 833–846

Konemann, H. (1981) Quantitative structure activity relationships in fish toxicity studies. Part I. Relationship for 50 industrial pollutants. Toxicology 19, 209–221

Kooijman, S.A.L.M. (1986a) Population dynamics on basis of budgets. In: Metz, J.A.J., Dickmann, O. (eds.) The Dynamics of Physiologicaly Structured Populations Lecture Notes in Biomathematics, vol. 68. Springer, Berlin Heidelberg New York London Paris Tokyo, pp. 266–297

Kooijman, S.A.L.M. (1986b) What the hen can tell about her eggs: Egg development on the basis of energy budgets. J. Math. Biol. 23, 163–185

Kooijman, S.A.L.M. (1986c) Energy budgets can explain body size relations. J. theor. Biol. 121, 269–282

Kooijman, S.A.L.M., Metz, J.A.J. (1984) On the dynamics of chemically stressed populations: The deduction of population consequences from effects on individuals. Ecotoxicology and Environmental Safety 8, 254–274

Kooijman, S.A.L.M., van der Hoeven, N., van der Werf, D.C. (1987) Population consequences of a physiological model for individuals (to appear)

Lassiter, R.R. (1986) A theoretical basis for modeling element cycling. In: Hallam, T.G., Levin, S.A. (eds.)

Mathematical Ecology: An Introduction. Biomathematics, vol. 17. Springer, Berlin Heidelberg New York London Paris Tokyo, pp. 341–380

Lassiter, R.R., Hallam, T.G. (1989) Survival of the fattest: A theory for assessing acute effects of hydrophobic, reversibly acting chemicals on populations. Envir. Toxicol. Chem. (to appear)

Li, Jia (1987) Persistence and extinction of populations. Dissertation. University of Tennessee, Knoxville, TN

Lynch, M., Weider, L.J., Lampert, W. (1986) Measurement of the carbon balance in Daphnia. Limnol. Oceanogr. *31*(1), 17–33

McKendrick, A.G. (1926) Application of mathematics to medical problems. Proc. Edinb. Math. Soc. *44*, 98–130.

Metz, J.A.J., Diekmann, O. (1986) The dynamics of physiologically structured populations. Lecture Notes in Biomathematics, vol. 68. Springer, Berlin Heidelberg New York London Paris Tokyo

Newsholme, E.A., Start, C. (1973) Regulation in Metabolism. Wiley, New York

Nisbet, R.M., Gurney, W.S.C. (1986) The formulation of age-structure models. In: Hallam, T.G., Levin, S.A. (eds.) Mathematical Ecology: An Introduction. Biomathematics, vol. 17. Springer, Berlin Heidelberg New York London Paris Tokyo, pp. 95–116

O'Neill, R.V., Gardner, R.H., Barnthouse, L.W., Suter, G.W., Hildebrand, S.G., Gehrs, C.W. (1982) Ecosystem risk analysis: A new methodology. Envir. Toxicol. Chem. *1*, 167–177

Paloheimo, J.E., Crabtree, S.J., Taylor, W.D. (1982) Growth model of Daphnia. Can. J. Fish. Aquat. Sci. *39*, 598–606

Randle, P.J., Garland, P.B., Hales, C.N., Newsholme, E.A. (1963) The glucose fattyacid cycle. Its role in insulin sensitivity and the metabolic disturbances of diabetes mellitins. Lancet *1*, 785

Rashevsky, N.F. (1959) Some remarks on the mathematical theory of nutrition of fishes. Bull. Math. Biophys. *21*, 161–183

Richman, S. (1958) The transformatin of energy by Daphnia pulex. Ecol. Monogr. *28*, 273–291

Ricker, W.E. (1954) Stock and recruitment. J. Fish Res. Bd. Can. *5*(15), 991–1006

Sinko, J.W., Streifer, W. (1967) A new model for age-size structure of a population. Ecology *48*, 910–918

Slooff, W. (1985) The role of multispecies testing in aquatic toxicology. In: Cairns, J. (ed.) Multispecies Toxicity Testing. Pergamon, New York

Smith, F.E. (1963) Population dynamics in Daphnia magna and a new model for population growth. Ecology *44*, 651–663

Swenson, W.A., Lloyd, L., Smith, Jr. (1973) Gastric digestion, food consumption, feeding periodicity, and food conversion efficiency in walleye (Stizostedion vitreum vitreum). J. Fish. Res. Bd. Can. *30*(9), 1327–1336

Taylor, B.E. (1986) Effects of food limitation on growth and reproduction of Daphnia. Arch. Hydrobiol. Beih. Ergebn. Limnol. *21*, 285–296

Tessier, A.J., Goulden, C.E. (1987) Cladoceran juvenile growth: Implications for competitive ability (to appear)

Thieme, H.R. (1986) A differential-integral equation modelling the dynamics of populations with a rank structure. In: Metz, J.A.J., Diekmann, O. (eds.) The Dynamics of Physiologically structured populations. Lecture Notes in Biomathematics, vol. 68. Springer, Berlin Heidelberg New York London Paris Tokyo, pp. 496–511

van Cappelleveen, E. (1987) Ecotoxicity of heavy metals for terrestrial isopods. Ph.D. Thesis, Free University, Amsterdam

Vaughan, B.E., et al. (1982) Pacific Northwest Laboratory Annual Report for 1981 to the DOE Office of Energy Research, Part 2. Ecological Research. PNL-4100, Pt. 2. 59–63. Battelle, Richland Washington

Veith, G.D., Call, D.J., Brooke, L.T. (1983) Structure-toxicity relationships for the fathead minnow, Pimephales promelas: narcotic industrial chemicals. Can. J. Fish. Aquat. Sci. *40*, 743–748

von Foerster, H. (1959) Some remarks on changing populations. In: Stohlmann, F. (ed.) The Kinetics of Cellular Proliferation. Grune and Stratton, New York

Wallis, I.G. (1975) Modelling the impact of waste on a stable fish population. Water Research *9*, 1025–1036

Zonneveld, C., Kooijman, S.A.L.M. (1987) The application of a dynamical energy budget model to the pondsnail Hymnaea stagnalis (to appear)

Part V. Demography and Population Biology

Mathematical Models in Plant Biology: An Overview

Louis J. Gross

Contents

Introduction

The study of plants could be undertaken at essentially every level of organization within biology, from that within a cell to the entire biosphere. To cover even the main theoretical questions on these diverse levels and the mathematical approaches used to analyse them would require several volumes. My objective here is to consider a small subset of the work that has been done, dealing only with the levels normally taken as being part of the purview of ecology and touching somewhat on a few more applied problems in agriculture. I will not discuss statistical analyses of plant community assemblages nor most aspects of plant-animal interactions, such as pollination biology (Real, 1983). Biophysical approaches were reviewed earlier (Gross, 1986b). Background references should be consulted for further details (e.g. France and Thornley, 1984; Givnish, 1986a; Gross and Miura, 1986; Jean, 1984; Rose and Charles–Edwards, 1981).

The importance of plant biology to life on this planet cannot be overestimated. Without the ability of plants to utilize the energy in sunlight to manufacture hydrocarbons through the process of photosynthesis, life as we know it could not have evolved. Green plants exist in virtually every habitat, terrestrial and aquatic, and serve as the base for all the intricate food webs which make up the earth's ecological systems. Many of our drugs are derived from plant compounds, our food supply depends on farmers' abilities to irrigate, fertilize and grow appropriate crops, and the beauty of our landscape is enhanced by the work of horticulturalists and plant breeders.

As in much of biology, mathematical models in plant biology are of relatively recent vintage, however the general theoretical development of plant biology lags in certain respects that of animal biology. For example much of the mathematical development of population biology has been undertaken with a specific animal bias, ignoring the uniquely different requirements of a general theory of plant populations. This lack of theory for plants is recognized and is now being addressed. The utility of modeling approaches in plant biology is made particularly clear when one considers global questions. For example attempts to analyze the global carbon cycle would be stymied without the recourse to mathematical models of photosynthesis and attempts to extrapolate them over wide areas (King et al., 1987).

My procedure will be to describe in detail only a couple of models at each level of organization, while providing references and some background on other models. The topic grouping is somewhat arbitrary. In particular it does not reflect the relative amount of progress to date—the work at the physiological level is far more complete than that at any of the others. This will be evident from the amount of space devoted to each of the levels discussed.

Plant Physiology

There are hosts of topics which might be included under this: cellular structure and growth, organogenesis, regulation by plant hormones, photosynthesis, respiration, transport of fluids, gas exchange and water relations, plant form, and allocation between plant parts (Rose and Charles–Edwards, 1981; Thornley, 1976). I will not cover cellular level problems except as they enter into the models for photosynthesis and water relations. However, with the recent rapid advances in tissue culture techniques for plant growth, I expect there to be much development of plant cell models in the near future.

Plant Architecture

A very active area of theory concerns the architecture or form of plants and associated questions of phyllotaxis (Horn, 1971; Thornley, 1977; Jean, 1984). Anyone who has walked through fields and forests cannot help but be amazed at the diverse structures that plants develop, though they are all based upon some type of branched form. The main theoretical questions concern the how and why: what are the physiological mechanisms that control plant form and how do they work, and why is there so much diversity in the forms that we see.

Phyllotaxis concerns the study of the arrangements of primordia (i.e. the locations of future leaves, scales, or other plant organs), the most common arrangement being spiral in form as in sunflowers and pineapples. For example, consider the observation of a leafy stem in which one chooses an initial leaf, then constructs a spiral by moving up the stem connecting the points of attachment of each successive leaf. Continue this until a leaf is reached that is approximately in the same angular position on the stem as the initial leaf. Then the ratio of the number of turns around the stem to the number of leaves encountered, excluding the first, is usually one of the fractions made from the Fibonacci sequence:

$$1/2, \ 1/3, \ 2/5, \ 3/8, \ 5/13, \ 8/21, \ldots$$

The Fibonacci sequence is generated by the recurrence relation

$$u_{n+1} = u_n + u_{n-1}$$

with the above case obtained from $u_0 = u_1 = 1$. For example, pear trees have a foliar cycle defined by the ratio 3/8 while apple trees have 2/5.

Closely connected with the above is the fact that the angle between consecutively emerging primordia, known as the divergence angle, generally converges to the Fibonacci angle of 137.51 degrees (the golden mean angle), or to a divergence angle generated from another Fibonacci sequence with different values for u_0 and u_1. To see how this arises, consider the polar coordinate map

$$r_n = bn, \quad \theta_n = nz$$

for the position of the nth primordia, where z is the angle between each successive primordia and b is a constant which generates a background spiral pattern. If z is a rational divisor of 2π, then this generates rays emanating from the origin; for example if $z = 4\pi/5$, then there are 5 equally spaced rays produced. If however z is an irrational divisor of 2π, then secondary spirals are generated. In particular, a Fibonacci spiral is obtained when $z = 2\pi/m$ or $137.5°$ where m is the golden mean given by

$$m = 1/\left(\lim_{n \to \infty} (u_{n+1}/u_n) - 1 \right) = (1 + \sqrt{5})/2.$$

These fascinating observations have long generated the interest of botanists (the earliest publications on this date from 1836), and there have been a number of biological theories proposed to explain them (see Jean, 1983 for a brief review of the mathematical approaches, or Jean, 1984 for a more complete overview). Mathematical models based on these theories have only been developed over the last two decades or so. However descriptive models aimed at a complete and experimentally relevant quantitative description of the observations have been around longer (e.g. Richards, 1951). Jean (1984) breaks down the explanatory models which have been developed into four types. I will merely mention them here, without going into their mathematical formulations.

The contact pressure model assumes that there are mutual pressures between adjacent primordia which are responsible for the observed patterns. Adler (1977) models this by allowing primordia to grow in diameter until they touch their nearest neighbor, then growth proceeds so as to maximize the minimal distance between the centers of primordia. The space-filling theory assumes that leaves arise in the largest gap between existing leaves, and do so as soon as a gap reaches some required minimum size and distance from the growing tip. Adler (1975) develops this theory.

The diffusion theory supposes that primordia produce a morphogen, interpreted as a growth-inhibiting hormone, which diffuses from the primordia. The exact chemical nature of the morphogen is never specified. The morphogen acts as an inhibitor of new primordia, so that a new primordium is initiated where the field, that is the concentration of the morphogen, is minimal. Thornley (1976) develops this theory, which is based on a steady-state analysis of a partial differential equation of reaction-diffusion type. For more on the relationship between this theory and the golden mean, see Marzec and Kappraff (1983). Finally the systemic approach (Jean, 1984) argues that the phenomenon of phyllotaxis is an amplification of branching patterns in early plants. The mathematical formulation is based on partially ordered systems of primordia, interacting in an aggregative fashion. The hierarchies thus generated are then analyzed so as to choose those which minimize an entropy measure of the growth pattern.

Regarding the question of why there is so much diversity in plant form, consider the following comparison. Imagine that you are an engineer charged with designing a device for collecting solar energy. Typically then, you would choose a form that you feel is optimal (according to some criterion you've chosen, and associated constraints on the material to be used, cost, etc.), and then just produce copies of this to utilize it over large areas. Although at some level plants are constructed this way, consisting of somewhat similar leaves, shoots, and needles repeated many times, at the same time there is tremendous diversity of form, both within species and between species.

The usual explanation for this diversity rests upon the great variability of environmental conditions at different locations and with height and time at a single location, in part due to the feedback effects of plants on themselves. As a plant grows in height for example, it shades the lower portions of itself, and this modification of its environment varies throughout the lifespan of the plant. The presumption then is that, even if there were a single criterion for plants to "optimize", environmental variability induces alternative optimal forms dependent upon the actual conditions at the site. In particular, if conditions are in part unpredictable through plant lifespan, then diverse forms may be equally well suited as measured by the expected value of the optimization criterion. Here, new criteria which take account of the variance and the underlying probabilistic nature of the environment may be appropriate, as for example have been included in risk-sensitive foraging theory (Real and Caraco, 1986). Alternative explanations for the diversity of form involve the potential for feedback effects to allow different forms to be equivalent, or for selective fitness components to be only weakly coupled to form. It is natural then to investigate the effects of environment upon optimal form under a variety of criteria which might apply to plants. For a recent compendium of work along these lines, see Givnish (1986a).

One of the criticisms of this approach concerns the assumption that natural selection acts to optimize plant processes according to certain a priori criteria, and that it is possible for us to ascertain the nature of these criteria from current observations and the limited historical information available to us. As Givnish (1986a) points out in his introduction, there have been well grounded criticisms of these assumptions. These are due in part to the inherent circularity of the approach (the constraints included are based upon observations which we then try to predict), and the fact that evolution is a historical process that works more as a tinkerer might to produce some kind of workable object from limited materials (Jacob, 1977). Evolution then leads to organisms that are not optimally tuned to environment, but rather ones that have been determined to be "workable" through the force of natural selection, as constrained by phylogenetic history. Despite these criticisms, the use of optimization models has led to a host of new hypotheses about plant form and function, some of which can be rigorously tested. Still the limitations of the approach must be acknowledged, and as our understanding of the mechanisms and genetic control of plant behavior improves, perhaps the use of these phenotypically-based approaches will decline.

Much of the work on the branching forms of plants involves simulation models constructed to mimic such attributes as number of branches, interbranch lengths, and branching ratios (Bell et al., 1979). Although it may be natural to think of plant branching patterns in terms of a stochastic branching process, this seems not to have been attempted. Investigations have been made of a variety of organizing principles for plant growth including maximizing effective leaf area (Honda and Fisher, 1978; Honda and Tomlinson, 1982), biomechanical limitations (Keller, 1980, looks at tendril growth this way), and a combination of various criteria (Niklas, 1982, 1986).

Niklas (1986) argues that one should view the evolution of plant architecture as a reconciliation of conflicting design requirements. These include: (a) the need for light interception, which is affected by branching angles and the probabilities of branch initiation; (b) mechanical constraints due to the maintenance of a usually upright form, which are affected by branching angles and bending moments of the shoot material; (c) hydraulic constraints due to the requirements for water transport throughout the plant and a limitation on the amount of dessication various tissues can withstand; and finally, (d) there is the need for reproduction, which may require the display and maintenance of flowers and fruits at particular locations.

Through the application of various physical laws, Niklas (1986) argues that the structure of plants is defined in part by the same mathematical constraints on abiotic form. He then concludes that there is no single "optimal" form—because there are more than two design criteria, the form predicted in a particular environment is intimately coupled with the environmental conditions at that site. In addition it is found that evolutionary patterns of change in plant architecture conform well to the hypothesis that there have been increasingly more efficient solutions to the problem of survival on land. Although this approach does not explain the diversity of plant form at any given site, it provides a template against which to test alternative hypotheses about the interaction between the multiple constraints acting on form as well as investigate evolutionary patterns.

Fluid Mechanics

An area of great physiological interest involves the flow of fluids through plants (Rand, 1983). Photosynthesis is driven by the gaseous diffusion of carbon dioxide into a leaf and the accompanying loss of water presents a major constraint on plant growth. Liquid-phase flow may be broken into separate compartments. There is the flow within the veinous structure of leaves, from which water diffuses into the mesophyll cells. From these cells there is gaseous phase diffusion of water into the intercellular air spaces from which it leaves the leaf through pores in the leaf surface, called stomata. Hydrocarbons manufactured in the mesophyll cells are translocated as a fluid to other parts of the plant through the vascular phloem tissue. Water is absorbed in the root tissue, along with certain nutrients, and carried up to the leaves via flow through the xylem, the transpiration stream. Each component of the flow system presents its own problems in fluid mechanics, and I will merely touch on them here.

It is usually assumed that the upward flow of water to leaves from the roots is due to a pressure difference induced by the evaporation of water from the leaves. The analysis of this is complicated by the anatomy of the tracheid cells which make up the xylem; flow between neighboring cells occurs through pits, or pores which have membrane coverings that allow the pit to be relatively open or closed to flow. Chapman et al. (1977) analyze this flow assuming an ideal fluid. The Munch hypothesis is that the downward flow in the phloem is driven by local concentration differences which are maintained by active transport mechanisms. The sieve cells which make up the phloem are alive, allowing active transport to be a possible mechanism, while the tracheids which make up the xylem die upon reaching maturity. The Munch hypothesis has been incorporated in a time-dependent partial differential equation model for the pressure and sugar concentration in sieve cells (Smith et al., 1980) and other approaches are considered in Rand and Cooke (1978) and Rand et al. (1980).

The movement of water in plants occurs due to gradients in chemical composition which bring about diffusive flow, gradients in hydrostatic pressure and those in gravitational potential. It has been found to be useful to combine the effects of these through a single measure—the water potential, ψ. This is defined by (Nobel, 1974)

$$\psi = p - RTc + \rho gz$$

where p = hydrostatic pressure (in bars)

R = gas constant = $83.141 \, cm^3 - bar/mole \, K$
T = temperature (K)
c = concentration of all solutes in an assumed dilute solution (mole/cm^3)
ρ = density of water (g/cm^3)
g = acceleration due to gravity = $980 \, cm/sec^2$
z = height (cm).

The water potential, ψ, is measured in bars, where 1 bar = $10^6 \, dyne/cm^2$ and is approximately one atmosphere. A variety of workers have used water potential as a basic variable in models for pressure within plant cells and between the symplasm (essentially living material) and apoplasm (essentially non-living regions). See Molz (1976) for a coupled diffusion equation model of flow in these pathways.

In contrast to the work on the phloem and xylem, flow in the roots has received relatively little attention. In general, below ground parts of plants have not been well studied, in part due to the difficulty of performing reliable experimental studies on them. Flow in roots is most often modeled using discrete versions of Fick's law, in which the pathway is broken up, rather arbitrarily, into pieces with separate resistances to flow. One alternative approach, due to Plant (1983), utilizes a continuum model for root growth assuming that growth is due to mechanical forces created by the flux of water into the root. For a cellular-level approach to root growth, using an age-structured model for the spatial patterns of cell length in the root apex, see Bertaud et al. (1986) and Bertaud and Gander (1986).

Another aspect of fluid mechanics concerns the flow external to the plant. Wind patterns in air affect many plant processes through convective heat exchange and diffusive resistance to gas exchange, as well as through direct effects on plant architecture in some habitats (Nobel, 1981). A striking example of this are the "flag trees" which exist at tree line at high elevations in Krummholz vegetation, in which the dessicating effects of wind and blowing ice crystals produce trees with branches only on the leeward side. An analogous situation exists for macroalgae, some of which have structural features which enhance their ability to maintain their substratal attachments in moving water. For a biophysical analysis of the trade-offs of various seaweed morphologies as affected by water flow, see Koehl (1986).

Photosynthesis

As a process that is basic to life on this planet, photosynthesis has been well studied from both an experimental and a theoretical perspective. The most complete summary to date on the modeling aspects is Hesketh and Jones (1980). Included are a variety of models to mimic the steady-state response of photosynthesis in plants with the three different main biochemical pathways of the process, the C_3, C_4, and CAM metabolisms. The environmental factors considered include light quantity and quality, carbon dioxide concentration, temperature, and humidity. One of the great difficulties involves attempting to combine the often conflicting experimental results from controlled laboratory conditions with those from field conditions. The host of interacting variables involved complicates the modeling process and has led to a multitude of models which take into account different environmental factors. For example, there are at least three major types of models derived to indicate photosynthetic response to temperature (Sharpe, 1983) and it is not clear which approach is most consistent with the available data.

Photosynthetic rate is defined as the net rate at which a plant or plant part exchanges carbon dioxide with the atmosphere. It is mainly measured at the leaf level in an ecological context, and is positive when there is a net flux of CO_2 into the leaf. It is negative when the respiratory processes in the leaf release more CO_2 than the photosynthetic processes absorb, as at night. Thus the study of photosynthesis requires an understanding of respiration, which is usually broken into that due to the normal metabolic activity of cells (dark respiration) and that which is due to the biochemistry of photosynthesis (photorespiration). It has been very difficult to separate these experimentally, and there is still not agreement as to their relative magnitudes or how they are coupled to alternative leaf forms. This is further complicated by the fact that respiration continues through the night, and total nightly respiration seems to be a function of the level of photosynthesis through the day.

Photosynthesis is driven by the gaseous diffusion of CO_2 into the leaf, mainly via pores in the leaf surface called stomata. There is an attendant loss of water from the leaf interior due to the gradient of concentration from the essentially water saturated condition internal to the leaf to the dry air external to a boundary layer surrounding the leaf. This loss of water presents a major constraint on plant growth, and thus plants have evolved the capacity to modulate it through control of the apertures of stomata. A number of workers have modeled stomatal movements (Cowan, 1972; Delwiche and Cooke, 1977; Cooke et al., 1977; Sharpe and Wu, 1978) from the point of view of biomechanics. A parameter study of these models indicates that Hopf bifurcations to a limit cycle can occur, giving oscillations in pressure within the cells surrounding the pore, and thereby producing oscillations in stomatal apertures (Rand et al., 1981; Rand et al., 1982). There is evidence that the oscillations so induced are of evolutionary advantage (Upadhyaya et al., 1981). These models are not closely coupled with the environmental factors controlling stomatal apertures in field conditions, however. For a dynamic model which takes into account light variations, see Kirschbaum et al. (1988). Recent work on coupling between stomata at different positions on the same leaf indicates how this can lead to a wave of stomatal opening or closing across the leaf (Rand and Ellenson, 1986).

A variety of other models of stomatal regulation of plant water loss are reviewed in Hall (1982). These involve discrete versions of Fick's law of diffusion, breaking up the pathway of diffusive flux into compartments as described in Gross (1986b). One alternative approach to modeling stomatal functioning concerns the optimization of a measure of water use relative to carbon gain (Cowan, 1977; Cowan and Farquhar, 1977) and has been quite successful in synthesizing the available data (Schulze and Hall, 1982; Cowan, 1982). The approach starts with the assimilation rate per unit leaf area, A, and transpiration rate, E. These depend on time and location on the leaf, and E is viewed as an implicit function of A. The problem involves calculus of variations if the assumed objective is to minimize transpiration over some time period subject to the constraint that total assimilation over that time is fixed at some level. Presumably this level is one which will assure adequate photosynthate production to meet plant needs for growth and reproduction. If s is the space variable, the requirement is that

$$\int_0^T \int_0^S [E(A, s, t) - \lambda A] \, ds \, dt$$

be minimized where λ is a Lagrange multiplier. Euler's equation leads to the result that

$$\left(\frac{\partial E}{\partial A}\right)_{s,t} = \lambda \quad \text{and} \quad \left(\frac{\partial^2 E}{\partial A^2}\right)_{s,t} > 0.$$

For elaboration of this theory see Cowan (1982). The key result above, concerning the constancy of the partial derivative of E with respect to A, has been found to hold approximately as stomatal aperture varies through a day, but not in all circumstances. A key problem with this approach is that it doesn't provide a mechanistic basis to determine the parameter λ. For more on the relationship between this theory and alternative formulations to take account of transpirational costs, see Givnish (1986b). Aside from its inherent assumptions about optimality, this approach has not taken into account possible constraints on the time response of stomata. Thus it should only be viewed as being applicable on time scales slow relative to the time constants of stomatal response to environmental changes. As these seem to be on the order of 10–20 minutes, this approach should probably be viewed as a strictly steady-state one. Givnish (1986b) discusses some of the difficulties involved in extending the theory to deal with dynamic environments.

Models have also been constructed for the investigation of flows of CO_2 and water internal to a leaf. Here the objective is to compare leaves of different anatomies to determine their effect on carbon gain. In part the interest in this lies in the fact that leaves of the same plant can have quite different internal structuring dependent upon the environmental conditions in which they developed. For example, leaves raised in high light tend to be tightly packed with mesophyll tissue in comparison to those raised in low light. Parkhurst (1977) presented a 3-dimensional reaction diffusion type partial differential equation model to analyze steady-state CO_2 and water concentration patterns within leaves of different morphologies. The model is

$$\frac{\partial C}{\partial t} = \frac{\partial}{\partial x}\left(p_1 D \frac{\partial C}{\partial x}\right) + \frac{\partial}{\partial y}\left(p_2 D \frac{\partial C}{\partial y}\right) + \frac{\partial}{\partial z}\left(p_3 D \frac{\partial C}{\partial z}\right) - U$$

where $C(x, y, z, t) = CO_2$ concentration at (x, y, z) at time t

$U(x, y, z, C) = CO_2$ uptake rate at (x, y, z) with concentration C
$D(x, y, z) = $ diffusivity of CO_2 at (x, y, z)
$p_i(x, y, z) = $ leaf porosity in the ith direction at (x, y, z).

The results of this sort of analysis allow for much more detailed investigation of leaf anatomy effects than the coarse approach of the analog-resistance types of models which are prevalent (Gates, 1980). Gross (1981) investigated the dynamics of a one-dimensional version of the above, and found that the time scale of response is rapid in comparison to observed photosynthetic dynamics, indicating that biochemistry and stomata must be controlling photosynthetic dynamics, not diffusion. Parkhurst (1986) carried out a finite element analysis of the steady state behavior of the full three dimensional model, to point out that many aspects of internal leaf structure may be viewed as adaptations to enhance CO_2 diffusion.

The biochemical background for most photosynthesis models tends to be highly oversimplified. Due to the great complexity of even the simplest form of the Calvin cycle it requires at least seventeen non-linear, stiff differential equations to describe, and just estimating the rate parameters is a formidable problem (Milstein and Bremerman, 1979). With the aim of simulating leaf carbon metabolism over a 24 hour period, Hahn (1984) constructed a model with 19 state variables, including the principal metabolites within a chloroplast, and the components of sucrose metabolism in the cytoplasm. By switching parameter values, light versus dark conditions are mimicked, and dynamic simulations run. The steady-state behavior is derived in Hahn (1986), but the approach is severely limited by the lack of data to estimate the large number of parameters. Despite some stochastic models of the photochemistry of photosynthesis (Goel and Richter-dyn, 1974; DiMichelle et al., 1978), the integration of photosystem models with biochemical ones is still in its formative stages. Most biochemical models merely assume a light dependent input of ATP and NADPH, the energy drivers for the reactions.

Another area in which there has been very little work concerns the dynamics of the process. Essentially all approaches have been steady-state, though some recent attempts have been made to investigate the effects of variable light and CO_2 levels on photosynthesis (Gross, 1982; Kaitala et al., 1982). These involve use of history dependent rate parameters due to the time lags in the system. The dynamics of photosynthesis can be viewed on three different time scales (Gross, 1986a): the physiological time scale (within a day), the acclimation time scale (days to weeks), and the evolutionary time scale.

Concerning the first of these time scales, perhaps the most influential modeling work to date has been by Farquhar and co-workers (Farquhar and von Caemmerer, 1982). This involves a biochemically-based model which views carbon assimilation as co-limited by enzyme activity, light and the associated electron transport limitations, and phosphate availability. The key variable is p_i, the partial pressure of intercellular CO_2. A main point of the model is to specify the joint effects of biochemical and photochemical processes in determining how assimilation of carbon varies with p_i. Since stomatal aperture also limits assimilation, the model determines the p_i appropriate to given environmental conditions by setting assimilation rate from the biochemical calculations equal to that obtained from a stomatal model and solving implicitly for p_i. The theory of Cowan (1982) described above is typically used for the stomatal portion of this.

This modeling approach has been used for example to investigate temperature effects on photosynthesis (Kirschbaum and Farquhar, 1984). The model has focussed a great deal of attention by experimentalists on several key components of the biochemistry, and provided a mechanism to combine the current best guesses of how parts of the complex reactions involved are coupled. In particular, the significance of p_i as a key variable has even led instrument manufacturers to develop photosynthetic meters which automatically calculate its value when taking leaf measurements. The basic model has been modified in a number of ways to take account of other factors, for example to include internal leaf gradients in p_i and light (Gutschick, 1984a, b).

Whole Plant and Population Level

Whole Plant

A variety of quite different approaches have been used to analyze whole plant growth. Two general schemes to investigate the partitioning of resources during plant growth are compartment models based on mass balance, and those based on optimization of a fitness measure. In the first of these, the physiological models discussed above may be combined in compartmental models which subdivide a plant into roots, shoot, leaves, fruits, etc. and the flow of material or energy between these compartments is analyzed (Thornley, 1976; Loomis et al., 1979; Reynolds and Thornley, 1982). The central difficulty is specifying the rule for partitioning, for example between root and shoot growth. A typical approach has been empirical, producing equations relating shoot and root dry weights to the relative strengths of sinks, and respiration rates (Barnes, 1979). The background here is that observations imply that under constant conditions there is a form of balanced growth, with a constant ratio of root to shoot material—in models this leads to balanced exponential growth, meaning that variables such as dry weights of plant parts increase exponentially at the same rates while the substrate concentrations (carbon and nitrogen) in the various plant parts stay constant. This gives constant carbon-to-nitrogen ratios in the plant.

Several models (Reynolds and Thornley, 1982; Johnson, 1985; Thornley and Johnson; 1986) use a "goal-seeking" model for partitioning, in that the parameters which determine the partitioning of new growth are functions of the concentrations of carbon and nitrogen in root and shoot. The partitioning is chosen so as to lead to balanced growth, either a fixed carbon:nitrogen or root-to-shoot ratio, after a transient initial period. There is also typically a decomposition of root and shoot into separate storage and structural components. This generally involves a number of parameters which may be difficult to estimate from available data, but strong arguments can be made that this approach presents great advantages over purely empirical ones (Thornley and Johnson, 1986). Makela and Sievanen (1987) carry out a comparison of two basic model formulations due to Thornley and coworkers, pointing out that one formulation is embedded in the other. For an application of this approach to woody plants, see Makela (1986). For a simplified model in the case of lichen, see Childress and Keller (1980).

The second approach to analyze whole-plant growth applies the notions of life history theory to study such things as the timing of vegetative versus reproductive growth (Cohen, 1971, 1976; Partridge and Denholm, 1974). The typical compartments used here are reproductive and non-reproductive. Here a plant is viewed as having control of the partitioning of photosynthate among plant parts. The models assume it is possible to control growth so as to maximize some measure of total reproduction (the presumed measure of selective fitness), a typical choice being the size of the reproductive compartment at the end of a fixed lifespan. The earlier papers, which used discrete models, indicated that for annuals the optimal control is to have a sudden complete switch to reproductive growth at some point in the season (i.e. bang-bang control). Further work on this uses continuous control theory and indicates that there can be cases of multiple switches and graded allocation between vegetative and reproductive growth (Vincent and Pulliam, 1980; King and Roughgarden, 1982a, b; Schaffer et al., 1982). The case in which the senescence rate of plant parts is allowed to be a function of plant size, leading to plant lifetime being a function of the plant's behavior, is investigated using control theory by Paltridge et al. (1984). An excellent review of the most recent work is given in Roughgarden (1986).

As an alternative to compartment type approaches for whole plant growth, one may assume the existence of underlying organizing principles of evolutionary origin, which specify the plant growth form in a particular environment. The biophysical approach of Gates (1980) allows one to determine the effects of heat loading and transpiration on leaf temperature, and coupling this with a model of photosynthesis, to determine optimal leaf forms. Alternative definitions of optimal have been applied with Parkhurst and Loucks (1972) using water use efficiency (ratio of photosynthesis to transpiration) while Givnish (1979) uses net carbon assimilation under the assumption that transpiration induces a photosynthetic "cost". The latter assumption is similar to that used by Cowan (1982) as mentioned above to analyze stomatal control. These approaches allow the derivation of optimal leaf shapes and sizes under a variety of environmental conditions which qualitatively accord well with the available data (Givnish and Vermeij, 1976), though they do not explain the diversity of these found at any given location.

Each of the above approaches has certain advantages—the mass balance compartment models contain more physiological detail about alternative plant structures than the optimization models, but lack the emphasis on deriving allocation patterns that may have been selected by evolution. The difference entails the alternative between the a priori method specifying optimization criteria and then seeing what this implies, versus assuming some pattern of growth is optimal (e.g. balanced growth) and determining the physiological details of how this might be met. The first approach lacks much of a physiological basis for testing its predictions, though the general trends which arise may be investigated in the field. The second approach is often limited by our lack of knowledge of details of physiological control, leading to functional dependences in the models that are chosen for convenience rather than due to a mechanistic understanding of the true form. Combining the two approaches is in part limited by the mathematical difficulties inherent in control problems associated with systems of more than a couple of nonlinear differential equations. It may well be most useful in the near future to deal with relatively simple formulations of the optimization problem, coupled with fairly detailed physiology, as for example Cowan (1986) discusses for the problem of carbon allocation between root and shoot. The whole area of allocation of carbon to defense from herbivores is also being tackled in this way (Gulman and Mooney, 1986) and leads to a host of fascinating theoretical questions regarding the timing and spatial distribution of defensive compounds throughout a plant.

Population Biology

The vast majority of theory in population biology has centered on animals, and it is only recently that there have been attempts to develop a comparable theory for plants. As Schaffer and Leigh (1976) point out, major reasons for the lack of progress to date include the plasticity of plant growth form, the high amount of spatial heterogeneity, and the size rather than age dependence of plant growth. In addition, Roughgarden (1986) argues that a theory of plant population dynamics needs to consider neighborhood interactions between nearby plants, the mass and energy flows needed for growth (emphasized above), pollination requirements, and the fact that many plants reproduce both vegetatively and sexually, so it becomes difficult to define what is meant by an individual. This problem rarely arises in animal population biology. Plant ecologists now differentiate between a genet, or the collection of plants arising from a single zygote, and a ramet, which is a vegetatively separate module witin a genet. For a recent review of theoretical approaches in plant population dynamics see Pacala (1989).

A key component of plant population theory is the competitive effects of plants on each other, since they are fixed in location for much of their lifespan and thus are limited by the availability of light, water and nutrients. In the case of monocultures, this has led to two different approaches for investigating competitive effects. One of these consists of empirically-motivated "laws" for how plant density affects individual growth, and the second are detailed physiologically-motivated models, often using simulation techniques, for analyzing individual growth.

A couple of different relationships have become standard for describing density effects in plant populations. One of these is the yield-density relation in which per-plant yield, y, is a function of density, D according to

$$y = W/(1 + qD^b)$$

where W is maximum yield at the sample time and q and b are empirically derived constants. See Vandermeer (1984) for a discussion of the derivation of this and biological interpretations of the parameters. Another basic tenet of plant population biology is the self-thinning rule, which defines the rate at which plants die due to competition in a dense monoculture, as a function of biomass accumulation. It applies at higher densities than those at which the previous yield law is presumed to hold. The form is

$$w = cN^{-x}$$

where w is the average weight of an adult plant (usually a genet), N is the adult density, and c and x are constants. The w and N vary in time as the population thins. The usual value for x is $3/2$ which has been justified mainly empirically, though several different theoretical models lead to values for x in this vicinity (Pickard, 1983, 1984).

Though this rule seems to hold fairly generally over some density ranges, a careful analysis of the statistical validity of the exact value of x (Weller, 1987) shows that the value of x does vary among populations so that all populations do not obey a single quantitative rule. Westoby (1981) points out that there is still no good dynamic model which adequately describes the phase portrait of plant populations in the w-N space, although there have been some attempts (Aikmann and Watkinson, 1980; Watkinson, 1980). A statistical model to describe the w-N trajectories is given by Lloyd and Harms (1986), with fits to data on loblolly pine. Benjamin and Hardwick (1986) review statistical models for the variance in w through time as functions of the variances and covariances of individual growth rates, lifespans, and size at emergence. See Roughgarden (1986) for more on recent work regarding analytically tractable population models. A particularly interesting new approach involves models which track the number of neighbors of an average individual in a population (Pacala and Silander, 1985; Pacala, 1989). This uses predictors of the plant's fecundity and survival as a function of the neighborhood structure to specify growth rates.

In addition to the yield laws mentioned above and relatively simple models to describe density effects, there are a range of models, typically derived for crops or forests, oriented towards predicting population-level effects of competition for light and nutrients. A great deal has been done on the canopy level in this regard, in which one models the population as a homogeneous canopy of leaves from which estimates of photosynthesis and growth rates may be obtained (Johnson and Thornley, 1984). Indeed this is critical in deciding optimal spacings of plantings for crops and forest plantations. Competition may be modeled by assuming a zone of competitive influence around each individual (Wixley, 1983), similar to the neighborhood method mentioned above (Pacala, 1989), though this has a more physiological basis. There are also detailed physiologically-based individual growth models coupled with rules for determining the effects of neighbouring individuals (Makela and Hari, 1986). Another approach uses a relatively simple individual growth model in which relative competitive ability is related to the amount of plant material in the competitive zone (i.e. crown volume for trees competing for light or root volume in the case of competition for water) (McMurtrie, 1981).

Although there is little agreement regarding the most appropriate models for analyzing unstructured plant population dynamics, there has been significant development in demographic types of models. As in animal ecology, these models are concerned with the substructure of populations, and how that structure changes through time and under alternative environmental conditions. These analyze the complex life cycles of plant populations utilizing matrix theory, and are an extension of the Leslie matrices used in animal demography. The inapplicability of most classical demographic approaches occurs due to the size rather than age dependence of plant growth. These matrix models break a population up into stage classes, based on size or plant form (Caswell and Werner, 1978). Due to this, it is possible for an individual to skip through one or more stage classes completely in the underlying time period chosen for the model, or to regress from one stage to a lower one. This leads to matrix models which are different in form from Leslie matrices. There are still quite general results however which allow one to define a reproductive value and easily find stable population structures utilizing the life cycle graph (Caswell, 1982a, b). See Caswell (1986) for an up-to-date review.

Alternative to the discrete approach of matrix theory, there has been a some work on continuous models of size and age structure in plant populations (Hara, 1984a, b). This is based on a model for individual size change through time as a stochastic diffusion process governed by infinitesimal mean growth rate and variance as functions of time and size as well as a size-dependent mortality rate. This has been coupled with a canopy photosynthesis model to look at density effects on size variability within a population, from which it is also possible to decouple the relative importance of above-versus below-ground competitive effects in determining plant size (Hara, 1986).

Large Scale Systems

Community Theory

Although much of community ecology has been driven by studies of plant communities, the early theoretical development of the field was perhaps stifled by an emphasis on elaborate classification schemes for natural systems. More recently there has been much work on statistical techniques for plant community analysis. In a similar vein, many of the modern questions posed by community ecology, dealing with niche partitioning and species diversity, are not limited to plants although plants do provide a major observational tool in the comparison of various theories. Roughgarden (1986) gives a brief review of studies in this regard.

As at the population level, competition is a key component of plant community models. Simple models based upon standard predator-prey models in which the prey variables are now viewed as measures of resources (light, nutrients, etc.), with a depletion rate due to utilization by the plant species densities, have dealt with the question of why several species can coexist on a few resources (Tilman, 1982). This approach averages over the spatial character of competitive interactions in plants. Different approaches have been taken to look at the localized nature of these interactions. Pacala (1986a, b, 1989) considers multi-species discrete-time models in which the population density dynamics of a species is governed by the probability that a randomly chosen individual of that species has a certain number of neighbors of each of the other species, times a function giving the individual's fecundity and survivorship with that set of neighbors. Under appropriate assumptions about dispersal and fecundity, this reduces to the Lotka-Volterra competition equations, but more realistic assumptions lead to interesting conclusions about coexistence and spatial effects.

Much of plant competition is mediated by seed dispersal. The evolution of seed dispersal patterns in patchy habitats has been investigated by Levin et al. (1983). General patch type models of competition for space are analyzed by Yodzis (1978) and may be applicable to forest situations. These models consider space as broken up into a large number of cells, with competitive interactions only between individuals in a given cell. The within-cell competition gives a rule for determining what species occupy that cell, and a usual form is the lottery competition case of Chesson and Warner (1981), in which a weighted random draw of juveniles present determines the winner when an adult within a cell dies. Weiner and Conte (1981) and Hubbell (1980) simulate local competition for space in annuals and tropical trees respectively to investigate competitive exclusion and the effects of dispersal on spatial pattern. Shmida and Ellner (1983) analyze a similar model and conclude, as do the simulations, that localized competition could allow for the extended coexistence of similar species.

Though it is realized that seedbanks form an important means by which a genotype can average temporally over environmental changes, relatively little work has been done on how this affects population growth. MacDonald and Watkinson (1981) build a non-linear difference equation model for annuals with a seedbank. In a random environment, these become stochastic difference equations, the long term behavior of which is analyzed in Ellner (1984). Associated with seedbanks are the germination and dormancy characteristics of the species, and an analysis of the evolutionarily stable germination fraction has been undertaken for logistic-type stochastic difference equation models (Ellner, 1985a). In general the evolutionarily stable fraction of seeds which germinate each year decreases with an increased survivorship of seeds in the soil and with increased variability of seed yields, though the appropriate measure of this variability depends on the circumstances (Ellner, 1985b). Templeton and Levin (1979) investigate some of the population genetics consequences of a seedbank.

There are a variety of modeling approaches used to analyze successional changes within plant communities. The successional state space approach considers a community to be in one of a number of states, and views the dynamics of the system as a Markov chain (Horn, 1975; Enright and Ogden, 1979; Acevedo, 1981; Runkle, 1981). The ecosystem approach attempts to include essentially all biotic interactions within a system and investigate the effects of abiotic driving variables. The approach involves enormously complicated models which are extremely difficult, if not impossible to validate in any reasonable manner.

One alternative applied to forest stands is to consider the dynamics of the system as being governed only by the relative competitive abilities of the tree species present. This approach, which follows the change in stand composition in a plot of fixed size via a Monte-Carlo type of simulation, has been applied to a host of different forest communities (Shugart, 1984). Given the great lack of physiological detail in the models, the results accord well with observations and the pollen record where it is available. This may be due however to the ability to "tune" these models to particular situations, producing qualitatively realistic results. Huston and Smith (1987) review a number of succession models however, and argue that the individual-based approach offers great hope for a unified view of successional dynamics. Indeed a fuller understanding of plant community structure can be tied to the choice of appropriate averages or filters, in time and space, to isolate the the system at the level of resolution required to investigate the questions being addressed (Allen and Wyleto, 1983).

Agriculture

An excellent review of mathematical models in agriculture is given by France and Thornley, 1984. Many of the above mentioned physiological models have been combined into intricate systems models to simulate the growth of entire crops (Loomis et al., 1979; Barrett and Peart, 1981). These have been developed for a wide variety of crops including cotton, soybean, rice, wheat and corn, each of which present their own difficulties in modeling. Complete documentation on these models is usually inaccessible in the open literature. Fairly complete versions are given in the series of Simulation Monographs of the Centre for Agricultural Publishing and Documentation (PUDOC) of the Netherlands, written by C.T. deWit and associates (e.g. deWit and Goudriaan, 1978). The models tend to be extremely complicated compendia of subprocess models which simulate the dynamics of such processes as respiration, photosynthesis, leaf growth, shoot elongation, and nutrient uptake using discretizations of compartment-type differential equation models. These tend to iterate on hourly or daily time scales and either require weather data as input or simulate weather patterns from previous years' data. Very elaborate geometrical models have also been constructed to ascertain the penetration and distribution of radiant energy through the crop canopy (Norman, 1975), though usually only simplified versions of these are used in the full crop models.

At their present stage of development, these models do not as accurately predict yields as considerably simpler regression models. Perhaps the greatest utility of the modeling effort to date has been to point out those areas of plant systems which are not adequately understood, and indicate what further experimental work is needed. The models do offer the advantage of being more general than regression models, whose coefficients must be fit with data from a given crop variety in given environmental circumstances and thus may not be predictive for other conditions. The systems models also present the potential for simulating the response of the crop to a range of management techniques, such as fertilization or irrigation schedules, which would be expensive and time-consuming to undertake in the field.

At an intermediate scale of model complexity, somewhat more general crop models have been developed along the lines of the mass-balance compartment models discussed earlier. These models can include effects of environmental variables on the allocation to various structural components in a plant (Thornley and Johnson, 1986), and can be coupled to other models, such as for animal intake to analyze effects of grazing (Johnson and Parsons, 1985). At this intermediate scale, it is possible to use the models to investigate alternative possible carbon allocation schemes and the implications for productivity, for example in forests (McMurtrie, 1986).

The techniques of operations research have been applied to a variety of economic problems in agriculture, including pest management, crop rotation, risk associated with poor weather, and farm management problems on a regional and national scale (Rovinsky and Shoemaker, 1981; Plant and Mangel, 1987). Currently a number of software packages using these techniques are available to farmers to help manage their crops. For example, one third of the cotton raised in Australia is managed in part through a computer model. An extremely elaborate software package is available to farmers all over the USA through Nebraska's AGNET. For the management of forest stands, Markov type models for projecting the diameter distributions of stands have been useful in setting harvesting schedules (Bruner and Moser, 1973; Peden et al., 1973). Another useful approach to aid in management of forests and crops is based on expert systems, in which the opinions of experts with extensive experience working on a particular crop are combined with simple empirical or mechanistic models to provide advice on specific management questions. These need not rely on the many, and often untested assumptions built into most detailed crop and forest growth models (Jones, 1985; McKinion and Lemmon, 1985), but it may not be easy to general these from one particular locale and crop system to another.

On a final topic, plant epidemiology, there are a number of modeling problems which arise that require somewhat different approaches from those utilized in animal epidemiology. Since plants are fixed in space for most of their lifespans, disease spread does not occur due to contacts between individuals, but rather through the direct dispersal of the pathogen. Also plants can have a continuum of resistance levels to any particular pathogen, so that an infection on an individual may range from non-existent to severe damage. Thus, the usual epidemiological classification of individuals as infected, succeptible, immune, or removed is inadequate. The theory of macroparasitic infections is more appropriate (Dobson, this volume), but has not as yet been utilized for plants. We are as yet far from a general theory of plant epidemiology, but for a recent survey see Gilligan (1985).

Acknowledgements. The initial draft of this review was written while I was supported by a Faculty Leave Award from the University of Tennessee, for which I am very grateful. Comments from the participants in the 1986 Trieste Course were very helpful in revising the initial draft. I especially thank Marc Mangel and Robert Pearcy of the University of California at Davis for their hospitality while the final version of this review was completed and Leah Edelstein-Keshet for many helpful suggestions.

References

Acevedo, M.F. (1981) On Horn's Markovian model of forest dynamics with particular reference to tropical forests. Th. Pop. Biol. *19*, 230–250

Adler, I. (1975) A model of space filling in phyllotaxis. J. Theor. Biol. *53*, 435–444

Adler, I. (1977) The consequences of contact pressure in phyllotaxis. J. Theor. Biol. *65*, 29–77

Aikman, D.P., Watkinson, A.R. (1980) A model for growth and self-thinning in even-aged monocultures of plants. Ann. Bot. *45*, 419–427

Allen, T.F.H., Wyleto, E.P. (1983) A hierarchical model for the complexity of plant communities. J. Theor. Biol. *101*, 529–540

Barnes, A. (1979) Vegetable plant part relationships. II. A quantitative hypothesis for shoot/storage root development. Ann. Bot. *43*, 487–499

Barrett, J.R., Peart, R.M. (1981) Systems simulation in U.S. agriculture. Progress in Modeling and Simulation. Academic Press, New York, pp. 39–59

Bell, A.D., Roberts, D., Smith, A. (1979) Branching patterns: the simulation of plant architecture. J. Theor. Biol. *81*, 351–375

Benjamin, L.R., Hardwick, R.C. (1986) Sources of variation and measures of variability in even-aged stands of plants. Ann. Bot. *58*, 757–778

Bertaud, D.S., Gander, P.W. (1986) A simulation model for cell proliferation in root apices. II. Patterns of cell proliferation. Ann. Bot. *58*, 303–320

Bertaud, D.S., Gander, P.W., Erickson, R.O., Ollivier, A.M. (1986). A simulation model for cell growth and proliferation in root apices. I. Structure of model and comparisons with observed data. Ann. Bot. *58*, 285–301

Bruner, H.D., Moser J.W. Jr., (1973) A Markov chain approach to the prediction of diameter distributions in uneven-aged forest stands. Can. J. For. Res. *3*, 409–417

Caswell, H. (1982a) Optimal life histories and the maximization of reproductive value: a general theorem for complex life cycles. Ecol. *63*, 1218–1222

Caswell, H. (1982b) Stable population structure and reproductive value for populations with complex life cycles. Ecol. *63*, 1223–1231

Caswell, H. (1986) Life cycle models for plants. Pages 171–233 in: L.J. Gross and R.M. Miura (eds.). Some Mathematical Questions in Biology—Plant Biology. Amer. Math. Soc., Providence

Caswell, H., Werner, P.A. (1978). Transient behavior and life history analysis of teasel (Dispsacus sylvestris Huds.) Ecol. *59*, 53–66

Chapman, D.C., Rand, R.H., Cooke, J.R. (1977). A hydrodynamical model of bordered pits in conifer tracheids. J. Theor. Biol. *67*, 11–24

Chesson, P.L., Warner, R.R. (1981) Environmental variability promotes coexistence in lottery competitive systems. Amer. Natur. *117*, 923–943

Childress, S., Keller, J.B. (1980) Lichen growth. J. Theor. Biol. *82*, 157–165

Cohen, D. (1971) Maximizing final yield when growth is limited by time or by limiting resources. J. Theor. Biol. *33*, 299–307

Cohen, D. (1976) The optimal timing of reproduction. Amer. Natur. *110*, 801–807.

Cooke, J.R., Rand, R.H., Mang, H.A., Debaerdemaeker, J.B. (1977) A non-linear finite element analysis of stomatal guard cells. Am. Soc. Agric. Eng. Paper #77–5511

Cowan, I.R. (1972) Oscillations in stomatal conductance and plant functioning associated with stomatal conductance: observations and a model. Planta *106*, 185–219

Cowan, I.R. (1977) Stomatal behavior and environment. Adv. Bot. Res. *4*, 117–228

Cowan, I.R. (1982) Regulation of water use in relation to carbon gain in higher plants. In: Lange, O.L., Nobel, P.S., Osmond, C.B., Ziegler, H. (eds.), Physiological Plant Ecology II, Springer-Verlag, Berlin, pp. 589–613

Cowan, I.R. (1986) Economics of carbon fixation in higher plants. In: Givnish, T.J. (ed.), ibid, pp. 133–170

Cowan, I.R., Farquhar, G.D. (1977) Stomatal function in relation to leaf metabolism and environment. Soc. Exp. Biol. Symp. *31*, 471–505

Delwiche, M.J., Cooke, J.R. (1977) An analytic model of the hydraulic aspects of stomatal dynamics. J. Theor. Biol. *69*, 113–141

deWit, C.T., Goudriaan, J. (1978) Simulation of Ecological Processes. PUDOC, Wageningen, the Netherlands

DiMichele, D.W., Sharpe, P.J.H., Goeschle, J.D. (1978) Towards the engineering of photosynthetic productivity. In: Critical Reviews in Bioengineering, CRC Press, Boca Raton, Florida, pp. 23–91

Ellner, S.P. (1984) Asymptotic behavior of some stochastic difference equation population models. J. Math. Biol. *19*, 169–200

Ellner, S.P. (1985a) ESS germination strategies in randomly varying environments. I. Logistic-type models. Theor. Pop. Biol. *28*, 50–79

Ellner, S.P. (1985b) ESS germination strategies in randomly varying environments. II. Reciprocal yield-law models. Theor. Pop. Biol. *28*, 80–116

Enright, N., Ogden, J. (1979) Applications of transition matrix models in forest dynamics: Araucaria in Papau, New Guinea and Nothofagus in New Zealand. Aust. J. Ecol. *4*, 3–23

France, J., Thornley, J.H.M. (1984) Mathematical Models in Agriculture. Butterworths, London

Gates, D.M. (1980) Biophysical Ecology. Springer-Verlag, New York

Givnish, T. (1979) On the adaptive significance of leaf form. P. 375–407 in Solbrig, O.T., Jain, S., Johnson, G.B., Raven, P.H. (eds.), Topics in Plant Population Biology. Columbia Univ. Press, New York

Givnish, T. (ed.). (1986a) On the Economy of Plant Form and Function. Cambridge University Press, Cambridge

Givnish, T. (1986b) Optimal stomatal conductance, allocation of energy between leaves and roots, and the marginal cost of transpiration. In: Givnish, T.J. (ed.), ibid, pp. 171–213

Givnish, T., Vermeij, G.J. (1976) Sizes and shapes of liane leaves. Amer. Natur. *110*, 743–776

Goel, N.S., Richter-Dyn., N. (1974) Stochastic Models in Biology. Academic Press, New York

Gross, L.J. (1981) On the dynamics of internal leaf carbon dioxide uptake. J. Math. Biol. *11*, 181–191

Gross, L.J. (1982) Photosynthetic dynamics in varying light environments: a model and its application to whole leaf carbon gain. Ecol. *63*, 84–93

Gross, L.J. (1986a) Photosynthetic dynamics and plant adaptation to environmental variability. In: Gross, L.J., Miura, R.M. (eds.), ibid, pp. 135–170

Gross, L.J. (1986b) Biophysical ecology: an introduction to organism response to environment. In: Hallam, T.G., Levin, S.A. (eds.), Mathematical Ecology. Springer-Verlag, Berlin, pp. 19–36

Gross, L.J., Miura, R.M. (eds.). (1986) Some Mathematical Questions in Biology—Plant Biology. American Math. Soc., Providence

Gulmon, S.L. and Mooney, H.A. (1986) Costs of defense and their effects on plant productivity. In T.J. Givnish (ed.), ibid, pp. 681–698

Gutschick, V.P. (1984a) Photosynthesis model for C_3 leaves incorporating CO_2 transport, propagation of radiation, and biochemistry 1. Kinetics and their parametrization. Photosynthetica *18*, 549–568

Gutschick, V.P. (1984b) Photosynthesis model for C_3 leaves incorporating CO_2 transport, propagation of radiation, and biochemistry 2. Ecological and agricultural utility. Photosynthetica *18*, 569–595

Hahn, B.D. (1984) A mathematical model of leaf carbon metabolism. Ann. Bot. *54*, 329–339

Hahn, B.D. (1986) A mathematical model of the Calvin cycle: analysis of the steady state. Ann. Bot. *57*, 639–653

Hall, A.E. (1982) Mathematical models of plant water loss and plant water relations. P. 231–261 in Lange, O.L. et al. (eds.), ibid

Hara, T. (1984a) A stochastic model and the moment dynamics of the growth and size distribution in plant populations. J. Theor. Biol. *109*, 173–190

Hara, T. (1984b) Dynamics of stand structure in plant monocultures. J. Theor. Biol. *110*, 223–239

Hara, T. (1986) Effects of density and extinction coefficient on size variability in plant populations. Ann. Bot. *57*, 885–892

Hesketh, J.D., Jones, J.W. (eds.). (1980) Predicting Photosynthesis for Ecosystems Models. CRC Press, Boca Raton, Florida.

Honda, H., Fisher, J.B. (1978) Tree branch angle: maximizing effective leaf area. Science *199*, 888–890

Honda, H., Fisher, J.B., Tomlinson, P.B. (1982) Two geometrical models of branching of botanical trees Ann. Bot. *49*, 1–11

Horn, H.S. (1971) The Adaptive Geometry of Trees. Princeton Univ. Press, Princeton, New Jersey

Horn, H.S. (1975) Markovian processes of forest succession. In: Cody, M.L., Diamond, J.M. (eds.), Ecology and Evolution of Communities. Cambridge, Massachusetts, pp. 196–211

Huston, M., Smith, T. (1987) Plant succession: life history and competition. Amer Natur. *130*, 168–198

Hubbell, S.P. (1980) Seed predation and the coexistence of tree species in tropical forests. Oikos *38*, 214–229

Jacob, F. (1977) Evolution and tinkering. Science *196*, 1161–1166

Jean, R. (1978) Phytomathematique. Univ. of Quebec Press, Quebec, Canada

Jean, R. (1983) Mathematical modeling in phyllotaxis: the state of the art. Math. Biosci. *64*, 1–27

Jean, R. (1984) Mathematical Approach to Pattern and Form in Plant Growth. Wiley, New York

Johnson, I.R. (1985) A model of the partitioning of growth between the shoots and roots of vegetative plants. Ann. Bot. *55*, 421–431

Johnson, I.R., Thornley, J.H.M. (1984) A model of instantaneous and daily canopy photosynthesis. J. Theor. Biol. *107*, 531–545

Johnson, I.R., Parsons, A.J. (1985) A theoretical analysis of grass growth under grazing. J. Theor. Biol. *112*, 345–367

Jones, J.W. (1985) Using expert systems in agricultural models. Agric. Engin. *66*, 21–23

Kaitala, V., Hari, P., Vapaavuori, E., Salminen, R. (1982) A dynamic model for photosynthesis. Ann. Bot. *50*, 385–396

Keller, J.B. (1980) Tendril shape and lichen growth. In: Oster, G.F. (ed.), Some Mathematical Questions in Biology, Vol. 13, Amer. Math. Soc., Providence, Rhode Island, pp. 257–274

King, A.W., DeAngelis, D.L. Post, W.M. (1987) The Seasonal Exchange of Carbon Dioxide Between the Atmosphere and the Terrestrial Biosphere: Extrapolation from Site-Specific Models to Regional Models. Environmental Sciences Division Publication #2988, Oak Ridge National Laboratory. ORNL/T-10570

King, D., Roughgarden, J. (1982a) Multiple switches between vegetative and reproductive growth in annual plants. Theor. Pop. Biol. *21*, 194–204

King, D., Roughgarden, J. (1982b) Graded allocation between vegetative and reproductive growth for annual plants in growing seasons of random length. Theor. Pop. Biol. *22*, 1–16

Kirschbaum, M.U.F., Farquhar, G.D. (1984) Temperature dependence of wholeleaf photosynthesis in Eucalyptus Pauciflora Sieb. ex Spreng. Aust. J. Plant. Physiol. *11*, 519–538

Kirschbaum, M.U.F., Gross, L.J., Pearcy, R.W. (1988) Observed and modelled stomatal responses to dynamic light environments in the shade plant Alocasia macrorrhiza. Plant, Cell and Environ. *11*, 111–121

Koehl, M.A.R. (1986) Seaweeds in moving water: form and mechanical function. In: Givnish, T.J. (ed.), ibid, pp. 603–634.

Levin, S.A., Cohen, D., Hastings, A. (1983) Dispersal strategies in patchy environments. Theor. Pop. Biol. *26*, 165–191

Lloyd, F.T., Harms, W.R. (1986) An individual stand growth model for mean plant size based on the rule of self-thinning. Ann. Bot. *57*, 681–688

Loomis, R.S., Rabbinge, R., Ng, E. (1979) Explanatory models in crop physiology. Ann. Rev. Plant Physiol. *30*, 339–367

MacDonald, N., Watkinson, A.R. (1981) Models of an annual plant population with a seed bank. J. Theor. Biol. *93*, 643–653

Makela, A. (1986) Implications of the pipe model theory on dry matter partitioning and height growth in trees. J. Theor. Biol. *123*, 103–120

Makela, A., Hari, P. (1986) Stand growth model based on carbon uptake and allocation in individual trees. Ecol. Model. *33*, 205–229

Makela, A., Sievanen, R.P. (1987) Comparison of two shoot-root partitioning models with respect to substrate utilization and functional balance. Ann Bot. *59*, 129–140

Marzec, C., Kappraff, J. (1983) Properties of maximal spacing on a circle related to phyllotaxis and the golden mean. J. Theor. Biol. *103*, 201–226

McKinion, J.M., Lemmon, H.E. (1985) Expert systems for agriculture. Computers and Electronics in Agric. *1*, 31–40

McMurtrie, R. (1981) Suppression and dominance of trees with overlapping crowns. J. Theor. Biol. *89*, 151–174

McMurtrie, R. (1986) Forest productivity in relation to carbon partitioning and nutrient cycling: a mathematical model. In: Cannell, M.G.R. and Jackson, J.E. (eds.), Attributes of Trees as Crop Plants. Inst. of Terrestrial Ecology, Monks Wood, Abbots Ripton, Hunts, UK, pp. 194–207

Milstein, J., Bremermann, H.J. (1979) Parameter identification of the Calvin photosynthesis cycle. J. Math. Biol. *7*, 99–116

Molz, F.J. (1976) Water transport through plant tissue: the apoplasm and symplasm pathways. J. Theor. Biol. *59*, 277–292

Niklas, K. (1982) Computer simulations of early land plant branching morphologies: canalization of patterns during evolution? Paleobiol. *8*, 196–210

Niklas, K. (1986) Computer simulations of branching-patterns and their implications on the evolution of plants. Pages 1–50 in: Gross, L.J., Miura, R.M. (eds.), ibid

Nobel, P.S. (1974) Introduction to Biophysical Plant Physiology. Freeman, San Francisco

Nobel, P.S. (1981) Wind as an ecological factor. In: Lange, O.L., Nobel, P.S., Osmond, C.B., Ziegler, H. (eds.), Physiological Plant Ecology I. Springer-Verlag, Berlin, pp. 475–500

Norman, J.M. (1975) Radiative transfer in vegetation. In: deVries, D.A., Afgan, N.H. (eds.), Heat and Mass Transfer in the Biosphere. Scripta, Washington, D.C., pp. 187–205

Pacala, S.W. (1986a) Neighborhood models of plant population dynamics. II. Multi-species models of annuals. Theor. Pop. Biol. *29*, 262–292

Pacala, S.W. (1986b) Neighborhood models of plant population dynamics. IV. Single and multi-species models of annuals with dormant seed. Amer. Natur. *128*, 859–878

Pacala, S.W. (1989) Plant population dynamic theory. In: May, R.M., Levin, S.A., Roughgarden, J. (eds.), Perspectives in Theoretical Ecology. Princeton Univ. Press, Princeton, NJ, pp. 54–67

Pacala, S.W., Silander Jr., J.A. (1985) Neighborhood models of plant population dynamics. I. Single-species models of annuals. Amer. Natur. *125*, 385–411

Paltridge, G.W., Denholm, J.V. (1974) Plant yield and the switch from vegetative to reproductive growth. J. Theor. Biol. *44*, 23–34

Paltridge, G.W., Denholm, J.V., Connor, D.J. (1984) Determinism, senescence and the yield of plants. J. Theor. Biol. *110*, 383–398

Parkhurst, D.F. (1977) A three dimensional model for CO_2 uptake by continuously distributed mesophyll in leaves. J. Theor. Biol. *67*, 471–488

Parkhurst, D.F. (1986) Internal leaf structure: a three-dimensional perspective. In: Givnish, T.J. (ed.), ibid, pp. 215–249

Parkhurst, D.F., Loucks, O.L. (1972) Optimal leaf size in relation to environment. J. Ecol. *60*, 505–537

Peden, L.M., Williams, J.S., Frayer, W.E. (1973) A Markov model for stand projection. For. Sci. *19*, 303–314

Pickard, W.F. (1983) Three interpretations of the self-thinning rule. Ann. Bot *51*, 749–757

Pickard, W.F. (1984) The self-thinning rule. J. Theor. Biol. *110*, 313–314

Plant, R.E. (1983) Analysis of a continuum model for root growth. J. Math. Biol. *16*, 261–268

Plant, R.E., Mangel, M. (1987) Modeling and simulation in agricultural pest management. SIAM Rev. *29*, 235–261

Rand, R.H. (1983) Fluid mechanics of green plants. Ann. Rev. Fluid Mech. *15*, 29–45

Rand, R.H., Cooke, J.R. (1978) Fluid dynamics of phloem flow: an axisymmetric model. Trans. Amer. Soc. Agric. Engin. *21*, 898–906

Rand, R.H., Upadhyaya, S.K., Cooke, J.R. (1980) Fluid dynamics of phloem flow: II. An approximate formula. Trans. Amer. Soc. Agric. Engin. *23*, 581–584

Rand, R.H., Upadhyaya, S.K., Cooke, J.R., Storti, D.W. (1981) Hopf bifurcation in a stomatal oscillator. J. Math. Biol. *12*, 1–11

Rand, R.H., Storti, D.W., Upadhyaya, S.K., Cooke, J.R., (1982) Dynamics of coupled stomatal oscillators. J. Math. Biol. *15*, 131–149

Rand, R.H., Ellenson, J.L. (1986) Dynamics of stomate fields in leaves. In: Gross, L.J., Miura, R.M. (eds.), ibid, pp. 51–86

Real, L. (ed.). (1983) Pollination Biology. Academic, Orlando

Real, L., Caraco, T. (1986) Risk and foraging in stochastic environments. Ann. Rev. Ecol. System. *17*, 371–390

Reynolds, J.F., Thornley, J.H.M. (1982) A shoot: root partitioning model. Ann. Bot. *49*, 585–597

Richards, F.J. (1951) Phyllotaxis: its quantitative expression and relation to growth in the apex. Phil. Trans. Roy. Soc. B *235*, 509–564

Rose, D.A., Charles-Edwards, D.A. (eds.). (1981) Mathematics and Plant Physiology. Academic, London

Roughgarden, J. (1986) The theoretical ecology of plants. In: Gross, L.J., Miura, R.M. (eds.), ibid, pp. 235–267

Rovinsky, R.B., Shoemaker, C. (1981) Operations research: applications in agriculture. Proc. Amer. Math. Soc. Sympos. Appl. Math. *25*, 151–174

Runkle, J.R. (1981) Gap regeneration in some old-growth forests of the eastern United States. Ecol. *62*, 1041–1051

Schaffer, W.M., Leigh, E.G. (1976) The prospective role of mathematical theory in plant ecology. Systematic Bot. *1*, 209–232

Schaffer, W.M., Leigh, E.G., Inonye, R.S., Whittam, T.S. (1982) Energy allocation by an annual plant when the effects of seasonality on growth and reproduction are decoupled. Amer. Natur. *120*, 787–815

Schmida, A., Ellner, S.P. (1984) Coexistence of plant species with similar niches. Vegetatio *58*, 29–55

Schulze, E.D., Hall, A.E. (1982) Stomatal responses, water loss and CO_2 assimilation rates of plants in contrasting environments. In: Lange, O.L. et al. (eds.), Physiological Plant Ecology II. Springer-Verlag, Berlin, pp. 181–230

Sharpe, P.J.H. (1983) Responses of photosynthesis and dark respiration to temperature. Ann. Bot. *52*, 325–343

Sharpe, P.J.H., Wu, H-I. (1978) Stomatal mechanics: volume changes during opening. Plant, Cell and Environ. *1*, 259–268

Shugart, H.H. (1984) A Theory of Forest Dynamics. Springer-Verlag, New York

Sinclair, T.R., Rand, R.H. (1979) Mathematical analysis of cell CO_2 exchange under high CO_2 concentrations. Photosynthetica *13*, 239–244

Smith, K.C., Magnuson, C.E., Goeschl, J.D., DiMichele, D.W. (1980) A time-dependent mathematical expression of the Munch hypothesis of phloem transport. J. Theor. Biol. *86*, 493–505

Templeton, A.R., Levin, D.A. (1979) Evolutionary consequences of seed pools. Amer. Natur. *114*, 232–249

Thornley, J.H.M. (1976) Mathematical Models in Plant Physiology. Academic Press, New York

Thornley, J.H.M. (1977) A model of apical bifurcation applicable to trees and other organisms. J. Theor. Biol. *64*, 165–176

Thornley, J.H.M., Johnson, I.R. (1986) Modelling plant processes and crop growth. In: Gross, L.J., Miura, R.M. (eds.) ibid, pp. 87–133

Tilman, D. (1982) Resource Competition and Community Structure. Princeton Univ. Press, Princeton, NJ

Upadhyaya, S.K., Rand, R.H., Cooke, J.R. (1981) The role of stomatal oscillations in plant productivity and water use efficiency. Amer. Soc. Agric. Engin. Paper #81-4017

Vandermeer, J. (1984) Plant competition and the yield-density relationship. J. Theor. Biol. *109*, 393–399

Vincent, T.L., Pulliam, H.R. (1980) Evolution of life history strategies for an asexual annual plant model. Theor. Pop. Biol. *17*, 215–231

Watkinson, A.R. (1980) Density-dependence in single-species populations of plants. J. Theor. Biol. *82*, 345–357

Weiner, J., Conte, P.T. (1981) Dispersal and neighborhood effects in an annual plant competition model. Ecol. Model. *13*, 131–147

Westoby, M. (1981) The place of the self thinning rule in population dynamics. Amer. Natur. *118*, 581–587

Wixley, R.A.J. (1983) An elliptical zone of influence model for uneven-aged row crops. Ann. Bot. *51*, 77–84

Yodzis, P. (1978) Competition for Space and the Structure of Ecological Communities. Lect. Notes in Biomath., Vol. 25. Springer-Verlag, Berlin

Stable Population Theory and Applications

John Impagliazzo

Contents

Introduction

Demography is the study of population, primarily human population, in terms of its growth and decay, its fertility and mortality, its relative mobility, and its composition, density and size. The impact on economic, political and social components of society has always been of interest to demographers. Early studies of demography as a science can be traced back to the Roman Empire, where in the year 225 A.D., the mortality schedules of Macer and Ulpian were the first published documents on the subject.

The mortality schedule of Ulpian evolved into what is better known as the life table. The life table models the life history of a cohort of people as if the people in the cohort lived every year of their lives in the base year of the table. A life table can only describe the mortality of a population at a given time. Life tables ignore the aspect of fertility and as a result, provide a poor picture of actual demographic patterns. At the turn of the twentieth century, the life table was the dominant model that was used to represent a population.

Other early attempts have been made to model populations. For example, at the end of the eighteenth century, Thomas Malthus [1798] hypothesized that food is necessary for existence with only a finite amount of land on which to grow it. In addition, he assumed that the rate of human reproduction remained a constant. From this it was deduced that the growth of food supply would increase arithmetically while the growth of population would increase geometrically. In mathematical terms this may be stated as, the rate of growth of a population is proportional to the population itself, and takes the familiar exponential form

$$P(t) = P_0 \exp\left(r(t - t_0)\right) \tag{1}$$

where $P(t)$ is the population at time t, P_0 is the population at time t_0; the rate of growth, r, is assumed constant over time.

Stable Theory of Population

Because of the interest of people in their own destiny, and the long-time availability of census data and record keeping, population studies tend to focus on human populations. A major problem in the study of a population is migration, the external flow of people into and out of a population. A population is said to be closed if it is devoid of migration. This is the first assumption of the stable theory of population, hereinafter called stable theory.

Experience shows that the rates of birth and death do vary with age as well as time. For example, for human populations, a woman of age 25 is more likely to produce a child than a woman of age 50; a man of age 70 is more likely to die in the next time period than a man of age 40. The age-specific rates of birth and death generate models that better reflect existing conditions of biological populations. Also, the rate of death may vary between different age groups of a population. Indeed, for a given age or age interval the death rate may vary over time also. The same holds true for birth rates. It is assumed in the classical formulation of stable theory that the rates of death and birth are not functions of time but functions of age alone.

The fundamental assumptions of stable population theory are that a population is closed, and that birth and death rates are time-invariant. Such models are usually confined to a single sex, the female, due to well-defined childbearing years, limited annual birth cycles, and the practice of registration of births by age of the mother. It is also assumed, usually for mathematical convenience, that there is an ultimate age of life w (also omega) and that reproduction occurs on a finite interval beginning at age a (also alpha) and ending at age b (also beta). These form the fundamental assumptions of stable theory in its classical sense.

Models of stable theory are of two types: deterministic and stochastic. In addition, they may be viewed from a discrete time or a continuous time approach. For purposes of discussion, a deterministic, discrete time model will be highlighted followed by the usual limiting considerations to generate a deterministic continuous model. For a stochastic approach to stable theory, consult Mode (1985), Chapter 7.

Development of a Deterministic, Discrete-Time Model

A cohort is a set of individuals having experienced a particular event during the same time period. Consider a population at three contiguous time intervals: $t - h$, t and $t + h$ as shown in Fig. 1. The birth cohort at time t with age category indexed by i will find itself in the age category indexed by $i + 1$ at time $t + h$. Let $F(t, i)$ represent the size of a female cohort with age index i at time t. At time $t + h$ this same cohort will number $F(t + h, i + 1)$.

However, not all the $F(t, i)$ may survive a time $t + h$. If g_{i+1} is defined to be the survivorship ratio between age index i and $i + 1$, then

$$F(t + h, i + 1) = g_{i+1} F(t, i). \tag{2}$$

If z is the index of the age category containing w, the last age of life, then it is clear that for $0 \le i \le z$,

$$0 \le g_i \le 1$$

where $g_{z+1} = 0$ since all members of the cohort will have died. Equivalently, the $F(t, i)$ members of the cohort at time t are the survivors of the $F(t - h, i - 1)$ members of the same cohort that were alive at time $t - h$, and

$$F(t, i) = g_i F(t - h, i - 1) \tag{3}$$

Combining (2) and (3) yields

$$F(t + h, i + 1) = g_{i+1} g_i F(t - h, i - 1).$$

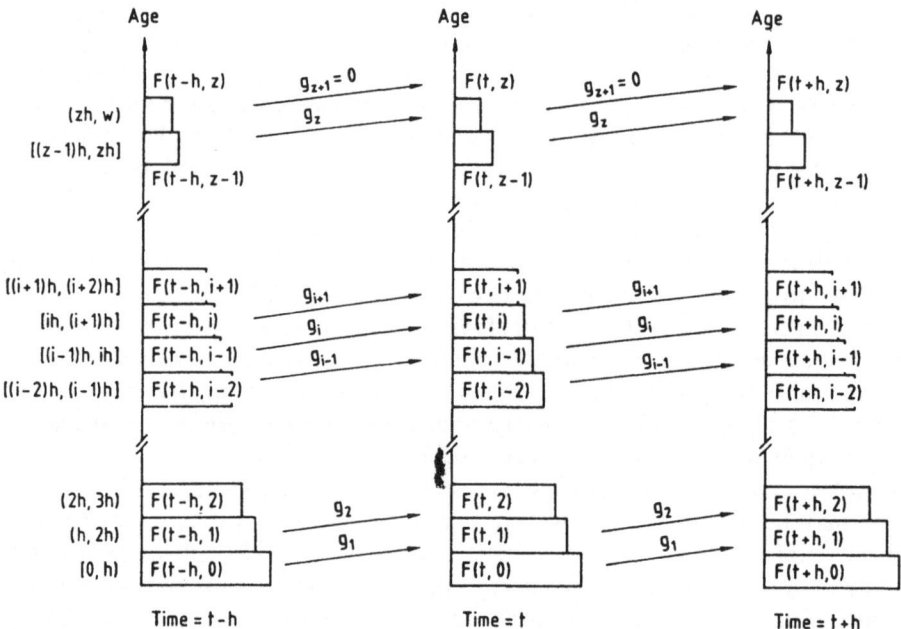

Fig. 1. Cohort mobility between three continuous time periods

In general, for $1 \leq k \leq z + 1$

$$F(t, k) = \left[\prod_{j=1}^{k} g_j \right] F(t - kh, 0) \tag{4}$$

where g_{z+1} is zero. Thus, (4) relates the size of a cohort of age index k at time t to its size at birth at time $t - kh$ on the assumption that survivorsip is not a function of time.

In order for a population to avoid extinction, it is necessary that it exhibit a reproductive or maternity factor. Indeed, the set of females on age interval $[0, h)$ at exact time t is composed of those females born in the time interval $(t - h, t]$ and who actually survive to time t. If $B(t, h)$ represents the number of female births on the time interval $(t - h, t]$ and $F(t, 0)$ represents the number of females on the age interval $[0, h)$ at exact time t, then for survivorship ratio g_0 at birth

$$F(t, 0) = g_0 B(t, h). \tag{5}$$

Combining (4) and (5) results in

$$F(t, k) = \left[\prod_{j=0}^{k} g_i \right] B(t - kh, h) \tag{6}$$

which states that the females in age category indexed by k at time t are the survivors of the cohort that was born on the time interval $(t - kh, t - (k - 1)h]$. Specifically, Fig. 2 shows a contribution to the births of females at time t generated from those females that were born on the time interval $(t - 4h, t - 3h]$ and represented by the birth function $B(t - 3h, h)$.

The total number of births occurring on the time interval $(t - h, t]$ results from many sources; specifically, each age category in the population at time $t - h$ makes a contribution to the births $B(t, h)$. Let $B(t, h; i)$ be the number of daughters on the time interval $(t - h, t]$ produced by those females in the age category indexed by i at time $t - h$. The unit maternity rate m_i for age category indexed by i is the non-negative

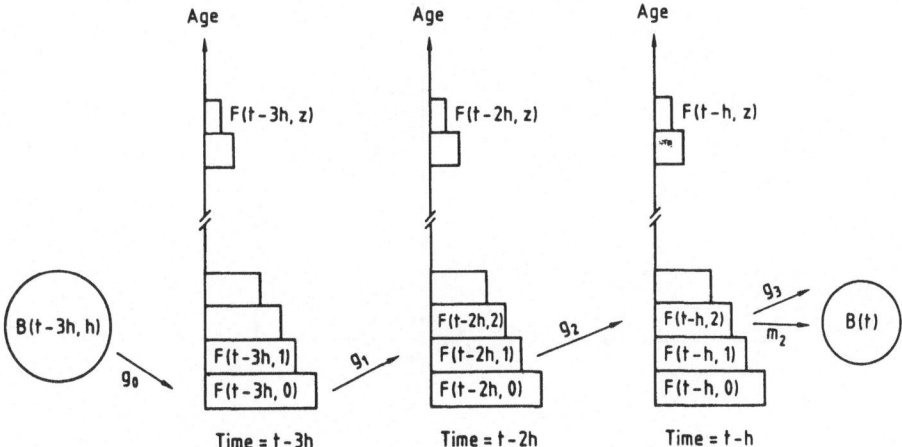

Fig. 2. Births at time t related to births at time $t - 3h$

ratio

$$m_i = B(t, h; i)/F(t - h, i)h \qquad (7)$$

In this sense m_i represents the annual maternity rate for the i-th age category when t is measured in years, and is called the age-specific maternity rate. Hence,

$$B(t, h; i) = m_i F(t - h, i)h \qquad (8)$$

and the total births $B(t, h)$ on the time interval $(t - h, t]$ is the aggregate of these contributions as shown in Fig. 3, and is described by

$$B(t, h) = \sum_{i=0}^{z} B(t, h; i). \qquad (9)$$

Substitution of (8) into (9) results in

$$B(t, h) = \sum_{i=0}^{z} m_i F(t - h, i)h. \qquad (10)$$

A generalization of this concept is shown in Fig. 4. In particular, the females on age interval $[ih, (i + 1)h]$ at time $t - h$ were born on the time interval $(t - (i + 2)h, t - (i + 1)h]$ and make a contribution to the births on the time interval $(t - h, t]$ numbering $B(t, h; i)$. The aggregate of these births form the total births $B(t)$ on the interval $(t - h, t]$.

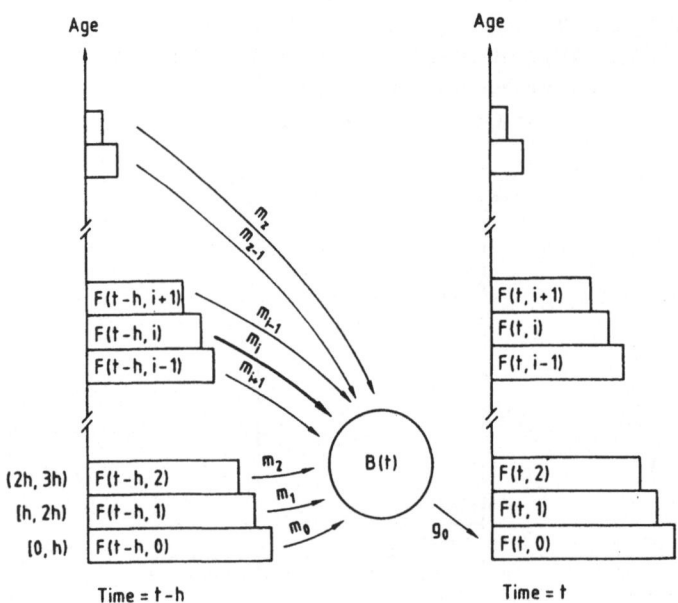

Fig. 3. Contributors to births at time t

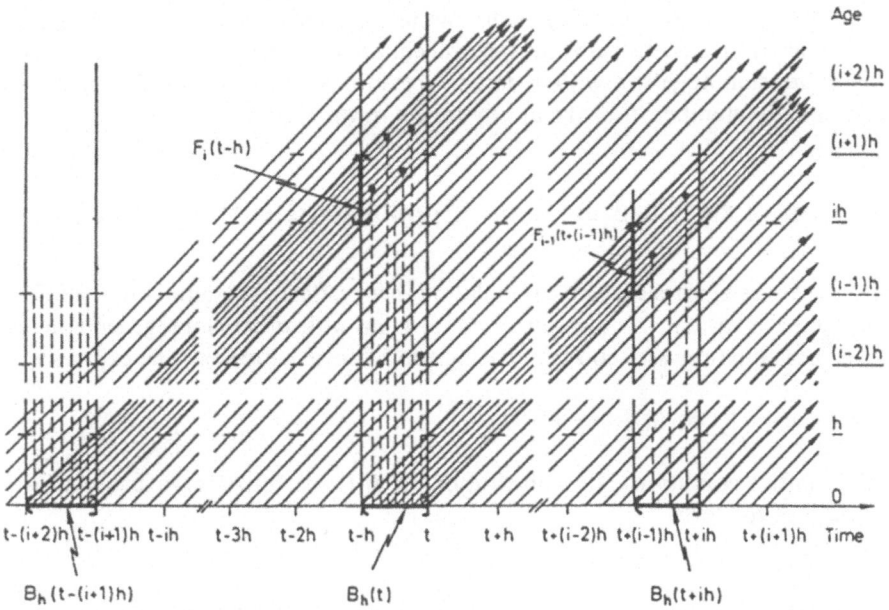

Fig. 4. Illustration showing two generations of births

When expression (6) is substituted into (8) the resulting equation becomes

$$B(t, h; i) = m_i \left[\prod_{j=0}^{i} g_j \right] B(t - (i + 1)h, h)h. \tag{11}$$

The net maternity function ϕ_i for age category indexed by i is defined as the product of the age-specific maternity rate and the total survivorship ratio. This is expressed as

$$\phi_{i+1} = m_i \left[\prod_{j=0}^{i} g_j \right]. \tag{12}$$

Substitution into (11) yields

$$B(t, h; i) = \phi_{i+1} B(t - (i + 1)h, h)h. \tag{13}$$

Substitution of (13) into (9) results in

$$B(t, h) = \sum_{i=0}^{z} \phi_{i+1} B(t - (i + 1)h, h)h. \tag{14}$$

The time (age) intervals of width h may be converted to unit time intervals. By setting $k = i + 1$, where k varies from 1 to n and where $B(t, h)$ becomes $B(t)$, Eq. (14) can be normalized and rewritten as

$$B(t) = \sum_{k=1}^{n} \phi_k B(t - k) \tag{15}$$

where $B(t)$ is the total number of births on the unit time interval $(t-1,t]$. Equation (15) is a deterministic, discrete time renewal equation in stable theory of population. This expression together with n initial conditions $B(t-1), B(t-2), \ldots, B(t-n)$ determine the number of births at any time t in the future.

Solution of the Discrete Time Population Renewal Equation

By the premise of classical stable population theory, the net maternity function ϕ of (15) is time invariant. As a result, Eq. (15) is an n-th order homogeneous linear recurrence equation with constant coefficients. This equation may be expressed as

$$B(t+n) = \sum_{k=1}^{n} \phi_k B(t+(n-k)) \tag{16}$$

with n initial conditions. By substitution of $B(t) = s^t$ into (16), a characteristic equation of the form

$$s^n - \phi_1 s^{n-1} - \cdots - \phi_n = 0 \tag{17}$$

is generated which is an n-th degree polynomial in s. It can be shown that a solution to (16) is

$$B(t) = \sum_{i=1}^{q} \left[\sum_{j=1}^{p(i)} a_{ij} t^{j-1} \right] s_i^t \tag{18}$$

where q is the number of distinct roots s_i of the characteristic Eq. (17) having multiplicity $p(i)$, and where the a_{ij} are constants. The non-negative demographic nature of the net maternity function may be employed to easily show (by Descartes' Rule of Signs) that the characteristic equation admits exactly one, simple, positive, real root. Let s_1 denote this principal root. Then it is clear that (18) becomes

$$B(t) = a_1 s_1^t + \sum_{i=2}^{q} \left[\sum_{j=1}^{p(i)} a_{ij} t^{j-1} \right] s_i^t \tag{19}$$

where the constants a_1 and a_{ij} are obtained by the initial conditions. Since the remaining $n-1$ roots must be complex [Impagliazzo, 1985], an alternate form of (18) can be demonstrated to be

$$B(t) = a_1 \exp(r_1 t) + \sum_{i=2}^{q} \left[\sum_{j=1}^{p(i)} a_{ij} t^{j-i} \right] \exp(u_i t) \cos(v_i t) \tag{20}$$

for $k \geq 2$ where $r_k = \ln(s_k) = u_k + iv_k$, a complex number. Thus, the solution to the population renewal equation consists of algebraic, exponential and trigonometric functions, a more encompassing expression when compared to the Malthusian model (1).

It is significant to note that the principal root of the characteristic equation of (17) dominates in magnitude all other roots of that equation. That is, for all $i = 2, 3, \ldots, q$

$$s_1 > |s_i| \quad \text{or} \quad r_1 > u_i = \text{Re}(r_i).$$

This is of paramount importance in that for large values of time t, the term $a_1 \exp(r_1 t)$ will overpower all other terms of the birth function described in (20). That is, as $t \to \infty$

$$B(t) \sim a_1 \exp(r_1 t). \tag{21}$$

This is an asymptotic result of stable population theory which parallels the Malthusian model. It has been shown by Meyer (1982) that a_1 must be positive and real.

The discrete time recursive model in the stable theory of population has received little mathematical attention. An advantage of this model is that problems may be formulated and solved by simple calculation without the use of standard analysis. A more detailed presentation of this model may be found in Impagliazzo (1985).

Discrete Time Matrix Population Models

A discrete time model that has received extensive use is the matrix model. This model was introduced by Bernardelli (1941) and later expanded by Lewis (1942) and Leslie (1945). Development of this model may be found in Keyfitz (1968), Impagliazzo (1985) and Frauenthal (1986). The following is an overview of this important aspect of stable population theory.

From the previous section, Eq. (2) is

$$F(t+h, i+1) = g_{i+1} F(t, i)$$

and Eq. (5) and (10) may be written as

$$F(t,0) = \sum_{i=0}^{z} g_0 m_i F(t-h, i) h. \tag{22}$$

Normalizing over unit time intervals and setting $u = i + 1$, allows these equations to be written as

$$F(t+1, 1) = \sum_{j=1}^{n} g_0 m_j F(t, j) \tag{23}$$

and

$$F(t+1, u+1) = g_u F(t, u) \tag{24}$$

for $u = 1, 2, \ldots, n$ where n is the index of the interval that contains the last age of childbearing and $g_n = 0$. Expressed as a matrix, Eq. (23) and (24) become

$$
\begin{bmatrix}
F(t+1, 1) \\
F(t+1, 2) \\
F(t+1, 3) \\
\vdots \\
F(t+1, n)
\end{bmatrix}
=
\begin{bmatrix}
g_0 m_1 & g_0 m_2 & g_0 m_3 & \cdots & g_0 m_{n-1} & g_0 m_n \\
g_1 & 0 & 0 & \cdots & 0 & 0 \\
0 & g_2 & 0 & \cdots & 0 & 0 \\
\vdots & \vdots & \vdots & & \vdots & \vdots \\
0 & 0 & 0 & \cdots & g_{n-1} & 0
\end{bmatrix}
\begin{bmatrix}
F(t, 1) \\
F(t, 2) \\
F(t, 3) \\
\vdots \\
F(t, n)
\end{bmatrix}
$$

$$\tag{25}$$

The $n \times n$ matrix of (25) is called the Leslie Projection Matrix. The column vector is called the age distribution vector. In symbolic form and with obvious equivalences, (25) may be expressed as

$$\mathbf{F}(t + 1) = \mathbf{Z}\mathbf{F}(t). \tag{26}$$

It is easy to show that this recursive matrix equation can be used to project the initial population distribution $\mathbf{F}(t)$ to k time intervals in the future, and that

$$\mathbf{F}(t + k) = \mathbf{Z}^k\mathbf{F}(t). \tag{27}$$

The concept of stability in mathematical demography implies that the ratio of females in a particular age category compared to the total number of females is asymptotically a constant. From the standpoint of matrices, the interpretation is that the age distribution at time $t + 1$ is proportional to the age distribution at time t, and

$$\mathbf{F}(t + 1) = s\mathbf{F}(t). \tag{28}$$

Substitution of (28) into (26) results in an eigenvalue problem with characteristic equation

$$\det(\mathbf{Z} - s\mathbf{I}) = 0 \tag{29}$$

It can be demonstrated (Gantmacher, 1959) that the non-negative projection matrix $\mathbf{Z}$ is irreducible and primitive. The characteristic equation (29) will admit roots such that only one of them, the principal eigenvalue, is simple, positive and dominates in magnitude all other eigenvalues.

When the eigenvalues s_j are distinct, the population renewal equation (27) results in a solution of the form

$$\mathbf{F}(t + k) = \sum_{j=1}^{n} a_j s_j^k \mathbf{Y}_j \tag{30}$$

where the $\mathbf{Y}_j$ are the eigenvectors corresponding to the s_j. If s_1 is the principal eigenvalue, then as k becomes unbounded, the age distribution vector has the asymptotic result expressed as

$$\mathbf{F}(t + k) \sim a_1 s_1^k \mathbf{Y}_1. \tag{31}$$

Because of this asymptotic result, the eigenvector $\mathbf{Y}_1$ is often called the principal stable eigenvector.

There is no guarantee that the eigenvalues of the characteristic equation will be distinct, although it appears to be true for human populations. When the eigenvalues are not distinct, however, it has been shown that a solution to the matrix population renewal equation exists. Furthermore, the solution obtained for the age distribution vector has an asymptotic result similar to (31). For a complete discussion of the matrix population renewal equation including the case for non-distinct eigenvalues, consult Impagliazzo (1985).

A Continuous Time Population Model

The discussion which follows is a continuous time alternative to the deterministic discrete time recursive approach. The extension to the continuous time model may be accomplished by letting h approach zero. If $kh \leq x < (k+1)h$, then it is easy to show that in the limit, age intervals $[kh, (k+1)h)$ approach exact age x and the net maternity function ϕ_i becomes $\phi(x)$. Based on these facts, Eq. (15) may now be written as

$$B(t) = \int_0^w \phi(x)B(t-x)dx \tag{32}$$

where B is a continuous function of time. A function of the form $c\exp(rt)$ may be substituted for $B(t)$ to obtain a solution. This function is a solution if and only if it satisfies the equation

$$\int_0^w \exp(-rt)\phi(x,t)dx = 1 \tag{33}$$

which is called the characteristic equation for the continuous time model. The left side of the equation is called the characteristic function.

It has been demonstrated by Lotka (1907), Pollard (1973) and others that a solution to this linear Volterra integral equation takes the form

$$B(t) = a_1\exp(r_1 t) + \sum_{i=2}^q \left[\sum_{j=1}^{p(i)} a_{ij}t^{j-1} \right]\exp(r_i t) \tag{34}$$

where r_1 is the dominant root of the characteristic equation, and for $i = 2, 3, \ldots, q$, r_i is a root of the characteristic equation of multiplicity $p(i)$. As with the discrete time model, the continuous birth function is asymptotic to an exponential function and is given by

$$B(t) \sim a_1\exp(r_1 t) \tag{35}$$

where a_1 is a positive, real constant and r_1 is the principal root of the characteristic

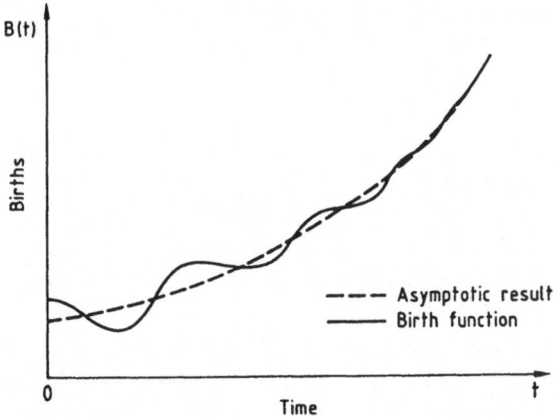

Asymptotic result
Birth function

Fig. 5. Graph demonstrating a typical solution to the discrete time recurrence model

equation (33). A typical illustration of the birth function is shown in Fig. 5. This once more illustrates the asymptotic behavior of the stable theory of population.

Crude Rates of Birth, Death and Increase

Consider the time interval $(t - h, t + h]$ of length $2h$ for $h > 0$. Let P_t be the total number of individuals of a population at time t. Let B_t represent the total number of live births and D_t the total number of deaths on this time interval. The ratios $b_c = B_t/P_t$ and $d_c = D_t/P_t$ are respectively the crude birth rate and the crude death rate of the population at time t. In practice, the time interval $2h$ is usually a calendar year causing P_t to be the mid-year population with b_c and d_c considered annual rates. The crude rate of increase is defined to be

$$r_c = b_c - d_c \tag{36}$$

It is possible to obtain an expression for the crude rate of birth, death and increase. In the discrete time mode, $F(t, i)$ was defined to be the number of females in age category indexed by i at time t. The continuous time analog of this is the number of females whose age lies on the interval $[x, x + dx)$ at time t which is represented by $F(t, x)dx$. The total number of females alive at time t is given by

$$F_T(t) = \int_0^w F(t, x)\,dx. \tag{37}$$

The number of daughters born to women of all ages at time t is given by

$$B(t) = \int_0^w m(x, t)F(t, x)\,dx \tag{38}$$

where m is the continuous time maternity function for age x at time t. The portion of the female population between ages x and $x + dx$ at time t is given by

$$c(x, t)dx = F(t, x)dx/F_T(t). \tag{39}$$

The number of females between ages x and y at time t is

$$\int_x^y c(x, t)dx = \int_x^y F(t, x)dx/F_T(t). \tag{40}$$

The crude birth rate can be expressed as

$$b_c = \int_0^w c(x, t)m(x, t)dx \Big/ \int_0^w c(x, t)\,dx. \tag{41}$$

Clearly,

$$\int_0^w c(x, t)\,dx = 1 \tag{42}$$

so the crude birth rate may be expressed also as

$$b_c = \int_0^w c(x, t)m(x, t)\,dx. \tag{43}$$

From the viewpoint of mortality, the parameter that expresses the annual rate of mortality at the precise moment of attaining age x at time t is $\mu(x, t)$ and is called the force of mortality. The parameter is expressed as the age derivative

$$\mu(x, t) = -\frac{1}{p(x, t)} \frac{d}{dx} [p(x, t)] = \frac{-d}{dx} [\ln (p(x, t))] \tag{44}$$

where $p(x, t)$ is the continuous time survivorship ratio for exact age x at time t. The number of female deaths for all ages at time t is

$$D(t) = \int_0^w \mu(x, t) F(t, x) dx \tag{45}$$

and the crude death rate at time t is

$$d_c = \int_0^w \mu(x, t) F(t, x) dx / F_T. \tag{46}$$

Following the same reasoning used in the crude birth rate, the crude death rate at time t may be expressed as

$$d_c = \int_0^w c(x, t) \mu(x, t) dx. \tag{47}$$

Combining (43), (47) with (36), the crude rate of increase at time t is

$$r_c = \int_0^w c(x, t) [m(x, t) - \mu(x, t)] dx. \tag{48}$$

Intrinsic Parameters of Stable Population Theory

There are certain results that are direct consequences of stable population theory which are useful in the analysis of populations and their models. The following illustrates the derivation of some of the more important parameters.

From (6) the number of females in age category indexed by k at time t is

$$F(t, k) = p_k B(t - kh, h). \tag{49}$$

The probability of surviving birth to the age category indexed by k is

$$p_k = \prod_{j=0}^k g_j. \tag{50}$$

For continuous time models, (49) may be expressed as

$$F(t, x) = p(x, t) B(t - x). \tag{51}$$

This represents the number of females alive who are exact age x at time t and are now the survivors of those females born at time $t - x$. Substitution of (37) and (51) into (39) results in

$$c(x, t) = p(x, t) B(t - x) / \int_0^w p(x, t) B(t - x) dx. \tag{52}$$

From the stable theory of population, substitution of (35) into (52) results in

$$c(x,t) \sim \frac{a_1 \exp(r_1 t) \exp(-r_1 x) p(x,t)}{a_1 \exp(r_1 t) \int\limits_0^w \exp(-r_1 x) p(x,t) dx} \tag{53}$$

from which one may write

$$c(x,t) \sim \frac{\exp(-r_1 x) p(x,t)}{\int\limits_0^w \exp(-r_1 x) p(x,t) dx} \tag{54}$$

Substitution of (54) into the crude birth rate (43) and taking the limit as $t \to \infty$ results in

$$b = \lim b_c = \frac{\int\limits_0^w \exp(-r_1 x) m(x,t) p(x,t) dx}{\int\limits_0^w \exp(-r_1 x) p(x,t) dx}. \tag{55}$$

Since the numerator of (55) is the characteristic function of the continuous time model and is equal to unity, then

$$b = 1 / \int\limits_0^w \exp(-r_1 x) p(x,t) dx \tag{56}$$

where b is called the intrinsic birth rate of stable population theory.

By similar reasoning substitution of (54) into the crude death rate (47) and taking the limit as $t \to \infty$ results in

$$d = \lim d_c = \frac{\int\limits_0^w \exp(-r_1 x) p(x,t) \mu(x,t) dx}{\int\limits_0^w \exp(-r_1 x) p(x,t) dx}. \tag{57}$$

Recall from (44) that

$$p(x,t) \mu(x,t) dx = -d(p(x,t)). \tag{58}$$

Substituting (58) into (57) and recalling that the denominator of (57) is $1/b$ results in

$$d = -b \int\limits_0^w \exp(r_1 x) d(p(x,t)) \tag{59}$$

where d is called the intrinsic death rate of stable population theory.

The difference between the intrinsic rates of birth and death may be expressed as

$$b - d = b[1 + \int\limits_0^w \exp(r_1 x) d(p(x,t))]. \tag{60}$$

Integration by parts results in

$$b - d = b[1 + \exp(-r_1 w) p(w,t) - p(0,t)$$
$$+ r \int\limits_0^w \exp(-r_1 x) p(x,t) dx]. \tag{61}$$

Since $p(w, t) = 0$ and $p(0, t) = 1$ then

$$b - d = b[1 - 1 + r_1/b] = r_1. \tag{62}$$

This illustrates a long-standing assumption under the stable theory of population that the intrinsic rate of increase is the difference between the intrinsic rates of birth and death and is equal to the principal characteristic root. Without loss of generality, the subscript may be dropped and

$$r = b - d. \tag{63}$$

The intrinsic rates r, b and d coupled with the age-specific functions of maternity m, mortality μ and the age distribution function c, form the fundamental parameters or functions of stable population theory. Although these functions have been expressed as generalized functions of time, it is understood that they are assumed to be true for all time. Therefore, they are simply written as functions of age alone as $m(x)$, $p(x)$, $\mu(x)$ and $c(x)$.

Gross and Net Reproductive Rates

The number of daughters that a female just born may expected to bear during her reproductive life, assuming she survives to the end of childbearing, is called the gross reproductive rate (GRR). If the reproductive age interval is $[a, b]$ then

$$GRR = \int_a^b m(x)dx \tag{64}$$

for all t. The gross reproductive rate assumes that the female survives to age b but in real life this is not necessarily the case. The probability that a female survives the age interval $[x, x + h]$ and gives birth to a daughter is the net maternity function ϕ as defined in (12). Then the number of daughters that a female just born may expect to bear during her reproductive life is called the net reproductive rate and is defines as

$$NRR = \int_a^b m(x)p(x)dx. \tag{65}$$

The characteristic function (33) of the continuous time model evaluated when the principal characteristic value (intrinsic rate of increase) r equals zero is simply NRR. That is,

$$NRR = \int_a^b \exp(-0x)m(x)p(x)dx = 1. \tag{66}$$

Clearly, the net reproductive rate is directly related to the intrinsic rate of increase. Specifically, based on the nature of the characteristic function $NRR = 1$ when $r = 0$, $NRR < 1$ when $r < 0$ and $NRR > 1$ when $r > 0$. The symbol R_0 is also used to represent the net reproductive rate.

A measure associated with the net reproductive rate is the length of a generation denoted T. The number of births at time t may be considered proportional to the number of births at time $t - T$. Let the constant of proportionality be R_0, the net reproductive rate. Then,

$$B(t) = R_0 B(t - T). \tag{67}$$

Dropping the subscript on the principal root of the characteristic equation, the stable solution for the birth function B is $a_1 \exp(rt)$ and when substituted into (67) gives

$$1 = R_0 \exp(-rT). \tag{68}$$

Solving for T yields

$$T = ln(R_0)/r \tag{69}$$

which is the length of a generation.

Population Substructures

It was already shown that the fraction of a population between ages x and $x + dx$ at time t for a stable population is

$$c(x)dx = b\exp(-rx)p(x)dx \tag{70}$$

where p is the probability of survival to age x from birth and b is the intrinsic birth rate. Let u and v be two ages between 0 and w such that $u \leq v$. The fraction of the population between ages u and v is simply

$$C(u, v) = \int_u^v c(x)dx$$

$$= b\int_u^v \exp(-rx)p(x)dx. \tag{71}$$

Clearly, the mean age of the population between age u and v at time t, denoted $a'_{u,v}$, is

$$a'_{u,v} = \frac{\int_u^v x\exp(-rx)p(x)dx}{\int_u^v \exp(-rx)p(x)dx}. \tag{72}$$

The mean age of a population over all ages of survival, denoted a', is $a'_{0,w}$ and is expressed as

$$a' = \frac{\int_0^w x\exp(-rx)p(x)dx}{\int_0^w \exp(-rx)p(x)dx}. \tag{73}$$

The integral of the denominator is simply $1/b$. Therefore, combining (56) and (70) with (73) gives

$$a' = \int_0^w xc(x)dx \tag{74}$$

Supply and Depletion of Monetary Funds

Consider a stable population where a portion of the population reap benefits of an annuity fund while another portion of the population makes contributions to the fund. Let N be the size of a population for all ages at some particular time. For simplicity, assume that all contributors make equal incomes of value M monetary units per year. Let the benefactors of this population be between ages v and w such that each person receives the same annuity of value A monetary units per year. For a balanced annuity plan, the total contributions to the fund must equal the total receipts from the fund. If f is the fraction of income that a contributor is required to make in order to maintain a balanced annuity fund, then

$$fMNC(u, v) = ANC(v, w) \tag{75}$$

where $C(x, y)$ represents the fraction of the population between ages x and y. Solving for f gives

$$f = (A/M)[C(v, w)/C(u, v)] \tag{76}$$

a value independent of population size.

An interesting aspect of this topic concerns the effects of a change in the intrinsic growth rate on the value of the contribution. The derivative of f with respect to r will determine this. Before addressing this issue, however, consider the derivative of $C(u, v)$ with respect to r. By using (71) and (56) it is easy to show that

$$\frac{d}{dr}C(u, v) = \frac{\int_u^v \exp(-rx)p(x)dx \int_0^w x\exp(-rx)p(x)dx - \int_0^w \exp(-rx)p(x)dx \int_u^v x\exp(-rx)p(x)dx}{\left[\int_0^w \exp(-rx)p(x)dx\right]^2}. \tag{77}$$

from which it is clear that

$$\frac{d}{dr}C(u, v) = (a' - a'_{u,v})C(u, v). \tag{78}$$

Returning to the expression for f expressed in (76) and taking the natural logarithm results in

$$\ln f = \ln(A/M) + \ln C(v, w) - \ln C(u, v) \tag{79}$$

Then the derivative with respect to r is

$$\frac{d}{dr}[\ln f] = [1/C(v, w)]\frac{d}{dr}C(v, w) - [1/C(u, v)]\frac{d}{dr}C(u, v) \tag{80}$$

where the derivative of A/M is zero. Substitution of (78) into (80) results in

$$\frac{d}{dr}\ln f = -(a'_b - a'_c) \tag{81}$$

where a_b' is the mean age for benefactors and a_c' is the mean age for contributors. In Western societies the difference between these two means is about 25 years and may be considered a constant. Based on this assumption the solution for f is simply

$$f = k \exp[-(a_b' - a_c')r] \tag{82}$$

for some constant k. Clearly, this demonstrates that an increase in r causes a decrease in f while a decrease in r causes an increase in f.

Estimation of the Intrinsic Rate of Growth

Consider the situation of a new population region or country having no previous demographic history. It is desired to estimate the intrinsic rate of growth. A single census is taken and the number of individuals aged x at time t, denoted $N(x, t)$, is recorded. With the use of a life table from a similar region or country, it is assumed that the individuals $N(x, t)$ are the survivors of those born x years ago under the influence of the average survivorship function L_x. For two different cohorts at time t

$$N(x, t) = L_x B(t - x)$$

and $\tag{83}$

$$N(y, t) = L_y B(t - y).$$

The births at times $t - x$ and $t - y$ are not known. However, from stable population theory, equation (35) may be used to express the ratio

$$\frac{B(t - x)}{B(t - y)} = \frac{N(x, t)/L_x}{N(y, t)/L_y} = \frac{a_1 \exp(r(t - x))}{a_1 \exp(r(t - y))} = \exp(r(y - x)). \tag{84}$$

Rewriting (84) results in

$$\exp(r(y - x)) = \frac{N(x, t)/N(y, t)}{L_x/L_y}$$

from which

$$r = (1/(y - x)) \ln\left[\frac{N(x, t)/N(y, t)}{L_x/L_y}\right]. \tag{85}$$

Equation (85) represents an estimate for the intrinsic growth rate r when a single census and a life table are used.

When two census years are available, a life table is not required and the guesswork of L_x/L_y is eliminated. Suppose a census is taken at time t' and another is taken at time t''. For a given age x, the population at times t' and t'' may be written as

$$N(x, t') = L_x B(t' - x)$$

and $\tag{86}$

$$N(x, t'') = L_x B(t'' - x).$$

Since the assumption of stability ensures that L_x is the same for time t' and t'', then (86) and (35) result in

$$N(x, t')/N(x, t'') = B(t' - x)/B(t'' - x) = \exp\left(r(t' - t'')\right)$$

from which

$$r = (1/(t' - t'')) \ln \left[N(x, t')/N(x, t'')\right]. \tag{87}$$

Equation (87) is a two-census estimate for the intrinsic rate of growth.

Estimation of the Intrinsic Rate of Birth

An extension of the previous discussion may be used to estimate the intrinsic rate of birth. From (71) the portion of the population between ages x and $x + n$ is

$$C(x, x + n) = b \int_{x}^{x+n} \exp\left(-rx\right)p(x)dx \tag{88}$$

For a small enough interval, the exponential may be evaluated at the midpoint of the interval and treated as a constant; hence (88) is approximated as

$$C(x, x + n) \doteq b \exp\left(-r(x + n/2)\right) \int_{x}^{x+n} p(x)dx. \tag{89}$$

The integral of (89) may be interpreted as $_nL_x$ with radix 1_0. Therefore, (89) becomes

$$\exp\left(r(x + n/2)\right) \doteq {_nL_x}b/C(x, x + n). \tag{90}$$

The intrinsic rate of growth is assumed to be constant over age. For two different ages x and y, (90) may be expressed as

$$\begin{aligned} r &\doteq (1/(x + n/2)) \ln \left[_nL_xb/(C(x, x + n))\right] \\ &\doteq (1/(y + n/2)) \ln \left[_nL_yb/(C(y, y + n))\right]. \end{aligned} \tag{91}$$

Rearrangement of the equality in (91) with exclusion of r yields

$$b^{(y - x)} \doteq [C(x, x + n)/_nL_x]^{y + n/2}/[C(y, y + n)/_nL_y]^{x + n/2}$$

from which

$$b \doteq \left[\frac{[C(x, x + n)/_nL_x]^{y + n/2}}{[C(y, y + n)/_nL_y]^{x + n/2}} \right]^{1/(y - x)} \tag{92}$$

Equation (92) indicates that under a stable population, the intrinsic birth rate may be approximated by a single census and a life table.

Conclusion

The continuous time population model is attributed to Lotka (1907). McKendrick (1926) and von Foerster (1959) have also made contributions to this model. Because of the ability to use the calculus, the continuous time model has received much exposure and attention. One should not lose sight that the elegance which is brought forth by the continuous time model is not always useful for actual calculations when applied to real population data. Herein lies the art of balancing the advantages of both discrete time and continuous time models.

Other models are also possible such as two-sex models and quasi-stable models as suggested by Coale (1963) and Lopez (1961) as well as stochastic models developed by Mode (1985). There are numerous applications that may be considered when dealing with population models. A sample includes the estimation of the intrinsic birth rate, the effects of migration on a population, as well as morbidity, kinship and marriage which may be found in Keyfitz (1977). In addition, there are numerous extensions and applications in mathematical ecology that lend themselves to population (not necessarily human population) modeling. Interested readers are directed to sources such as Levin (1981), Hallam (1986) and Gross (1986) for further information in these areas.

References

Bernardelli, H. (1941) Population waves. Journal of the Burma Research Society, vol. *31*, part 1

Coale, A.J. (1963) Estimates of various measures through the quasi-stable age distribution. Emerging techniques in population research. Thirty-Ninth Annual Conference of the Milbank Memorial Fund, New York

Feller, W. (1941) On the integral equation of renewal theory. Annals of Mathematical Statistics, vol. *12*

Foerster, H. von (1959) In: Stohlman, F., Jr. (ed.) The kinetics of cellular proliferation. Grune and Stratton, New York

Frauenthal, J.C. (1986) Analysis of age-structure models. In: Hallam, T.G., Levin, S.A. (eds.) Mathematical ecology. Biomathematics, vol. *17*. Springer, Berlin Heidelberg New York London Paris Tokyo, pp. 117–147

Gantmacher, Feliks, R. (1959) Applications of the theory of matrices. Interscience Publishers, New York

Goldberg, S. (1958) Introduction to difference equations with illustrative examples from economics, psychology and sociology. John Wiley and Sons, New York

Gross, L.J. (1986) Ecology: An idiosyncratic overview. In: Hallam, T.G., Levin, S.A. (eds.) Mathematical ecology. Biomathematics, vol. 17. Springer, Berlin Heidelberg New York London Paris Tokyo, pp. 3–15.

Hallam, T.G., Levin, S.A. (1986) Mathematical ecology. Biomathematics, vol. 17. Springer, Berlin Heidelberg New York London Paris Tokyo

Impagliazzo, J. (1985) Deterministic aspects of mathematical demography. Biomathematics, vol. *13*. Springer, Berlin Heidelberg New York London Paris Tokyo

Keyfitz, N. (1968) Introduction to the mathematics of population. Addison-Wesley, Reading, Mass

Keyfitz, N. (1977) Applied mathematical demography. Wiley and Sons, New York

Leslie, P.H. (1945) On the use of matrices in certain population mathematics. Biometrika, vol. 33

Levin, S.A. (1981) Age structure and stability in multiple-age spawning populations. In: Vincent, T.L., Skowrouski, J.M. (eds.) Renewable Resource Management. Lecture Notes in Biomathematics, vol. 39. Springer, Berlin Heidelberg New York London Paris Tokyo

Lewis, E.G. (1942) On the generation and growth of a population. Sankhya, vol. 6

Lopez, A. (1961) Problems in stable population theory. Office of Population Research, Princeton University, Princeton

Lotka, A.J. (1907) Relation between birth rates and death rates. Scinece N.S., vol. *26*

Malthus, T.R. (1798) An essay on the principle of population as it affects the future improvement of society with remarks on the speculations of Mr. Godwin, Condorcet M. and other writers. Reprinted 1926. Macmillan, London

McKendrick, A.G. (1926) Applications of mathematics to medical problems. Proceedings of the Edinburgh Mathematical Society, vol. 44, series 1

Meyer, W. (1982) Asymptotic birth trajectories in the discrete form of stable population theory. Theoretical Population Biology, vol. 21, no. 2

Mode, C.J. (1985) Stochastic processes in demography and their computer implementation. Biomathematics, vol. 14. Springer, Berlin Heidelberg New York London Paris Tokyo

Pollard, J.H. (1973) Mathematical models for the growth of human populations. Cambridge University Press, London

Smith, D., Keyfitz, N. (1977) Mathematical demography: selected papers. Biomathematics, vol. 6. Springer, Berlin Heidelberg New York

Stage Structure Models Applied in Evolutionary Ecology

R.M. Nisbet, W.S.C. Gurney, and J.A.J. Metz

Contents

1. Introduction

Mathematical models of physiologically structured populations are now well established within the mainstream of theoretical ecology (Metz and Diekmann, 1986, and references therein), but to date their utilisation in many areas of ecology has been restricted by two types of difficulty. First, the numerical solution of the partial differential equations that arise naturally in the description of many structured populations is far from straightforward, and although promising methods are currently being developed (e.g. de Roos, 1988), the numerically unsophisticated worker does not have ready access to well-tested "off-the-shelf" computer packages such as are available for models posed in terms of ordinary differential equations, difference equations, or Leslie matrices. Second, practical applications of structured models demand large quantities of biological information, and it is seldom easy to formulate models that only require parameters which can be calculated from existing data.

These twin demands—that models be computationally tractable and parametrised with an eye on available data—motivated the development of "stage structure" models. These models are applicable to species whose life history consists of a succession of distinct stages (e.g. insect instars) with a single "development index" or "physiological age" determining the transition from one stage to its successor. The ground rules for the systematic formulation of simple stage structure models are by now well established (see for example Nisbet and Gurney, 1986: our paper in the preceding volume; also Gurney, Nisbet and Lawton, 1983; Nisbet and Gurney, 1983), and a number of elaborations of the basic techniques have been developed (Blythe, Nisbet and Gurney, 1984; Nisbet et al., 1986; Gurney, Nisbet and Blythe, 1986). The end point of this work is that a wide range of structured models may be formulated in terms of sets of coupled delay differential equations with parameters specific to particular life history stages, the price paid for this simplification being partial or total neglect of within stage variability. Such models have found application in "strategic" studies (e.g. Gurney and Nisbet, 1985; Murdoch et al., 1987) and in more detailed simulations of natural populations (e.g. Crowley et al., 1987).

In parallel with these developments, there has been a growing awareness of the relevance of non-linear structured population models to life history theory and to population genetics. The most elementary "textbook" models (discussed for example in Crow, 1986; Hedrick, 1983) for gene frequency change through selection in large populations assume constant (i.e. density independent) relative individual fitnesses, and consider differential survival prior to reproduction as the only selective force. Developments over the past one and a half decades have seen the development of theory for evolution in age structured populations (Charlesworth, 1980) and of systematic recipes for the formulation of models of density—and frequency-dependent selection (Roughgarden, 1979), advances which have freed population genetics from the straitjacket of assumed exponential growth and which provide insight on evolutionary pressures in populations with nonlinear dependence of vital rate parameters on population density or environmental factors. Yet there remain significant gaps in our understanding of density-dependent selection in structured populations, for essentially the reasons identified above: mathematical or computational intractability, and a mismatch between model parameters and available data. Specifically, we have rather little understanding of invasibility conditions or of the rate of gene substitution in cycling age-structured populations, and know of no quantitative analysis of any long term population experiments that take account of age structure.

Our aim in this chapter is to explore the utility of the compromise offered by stage structure models in modelling evolution in structured populations. The models are motivated by the classic experiments of Nicholson (1957) on laboratory populations of the sheep blowfly *Lucilia cuprina*, in which the nature of the fluctuations in population size changed dramatically in the course of a two-year experiment, a change Nicholson attributed to the development in adult females of the ability to produce the first eggs without a protein meal. Shorter experiments (Nicholson, 1954) lasting about one year on the same system have been successfully modelled by a very simple stage structure model (Gurney, Blythe and Nisbet, 1980; Nisbet and Gurney, 1982); all the models in the present paper are elaborations on that basic model.

In Sect. 2, as a prerequisite to what follows, we review some known results on models of evolution in unstructured populations, with particular emphasis on invasibility. In the following section we review the relevant experimental results from Nicholson's papers, outline the key steps in the formulation and testing of the simple model, and then interpret the results of the "long" (1957) experiments as "drift" in the values of the parameters in that model (Stokes, 1985; Stokes et al., 1988). In Sect. 4 we investigate the generation structure of the cycling populations, this being relevant to our modelling strategy in the remainder of the paper. We then study, in Sect. 5, an exceedingly simple "clonal" model, for which we determine invasibility conditions analytically for situations in which the established population is stable, and perform numerical simulations for the situation where (as in Nicholson's experiments) the population is executing large amplitude limit cycles. Unfortunately, no amount of special pleading can justify a clonal model as a representation of blowflies, so in Sect. 6 we make similar investigations for a one-locus, two-allele diploid model.

2. Some Basic Evolutionary Genetics

Two major aims of simple models in evolutionary genetics are to predict the speed of change through time of the frequencies of different genes already present in a population (gene substitution), and to determine the possibility of establishment of a rare mutant in a genetically homogeneous resident population (invasion).

The simplest idealised model for selection in large populations relies on a number of assumptions:

(1) one locus, two alleles,
(2) no mutations,
(3) random mating,
(4) nonoverlapping generations,
(5) differential survival prior to reproduction ("viability") as the only selective force,
(6) constant (i.e. *inter alia* density independent) relative individual fitnesses.

If the two alleles are represented by A and a, and if P, Q, and R represent the frequencies of the zygotic genotypes AA, Aa and aa respectively, then p and q, the relative zygotic frequencies of the A and a alleles, are (by definition) given by

$$p = P + Q/2 \quad q = 1 - p. \tag{1}$$

If assumptions (3) to (5) hold, then the zygotic genotype frequencies are given by a relationship known as Hardy–Weinberg equilibrium:

$$P = p^2, \quad Q = 2pq, \quad R = q^2. \tag{2}$$

Consequently, in this model, one zygotic gene frequency (p or q) suffices to specify the genetic structure of the population.

An allele A is said to be *fixed* in the population if the population is made up exclusively of the genotype AA ($p = P = 1$); a population in which A and a persist (with p bounded away from both zero and one) is said to exhibit *protected polymorphism*. If v v_2, v_3 denote the viabilities of genotypes AA, Aa, and aa respectively, then in the simple model referred to above, the outcome of selection may be summarised as follows:

(a) If $v_1 \geq v_2 \geq v_3$, with at least one inequality, then A goes to fixation for any initial gene frequency $p_0 > 0$.

(b) If $v_1 < v_2 > v_3$ (heterozygote superiority), then there is a stable and globally attractive internal equilibrium with non-zero frequencies for both alleles,

(c) If $v_1 > v_2 < v_3$ (heterozygote inferiority), then either A or a goes to fixation depending on whether the initial gene frequency p_0 is smaller or larger than the (now unstable) equilibrium value.

Three further results are relevant for investigating invasibility and the speed of gene substitution subsequent to invasion, namely:

(e) If $v_2 > v_3$ or $v_1 > v_2 = v_3$, then a population made up exclusively of a is invasible by A. The initial population growth is exponential if $v_2 > v_3$, and a is ultimately removed from the population at an exponential rate if $v_1 > v_2$.

(f) If $v_1 > v_2 = v_3$ (allele A fully recessive), and the initial value (p_0) of p is very small, then p initially increases very slowly; after t generations, to second order in p_0 (and to first order in t):

$$p(t) = p_0 + (v_1 - v_2)v_2^{-1}p_0^2 t, \tag{3}$$

so that p increases linearly (rather than exponentially) and at a rate proportional to p_0^2.

(g) If $v_1 = v_2 > v_3$ (allele A fully dominant), then the asymptotic approach of q towards zero is very slow; if t is large, then after t generations

$$q(t) = v_2/[(v_2 - v_3)(C + t)] \tag{4}$$

where C is a constant determined by the initial conditions.

Many of the above results have their counterparts in more complex models: this is true for example with (f) and (g) which reflect the fact that the frequency of homozygotes of a relatively rare gene varies as the square of the gene frequency, while the frequency of heterozygotes is to a first approximation linear in the gene frequency, so it is the performance of heterozygotes that determines the rates of invasion or fixation.

If any of the idealisations (3)–(5) cannot be made, then we cannot assume Hardy–Weinberg equilibrium and have to keep track of *two* genotype frequencies P and Q (remembering that $R = 1 - P - Q$). If, additionally, individual properties such as viability or fertility vary with overall population density (denoted by N), we also have to include this quantity in our description of population state; alternatively we may parametrize the population state by three *genotype densities*: $N_1 = NP$, $N_2 = NQ$, and $N_3 = NR$. In the still more complex case of a structured population where an individual can be in one of a number of states (e.g. age classes), the minimal description involves densities of all possible genotypes for each individual state; thus in a one-locus, two-allele stage structure model we have to work with three genotype densities for each stage. The population balance equations can be formed according to the usual book-keeping considerations, except that in the birth terms we have to account for the rules of pair formation (for example by assuming random mating) and the Mendelian ratios of the corresponding neonate outputs. An example of this procedure will be worked through in detail in Sect. 6.

Clearly there is scope for rapid escalation in complexity with the associated problems of parameter richness (the unconvinced reader might try to write down the equations for a two-locus, two-stage model!), and much published work makes further approximations or idealisations in order to achieve an acceptable level of mathematical complexity. One such simplification results in asexual or *clonal* models in which the assumption of Mendelian inheritance is dropped and competition is assumed to take place between identical reproducing clones. This ploy can be justified *a posteriori* from the observation that many essential properties, like initial rate of increase, tend to be shared by both types of model. This is commonly the case when genotypes differ only in their survival probabilities, but considerable care is needed for example when, as is usual for fertility selection, vital parameters differ between the sexes.

Where clonal models are inappropriate, an alternative ploy is to study the fate of a rare mutant invading a stable, genetically homogeneous, resident population. As initially selection operates mainly through the heterozygotes, and overall population density is determined by the resident, the equations describing the immediate fate of the invader are almost always linear (an obvious exception being the case of a fully recessive invader—see result (f) above). Quite a lot of information can be gathered from such invasibility calculations. For example, if *a* can invade a population in which A is fixed but not vice versa, then there is unlikely to be long term polymorphism with A and a coexisting for ever. Conversely if A and a can mutually invade, we have a protected polymorphism, whether in stable equilibrium or with persisting fluctuations.

Both simplifications will be used in this chapter in our modelling on Nicholson's blowflies. When we investigate invasibility in the narrow context of the blowfly modelling, we of course do not envisage the spontaneous appearance of an advantageous mutant, a singularly unlikely outcome in an experiment lasting only a few tens of generations, made even more unlikely by the fact that the observed drift in population behaviour occurred over roughly the same amount of time in different experiments (Nicholson, 1957). Rather we recognise that an individual introduced to Nicholon's laboratory conditions experienced very different selective pressures from those experienced by wild type files, and model the progress of substitution of a gene with a very low initial density. Mathematically this is equivalent to modelling invasion.

This program offers the possibility of insight with wider applicability than the narrow aim of "postdicting" the properties of Nicholson's blowflies, since relative invasibility is probably the most important indicator of the direction of evolutionary progress. One scenario is that advantageous mutants occur only very rarely (advantageous being defined as "potentially invading"!), and moreover are often lost due to random fluctuations in the very early stages when there are still a small number of copies around. In that case, a population will for any particular adaptive trait usually go through relatively short spells of transient polymorphism interspersed with long waiting periods before the next advantageous mutation comes along. Which mutations are going to make it in the next round, and therefore which direction adaptation is driving the system depends on relative invasibility.

Modelling invasibility would be particularly straightforward if there were many situations in which we could find a function F such that any population of individuals with some trait combination x could be invaded by individuals with trait combination y if and only if $F(y) > F(x)$. In that case evolution will cause F to increase subject only to local constraints. The long term effect of evolution is then to maximise the function F subject to the appropriate constraint(s). The clonal model in Sect. 5 of this chapter provides an example of such a maximum principle.

Of course life is not as simple as that (see e.g. Hastings, 1984). From the invasion condition calculated in Sect. 6 of the present paper (Eq. 34) we shall see that even for a particularly simple model such as is developed in that section, there need not be a simple function that will be maximised by evolution. Understanding of possible evolutionary mechanisms can therefore only proceed by painstaking study of simplified models peretaining to specific biological situations. Our aim in the remainder of this chapter is to demonstrate that stage structure models can play some part in this program.

3. Modelling Nicholson's Blowflies

Short Experiments

Nicholson (1954) found large amplitude cycles in laboratory populations of the sheep blowfly *Lucilia cuprina* (Wied), kept under conditions where the population was regulated by availability of protein for the adult flies, although there was ample sugar and water to meet maintenance requirements. One example of these data is shown in Fig. 1a. A number of models have been proposed to explain the appearance of these cycles (e.g. Varley, Gradwell and Hassell, 1973; May, 1974; Maynard Smith, 1974; Oster, 1976; Gurney, Blythe and Nisbet, 1980; Readshaw and Cuff, 1980; Gurney, Nisbet and Lawton, 1983), and there is broad agreement that the primary

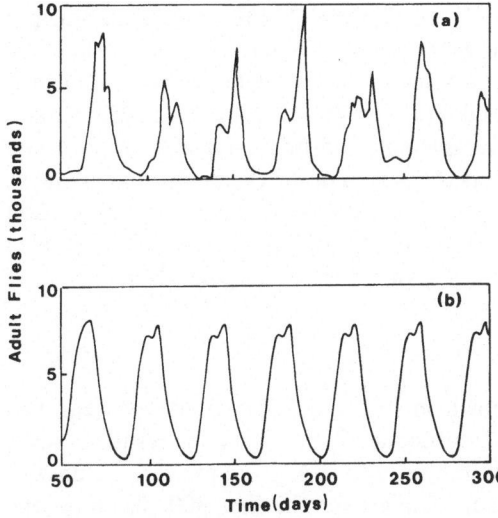

Fig. 1a, b. Fluctuations in Nicholson's blowfly populations: (a) Experimental results for a population limited by the availability of protein for adults (Nicholson, 1954), (b) predictions of the simple model (Eqs. 5 and 6) with $r = 6$ day^{-1}, $N_0 = 850$ flies, $m = 0.2$ day^{-1}, $T = 15$ day

cause is the delay between the immediate reduction in fecundity in flies experiencing shortage of food, and any consequent reduction in the rate of recruitment to the adult population.

The simplest model (Gurney et al., 1980; Gurney et al., 1983; Stokes et al., 1988) that is consistent with the known biology of individual flies, and in reasonable quantitative agreement with the observed population fluctuations, is based on a balance equation for the population $N(t)$ of mature adult flies

$$dN(t)/dt = E(t - T) - mN(t) \qquad (5)$$

in which m represents the age- and density-independent *per capita* death rate of adult flies, T the (assumed constant) development time from egg to mature adult, and $E(t)$ the total rate of "viable" egg production, an egg being regarded as viable if it eventually survives to become a mature adult. The total viable egg production rate is assumed to depend on N according to the relationship

$$E(t) = rN(t)\exp(-N(t)/N_0), \qquad (6)$$

which can be justified either as a plausible fit to data of Readshaw and van Gerwen (1983) on egg production as a function of food supply, or by statistical fits to segments of Nicholson's population runs (Stokes, 1985). In Eq. (6), r represents the product of the maximum possible fecundity and the (density-independent) survival from egg to mature adult, while N_0 is a measure of the effect of adult population size on fecundity, with very little egg production occurring if N is substantially in excess of N_0.

The model as now formulated has four parameters: r, N_0, m, and T. The first two can be estimated either from experiments on individual flies (published by Readshaw and Cuff, 1980), or alternatively from parts of the blowfly population data (e.g. by estimating N_0 from the population sizes at peak egg production: Gurney et al., 1980). Fortunately, there is reasonable consistency in the results whichever approach is used. The mortality rate can be estimated from the decline phase of the population cycles (Gurney et al., 1980), or from the results of survival experiments by Readshaw and van Gerwen (1983). The developmental delay was measured by Nicholson (1954). Details of the different methods of estimating the parameters, and a variety of tests of the model have been published elsewhere (Gurney et al., 1980; Nisbet and Gurney, 1982; Gurney et al., 1983; Stokes, 1985), but we illustrate in Fig. 1b a sample solution of Eq. (1) with appropriate parameters. The model correctly predicts the period of the cycles (around 40 days), and makes a reasonable estimate of the peak populations, the error in this prediction being within the range attributable to uncertainty in N_0. The model cycles also exhibit the "double peak" structure that is evident in the original data.

Long Experiments

The experiment whose results are shown in Fig. 1 ran for about one year and produced seven cycles. Nicholson (1957) subsequently published the results of some even longer experiments, one of which (reproduced in Fig. 2a) ran for 722 days— almost twice as long as its predecessor. While over the first 300–400 days the

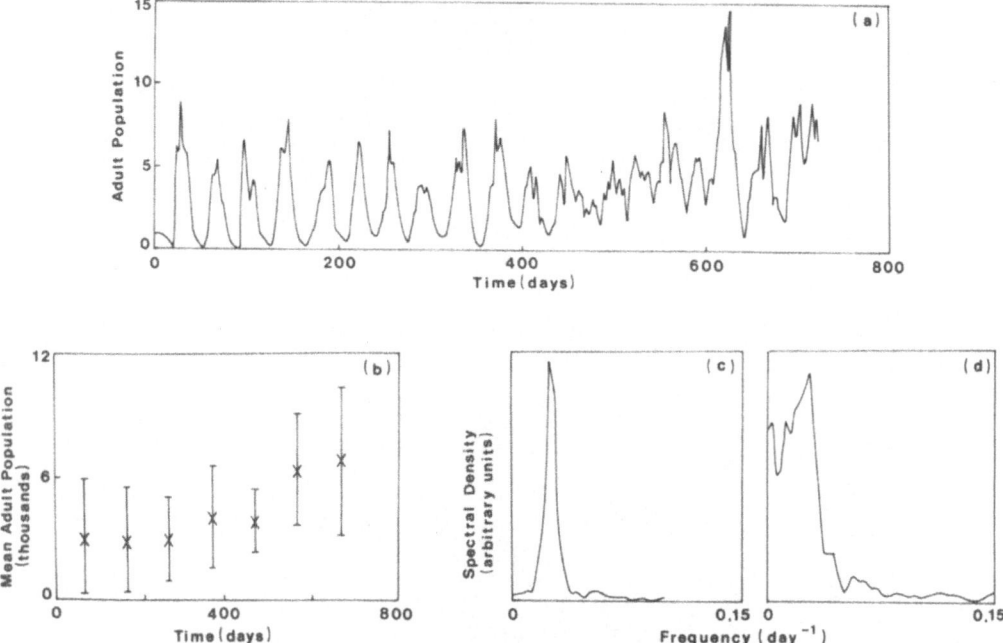

Fig. 2a–d. Results of Nicholson's "long" (1957) experiment on blowflies: **(a)** Adult population fluctuations, **(b)** 100 day means (with error bars), **(c)** spectral density for first 400 days only, **(d)** spectral density for remainder of data. [Based on analysis in Stokes et al., 1988]

population fluctuations in this second experiment are consistent with expectations based on the earlier experiments, the second half of the data set develops in an unexpected way which is inconsistent with the predictions of the simple model represented in Eq. (5) and (6). The mean population begins to drift upwards and the fluctuations lose their distinctive quasi-cyclic character (see Figs. 2b–d).

Nicholson's data on death and recruitment from which the population fluctuations shown in Fig. 2a were constructed are available in a paper by Brillinger, Guckenheimer, Guttorp and Oster (1980). Using these data Stokes (1985) and Stokes et al. (1988) estimated the best fitting values of the parameters of the simple model *for successive 100-day blocks of data.* There was evidence of substantial "drift" in the values of r (which dropped from 6 to 1 day^{-1}) and N_0 (which increased from 850 to 2500 flies). The drift in the estimates of the *per capita* death rate m was less well defined and scarcely statistically significant. No attempt was made to estimate drift in the developmental delay T.

Having identified significant drift in the parameters r and N_0, it is possible to regard these quantities as continuous functions of time, chosen to fit the observed drift, and then to compute a numerical solution to Eqs. (5) and (6). The result of this exercise is shown in Fig. 3a, and indicates that the observed parameter drift is indeed consistent with the observation that the cycles disappear after around 400 days. However the experimental population continued to exhibit rather large, if non-cyclic, fluctuations for the remainder of the experiment; Stokes et al. argued that

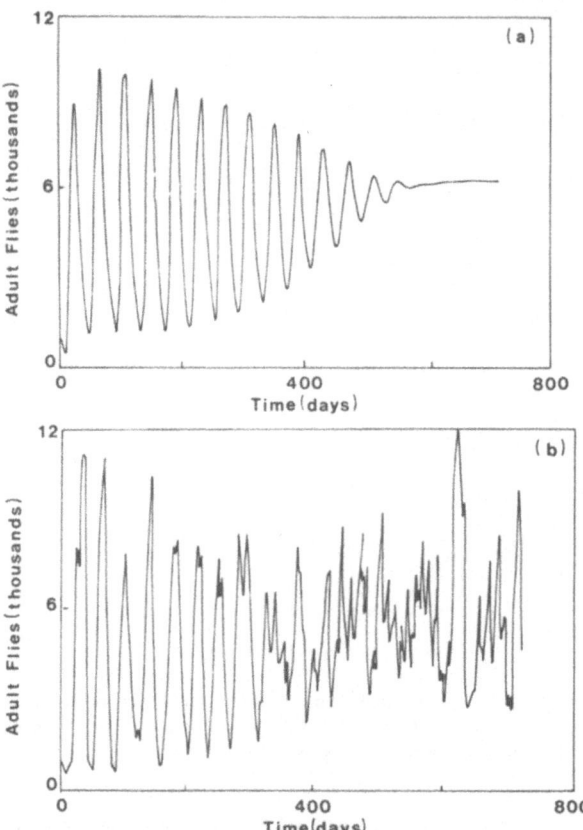

Fig. 3a, b. Simulations by Stokes et al. (1988) of Nicholson's blow-flies based on simple model, but with parameters varying with time: (**a**) deterministic, (**b**) with added noise

these could be understood in terms of the response of the system to some noise source. This would have little effect on the rather robust limit cycles present at the start of the experiment, and a much greater effect as the instantaneous values of r dropped below the critical value for local stability. They were not able to identify the biological origin of this noise, but from the scatter in the regression used to fit the parameters were able to estimate its intensity and spectrum, which had sufficient structure to indicate some systematic contributory factor not included in the model. To assess the importance of this systematic error, they carried out a modelling exercise in which *white* noise of the appropriate intensity was added to the model, and the resulting stochastic delay-differential equation interpreted in the Ito sense (as was done in the analysis of the data). The outcome was that the resulting population fluctuations were of comparable magnitude and form to those observed (Fig. 3b).

4. Generation Structure

In view of the vital role played by discrete generation models in underpinning the traditional models of selection and invasion, it is of interest to explore the pattern of successive generations in the blowfly cycles, with particular emphasis on the extent of overlap. Naively we might regard each population "burst" as one generation; alternatively we might regard the "double peaks" in egg production as distinct generations. With more complex patterns of fluctuation, the possibilities are limitless and some systematic procedure for identifying generations is clearly needed.

Suppose there are x_0 adults present at some specified time $(t = 0)$, and we call these individuals generation 0. The population of generation zero at time t is then $x_0 \exp(-mt)$. If we denote by $x_1(t)$ the total number of adult offspring of these adults present at time t, and define an instantaneous fecundity by

$$f(t) = E(t)/N(t) = r \exp(-N(t)/N_0), \tag{7}$$

then the population of the first generation $x_1(t)$ varies as

$$\dot{x}_1(t) = f(t-T)x_0 \exp(-m(t-T)) - mx_1 \quad (t > T) \tag{8}$$

with $x_1 = 0$ for $t < T$. By similar reasoning, the numbers alive at time t in the ith generation are determined by the delay-differential equation

$$\dot{x}_i(t) = x_{i-1}(t-T)f(t-T) - mx_i, \quad (t > iT) \tag{9}$$

while the *total* size of the ith generation is just the total number of dead flies produced by individuals in that generation, i.e.

$$G_i = m \int_0^\infty x_i(t')dt'. \tag{10}$$

This can also be computed from a delay differential equation by defining

$$G_i(t) = m \int_0^t x_i(t')dt', \tag{11}$$

so that

$$\dot{G}_i(t) = mx_i(t), \quad G_i(0) = 0. \tag{12}$$

Eq. (12) is integrated until a steady state is approached to within some specified accuracy.

Figure 4 illustrates the sequence of generations in a run with the same parameters as were used to generate Fig. 1b. We arbitrarily chose those adults present at time 50 days (during a decline phase in population) as generation zero and studied the next five generations over a time interval which contained five large population peaks. The total *size* of successive generations quickly stabilises, but their distribution over time is more complex. Effectively all the members of the first generation fall within the next population peak (unsurprisingly), but this peak also contains a large number of members of generation 2, a feature reflected in the "double peaked" pattern of egg production (Fig. 1a). The generations soon smear out among the population peaks, with generation four being split between three peaks.

Both main features noted in Fig. 4—smearing of generations and constant asymptotic generation size—can be proved to hold for a wide range of population

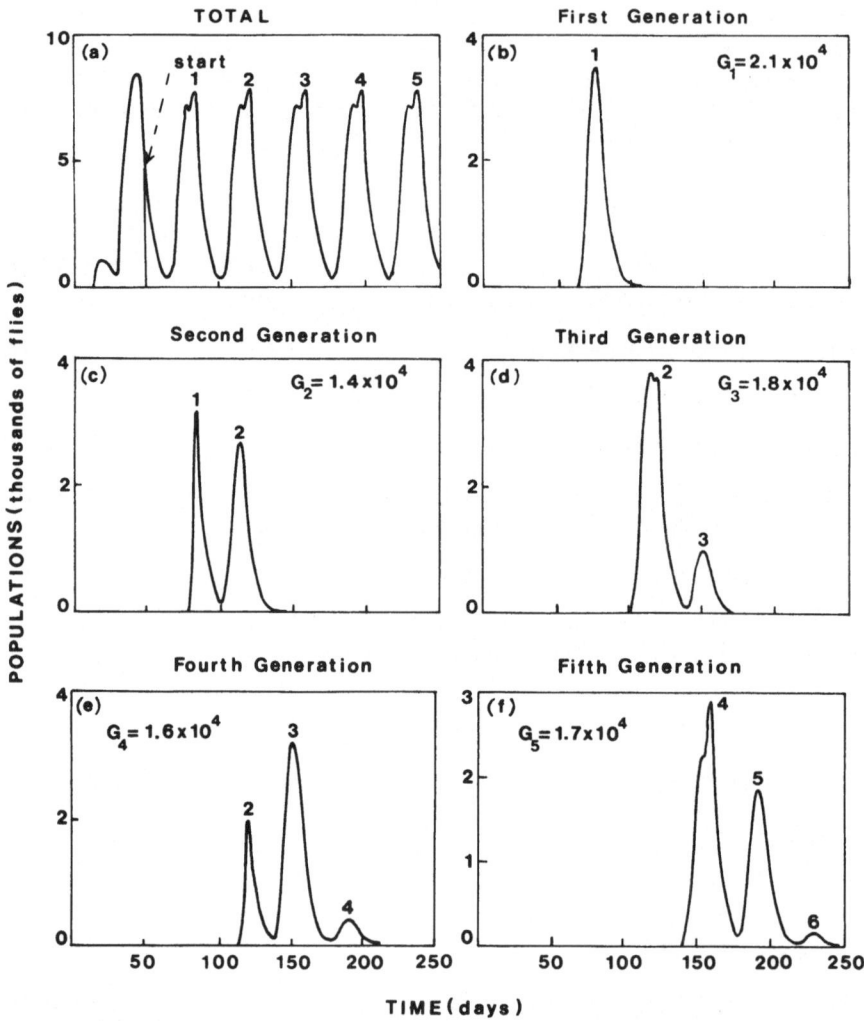

Fig. 4a–f. Structure of successive generations predicted by the model of Nicholson's blowflies with the same parameters as were used to generate Fig. 1b. Adults in the population at $t = 50$ days were arbitrarily designated to be generation zero. (**a**) Total adult population, (**b**)–(**f**) numbers in generations 2–5. The peaks in total population are labelled 1–5 and marked on the generation plots, as are the total generation sizes (defined in Eq. (10) of the main text)

models with periodic solutions (J.A.J. Metz—unpublished work), and we can speculate that the results remain true with chaotic solutions. Figure 5 illustrates the successive generation sizes, and the distribution over time of the fortieth generation for two further sets of parameters—one where the original equations have a "period-doubled" solution, the other where there is apparently chaos.

From the distribution of generations shown in Figs. 4 and 5, it is clear that, in spite of the distinctive population peaks, there is no prospect of meaningfully

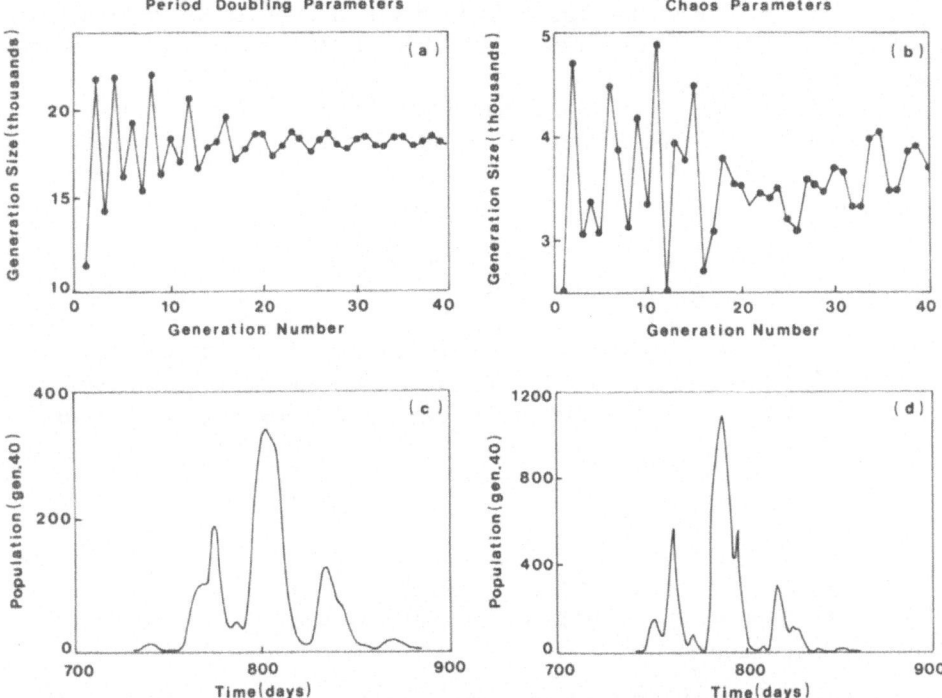

Fig. 5a–d. Total size of successive generations predicted by the blowfly model for (**a**) parameters in which the population fluctuations have passed through one period doubling (PD) bifurcation, and (**b**) parameters for which the population fluctuations are chaotic. Also shown (figs. **c** and **d**) is the variation with time of the instantaneous size, $x_{40}(t)$, of the fortieth generation in each case.

The parameters used were selected with the aid of Fig. 10 of Blythe et al. (1982) and were:
$r = 6\,\mathrm{day}^{-1}$, $N_0 = 1000$ flies, $m = 0.333\,\mathrm{day}^{-1}$, $T = 15\,\mathrm{day}$—PD,
$r = 20\,\mathrm{day}^{-1}$, $N_0 = 1000$ flies, $m = 0.5\,\mathrm{day}^{-1}$, $T = 15\,\mathrm{day}$—chaos

modelling density dependent selection in Nicholson's blowflies within any discrete generation framework.

5. Clonal Model

We now construct a particularly simple stage-structure model which can be used to describe the replacement of a wild type by an invader. The model is a natural extension of the blowfly model used in the previous two sections, and we use it to demonstrate that the pace of selection within the blowfly population *could* be fast enough to produce the parameter drift highlighted in Sect. 3.

Following Stokes et al. (1988), we imagine that at the start of the blowfly experiments, the population consisted predominantly of wild-type individuals characterised by a set of parameters r_w, N_{ow}, m_w, and T_w, together with a very few individuals of a rare type characterised by parameters r_i, N_{oi}, m_i, and T_i. We assume scramble competition among *all* adults for food with equal success for adults of both types, but that adults of the two types differ in their ability to produce eggs at a given level of protein input.

The essential assumption of a clonal model is that the offspring of wild type females are all of wild type; likewise the offspring of the initially rare type females are also of the initially rare type. The two viable egg production rates can then plausibly be assumed to take a form analogous to eq. (6). namely

$$E_w(t) = r_w N_w(t) \exp\left[-(N_w(t) + N_i(t))/N_{ow}\right] \tag{13}$$

$$E_i(t) = r_i N_i(t) \exp\left[-(N_w(t) + N_i(t))/N_{oi}\right] \tag{14}$$

in which $N_w(t)$ and $N_i(t)$ are the individual populations of wild and rare type respectively. Provided the maturation delay (T) is the same for each type, the adult population dynamics are now given by two equations analogous to eq. (5), namely

$$\dot{N}_w(t) = E_w(t - T) - mN_w(t), \tag{15}$$

$$\dot{N}_i(t) = E_i(t - T) - mN_i(t). \tag{16}$$

It would clearly be of interest to have analytic conditions for takeover and the initial rate of growth of an invading population. The first step in such a derivation is to note that if the population of the invader $N_i(t)$ is sufficiently small, then Eqs. (15) and (16) take the form

$$\dot{N}_w(t) = r_w N_w(t - T) \exp\left(-N_w(t - T)/N_{ow}\right) - m_w N_w(t), \tag{17}$$

$$\dot{N}_i(t) = r_i N_i(t - T) \exp\left(-N_w(t - T)/N_{oi}\right) - m_w N_i(t). \tag{18}$$

Thus to this order of approximation, the dynamics of the resident population is unaffected by the invader, while that of the invader is given by a *linear* delay-differential equation with, in general, a time dependent forcing term multiplying the delayed argument. The form of this forcing term is determined by the fluctuations in the resident population. If, however, the resident population is at or near the stable equilibrium

$$N_w^* = N_{ow} \ln\left(r_w/m_w\right) \tag{19}$$

then the invasion problem reduces to one of calculating the roots of a characteristic equation, derived by assuming a solution of the form $N_i \propto \exp(\lambda t)$ for Eq. (18). The result is

$$m_i R_0 \exp\left(-\lambda T\right) = m_i + \lambda, \tag{20a}$$

where

$$R_0 = r_i (m_w/r_w)^{N_{ow}/N_{oi}}/m_i \tag{20b}$$

is the mean number of viable offspring produced by an invader as long as invader density remains low.

Since R_0 and m_i are positive, it can be shown that Eq. (19) has a dominant real root with the property that all other roots have smaller real parts. This dominant root is positive if and only if $R_0 > 1$, which can only occur if

$$N_{oi} \ln\left(r_i/m_i\right) > N_{ow} \ln\left(r_w/m_w\right), \tag{21}$$

suggesting (cf. Eq. 19) that evolution maximises equilibrium population size. Since it is a tenet of faith that natural selection operates at the level of individuals, this is more appropriately expressed as: evolution maximises the density of conspecifics that an individual can bear in its neighbourhood, while still producing on average one viable offspring in its lifetime. We cannot be more specific without supplying some physiological details about the mechanism of competition. However in the present model we envisage protein shortage as the cause of reduced fecundity and can thus reinterpret the condition (21) in terms which truly relate to individual physiology as: evolution minimises the level of resource (protein) per individual at which an individual produces on average one viable offspring during its lifetime.

If the resident population is not in stable equilibrium but fluctuates in a regular limit cycle, Floquet theory (see e.g. Hale, 1977; chapter 8) combined with the natural positivity conditions implies that the long term overall increase or decrease of the solution to (18) will be exponential. However, practical means for calculating the dominant Floquet exponent, other than through direct numerical solution of the delay differential equation, are unavailable. From a limited number of numerical solutions of Eqs. (15) and (16), it appears that Eq. (21) is a reasonable approximation

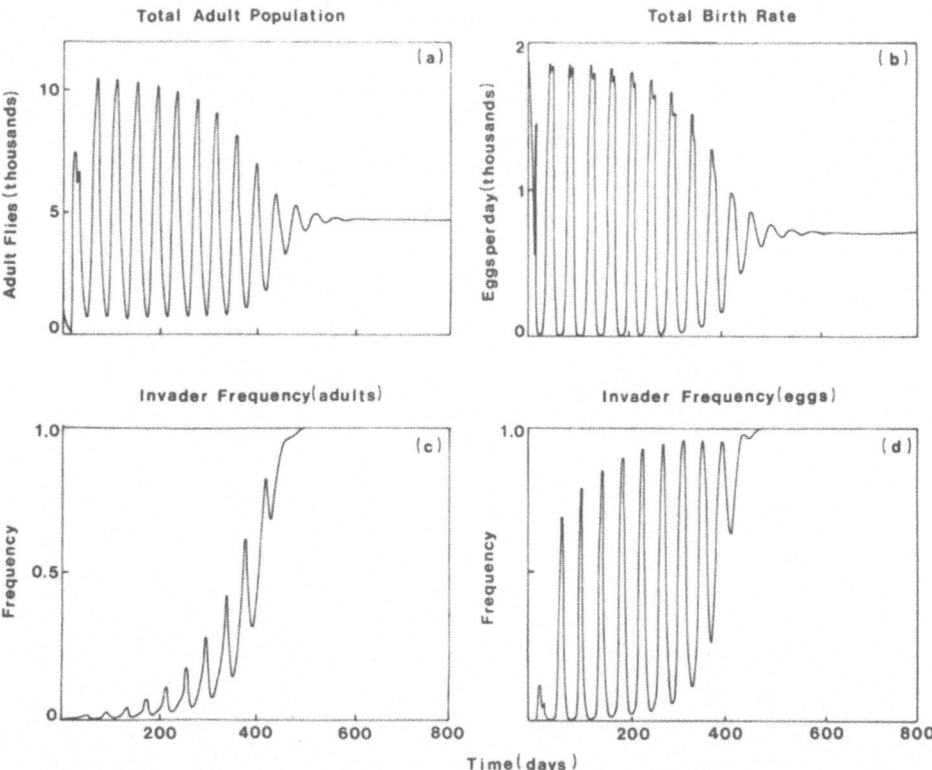

Fig. 6a–d. Solution of Eqs. (15)/(16) with parameters as specified in the text. The runs were started with a population of 1000 adult flies of which 1% were rare type. Figs. **a** and **b** show the development through time of the total population and total birth rate respectively while (**c**) and (**d**) show the instantaneous *proportions* of rare type in the adults and eggs respectively

to the invasion condition even if the resident population is cycling. However the detailed behaviour shows evidence of considerable dynamic subtlety which merits further research.

With parameters appropriate to Nicholson's blowflies, the resident population is clearly not at or near a stable equilibrium, at least in the early part of the experiment, so further study of this system must proceed numerically. If we guess that at the start of the experiment, the population was overwhelmingly made up of wild type flies,

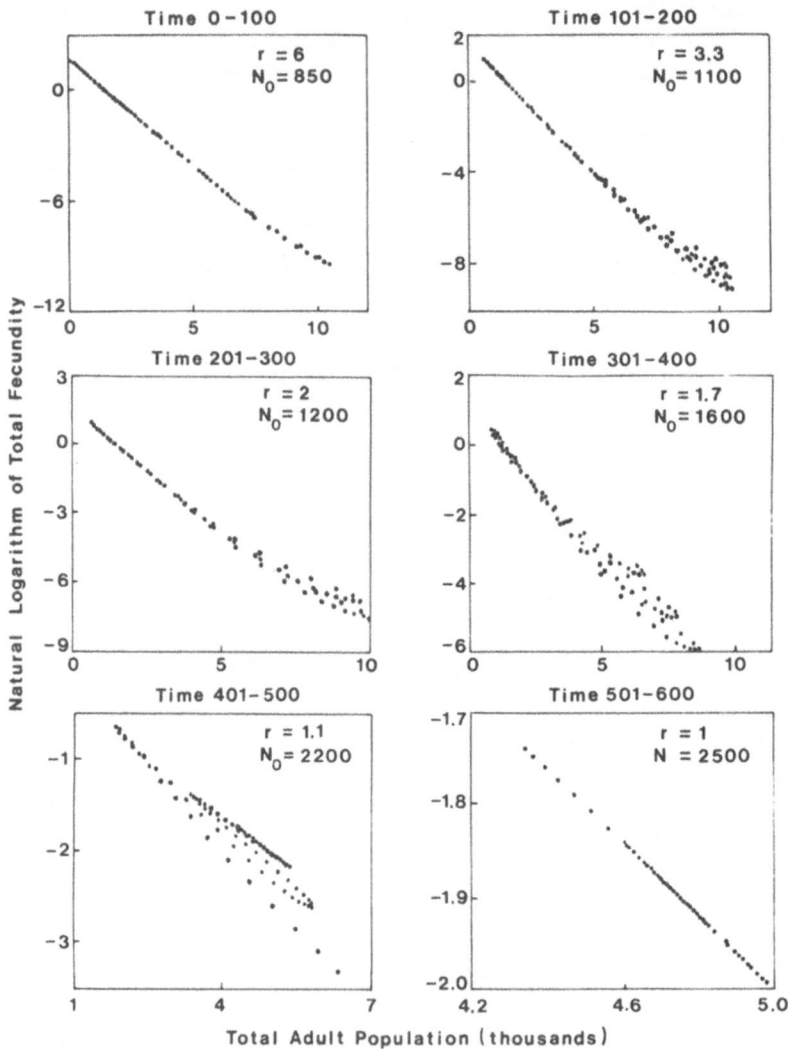

Fig. 7. The clonal model predictions of variation in logarithm of total fecundity with total population for successive blocks of 100 days. The results may be reconciled with the observed "parameter drift", noted by Stokes (1985) and Stokes et al. (1988) for the simple one-strain model, by "eyeballing" straight lines to the plots and inferring values of r and N_0 for the relevant interval of time. This was an exercise at the Trieste workshops; typical results are noted on each figure

and that the takeover by the initially rare type is essentially completed by the end of the run, then the parameter values in the egg production functions for the two types can be inferred from the fits by Stokes (1985) to egg production in the first and last 100 days. A run with these parameters, a fixed death rate of 0.15 day^{-1}, and a fixed delay of 15 days is illustrated in Fig. 6, from which it is clear that replacement of the wild type by the invader could be largely complete within a few hundred days. Further computations reveal that the timescale of takeover depends only weakly (approximately logarithmically) on the initial density of the rare type. Figures 6c and 6d illustrate the mechanism underlying this takeover: during the troughs of the cycles in total birth rate the larger value of N_o for the invader results in a very high proportion of eggs being of rare type (Fig. 6d). As a result, during the troughs in the adult population cycles the proportion of the invader increases until large enough to reduce the effective fecundity of the total population sufficiently to produce stability and eventually complete takeover.

It is an instructive exercise (which was part of the Trieste workshop course) to compare the "parameter drift" predicted by this simple two-strain model with that observed experimentally and discussed in Sect. 3. For the same run as was used to construct Fig. 6, we plot natural logarithm of fecundity (defined as total egg production rate over total mature adult population) against total mature adult population, *for successive 100 day blocks of data*. The results are shown in Figs. 7a-f from which we can easily infer the variation with time of the parameters r and N_0. The results of a rough exercise in "eyeballing" straight line fits to these graphs are noted in the figure, and show that the behaviour of the clonal model is at least qualitatively consistent with the observed parameter drift.

6. Diploid Model

Of course real blowflies are diploid, sexually reproducing creatures. The main difference between a clonal model and a diploid one-locus model is that in the latter we have to deal with three genotypes AA Aa and aa. We denote by N_1, N_2, and N_3 the densities of the three genotypes and further introduce:

$N = N_1 + N_2 + N_3$: total adult density
$P = N_1/N, \quad Q = N_2/N, \quad R = N_3/N$: adult genotype frequencies
$p = P + 0.5Q, \quad q = 0.5Q + R$: adult gene frequencies.

We assume as before that any density effects on reproduction are mediated through the total population density only. Then assuming a fixed sex ratio, we postulate that the "viable egg production" for the three types of females is given by

$$E_i(t) = r_i N_i(t) \exp[-N(t)/N_{oi}]. \quad i = 1, 2, 3 \tag{22}$$

To calculate the contribution of each of these egg outputs to the change in adult genotype densities after the appropriate development times have elapsed, we note that a fraction $P(t)$ of E_1 comes from insemination by AA males and is therefore

100% AA, a fraction $Q(t)$ comes from Aa males and is 50% AA and 50% Aa, while a fraction $R(t)$ comes from aa males and is therefore 100% Aa. Comparable arguments apply to E_2 and E_3. We conclude that the rate of viable egg input at time t of genotype AA is:

$$B_1(t) = [P(t) + 0.5Q(t)]E_1(t) + [0.5P(t) + 0.25Q(t)]E_2(t)$$

$$= p(t)[E_1(t) + 0.5E_2(t)]. \tag{23}$$

Similar arguments show that the instantaneous viable egg input for genotype Aa is

$$B_2(t) = q(t)E_1(t) + 0.5E_2(t) + p(t)E_3(t), \tag{24}$$

and for aa is

$$B_3(t) = q(t)[0.5E_2(t) + E_3(t)]; \tag{25}$$

the reader may find it instructive to check that with these expressions the sum of the egg inputs to the three genotypes indeed equals the total egg production. Finally we observe that if T_i is the maturation delay for genotype i, then the adult recruitment rate at time t is just the viable egg input at time $t - T_i$; consequently

$$\dot{N}_i(t) = B_i(t - T_1) - m_i N_i(t), \quad i = 1, 2, 3. \tag{26}$$

As a first step in the analysis of Eq. (26) we investigate the conditions under which A can invade into an aa population. When p is very small, both N_1 and N_2 will be small relative to N_3, and in addition N_1 will be small relative to N_2. Thus, to first order of approximation:

$$N = N_3, \tag{27}$$

$$q = 1, \quad 0.5E_2 + E_3 = E_3, \tag{28}$$

$$qE_1 + 0.5E_2 + pE_3 = 0.5E_2 + pE_3, \quad p = 0.5N_2/N_3, \tag{29}$$

implying

$$\dot{N}_3(t) = r_3 N_3(t - T_3)\exp[-N_3(t - T_3)/N_{o3}] - m_3 N_3, \tag{30}$$

$$\dot{N}_2(t) = 0.5r_2 N_2(t - T_2)\exp[-N_3(t - T_2)/N_{o2}]$$

$$+ 0.5r_3 N_2(t - T_2)\exp[-N_3(t - T_2)/N_{o3}] - m_2 N_2. \tag{31}$$

Barring certain very special cases such as the case of a fully recessive mutant investigated later, Eqs. (30) and (31) determine both the conditions under which A can invade and the initial relative rate at which invasion takes place (and hence the apparent overall rate of gene substitution).

As in the clonal case, the linearised Eq. (31) only admits straightforward analysis in circumstances where N_3 can justifiably be assumed to be constant; in other words when the steady state solution

$$N_3^* = N_{o3}\ln(r_3/m_3) \tag{32}$$

of (30) is stable. Then (31) reduces to

$$\dot{N}_2(t) = 0.5[r_2(m_3/r_3)^{N_{o3}/N_{o2}} + m_3]N_2(t - T_2) - m_2 N_2(t - T_2). \tag{33}$$

the form of which clearly reflects our assumption that the fertility differences among genotypes are an exclusively female property: the first term in the square brackets expressing the contribution through the female line, the second the contribution through the male line.

In the same manner as for the clonal model we conclude that A can invade only if R_0, given by

$$R_0 = 0.5[r_2(m_3/r_3)^{N_{o3}/N_{o2}} + m_3]/m_2 > 1. \tag{34}$$

If $R_0 = 1$, invasibility is determined by the properties of the homozygote AA. In the special case that $m_2 = m_3 = m$, we have that $R_0 > 1$ if and only if

$$N_{o2} \ln (r_2/m) > N_{o3} \ln (r_3/m). \tag{35}$$

indicating that if the death rates are the same, the invasibility condition are identical in the clonal and diploid cases.

The speed of invasion can again be determined from the characteristic equation

$$m_2 R_0 \exp(-\lambda T_2) = \lambda + m_2 \tag{36}$$

which has the same form as in the clonal model (Eq. 19), though the expression defining R_0 is different (Eq. 34). If there is no selection due to differential mortality $(m_2 = m_3)$, then with appropriate identification of parameters

$$R_{0,\text{diploid}} = (R_{0,\text{clonal}} + 1)/2 < R_{0,\text{clonal}} \tag{37}$$

provided $R_0 > 1$ in both cases, from which it is easy to show using Eq. (36) that initial gene substitution goes slower in the diploid than in the clonal case. Indeed a perturbation expansion in the quantity $(R_0 - 1)$ shows that in the case of sufficiently slow selection

$$\lambda_{\text{clonal}} = 2\lambda_{\text{diploid}}. \tag{38}$$

As in the clonal case, we have no analytic results for invasion of a population which is not at or near a stable equilibrium. We have however done some numerical investigations, with parameters based on Nicholson's data. Figures 8 and 9 contain the results from two of these computations. In both cases we assume that the parameters of the aa genotype are those of the starting phase of Nicholson's experiments, and that the parameters of the AA genotype are those of the final phase. The difference is in the parameters selected for the heterozygote Aa: in the first run (Fig. 8) we assume A to be fully dominant, in the second (Fig. 9) A is fully recessive.

In the dominant case, the oscillations effectively disappear between 400 and 500 days, much as with the clonal model. This indicates the importance of the population oscillations, as the parameters are such that if we were dealing with an invasion into the equilibrium population we would be well within the region of applicability of Eq. (38). Another noteworthy feature of the results is that the adult gene frequecies remain close to Hardy-Weinberg equilibrium, in sharp contrast to the gene frequencies among eggs. This is due to the fact that the adult population consists of a mix of cohorts, born at different times and with different gene frequencies. The effect of this is to enhance the relative numbers of heterozygotes (the "Wahlund effect"), whereas the selection regime (Fig. 9d) produced the opposite effect among the neonates. Apparently the two effects roughly cancel for much of the time.

The recessive case, depicted in Fig. 9, differs from the dominant case in the way anticipated in Sect. 2 on the basis of the discrete generation models—an exceedingly long (5000 days) time for the invasion to get started. All other features, such as deviation from Hardy-Weinberg equilibrium, are very much comparable to those for the dominant case.

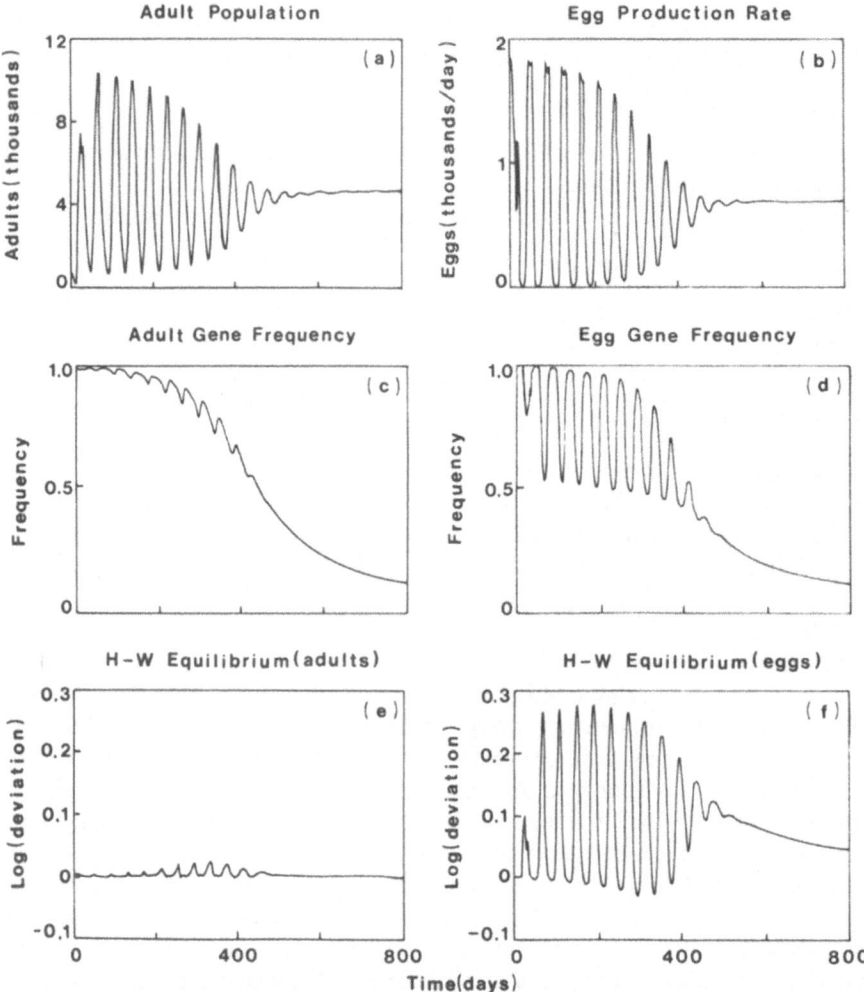

Fig. 8a–f. Numerical solutions for the diploid model with the "invader" gene fully dominant. The parameters used were: $r_1 = r_2 = 1$ day^{-1}, $r_3 = 6$ day^{-1}; $N_{01} = N_{02} = 2500$ flies, $N_{03} = 850$ flies; $m = 0.2$ day^{-1}, $T = 15$ day (all genotypes). The initial population of 1000 was made up of 980 in adult genotype 3 and 20 in adult genotype 2, corresponding to a gene frequency of 1% for the invader.

Plots (a) and (b) give the trajectory of the total population and birth rate, and plots (c) and (d) contain the variation of (wild type) gene frequency in respectively the adult and egg populations. Plots (e) and (f) give (on a logarithmic scale) the ratio of observed gene frequency (for adults and eggs respectively) to that which would occur if Hardy-Weinberg equilibrium prevailed.

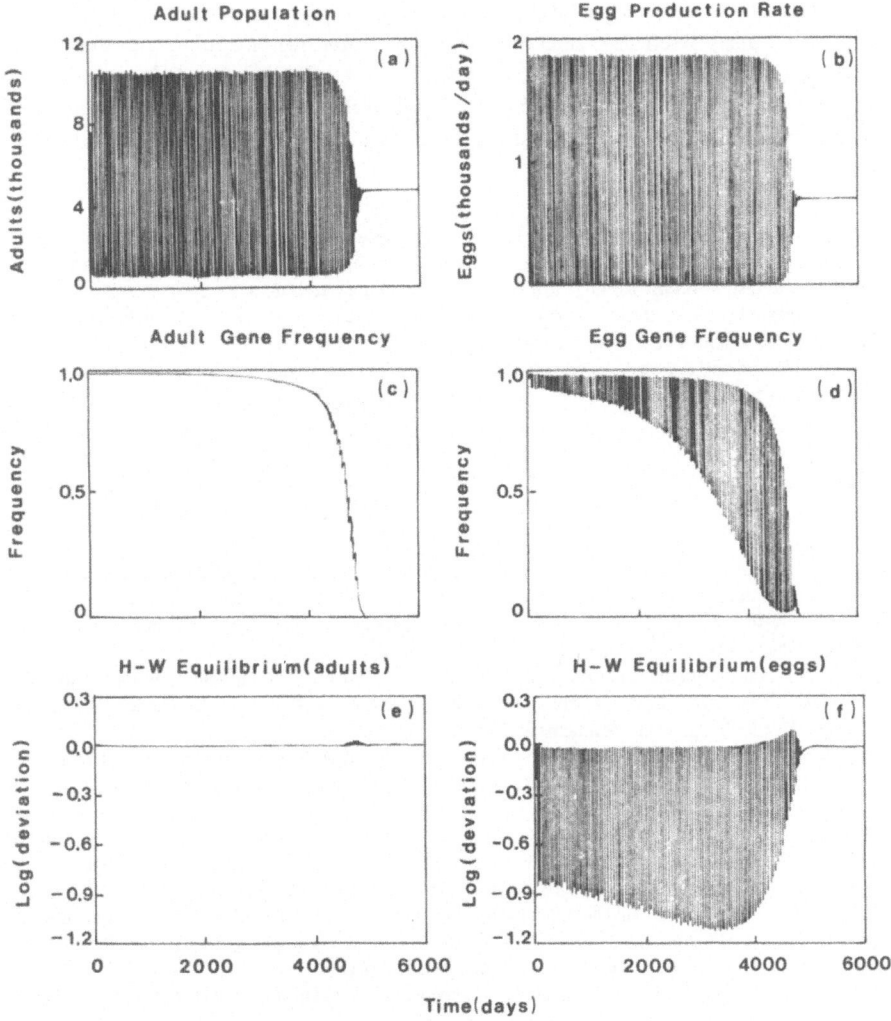

Fig. 9. Same as Fig. 8 but with the invader fully recessive

7. Concluding Remarks

The potential for using stage structure models in population genetics is clearly illustrated by the diploid model investigated in this paper. The arguments used to formulate the diploid model admit extension to more elaborate models; in particular models with various mixes of fertility and viability selection and with density dependent mortality in one or more life history stages can clearly be developed. However, the complexity of such models will certainly preclude any complete non-linear analysis, and it is likely that the fastest route to insight on the dynamics will be to concentrate on derivation of invasibility conditions for populations in stable equilibrium, supported by selective computations for populations exhibiting persisting fluctuations (limit cycles or chaos). Looking further ahead, deeper analytic study is required on the conditions for invasibility of cycling stage-structured populations, and on the generation structure of the invader and resident populations.

Acknowledgements. We thank Steve Blythe for suggesting an earlier "diploid" model and for comments on a previous draft of this work, Kevin Stokes for discussions on the "clonal" model, and Stefan Gerritz for remarks and insight on invasibility. Robert Beaton did some pareliminary work on the clonal model as an undergraduate project at the University of Strathclyde. The paper is based on modelling workshops at the 1986 Trieste school and on similar workshops held at the University of Leiden in May 1987; we thank Simon Wood and Andre de Roos for assistance with this teaching. Above all, we thank the students participating in the two courses for their interest and insight.

References

Blythe, S.P., Nisbet, R.M., Gurney, W.S.C. (1982) Instability and complex dynamic behaviour in population models with long time delays. Theor. Pop. Biol. *22*, 147–176.

Blythe, S.P., Nisbet, R.M., Gurney, W.S.C. (1984) Population models with distributed maturation periods. Theor. Pop. Biol. *25*, 289–311.

Brillinger, D.R., Guckenheimer, J., Guttorp, P., Oster, G. (1980) Empirical modelling of population time series data: The case of age and density dependent vital rates. Lectures on Mathematics in the Life Sciences *13*, 65–90. American Mathematical Society

Charlesworth, B. (1980) Evolution in Age-Structured Populations. Cambridge University Press

Crow, J.F. (1986) Basic Concepts in Population, Quantitative, and Evolutionary Genetics. Freeman, New York

Crowley, P.H., Nisbet, R.M., Gurney, W.S.C., Lawton, J.W. (1987) Population regulation in animals with complex life histories: Formulation and analysis of a damselfly model. Adv. Ecol. Res. *17*, 1–59

Gurney, W.S.C., Blythe, S.P. Nisbet, R.M. (1980) Nicholson's blowflies revisited. Nature *187*, 17–21

Gurney, W.S.C., Nisbet, R.M., Blythe, S.P. (1986) The systematic formulation of stage structure models. In: Metz. J.A.J., Diekmann, O. (eds.) The Dynamics of Physiologically Structured Populations, Springer, Berlin Heidelberg New York London Paris Tokyo

Gurney, W.S.C., Nisbet, R.M. (1985) Generation separation, fluctuation periodicity and the expression of larval competition. Theor. Pop. Biol. *28*, 150–180

Gurney, W.S.C., Nisbet, R.M., Lawton, J.H. (1983) The systematic formulation of tractable single species population models including age-structure. J. Anim. Ecol. *52*, 479–495.

Hale, J. (1977) Theory of Functional Differential Equations. Springer, Berlin Heidelberg New York Tokyo

Hastings, A. (1984) Evolution in a seasonal environment: simplicity lost? Evolution *38*, 350–358.

Hedrick, P.W. (1983) Genetics of Populations. Science Books International, Boston

May, R.M. (1974) Stability and Complexity in Model Ecosystems. Princeton University Press, Princeton, N.J.

Maynard Smith, J. (1974) Models in Ecology. Cambridge University Press, Cambridge, England

Metz, J.A.J., Diekmann, O. (eds.) (1986) The Dynamics of Physiologically Structured Populations. Springer, Berlin Heidelberg New York London Paris Tokyo

Murdoch, W.W., Nisbet, R.M., Blythe, S.P., Gurney, W.S.C., Reeve, J.D. (1987) An invulnerable age class and stability in delay-differential parasitoid-host models. Am. Nat. *129*, 263–282.

Nicholson, A.J. (1954) An outline of the dynamics of animal populations. Aust. J. Zool. *2*, 9–65

Nicholson, A.J. (1957) The self adjustment of populations to change. Cold Spring Harbour Symposium on Quantitative Biology *22*, 153–173

Nisbet, R.M., Blythe, S.P., Gurney, W.S.C., Metz, J.A.J. (1986) Stage structure models with distinct growth and development processes. IMA J. Math. Appl. Med. and Biol. *2*, 57–68

Nisbet, R.M., Gurney, W.S.C. (1982) Modelling Fluctuating Populations. Wiley and Sons, London New York Chichester

Nisbet, R.M., Gurney, W.S.C. (1983) The systematic formulation of population models for insects with dynamically varying instar duration. Theor. Pop. Biol. *23*, 114–135

Nisbet, R.M., Gurney, W.S.C. (1986) Age structure models. In: Hallam, T.G., Levin, S.A. (eds.) Mathematical ecology. Springer, Berlin Heidelberg New York London Paris Tokyo

Oster, G. (1976) Internal variables in population dynamics. In: Lectures in Mathematics in the Life Sciences *8*, 37–68

Readshaw, J.L., Cuff, W.R. (1980) A model of Nicholson's blowfly cycles and its relevance to predation theory. J. Anim. Ecol. *49*, 1005–1010

Readshaw, J.L., van Gerwen, A.C.M. (1983) Age specific survival, fecundity, and fertility of the adult blowfly Lucilia cuprina in relation to crowding, protein food and population cycles. J. Anim. Ecol. *52*, 879–888

de Roos, A. (1988) Numerical methods for structured population models: The escalator boxcar train. Numerical Methods for Partial Differential Equations *4*, 173–195

Roughgarden, J. (1979) Theory of population genetics and evolutionary Ecology: an Introduction. Macmillan, New York

Stokes, T.K. (1985) PhD thesis University of Strathclyde, Glasgow UK

Stokes, T.K., Gurney, W.S.C., Nisbet, R.M., Blythe, S.P. (1988) Parameter evolution in a laboratory insect population. Theor. Pop. Biol. *34*, 248–265.

Varley, G.C., Gradwell, G.R., Hassell, M.P. (1973) Insect Population Ecology. Blackwell Scientific Publications, London

Some Applications of Structured Models in Population Dynamics

Carlos Castillo-Chavez

Contents

It is now a well-known fact that age or size structure often affects qualitative changes in the dynamics of population models (see Nisbet and Gurney in *Mathematical Ecology. An Introduction*, eds. T.G. Hallam and S.A. Levin 1986). However the incorporation of age or size structure leads to infinite dimensional dynamical systems that are difficult to analyze. Furthermore, in some cases increased detail may reduce predictive capability because of problem of parameter estimation and error propagation. Because of this difficulty, what I call the "science of biological aggregation" has responded with systematic attempts to develop models of reduced mathematical complexity that do not sacrifice biological realism. In many instances, a minimal level of detail is required; further aggregation results in the loss of vital information and may lead to erroneous conclusions. Successful simplified realistic models have to be less aggregated.

In this chapter, I discuss some structured models that have been used in ecology and epidemiology. As these notes concentrate on some of the most basic mathematical techniques used in the analysis of these models as well as on the modelling approaches involved in their construction, the emphasis is on a hands-on approach. Topics are covered as a sequence of exercises, most of them simple, others requiring more thought, and a few which may lead to research articles. References to the literature indicate where more detailed discussions can be found. The selection of the applications is not totally arbitrary; my intention is to provide a few applications and expansions of some of the models in Hallam and Levin (1986). I have chosen applications that do not demand a sophisticated mathematical background and that yet give a feeling of active areas of research. Throughout these notes there will be references to some of the articles in Hallam and Levin (1986); however, I include a short review of age-structure models.

These notes are organized as follows: Sction I provides a review of the linear theory with age-dependence, a brief examination of the effects of harvesting strategies on the age-structure of a single population, and a short discussion of the relationship between discrete and continuos models. Section II discusses the classical nonlinear demographic model for a single population with age structure and linear and nonlinear size-structured models with time delays. Section III looks at the use of age-structured and sociologically-structured models in epidemiology that have been used to model the dynamics of influenza, AIDS, and hepatitis.

Section I

It was a physician, Lieut. Col. A.G. MacKendrick (1926), who first introduced age into the description of the dynamics of a one-sex population. The linear theory is based on the MacKendrick equation, which is most often referred to as the Von Foerster (1959) equation since MacKendrick's earlier work was not well known. Fortunately, Hoppensteadt (1984) has made the contributions of MacKendrick widely available to the scientific community.

The MacKendrick model assumes that the female population can be described by a density function $f(a, t)$ of chronological age a and time t, both to be measured on the same scale. Hence $\int_I f(a,t)da$ denotes the number of members of the population at time t that have ages in the age-interval $I = [c, d]$. Let $\theta(a, t)$ denote the age-specific death rate, so that $\theta(a, t) f(a, t)\Delta a$ denotes the number of individuals per unit time dying at time t with ages in $[a, a + \Delta a]$. Let $\lambda(a, t)$ denote the age-specific fertility rate, so that $\Delta t \int_I \lambda(a, t)f(a, t)da$ (where $I = (0, \infty)$) gives the total number of births in the time interval t to $t + \Delta t$. Finally, let $f_0(a)$ denote the initial age density and assume that changes in $f(a, t)$ are due only to individuals getting older or dying. More specifically, those individuals aged between a and $a + \Delta a$ that survive mature to age class $a + \Delta a$ to $a + 2\Delta a$. This assumption leads to the following initial boundary value problem as a first order approximation (see Gurney and Nisbet 1982):

$$\frac{\partial f(a, t)}{\partial a} + \frac{\partial f(a, t)}{\partial t} + \theta(a, t)f(a, t) = 0, \tag{1.1}$$

$$B(t) = f(0, t) = \int_0^\infty \lambda(a, t)f(a, t)da, \tag{1.2}$$

$$f(a, 0) = f_0(a). \tag{1.3}$$

Exercise 1.1. Derive the initial boundary value problem (1.1)–(1.3) from first principles. Hint: set up a continuity equation by determining in two ways the number of individuals of age a at time $t + \Delta t$, equate both expressions, expand them using appropriate mean value theorems, and drop the terms of order higher than one. For a derivation along these lines see Gurney and Nisbet (1982).

The solution to (1.1)–(1.3) can be found by several approaches (Nisbet and Gurney, 1986). One approach consists of integrating formally along characteristics on the positive quadrant of the a-t plane (for details see Hoppensteadt, 1975). By following this approach we obtain the formal solution given by

$$
f(a, t) = \begin{cases} f_0(a - t) \exp\left[-\int_0^t \theta(a - t + s, s)ds \right] & \text{if } a \geq t \\ B(t - a) \exp\left[-\int_0^a \theta(s, t - a + s)ds \right] & \text{if } a < t. \end{cases}
\tag{1.4}
$$

Note that in the above equation the only unknown on the right had side is given by B. If we substitute this formal solution into Eq. (1.2) we reduce the problem to the solution of an integral equation for B. Furthermore, when λ and θ are time-independent, we simply get a convolution equation.

Exerice 1.2. Formally substitute the solution (1.4) into Eq. (1.2) and arrive at an integral equation of the form

$$
B(t) = F(t) + \int_0^t K(a, t)B(t - a)da,
\tag{1.5}
$$

where $F(t)$ and $K(a, t)$ involve known data.

Exercise 1.3. Let θ and λ be time-independent. Then show that Eq. (1.5) reduces to the well-known renewal equation

$$
B(t) = F(t) + \int_0^t K(t - a)B(a)da.
\tag{1.6}
$$

Equation (1.6) was first derived directly by Sharpe and Lotka (1911) and has since been widely studied. If we assume that in Eq. (1.6) θ and λ are non-negative, piecewise continuous, and that λ has compact support, it then follows that $B(t)$ is Laplace transformable and that

$$
L\{B(t)\}(p) = \frac{L\{F(t)\}(p)}{1 - L\{K(a)\}(p)}
\tag{1.7}
$$

exists for all p with $\text{Re}(p) > p^*$ where p^* is the unique real solution to Lotka's characteristic equation

$$
k(p) = 1 \quad \text{where } k(p) = \int_0^\infty e^{-pa} K(a)da,
\tag{1.8}
$$

and

$$
K(a) = \lambda(a)e^{-\int_0^a \theta(u)du}
$$

All other roots (complex conjugate pairs) of Eq. (1.8) satisfy $\text{Re}(p) < p^*$ and hence the sign of p^* determines the stability of all solutions.

Exercise 1.4. Show that Eq. (1.8) has a unique real root p^*. Hint: note that $k(p)$ is a monotone function of p. Also show that all other roots (complex conjugate pairs) satisfy $\text{Re}(p) < p^*$.

Furthermore, it can be shown that $B(t) = O(e^{pt})$ for p's satisfying $\text{Re}(p) > p^*$ and $B(t) = B_0 e^{p^*t} + o(e^{p^*t})$ as $t \to \infty$, where B_0 is the residue of $L\{B(t)\}(p)$ at $p = p^*$. We also have that

$$e^{-p^*t} f(a, t) \to B_0 e^{-p^*a - \int_0^a \theta(u)du} \tag{1.9}$$

where the convergence is uniform as $t \to \infty$, and with the right side of (1.9) giving the expression for the "stable age distribution". In this context $k(0)$ represents the expected number of offspring during the lifetime of an individual, and is called the *reproductive number*; hence if $k(0) > 1$ the population will grow and if $k(0) < 1$ the population will die out. To confirm this, we observe that $p^* > 0$ whenever $k(0) > 1$ and $p^* < 0$ whenever $k(0) < 1$. The above analysis can be found in Hoppensteadt (1975), Keyfitz (1977), or Frauenthal (1986).

There are numerous of studies on the effects of harvesting on population growth. In the context of predator-prey interactions, some studies that treat the population being harvested as a homogeneous resource include those of Brauer and Sánchez (1975), Brauer et al. (1976), Yodzis (1976), Brauer and Soudack (1979a, b), and Brauer (1984). For a first look at the problem of harvesting from a bioeconomic or control theory point of view see the works of Clark (1976, and this volume) and Goh et al. (1974).

Among the first papers to study the effects of harvesting on an age-structured population are those of Sánchez (1978, 1980). The following discussion and series of exercises are based on his results. We look at an age-structured population, and assume that its dynamics is governed by the MacKendrick model just described. We study the effects of population growth and age structure of the following simple harvesting strategy; we are allowed to harvest at a constant rate $\alpha > 0$ the individuals of age $a > c$. This harvesting strategy only modifies equation (1.1) which now becomes

$$\frac{\partial f(a, t)}{\partial a} + \frac{\partial f(a, t)}{\partial t} + \theta(a, t)f(a, t) + \alpha\chi_{[c, \infty]}(a)f(a, t) = 0, \tag{1.10}$$

where $\chi_{[c, \infty]}(a)$ denotes the characteristic function of the age-interval $[c, \infty]$; that is, it equals 1 if $c \geq a$, and zero if $a < c$. The solution of the initial boundary value problem given by (1.2), (1.3), and (1.10) is given formally by expression (1.4) when $\theta(a, t)$ is replaced by $\mu(a, t)$ where $\mu(a, t) = \theta(a, t) + \alpha\chi_{[c, \infty]}(a)$. We can reduce the solution of this initial boundary value problem to the solution of a single integral equation, in the following exercise:

Exercise 1.5. Let

$$E(a, t) = \begin{cases} 1 & \text{if } a \leq c \\ e^{-\alpha(a-c)} & \text{if } c < a \leq c + t, \\ e^{-\alpha} & \text{if } c + t < a, \end{cases} \tag{1.11}$$

then find the modified $B(t)$, which we call $B^*(t)$, that results from this harvesting strategy. Compare the expression for $B^*(t)$ with Eq. (1.5).

If we assume now that λ and θ are time-independent and that $\lambda(a) = 0$ for $a > A$, then for $t > c$ we have the following expression for $B^*(t)$:

$$B^*(t) = F^*(t) + \int_0^t B^*(t - a)E(a)K(a)da, \tag{1.12}$$

where $F^*(t)$ and $K(a)$ involve known data, and $E(a) = 1$, $0 \leq a < c$, $E(a) = e^{-\alpha(a-c)}$ when $a > c$. Therefore, $B^*(t)$ is Laplace transformable and we can proceed to study the effects of harvesting on this population. The important question of what rate of harvesting will drive the population to extinction under this harvesting regime can be easily answered through the next exercise.

Exercise 1.6. Show that there exists a unique real number p_1 such that the Laplace transform of $B^*(t)$, $L\{B^*(t)\}(p)$ exists for p's with $\mathrm{Re}(p) > p_1$. What is the reproductive number in this case? What is the relationship between p^* in exercise (1.4) and p_1? Can the previous question be answered in terms of their respective reproductive numbers? What is the characteristic equation? Using the previous discussion on the asymptotic behavior of $B(t)$ as a guide, what would you think are the asymptotic properties of $B^*(t)$?

We now proceed with a simple exercise where direct computations are actually possible.

Exercise 1.7. Assume that λ and θ are constant, and that $f_0(a) = \delta_0(a)$, the Dirac delta function.

(1) When $\alpha = 0$ (unharvested case), show that $B(t) = \lambda e^{(\lambda - \theta)t}$, $p^* = \lambda - \theta$, and the "stable age distribution" is given by $\lambda e^{-\lambda a}$.
(2) When $\alpha \neq 0$, show that

$$B^*(t) = \lambda e^{-\theta t - \alpha(t - c)} + \lambda \int_0^t B^*(t - a)e^{-\theta a}E(a)da,$$

and that the characteristic equation is given by

$$\frac{\lambda}{p + \theta} + \lambda e^{-(p + \theta)c}\left[\frac{1}{p + \theta + \alpha} - \frac{1}{p + \theta}\right] = 0,$$

where p_1 (the dominant eigenvalue) is the unique root of this equation. Furthermore, we note that it can be shown that $p_1 = \lambda - \theta - e^{-\lambda c} + O(\alpha^2)$. What is the "stable age distribution" in this case?

In the following exercise the reader is encouraged to explore a more general harvesting strategy; for further extensions see the papers by Sánchez (1978, 1980).

Project 1.1. Suppose that we now harvest at a constant rate $\alpha > 0$, but only among those individuals with ages in the age interval $c \leq a \leq c + n$, where n is a positive integer. Carry out the analysis described in Exercises 1.5–1.7 in this case.

There are also discrete versions of the MacKendrick-Von Foerster model. The most common version is that of Leslie (1945; for a detailed exposition see Frauenthal, 1986). Leslie's model is derived (as is the MacKendrick-Von Foerster model) from first principles, and it has been generalized in a biological contexts by Caswell (1986) among others. The next project will explore the important mathematical question of the possible connections between continuous and discrete time models for the situation when the discretization that gives rise to a Leslie-type model converges to the solutions of the MacKendrick-Von Foerster model.

Project 1.2. Recently Saints (1987) has developed stable and convergent finite difference schemes for the MacKendrick-Von Foerster model that that gives rise to the Leslie model. He considers the version of the MacKendrick-Von Foerster model for which θ and λ are only functions of age. In addition he assumes that individuals do not live past the maximum age A. This is accomplished by imposing the condition $\int_I \theta(a)da = \infty$, where $I = [0, A]$. To apply the theory of infinite difference methods Saints rewrites the MacKendrick-Von Foerster model as follows:

$$Lf(a, t) = 0, \quad (a, t)\varepsilon G,$$
$$f(a, 0) = f_0(a) \quad 0 \leq a \leq A, \tag{1.13}$$

where L is the following operator

$$Lf(a, t) = \begin{cases} f(0, t) - \int_0^A \lambda(a)f(a, t)da, & a = 0, \\ \dfrac{\partial f(a, t)}{\partial a} + \dfrac{\partial f(a, t)}{\partial t} + \theta(a)f(a, t), & a > 0, \end{cases} \tag{1.14}$$

and the domain G is given by

$$G = \{(a, t): 0 \leq a \leq A, 0 \leq t \leq T\}.$$

An approximate solution f_Δ to the solution of the system (1.13) is developed on the grid

$$G_\Delta = \{(a_i, t_n): i = 0, 1, 2, \ldots, v, n = 0, 1, 2, \ldots, N\},$$

where $a_i = i\Delta$, $i = 0, 1, 2, \ldots, N$, $v = [A/\Delta]$, and $N = [T/\Delta]$. This approximation is done through the difference operator L_Δ. Therefore we need to solve the following approximate problem:

$$L_\Delta f_\Delta(a, t) = 0, \quad (a, t)\varepsilon G_\Delta, \tag{1.15}$$
$$f_\Delta(a, 0) = f_0(a), \quad (a, 0)\varepsilon B_\Delta,$$

where

$$B_\Delta = \{(a_i, 0): i = 0, 1, 2, \ldots, v\}.$$

See Saints (1987) for the sense in which the operator L_Δ approximates L (as $\Delta \to 0$), for a definition of the stability of (1.15), and a definition of the convergence of (1.15) to (1.13).

His convergent scheme is given by

$$L_\Delta f_\Delta(a, t) = \begin{cases} f_\Delta(0, t) - \displaystyle\sum_{i=1}^v \Delta\alpha_i\lambda(a_i)f_\Delta(a_i, t), & a = 0, \\ \dfrac{f_\Delta(a, t) - f_\Delta(a - \Delta, t - \Delta)\exp\left(-\displaystyle\int_{a - \Delta a}^a \theta(a)da\right)}{\Delta}, & a > 0. \end{cases}$$

The coefficients α_i are given by a quadrature method for approximating integrals. When $f_\Delta(a_{i+1}, t_{n+1}) = p_i f_\Delta(a_i, t_n)$, where

$$p_i = \exp\left(- \int_{a_i}^{a_{i+1}} \theta(a) da \right),$$

show that the solution of problem (1.15) with the scheme given by (1.16) is equivalent to the solution of the discrete Leslie model.

Saints notes that the difference scheme

$$L_\Delta f_\Delta(a, t) = \begin{cases} f_\Delta(0, t) - \sum\limits_{i=1}^{\nu} \Delta\alpha_i \lambda(a_i) f_\Delta(a_i, t), & a = 0, \\ \dfrac{f_\Delta(a, t) - f_\Delta(a - \Delta, t - \Delta)}{\Delta} + \theta(a) f_\Delta(a - \Delta, t - \Delta), & a > 0, \end{cases}$$

given by Oster (1978; formally equivalent to a Leslie model) does not in general converge to the solution of the MacKendrick-Von Foerster model.

Section II

A strong objection to the MacKendrick model derives from its assumption that the birth and death moduli are independent of the total population P. If we modify it and assume that these moduli depend on total population size, we arrive at the Gurtin and MacCamy (1974) model:

$$\frac{\partial f(a, t)}{\partial a} + \frac{\partial f(a, t)}{\partial t} + \theta(a, P) f(a, t) = 0, \tag{2.1}$$

$$P(t) = \int_0^\infty f(a, t) da, \tag{2.2}$$

$$B(t) = f(0, t) = \int_0^\infty \lambda(a, P) f(a, t) da \tag{2.3}$$

$$f(a, 0) = f_0(a). \tag{2.4}$$

By formally solving (2.1) and (2.3) along characteristics and substituting this formal solution into Eq. (2.2) and (2.3), Gurtin and MacCamy (1974) were able to reduce the solution of (2.1)–(2.4) to the solution of the following pair of coupled nonlinear Volterra integral equations for P and B:

$$P(t) = \int_0^t K(t - a, t; P) B(a) da + \int_0^\infty L(a, t; P) f_0(a) da, \tag{2.5}$$

$$B(t) = \int_0^t \lambda(t - a, P(t)) K(t - a, t; P) B(a) da$$

$$+ \int_0^\infty \lambda(a + t, P(t)) L(a, t; P) f_0(a) da, \tag{2.6}$$

where the nonlinear functionals K and L are given by

$$K(\alpha, t; P) = e^{-\int_{t-\alpha}^{t} \theta(\alpha+s-t, P(s))ds}, \qquad (2.7)$$

$$L(\alpha, t; P) = e^{-\int_{t-\alpha}^{t} \theta(s+\alpha, P(s))ds}. \qquad (2.8)$$

Project 2.1. See Hoppensteadt (1975) and use the development in his first chapter as a model to complete the above reduction for the Gurtin MacCamy initial boundary value problem. The complete analysis can be found in Gurtin and MacCamy's original paper.

Under appropriate conditions, Gurtin and MacCamy show that their initial boundary value problems (2.1)–(2.4) has a unique positive solution for all positive time. Furthermore, they also obtain local stability results for this model. To explain their results, let

$$\pi(a, P) = e^{-\int_{0}^{a} \theta(s, P)ds} \qquad (2.9)$$

denote the probability that an individual survives to age a given that the population size is P, and let

$$R(P) = \int_{0}^{\infty} \lambda(a, P)\pi(a, P)da \qquad (2.10)$$

denote the expected number of offspring that an individual has over its life-span given that the population size is P; that is, $R(P)$ is the *reproductive number*. Under appropriate conditions, a necessary and sufficient condition for the existence of a steady-stage age distribution $[f(a, t) \equiv f(a)]$ with total population $P > 0$ is that the reproductive number, $R(P) = 1$ (i.e. the population effectively replaces itself). To study the local exponential stability of the steady-state age distribution f_E, they consider perturbations from this steady-state of the form

$$\psi(a, t) = \psi^*(a)e^{\gamma t}, \quad \text{with} \quad \psi^*(a) \text{ and } \gamma \text{ complex}, \qquad (2.11)$$

and show that solutions of the form (2.11) exist if and only if γ satisfies the following transcendental equation:

$$1 = \int_{0}^{\infty} r(a)e^{-\gamma a}da + g_{\gamma}\left[\frac{K}{B_E} - \int_{0}^{\infty} r(a)h_{\gamma}(a)da\right], \qquad (2.12)$$

where $f_E(a)$ is the given steady-state age distribution with corresponding birth rate B_E and total population P_E, and where the other quantities in (2.12) are defines as follows:

$$r = \pi_E \theta_E, \qquad (2.13)$$

$$h_{\gamma} = \int_{0}^{a} \exp[-\gamma(a-s)]\theta'_E(s)ds, \qquad (2.14)$$

$$g_\gamma = \frac{B_E \int\limits_0^\infty \exp[-\gamma a]\pi_E(s)ds}{1 + B_E \int\limits_0^\infty \pi_E(a)h_\gamma(a)da}, \tag{2.15}$$

with

$$K = B_0 \int\limits_0^\infty \lambda'_E(a)\pi_E(a)da, \quad B_E = \frac{P_E}{\int\limits_0^\infty \pi_E(a)da},$$

$$\theta_E(a) = \theta(a, P_E), \quad \lambda_E(a) = \lambda(a, P_E), \quad \pi_E(a) = \pi(a, P_E),$$

$$\theta'_E(a) = \theta_P(a, P_E), \quad \lambda'_E(a) = \lambda_P(a, P_E), \quad P_E = \int\limits_0^\infty f_E(a)da.$$

They arrive at the following important result:

Theorem (*Gurtin-MacCamy*). *Assume that Eq. (2.12) has no solution γ with $\mathrm{Re}(\gamma) \geqq 0$. The there exist numbers $\delta > 0$ and $\varepsilon > 0$ such that given initial data f_0 with $\|f_0 - f_E\|_1 < \delta$ (L_1 norm), the corresponding solution of the population problem, if it exists for all positive time, satisfies*

$$P(t) - P_E = O(\exp[-\varepsilon t])$$
$$f(a, t) - f_E(a) = O(\exp[-\varepsilon t]) \quad \text{(for each } a\text{)}$$
$$\text{as } t \to \infty.$$

In reference to insect populations, Nisbet and Gurney (1983) note: "Physiologically... for most insect species it is not chronological age but weight gain that triggers the various moults, a doubling of weight during an instar being typical (Dyar's 'law', Chapman (1969))." Therefore if we wish to understand the mechanisms responsible for the observed dynamics of biological populations with complex life cycles, in some instances we have to incorporate physiological characters such as size or weight into the description of a population. The next modelling exercises introduce Gurney and Nisbet's linear size-structured model with a time lag due to an egg-stage duration (1982, 1983, 1986), as well as a natural nonlinear extension of that model. The basic assumptions underlying the dynamics of this model are that individuals in this population may reproduce, grow in size, get older, or die. Furthermore, it is assumed that individuals in this population produce eggs that remain (with a certain probability) as eggs for a fixed length of time. The nonlinear extension just assumes that the birth and death processes depend on total population size.

Exercise 2.1. Let $f(a, m, t)$ be a density function, where a denotes age, m denotes size or mass; furthermore, assume that it describes the size-structure of a population at time t. Let us assume that changes in $f(a, m, t)$ are due to individuals maturing and growing to an older age class and a bigger size class or dying. More specifically, individuals aged between a and $a + \Delta a$ and with sizes between m and $m + \Delta m$ who survive enter the age class $a + \Delta a$ to $a + \Delta a + \Delta t$ and enter the size class $m + g(a, m)\Delta t$, to $m + \Delta m + g(a, m + \Delta m)\Delta t$ where $g(a, m)$ denotes the size-specific growth rate of an individual of age a and size m. Let $\theta(a, m)$ denote the size-specific death rate of an individual of age a and size m; $\lambda(a, m, m')$, the size-specific fertility rate of an individual of age a and size m, where m' denotes the size of the offspring, and $f_0(a, m)$ the initial age-size distribution. Derive the following initial boundary value problem by using the approach of exercise (1.1):

$$\frac{\partial(fg)}{\partial m} + \frac{\partial f}{\partial a} + \frac{\partial f}{\partial t} + \theta f = 0, \tag{2.16}$$

$$f(0, m, t) = \int_0^\infty \int_0^\infty \lambda(a, m', m, t) f(a, m', t)\, dm'\, dt, \quad t > 0 \tag{2.17}$$

$$f(a, m, 0) = f_0(a, m). \tag{2.18}$$

A detailed derivation can be found in Castillo-Chavez (1987a).

The main difficulty in this type of models is that the sought density, in this case $f(a, m, t)$, forms part of the boundary condition through the reproduction process. In order to obtain some basic results, we proceed to derive in the next two exercises an age-independent version with a simplified "birth law".

Exercise 2.2. Assume that g, λ, and θ are age independent and let $\rho(m, t) = \int_0^\infty f(am, t)\, da$; that is, $\int_m^{m+\Delta m} \rho(m', t)\, dm$ denotes the number of individuals with sizes in the range m to $m + \Delta m$, irrespective of age. Integrate Eq. (2.16) formally from 0 to ∞ on the age variable, under the assumption that $f(\infty, m, t) = 0$, to arrive at the following initial boundary value problem:

$$\frac{\partial \rho}{\partial t} + \frac{\partial(g\rho)}{\partial m} + \theta \rho = f(0, m, t), \tag{2.19}$$

$$f(0, m, t) = \int_0^\infty \lambda(m, m', t)\rho(m', t)\, dm', \quad t > 0 \tag{2.20}$$

$$\rho(m, t) = \int_0^\infty f_0(a, m)\, da. \tag{2.21}$$

Exercise 2.3. Assume that for the biological population of interest the following assumptions regarding its reproductive process are reasonable:

(a) All 'eggs' have the same size at birth, m_1.
(b) The egg-state duration is always the same, a constant t_E.
(c) The probability of survivorship P_E from the egg-stage is constant.
(d) The population does not exist for $t < 0$, and the system is "seeded" instantaneously at time $t = 0$ by means of an arbitrary distribution $\rho_0(m)$.

Let $R(t)$ denote the total rate of recruitment at time t from the egg-stage and $E(t)$ the total egg-production rate at time t. Use the relation $R(t) = P_E E(t - t_E)$ to replace the initial boundary value problem (2.19)–(2.21) by the following:

$$\frac{\partial \rho}{\partial t} + g\frac{\partial \rho}{\partial m} + \left[\theta + \frac{\partial g}{\partial m}\right]\rho = 0, \quad m > m_1 \tag{2.22}$$

$$g(m_1)\rho(m_1, t) = \int_{m_1}^\infty \lambda(m)\rho(m, t - t_E)\, dm = B(t), \quad t > t_E \tag{2.23}$$

$$\rho(m, 0) = \rho_0(m), \tag{2.24}$$

$$\rho(m, t) = 0, \quad \text{for all } t < 0. \tag{2.25}$$

Under very general conditions the characteristic curves associated with the above model do not intersect. This situation corresponds to a lack of interference (in growth rates) between individuals. In this case, the above model belongs to a generalized class of MacKendrick models with time delays. The following exercise shows this relationship:

Exercise 2.4. Show that the change of variable provided by $J(m) = \int_{m_1}^{m} (ds/g(s))$ reduces the model (34)–(37) to a MacKendrick type-model with constant time delay.

Exercise 2.5. Denote the probability that an individual will grow to size m by

$$\pi(m) = \frac{P_E}{g(m_1)} e^{-\int_{m_1}^{m} \frac{\theta(u)}{g(u)} du}.$$

(a) Show that the reproductive number RN (i.e., the number of expected offspring over the life-span of an average individual) is given by

$$RN = \int_{m_1}^{\infty} \lambda(m)\pi(m)\, dm.$$

(b) Show that the Lotka-type characteristic ($RN = 1$) equation is given by

$$1 = \int_{m_1}^{\infty} \pi(m)\lambda(m)e^{-s\left(\tau_E + \int_{m_1}^{m} \frac{du}{g(u)}\right)} dm. \tag{2.26}$$

(c) It should be clear biologically that if $RN > 1$, then the population grows; and if $RN < 1$ then it decreases. Show that this is the case by studying the nature of the roots of Eq. (2.26).
(d) What effects does the delay play on the growth of the population? Compare Eq. (2.26) with the characteristic equation associated with Exercise 1.2.
(e) Determine the "stable size distribution" for this model.
For an elaboration of the ideas in Exercises 2.4 and 2.5 see Castillo-Chavez (1987a).

Project 2.2. An interesting economic problem is the determination of the optimal sustainable yield in an age-specific harvesting schedule. For discrete populations, this problem has been reduced by Beddington and Taylor (1973), Rorres and Fair (1975), Doubleday (1975), and Rorres (1976) to the solution of a linear programming problem. By assuming that the economic value of females is a function of age, and by taking into consideration the capital needed to harvest or raise such a population, Rorres and Fair (1980) were able to determine the age-specific harvest rate that maximizes the yield of the resulting sustainable harvest. They showed that if no upper bound is imposed on the age-specific harvest rate then the optimal harvesting rate is impulsive and bimodal. More specifically: at most two ages should be harvested, with the older being harvested to extinction.

Look at the paper by Rorres and Fair and see if their results can be extended in the context of the size-structured population described by the model (2.7)–(2.10). For further extensions that involve the use of size-structured populations in the determination of an optimal fishery policy see the works of Botsford (1981, 1985).

If we now assume that mortality and fertility rates depend on total population size P, we arrive at the following initial boundary value problem:

$$\frac{\partial \rho}{\partial t} + g\frac{\partial \rho}{\partial m} + \left[\frac{\partial g}{\partial m} + \theta(m, P)\right]\rho = 0, \quad m > m_1 \tag{2.27}$$

$$P(t) = \int_{m_1}^{\infty} \rho(m, t)\,dm, \tag{2.28}$$

$$B(t) = g(m_1)\rho(m_1, t) = P_E \int_{m_1}^{\infty} \lambda(m, P)\rho(m, t - t_E)\,dm, \quad t > t_E \tag{2.29}$$

$$B(t_E) \equiv 0, \tag{2.30}$$

$$\rho(m,) = \rho_0(m) \quad m > m_1, \tag{2.31}$$

$$\rho(m, t) = 0 \quad t < 0. \tag{2.32}$$

Project 2.3. Using the change of variable of the previous exercise, reduce the initial boundary value problem above to a Gurtin-MacCamy type model with constant time delay. Study this transformed model by following the analysis of the Gurtin-MacCamy model. For details and extensions see Castillo-Chavez (1987b).

Project 2.4. In the derivation of Nisbet-Gurney type models (2.22–2.25 or 2.27–2.32), it was assumed that the egg-stage duration is constant. In general, the maturation time will be different for different individuals and hence the egg-stage duration will not be constant. Assume that the egg-stage duration is given by a distribution $Q(t)$ and derive the initial boundary value problem corresponding to this situation. How far can you carry the mathematical analysis?

Project 2.5. In the nonlinear version of the Nisbet-Gurney model (2.27–2.32), it was assumed that the growth rate does not depend on the total population size P. How would the conclusions of the discussion on this model will be affected if we assume that $g = g(m, P)$?

Project 2.6. Using the approach of Saints (1987), or any other approach develop, stable and convergent difference schemes to the Nisbet-Gurney model (2.22)–(2.25). Can a connection be made between the matrices used by Caswell (1986) to model the life cycle of plants and the difference schemes that you develop?

Section III

The role that the age structure of a population plays in the dynamics of some infectious diseases has been a very active area of research (see Anderson and May, 1984; Dietz and Schenzle, 1985; Shenzle, 1984, 1985; May, 1986; Castillo-Chavez et al., 1988, 1989). This is, therefore, an area where mathematicians and biologists have been interacting successfully.

In this section we show the mathematical rudiments involved in the analysis of some typical epidemiology models with age-structure. We first introduce a model for a single population facing an infection produced by a single agent. Then we introduce a model that incorporates the effects of co-circulating variants of the same etiological agent on the dynamics of this population, and finally we look at a model that has been introduced recently in order to study the possible demographic effects of the AIDS epidemic on human populations.

We first assume that we are dealing with a population that is at equilibrium. We divide our population into 3 classes: the susceptible (S), the infected (I), and the recovered (R), where $s(a, t)$, $i(a, t)$, and $r(a, t)$ denote the densities of each respective class. For this SIR epidemiological model, $\int_J s(a, t) da$, $\int_J i(a, t) da$, and $\int_J r(a, t) da$ denote the number of individuals in each class that have ages in the interval J at time t. The individuals are assumed to "move" according to the following transfer diagram

TRANSFER DIAGRAM 1

$$\varepsilon \longrightarrow I \longrightarrow R.$$

S Susceptible individuals, I Infected individuals, R Recovered individuals.

We assume that births and deaths occur at equal rates, that all newborn individuals are susceptible, and that the transfer of infection is due to an age-dependent, proportionately-mixed, bilinear incidence rate. That is, if we let $b(a)$ denote the age-specific contact rate then a particular case of the proportionate mixing assumption states that the contact rate between a susceptible person of age a and an infective person of age a' is proportional to $b(a)b(a')$ (see Barbour, 1978; Nold, 1980; Hethcote and Yorke, 1984; and Dietz and Schenzle, 1985). Based on these assumptions and with the aid of the transfer diagram, we arrive at the following initial boundary value prolem (see Hoppensteadt, 1974; Dietz, 1975; Anderson and May, 1983; Dietz and Schenzle, 1985; Webb, 1985; May, 1986; Castillo-Chavez et al., 1988, 1989):

$$\frac{\partial s(a, t)}{\partial a} + \frac{\partial s(a, t)}{\partial t} = -\lambda(t)b(a)s(a, t) - \mu s(a, t), \tag{3.1}$$

$$\frac{\partial i(a, t)}{\partial a} + \frac{\partial i(a, t)}{\partial t} = \lambda(t)b(a)i(a, t) - (\mu + \gamma)i(a, t), \tag{3.2}$$

$$\frac{\partial r(a, t)}{\partial a} + \frac{\partial r(a, t)}{\partial t} = \gamma i(a, t) - \mu r(a, t), \tag{3.3}$$

$$\lambda(t) = \beta \int_0^\infty b(a)i(a, t) da, \tag{3.4}$$

$$s(a, 0) = s_0(a), i(a, 0) = i_0(a), r(a, 0) = r_0(a), \tag{3.5}$$

$$s(0, t) = \mu, i(0, t) = 0, r(0, t) = 0, \tag{3.6}$$

where μ and γ are, respectively, the constant mortality and recovery rates, $\lambda(t)$ is the "force of infection" at time t, and β is transmission scaling factor. The initial age distributions are zero beyond a maximum age.

We compute a threshold condition based on the work (in a slightly more general context) of Webb (1985), Dietz and Schenzle (1985), and Castillo-Chavez et al. (1989). A *threshold condition*, or *basic reproductive number* in epidemiology, identifies a quantity that must exceed one for the disease to remain endemic, and also can be interpreted as the number of secondary infections produced by an infectious individual in a population of susceptibles. Let us assume that steady-state age distributions are reached as time approaches infinity. Hence, at infinity λ becomes the constant $\lambda^* = \beta \int_J b(a) y^*(a) da$ where $J = (0, \infty)$. It is now a straightforward exercise in elementary ordinary differential equations to determine the steady states. This is done by setting the time derivatives in (3.1)–(3.3) equal to zero and integrating formally the system of ordinary differential equations. The steady states are given by the expressions (3.7)–(3.9).

Exercise 3.1. By setting the time derivatives equal to zero in (3.1)–(3.3), show that the steady-state age distributions are given by the following expressions:

$$s^*(a) = \mu e^{-\mu a - \lambda^* \int_0^a b(\alpha) d\alpha}, \tag{3.7}$$

$$i^*(a) = \mu e^{-(\mu + \gamma)} \int_0^a \lambda^* b(\theta) e^{\gamma \theta - \lambda^* \int_0^\theta b(\alpha) d\alpha} d\theta, \tag{3.8}$$

$$r^*(a) = \mu e^{-\mu a} - s^*(a) - i^*(a). \tag{3.9}$$

Observe that the equation for r^* is redundant. It we now substitute $i^*(a)$ into the expression for λ^*, we find that either $\lambda^* = 0$ or λ^* satisfies the following characteristic equation:

$$1 = \beta \int_0^\infty b(a) \mu e^{-(\mu + \gamma) a} \int_0^a b(\theta) e^{\gamma \theta - \lambda^* \int_0^\theta b(\alpha) d\alpha} d\theta \, da. \tag{3.10}$$

If we let

$$H(\lambda) = \beta \int_0^\infty b(a) \mu e^{-(\mu + \gamma) a} \int_0^a b(\theta) e^{\gamma \theta - \lambda \int_0^\theta b(\alpha) d\alpha} d\theta \, da,$$

then the characteristic Eq. (3.10) has a unique positive solution λ^* if and only if $H(0) > 1$. Note that $H(0)$ is the reproductive number, and denotes the number of secondary infections produced by an infectious individual in a purely susceptible population. When this threshold condition ($H(0) > 1$) is satisfied, an endemic (persistent) steady-state age distribution given by the expressions (3.7)–(3.9) exists. If $H(0) < 1$ (i.e., below the threshold) the disease dies out and the steady-state distributions are given by the same expressions but with $\lambda^* = 0$ (no infected or recovered individuals remain).

Exercise 3.2. Show that Eq. (3.10) has a unique real root. Hint: first show that $H(\lambda)$ is a monotone function of λ.

If $H(0) > 1$ (i.e., above the threshold), then the "trivial" steady state distribution—the infection-free state—is unstable, so that the disease persists. In order to show this, we take the following particular perturbations of the infection-free state ($\lambda^* = 0$):

$$s(a, t) = s^*(a) + \xi(a)e^{pt}, \tag{3.11}$$

$$i(a, t) = \eta(a)e^{pt}, \tag{3.12}$$

$$\lambda(t) = \theta e^{pt}. \tag{3.13}$$

These expressions are substituted back into (3.1)–(3.6), and then we proceed to linearize around this equilibrium; that is we approximate the original nonlinear model locally by a linear model. This approach allows us to study the local stability of the original system by looking at the nature of the admissible p's (the eigenvalues).

Exercise 3.3. Substitute the expressions (3.11)–(3.13) into the initial boundary value problems (3.1)–(3.6), and drop the terms of higher order; that is, linearize the initial boundary value problem around the infection-free state. Due to the nature of the perturbations, we can "separate variables" in the linearization obtained in Exercise 3.3. This allows us to determine the "stability" or "p" equation (see expression (3.16) below).

Exercise 3.4. Show that $\eta(a)$ is given by the following expression:

$$\eta(a) = \theta\mu e^{-\mu a} \int_0^a b(\phi)e^{p(\phi-a) + \gamma(\phi-a)} d\phi. \tag{3.14}$$

Moreover, since

$$\theta = \beta \int_0^\infty b(a)\eta(a)da, \tag{3.15}$$

by substituting the expression for $\eta(a)$ obtained in Exercise 3.3 into Eq. (3.15), we arrive at the following stability equation (p-equation):

$$1 = \beta \int_0^\infty b(a)\mu e^{-\mu a} \int_0^a b(\phi)e^{-\gamma(a-\phi) - p(a-\phi)} d\phi\, da. \tag{3.16}$$

Exercise 3.5. Show that if we are above the threshold ($H(0) > 1$), then Eq. (3.16) has a unique real positive root so that the trivial steady-state age distribution is unstable.

The study of the possible outcomes generated by the interplay between a heterogeneous host population and multiple strains of an etiological agent has only received attention recently. The following modelling project deals with a situation when competing strains of an etiological agent provide a degree of protection, or *cross-immunity*, against closely-related strains.

Project 3.1. (a) Assume a homogeneous population is facing two H1N1 viral strains of influenza type A. Assume that an individual cannot be simultaneously infected by both strains, and that an individual that has been infected with one of the strains will have a certain degree of cross-immunity to the other viral strain. Furthermore, assume that this degree of cross-immunity is measured by a susceptibility coefficient that reduces the transmission coefficient associated with new but related strains of the same subtype (in this case H1N1). Divide this population into eight classes: S (fraction susceptible), Y_i (fraction infected with strain $i, i = 1, 2$), Z_i (fraction recovered from the other strain), V_i (fraction infected strain i after recovery from the other strain), and W (recovered from both strains). The dynamics of the infection is then represented by the following transfer diagram.

TRANSFER DIAGRAM 2

X Susceptible individuals, Y_i individuals infected strain i, Z_i individuals recovered from strain i but susceptible to the other strain, V_i individuals infected with strain i but recovered from the other strain, W individuals recovered from both strains.

Let β_i denote the transmission coefficient of strain i, and define σ_j (where $j = 3 - i, i = 1, 2$) to be the "susceptibility factor", that is, the degree of cross-immunity of types Z_i and X to new but related strain j. We assume that σ_j is between 0 and 1. Let γ_i denote the recovery rate from strain i, and μ denote the constant natural mortality rate. Use the above transfer diagram, the "mass-action" law, and homogeneous mixing to arrive at the following set of ordinary differential equations governing the dynamics of influenza for this population:

$$S'(t) = - [\beta_1(Y_1 + V_1) + \beta_2(Y_2 + V_2) - \mu] S + \mu$$

$$Y_i'(t) = \beta_i(Y_i + V_i)S - (\gamma_i + \mu) Y_i, \qquad \text{for } i = 1, 2,$$

$$Z_i'(t) = \gamma_i Y_i - [\sigma_j \beta_j(Y_j + V_j) + \mu]Z_i, \qquad \text{for } i = 1, 2,$$

$$V_i'(t) = \sigma_i \beta_i(Y_i + V_i)Z_j - (\gamma_i + \mu)V_i, \qquad \text{for } i = 1, 2,$$

$$W'(t) = \gamma_1 V_1 + \gamma_2 V_2 - \mu W,$$

$$S(0) = X_0, Y_i(0) = Y_{i0}, Z_i(0) = Z_{i0}, V_i(0) = V_{i0}, W(0) = W_0, \qquad \text{for } i = 1, 2.$$

Recall that $j = 3 - i$; that is, $i = 1, j = 2$, or $i = 2, j = 1$.

(b) Find the equilibria of this model and study their local stability.

(c) Assume that you are dealing with an age-structured population. Follow the transfer diagram and use the same assumptions as for the one-strain model to arrive at the following two-strain epidemiological model with proportionate mixing and cross-immunity:

$$\frac{\partial s(a,t)}{\partial a} + \frac{\partial s(a,t)}{\partial t} = -(\lambda_1(t)b(a) + \lambda_2(t)b(a) + \mu(a))s(a,t),$$

$$\frac{\partial y_i(a,t)}{\partial a} + \frac{\partial y_i(a,t)}{\partial t} = \lambda_i(t)b(a)s(a,t) - (\gamma_i + \mu(a))y_i(a,t), \quad i = 1,2$$

$$\frac{\partial z_i(a,t)}{\partial a} + \frac{\partial z_i(a,t)}{\partial t} = \gamma_i y_i(a,t) - \sigma_j \lambda_j(t)b(a)z_i(a,t) - \mu(a)z_i(a,t), \quad i = 1,2$$

$$\frac{\partial v_i(a,t)}{\partial a} + \frac{\partial v_i(a,t)}{\partial t} = \sigma_i \lambda_i(t)b(a)z_j(a,t) - (\gamma_i + \mu(a))v_i(a,t), \quad i = 1,2$$

$$\frac{\partial w(a,t)}{\partial a} + \frac{\partial w(a,t)}{\partial t} = \gamma_1 v_1^{(a,t)} + \gamma_2 v_2^{(a,t)} - \mu(a)w(a,t),$$

$$\lambda_i(t) = \beta_i \int_0^\infty b(a')[y_i(a',t) + v_i(a',t)]\,da',$$

$$s(0,t) = \rho, \, y_i(0,t) = 0, \, z_i(0,t) = 0, \, v_i(0,t) = 0, \, w(0,t) = 0,$$

$$s(a,0) = s_0(a), \, y_i(a,0) = y_{0i}(a), \, z_i(a,0) = z_{0i}(a), \, v_i(a,0) = v_{0i}(a), \, r(0,t) = w_0(a),$$

$$\rho = \left[\int_0^\infty e^{-M(a)}\,da\right]^{-1} \quad \text{where } M(a) = \int_0^a \mu(\alpha)\,d\alpha. \tag{2.9}$$

How far can you carry out the analysis of part (a) in this case? Some mathematical results dealing with Project 4 can be found in Castillo-Chavez et al. (1988, 1989).

Elsewhere in this volume May and Anderson discuss their extensive work modelling the AIDS epidemic. In their discussion they introduce a model to investigate the demographic consequences of the spread of AIDS through horizontal and vertical transmission. This model is extended to age-structured populations in order to study the effects of the human immunodeficiency virus (HIV) transmission on the age-profiles of the population under consideration.

May and Anderson make the simplifying assumption that male-to-female and female-to-male transmission of HIV takes place at equal rates. This assumption allows them to deal effectively with one population rather than with separate male and female populations. They further assume that sex-ratios are 50:50 at all ages. To introduce their model, denote by $s(a,t)$ the density of susceptible individuals that at time t have age a, by $i(a,t)$ the corresponding density of infected (assumed infectious) individuals and by $n(a,t)$ the corresponding population density. The average incubation period after which individuals are assumed to be removed effectively from the population is given by the constant $(1/v)$; $\lambda(a,t)$ denotes the age-dependent probability per unit time that a given susceptible will acquire infection; $\mu(a)$ denotes the age-specific mortality rate; $m(a)$ denotes the age-specific fertility rate; ε denotes the fraction of the offspring of infected mothers who survive (that is, do not die from AIDS). Using the above definitions May et al. (1988, 1989) arrive at the following model:

$$\frac{\partial s}{\partial a} + \frac{\partial s}{\partial t} = -[\lambda(a,t) + \mu(a)]s, \tag{3.17}$$

$$\frac{\partial i}{\partial a} + \frac{\partial i}{\partial t} = \lambda(a,t)s - [v + \mu(a)]i, \tag{3.18}$$

$$\frac{\partial n}{\partial a} + \frac{\partial n}{\partial t} = -\mu(a)n - vi, \tag{3.19}$$

$$s(0,t) = n(0,t) = B(t) = \int_A^\infty m(a)[n(a,t) - (1-\varepsilon)i(a,t)]\,da, \tag{3.20}$$

$$s(a,0) = s_0(a), i(a,0) = i_0(a), n(a,0) = n_0(a), \tag{3.21}$$

$$\lambda(a,t) = \beta c \frac{\int_L^\infty p(a,a')i(a',t)\,da'}{\int_L^\infty p(a,a')n(a',t)\,da'}, \tag{3.22}$$

where β and c denote the transmission probability and the mean rate of acquiring new sexual partners. Let $p(a,a')$ denote the probability that a susceptible of age a will choose a partner of age a', and let A and L denote appropriate limits of integration (age of first reproduction and age at which an individual becomes sexually active).

Although May et al. (1988, 1989) have analyzed this model through the use of analytical and numerical techniques, they have not exhausted all the possibilities. A useful project would involve the development of stable and convergent difference schemes to solve it numerically so as to study the transient dynamics of the epidemic.

Project 3.2. Using the approach of Saints (1987), or any other approach, develop difference schemes for the above age-structured model. Using the information and data provided in May and Anderson's chapter (this volume), implement your schemes to study the transient dynamics of this model. Relax May and Anderson's assumptions and work with a full two-sex model.

A particularly new and interesting approach has been developed by Sattenspiel (1987) to study the geographical spread of hepatitis among day care centers in New Mexico. The fact that host populations are distributed in space makes the random mixing assumption invalid. A common approach has been to divide the host population into multiple interacting groups. However, once a population is divided into subpopulations, it is usually assumed that there is random mixing within each subpopulation with the amount of interaction usually greater if individuals are from the same subpopulation. The model developed by Sattenspiel differs because it considers two types of interactions between individuals:

(i) Interactions between individuals because of geographic proximity, and
(ii) interactions between individuals of the same or different subpopulation (s) because of attendance to common social functions.

This hierarchical model then incorporates a higher degree of reality into models for the spread of disease. The model is given by a system of $(6n)$ differential equations that describe the change in the number of each class [susceptible (social and nonsocial), infectious (social and nonsocial), and removed (social and nonsocial)] from a population subdivided into n neighborhoods. If o denotes nonsocial and s denotes social then the system of equations is:

$$\frac{dS_{oi}}{dt} = b_i N_{oi} - b_i S_{oi} - \beta \sigma_i (S_{oi} I_{oi} + S_{oi} I_{si}) - \beta S_{oi} (MM^T)_{oi} Y,$$

$$\frac{dS_{si}}{dt} = b_i N_{si} - b_i S_{si} - \beta \sigma_i (S_{si} I_{si} + S_{si} I_{si}) - \beta S_{si} (MM^T)_{si} Y,$$

$$\frac{dI_{oi}}{dt} = \beta \sigma_i (S_{oi} I_{oi} + S_{oi} I_{oi}) + \beta S_{oi} (MM^T)_{oi} Y - \gamma_i I_{oi} - b_i I_{oi},$$

$$\frac{dI_{si}}{dt} = \beta \sigma_i (S_{si} I_{oi} + S_{si} I_{si}) + \beta S_{si} (MM^T)_{si} Y - \gamma_i I_{si} - b_i I_{si},$$

$$\frac{dR_{oi}}{dt} = \gamma_i I_{oi} - b_i R_{oi},$$

$$\frac{dR_{si}}{dt} = \gamma_i I_{si} - b_i R_{si}, \quad i = 1, 2, 3, \ldots, n.$$

β denotes the transmission rate per unit contact, σ_i is a neighborhood-specific adjustment of the transmission rate, b_i is the birth and death rate as the population remains constant (even within a neighborhood), M is the "generalized" movement matrix, g_i is the recovery rate in neighborhood i, and $S_{oi}, S_{si}, I_{oi}, I_{si}, R_{oi},$ and R_{si} are the numbers of individuals in each class in neighborhood i. Y is a $2n \times 1$ vector with elements $(I_{o1}, \ldots, I_{on} I_{s1}, \ldots, I_{sn})$. The total number of births in each population is $b_i N_i$. All "newborns" are susceptible. Deaths in each class are given by $b_i S_{oi}$, etc. The next project deals with the mathematical analysis of this model.

Project 3.3. (a) Do the stability analysis of the infection-free state. (b) Experiment with different movement matrices in order to explore the spread of the disease.

A partial mathematical analysis of the above model can be found in Sattenspiel (1987) and Sattenspiel and Simon (1988).

Project 3.4. An excellent research project would consist in the utilization of this approach in modelling the sexual and drug-related transmission of the HIV.

(a) Modify the above model to allow for variable populations size.
(b) Derive a model of the above type for the sexual transmission of AIDS.
(c) Incorporate social drug groups into Sattenspiel's model.

Acknowledgements. This work has been partially supported by NSF grant DMS-8406472 to Simon A. Levin, by The Center for Applied Mathematics and the Office of the Provost at Cornell University, as well as by a Ford Foundation Postdoctoral Fellowship for Minorities. I give my thanks to all of them.

References

Anderson, R.M., May, R.M. (1983) Vaccination against rubella and measles: quantitative investigations of different policies. J. Hyg. *90*, 259–325

Beddington, J.R., Taylor, D.B. (1973) Optimal age specific harvesting of a population. Biometrics *29*, 801–809

Barbour, A.D. (1978) Macdonald's model and the transmission of bilharzia. Trans. Royal Soc. Trop. Med. Hyg. *72*, 6–15

Bostford, L.W. (1981) Optimal fishery policy for size-specific, density-dependent population models. J. Math. Biol. *12*, 265–93

Botsford, L.W., Wainwright, T.C. (1985) Optimal fishery policy: An equilibrium solution with irreversible investment. J. Math. Biol. *21*, 317–327

Brauer, F. (1984) Constant yield harvesting of population systems. In: Hallam, T.G., Levin, S.A. (eds.) Mathematical Ecology Proceedings of the Conference, Trieste, Italy, Lecture Notes in Biomathematics, vol. 54. Springer, Berlin Heidelberg New York London Paris Tokyo, pp. 234–242

Brauer, F., Sánchez, D.A. (1975) Constant rate population harvesting: equilibrium and stability. Theor. Pop. Biol. *8*, 12–30

Brauer, F., Soudack, A.C., Jarosh, H.S. (1976) Stabilization and destabilization of predator-prey systems under harvesting and nutrient enrichment. Int. J. Control *23*, 553–573

Brauer, F., Soudack, A.C. (1979a) Stability regions and transition phenomena for harvested predator-prey systems. J. Math. Biol. *7*, 319–337

Brauer, F., Soudack, A.C. (1979b) Stability regions in predator-prey systems with constant rate harvesting. J. Math. Biol. *8*, 55–71

Castillo-Chavez, C. (1987a) Linear character dependent models with constant time delay in population dynamics. Int. J. of Math. Mod. *9*, 821–36

Castillo-Chavez, C. (1987b) Nonlinear character dependent models with constant time delay in population dynamics. J. Math. Anal. and Appl. *128*, 1–29

Castillo-Chavez, C., Hethcote, H.W., Andreasen, V., Levin, S.A., Liu, W-m (1988) Cross-immunity in the dynamics of homogeneous and heterogeneous populations. In: Hallam, T.G., Gross, L.G., Levin, S.A. (eds.) Proc. of the Research Conference, Second Autumn Course on Mathematical Ecology, Trieste, Italy, 1986. Singapore: World Scientific Publishing Co., pp. 303–316

Castillo-Chavez, C., Hethcote, H.W., Andreasen, V., Levin, S.A. Liu, W-m (1989) Epidemiological models with age structure and proportionate mixing J. Math Biol. (in press)

Caswell, H. (1986) Life cycle models for plants. In: Some Mathematical Questions in Biology, vol. XVIII. Lectures on Mathematics in the Life Sciences

Clark, C.W. (1976) Mathematical Bioeconomics, Wiley & Sons, New York

Dietz, K. (1975) Transmission and control of arbovirus diseases. In: Cooke, K.L. (ed.) Epidemiology. Society for Industrial and Applied Mathematics, Philadelphia, pp. 104–121

Dietz, K., Schenzle, D. (1985) Proportionate mixing models for age-dependent infection transmission. J. Math. Biol. *22*, 117–120

Doubleday, W.G. (1975) Harvesting in matrix population models. Biometrics *31*, 189–200.

Frauenthal, J. C. (1986) Analysis of age structured models. In: Hallam, T., Levin, S.A. (eds.) Mathematical Ecology. An Introduction. Biomathematics, vol. 17. Springer, Berlin Heidelberg New York London Paris Tokyo, pp. 117–147

Goh, B.S., Leitmann, G., Vincent, T.L. (1974) Optimal control of a prey-predator system. Math. Biosci. *19*, 263–286

Gurney, W.S.C., Nisbet, R.M. (1982) Modelling Fluctuating Populations. Wiley & Sons, New York

Gurtin, M.E., MacCamy, R.C. (1974) Non-linear age dependent population dynamics. Arch. for Rat. Mech. and Anal. *54*(3), 281–300

Hallam, T.G., Levin, S.A. (eds.) (1986) Mathematical Ecology. An Introduction. Biomathematics, vol. 17. Springer, Berlin Heidelberg New York London Paris Tokyo

Hethcote, H.W., Yorke, J.A. (1984) Gonorrhea Transmission Dynamics, and Control. Lecture Notes in Biomathematics, vol. 56. Springer, Berlin Heidelberg New York London Paris Tokyo

Hoppensteadt, F. (1974) An age dependent epidemic model. J. of Franklin Institute *297*, 325–333

Hoppensteadt, F. (1975) Mathematical Theories of Populations: Demographics, Genetics and Epidemics. SIAM Regional Conference Series in Applied Math., No. 20, Philadelphia

Hoppensteadt, F. (1984) Some influences of population biology in mathematics. In: Essays in the History of Mathematics. Memoirs of the AMS 48(298), 25–29

Keyfitz, N. (1977) Introduction to the Mathematics of Populations. Addison-Wesley, Reading, Mass

Leslie, P.H. (1945) On the use of matrices in certain population mathematics. Biometrica 33, 183–212

MacKendrick, A.G. (1926) Applications of mathematics to medical problems. Proc. Edinburgh Math. Soc. 44, 98–130

May, R.M. (1986) Population biology of microparasitic infections, In: Hallam, T., Levin, S.A. (eds.) Mathematical Ecology. An Introduction. Biomathematics, vol. 17. Springer, Berlin Heidelberg New York London Paris Tokyo, pp. 405–442

May, R.M., Anderson, R.M., McLean, A.R. (1988) Possible demographic consequences of HIV/AIDS: I, assuming HIV infection always leads to AIDS. Math. Biosci. (in press)

May, R.M., Anderson, R.M., McLean, A.R. (1989) Possible demographic consequences of HIV/AIDS: II, assuming HIV infection does not necessarily lead to AIDS. In: Castillo-Chavez, C., Levin, S.A., Shoemaker, C. (eds.) Mathematical Approaches to Resource Management and Epidemiology. Lecture Notes in Biomathematics (in press)

Nisbet, R.M., Gurney, W.S.C. (1983) The systematic formulation of population models for insects with dynamically varying instar duration. Theor. Pop. Biol. 23, 114–135

Nisbet, R.M., Gurney, W.S.C. (1986) The formulation of age-structure models. In: Hallam, T., Levin, S.A. (eds.) Mathematical Ecology. An Introduction. Biomathematics, vol. 17. Springer, Berlin Heidelberg New York London Paris Tokyo, pp. 95–115.

Nold, A. (1980) Heterogeneity in diseases-transmission modeling. Math Biosci. 52, 227–240

Oster, . (1978) The dynamics of nonlinear models with age structure. In: Levin, S.A. (ed.) Studies in Mathematical Biology, vol. 16, part II: Populations and Communities. M.A.A., pp. 411–438

Rorres, C. (1976) Optimal sustainable yield of a renewable resource. Biometrics 32, 945–948

Rorres, C., Fair, W. (1975) Optimal harvesting policy for an age-specific population. Math. Biosc. 24, 31–47

Saints, K. (1987) Discrete and continuous models of age-structured population dynamics. Senior Research Report, Harvey Mudd College, Claremont, CA

Sánchez, D. (1978) Linear age-dependent population growth with harvesting. Bull. Math. Biol. 40, 377–385

Sánchez, D. (1980) Linear age-dependent population growth will seasonal harvesting. J. Math. Biol. 9, 361–368

Sattenspiel, L. (1987) Population structure and the spread of disease. Human Biol. 59, 411–438

Sattenspiel, L., Simon, C.P. (1988). The spread and persistence of infectious diseases in structured populations. Math. Biosci. (in press)

Schenzle, D. (1985) Control of virus transmission in age-structured populations. In: Capasso, V., Grosso, E., Paveri-Fontana, S.L. (eds.) Mathematics in Biology and Medicine. Lecture Notes in Biomathematics, vol. 57. Springer, Berlin Heidelberg New York London Paris Tokyo, pp. 171–178

Sharpe, F.R., Lotka, A.J. (1911) A problem in age distribution. Phil. Mag. 21, 435–438

Von Foerster, H. (1959) In: Stohlmann, F. (ed.) The Kinetics of Cellular Proliferation Grune and Stratton, New York

Webb, G.F. (1985) Theory of Nonlinear Age-Dependent Population Dynamics. Marcel Dekker, Inc., New York

Yodzis, P. (1976) Effects of harvesting on competitive systems. Bull. Math. Biol. 38, 97–109

Author Index

Subject Index

Bio-mathematics

Managing Editor: S. A. Levin

Editorial Board: M. Arbib, H. J. Bremermann,
J. Cowan, W. M. Hirsch, S. Karlin, J. Keller,
K. Krickeberg, R. C. Lewontin, R. M. May,
J. D. Murray, A. Perelson, T. Poggio, L. A. Segel

Springer-Verlag Berlin
Heidelberg New York London
Paris Tokyo Hong Kong

Springer

Bio-mathematics

Managing Editor: S. A. Levin

Editorial Board: M. Arbib, H. J. Bremermann, J. Cowan, W. M. Hirsch, S. Karlin, J. Keller, K. Krickeberg, R. C. Lewontin, R. M. May, J. D. Murray, A. Perelson, T. Poggio, L. A. Segel

Springer-Verlag Berlin
Heidelberg New York London
Paris Tokyo Hong Kong